QUANTITATIVE STUDIES IN THE GEOLOGICAL SCIENCES

The Geological Society of America, Inc.
Memoir 142

Quantitative Studies in the Geological Sciences

A Memoir in Honor of William C. Krumbein

Edited by

E. H. TIMOTHY WHITTEN

1975

Copyright 1975 by The Geological Society of America, Inc.
Copyright is not claimed on any material
prepared by U.S. Government employees
within the scope of their employment.
Library of Congress Catalog Card Number 74-15932
I.S.B.N. 0-8137-1142-8

Published by
THE GEOLOGICAL SOCIETY OF AMERICA, INC.
3300 Penrose Place
Boulder, Colorado 80301

Printed in the United States of America

*The Memoir series was originally made possible
through the bequest of
Richard Alexander Fullerton Penrose, Jr.*

Contents

Foreword . . . . *E. H. Timothy Whitten*   ix
William C. Krumbein: The making of a methodologist . . . *Arthur L. Howland*   xiii

STRATIGRAPHIC, SEDIMENTARY, AND PALEONTOLOGICAL TOPICS

Phenetic variation in some Middle Ordovician strophomenid brachiopods
. . . . *Peter W. Bretsky and Sara S. Bretsky*   3
Laws of distribution applied to sand sizes . . . . *E. C. Dapples*   37
Burrowing depths of some Neogene heterodont bivalves
. . . . *David E. Ogren, Keewhan Choi, and Martin D. Fraser*   63
Geometry of sedimentary basins: Applications to Devonian of North America
and Europe . . . . *L. L. Sloss and Wolfgang Scherer*   71
Cross-bed variability in a single sand body . . . . *R. Steinmetz*   89

SEDIMENT TRANSPORT

Model of recurring random walks for sediment transport . . *Michael F. Dacey*   105
Multilayer Markov mixing models for studies of coastal contamination
. . . . *William R. James*   121
Influence of grain shape on the selective sorting of sand waves
. . . . *David Poché and H. G. Goodell*   137

GEOPHYSICS AND HYDRODYNAMICS

Generation of standing edge waves through nonlinear subharmonic resonance
. . . . *G. E. Birchfield and Cyril J. Galvin, Jr.*   151
Lithospheric flexure as shown by deformation of glacial lake shorelines in southern
British Columbia . . . . *R. J. Fulton and R. I. Walcott*   163
Theory of velocity of earthquake dislocations . . . . . . *Johannes Weertman*   175

GEOCHEMISTRY

Diagenetic reactions as stochastic processes: Application to the Bermudian
eolianites . . . . *G. Michel Lafon and H. L. Vacher*   187
Modeling of geochemical cycles: Phosphorus as an example
. . . . *A. Lerman, F. T. Mackenzie, and R. M. Garrels*   205
Mixing of sea water with calcium carbonate ground water . . . *L. N. Plummer*   219

PETROLOGY

Three-dimensional polynomial trend analysis applied to igneous petrogenesis
. . . . *Geoffrey W. Mathews, J. Allan Cain, and Philip O. Banks*   239

Petrogenetic significance of grain-transition probabilities, Cornelia pluton, Ajo, Arizona .... *William B. Wadsworth*   257
Appropriate units for expressing chemical composition of igneous rocks
.... *E. H. Timothy Whitten*   283

STATISTICAL METHODOLOGY

Segmentation of discrete sequences of geologic data
.... *D. M. Hawkins and D. F. Merriam*   311
Determination of important parameters in a classification scheme
.... *Thomas A. Jones and Robert A. Baker*   317
Variograms and variance components in geochemistry and ore evaluation
.... *A. T. Miesch*   333
Effect of closure on the comparison of means in ternary systems
.... *P. G. Sutterlin and R. W. May*   341
Comparison of fan-pass spatial filtering and polynomial surface-fitting models for numerical map analysis .... *Michael D. Wilson*   351
Texture analysis .... *Geoffrey S. Watson*   367
Author index 393
Subject index 401

Foreword

This volume is dedicated to William C. Krumbein in recognition of his stimulating teaching and guidance over many years and his continuing leadership and research as Emeritus William Deering Professor of Geological Sciences at Northwestern University.

In the middle of January 1973, a small group of people, long associated with William C. Krumbein, conceived the idea of preparing a monograph to honor both his past leadership in quantitative geology and his continuing stimulus to colleagues and students. For the past two decades, Krumbein has been especially instrumental in the transition from the traditional, qualitative, and often subjective, descriptive approach to geologic problems to the more vigorous and objective analysis of earth science problems. With this in mind, on January 22, 1973, all of Krumbein's past and present faculty colleagues and graduate students, together with a few scientists who have closely collaborated or have been intimately associated with him, were each encouraged to contribute an article under the unifying theme "Quantitative Models in the Earth Sciences." Immediately, promises of an overwhelming number of papers were received. As editor of the volume, I was determined to see it in print in 1974 if humanly possible. This necessitated adherence to firm deadlines. I am personally grateful to all who so assiduously met deadline dates, but I must also apologize to the many potential authors who, for a variety of reasons in a busy world, were unable to complete their manuscripts in time for inclusion in the volume.

Each paper in this volume reflects the impact of some different aspects of Krumbein's research and teaching. The papers are loosely grouped under six headings: (1) stratigraphic, sedimentary, and paleontological topics; (2) sediment transport; (3) geophysics and hydrodynamics; (4) geochemistry; (5) petrology; and (6) statistical methodology. The range of subject matter is large. Krumbein was frequently at pains to emphasize that most quantitative techniques, although developed to cope with a specific problem, commonly find ready application to many other domains. Hence, it is particularly appropriate that this volume include both a selection of specific *applications* of objective quantitative techniques to clearly defined earth science problems and a group of *methodological* papers. Of particular interest in the latter group is Watson's paper, which reviews and develops in detail the mathematics underlying the exciting new work in stereology; recent developments (mainly by French scientists and their texture analyzer) have demonstrated the importance of stereological techniques to the quantitative study of sedimentary and other particulate rocks. This work and other methodological research are likely to be forerunners of the next cycle of advance in objective, quantitative and qualitative, earth science; that is, in a continuation of the trends so ably fostered by William C. Krumbein.

E. H. Timothy Whitten

Evanston, Illinois
March 1974

William C. Krumbein

William C. Krumbein: The Making of a Methodologist

The nature of a career is determined by initial factors of heredity and early training which are difficult or impossible to discern. Although it is said that Mozart and John Stuart Mill were identified in childhood, for most careers the record begins at the stage when character has been formed and when its interactions with the environment in which it exists can be recognized. Then, watching the progress of that career, we can see the relationship between the choices that were made and the accomplishments that resulted.

The distinguished career of William C. Krumbein as scientist and teacher began, then, with his entering the School of Administration of the University of Chicago. His having been awarded a scholarship to the university indicates that his ability already had been recognized. He received his Ph.B. in 1926 and entered the business world, working for several years in the field of insurance adjustments with a finance company. As an undergraduate, he had taken an elective course in geology with Paul MacClintock. Whether or not this election was a deterministic or a probabilistic event, it proved to be of major importance for his future and for the future of geology. Krumbein found MacClintock to be a stimulating teacher and counselor, and under his guidance, Krumbein took several part-time courses in geology in the following years. In 1929, perhaps with rare foresight of the approaching stock market crash, he decided to leave business and return to the University of Chicago for graduate work in geology. He received an M.S. degree in 1930 and a Ph.D. under J Harlan Bretz in 1932.

In 1928 MacClintock left Chicago for Princeton, but other influences were at work. Francis Pettijohn had just joined the Chicago faculty, and Krumbein was a student in his first course in sedimentation. He immediately recognized the potentialities for the application of statistics, a field familiar to him from business school, to the analysis of sedimentological data. A paper on some topic in the history of geology was an assignment for doctoral candidates, and Krumbein's choice was "A History of the Principles and Methods of Mechanical Analysis," published in the *Journal of Sedimentary Petrology* in 1932 as his first paper. But a historical review was only one step into the subject, and current aspects were also occupying his attention, for a paper on "Mechanical Analysis of Fine-Grained Sediments" followed in the same volume. Thus was launched a lifelong study not only of sediments but, in particular, of how to study them.

In 1933 Krumbein joined the staff of the University of Chicago to teach in the physical science survey course which formed part of the common curriculum that had recently been initiated by Robert Hutchins. As a graduate student, he had shared an office with M. King Hubbert. The interchange between Hubbert and Krumbein reinforced their belief in the importance of basing geologic studies on a strong foundation of mathematics and the principles of physics and chemistry. The teaching in the physical science survey course, therefore, had congenial aspects for Krumbein, because it was based on an integrated approach in science. From his collaboration with Carey Croneis in this course, came a new approach to geology texts. *Down to Earth* was written in 1935 in lively style and stressed analytical treatment and evolution of concept. This book, intended for their students, has sold over 33,000 copies and is still in print nearly 40 years later.

As an instructor Krumbein had free entry into university courses, and over the next several years he used this opportunity to advance his knowledge in mathematics and physics. At the same time, his studies in sedimentation and sediment analysis were proceeding with the publication of a series of papers that dealt with methods for mechanical analysis, sampling problems, and the representation of data. Familiar to all students of sediments is the *phi* scale, which Krumbein proposed as a logarithmic transform for the treatment of particle-size distributions. This scale was identified by Karl Eckart, a colleague in physics, as equal in importance to the pH and star-magnitude scales. This and continuing work in collaboration with Francis Pettijohn led to the *Manual of Sedimentary Petrology*, published in 1938, which at once became an authoritative reference book for all who would deal with sediments and sedimentary rocks.

Successful application of statistical procedures, revealing the nature of distributions, led Krumbein inevitably to a search for physical causes and the character of the dynamic processes controlling particle movements. A series of papers on sedimentary environments began with "The Sediments of Barataria Bay" by Aberdeen and Krumbein in 1937. One can follow his interest in this line of investigation over the years through his publications on the flood gravels of San Gabriel Canyon and Arroyo Seco and on lake and ocean beaches.

In 1941–42 Krumbein was awarded a Guggenheim fellowship, and characteristically, he employed it to become acquainted with developments in hydraulics and fluid dynamics. He spent part of the year in flume studies with Hunter Rouse at the University of Iowa, and part with M. P. O'Brien, Dean of the School of Engineering at the University of California, working with wave tanks and on beach studies.

It was on the basis of reports prepared at Berkeley that Krumbein was called to Washington in 1942 to join the Landing Beach Intelligence Group of the Beach Erosion Board of the U.S. Army Corps of Engineers. This group worked closely with the Military Geology Unit of the U.S. Geological Survey, directed initially by W. H. Bradley, and applied geologic expertise to the planning for military operations in North Africa, Sicily, and on many Pacific beaches.

Invited to lecture on the American Association of Petroleum Geologists tour, Krumbein chose as his topic "Recent Sediments and the Search for Petroleum." Not surprisingly, this led to an invitation in the fall of 1945 to join Gulf Research and Development Company, the geological section of which was under the direction of Ben Cox. Here again Krumbein took advantage of the opportunities to exchange ideas and expand his interests through contact with a new and lively set of associates. In Tulsa, Charles Ryniker discoursed on limestone and pointed out the uses of facies mapping. His attention was directed by Roy Hazzard in Shreveport and by Sig Hammer and others in Pittsburgh to subsurface and to geophysical studies.

In 1946, to the great delight of many of his friends, Krumbein returned to the academic world by joining the faculty at Northwestern University. The carry-over of interests and ideas from his work with Gulf is apparent, for in the next few years in a series of papers, he developed the concepts of facies mapping and the use of measures such as clastic ratio and sand-shale ratio, always emphasizing the importance of subsurface data now so abundantly available through drilling and seismic studies. A fertile collaboration with his Northwestern colleagues, E.

C. Dapples and L. L. Sloss, produced a series of joint papers on facies analysis and sedimentary tectonics. Collaboration in teaching was as natural as in research. In 1951 Krumbein and Sloss published *Stratigraphy and Sedimentation* as one of the fruits of this collaboration. A second edition appeared in 1963. An *Atlas of Lithofacies Maps*, compiling the work of many of their students, was published in 1960 by Dapples, Krumbein, and Sloss.

Statistical methods for analyzing geologic data have never ceased to be one of Krumbein's major interests, and the advent of the electronic digital computer opened new avenues. The high-speed digital computer first appeared in one of his titles in 1958 in a joint paper with L. L. Sloss on its use in stratigraphic and facies analysis. Krumbein used his vice-presidential address to Section E of the American Association for the Advancement of Science in 1961 to discuss the computer in geology, and his publications since then contain many contributions advancing the application of computer methods to a variety of geologic problems. Churchill Eisenhart of the National Bureau of Standards; John Tukey, Professor of Mathematics at Princeton; and Franklin Graybill, Professor of Statistics at Colorado State University, were particular sources of stimulation and discussion during the period, and in 1965 Krumbein and Graybill published their book *Introduction to Statistical Models in Geology*.

Recognition has come to Krumbein in many forms. He was elected to The American Academy of Arts and Sciences and given the William Deering Chair in Geology at Northwestern University. He has held appointments as Guggenheim Fellow, Fulbright Lecturer, and President's Fellow at Northwestern, and has been elected Vice-President of Section E of the American Association for the Advancement of Science, and President of the Society of Economic Paleontologists and Mineralogists. At its organizational meeting, the International Association for Mathematical Geology honored him by immediately electing him Past President, a fitting recognition of the influence of his work. His service on many professional committees, as Editor of the Bulletin of the American Association of Petroleum Geologists, and as Councilor of the Geological Society of America make clear how much his professional colleagues have valued his wisdom and advice.

Even in a brief review such as this of Bill Krumbein's career, certain points stand out. One, I think, is the clarity and consistency of his vision: the early recognition of the need for quantitative measurements, for sound physical and mathematical foundations, for the application of general principles to particular problems, and for new methodology to reveal relationships. Another is his steady productivity which continues today without any recognizable break, although he has been Professor Emeritus since 1970. One aspect of his career, implicit throughout what has been said, deserves a few further words of emphasis: his teaching. Never has it been routine or static, whether aimed at freshmen in physical science or at graduate students grappling with statistical models. If one had a series of his course outlines, one could follow the progress of geological science for the last four decades. One of his principles has been that his office door is never closed. Student or colleague, old acquaintance or stranger is welcome and will have his attention and help.

Those who have had a part in the making of this volume take pleasure in saluting him. Recognizing no perturbation in his scientific career from the addition of "emeritus" to his title, we look forward to continuing fruitful association and can repeat with Pip and Joe in another tale of great expectations their same words of anticipation: "What larks!"

Arthur L. Howland

BIBLIOGRAPHY OF WILLIAM C. KRUMBEIN

1932 A history of the principles and methods of mechanical analysis: Jour. Sed. Petrology, v. 2, p. 89-124.

—— The mechanical analysis of fine-grained sediments: Jour. Sed. Petrology, v. 2, p. 140-149.

1933 Textural and lithological variations in glacial till: Jour. Geology, v. 41, p. 382-408.

—— The dispersion of fine-grained sediments for mechanical analysis: Jour. Sed. Petrology, v. 3, p. 121-135.

1934 The probable error of sampling sediments for mechanical analysis: Am. Jour. Sci., v. 227, p. 204-214.

—— Size frequency distributions of sediments: Jour. Sed. Petrology, v. 4, p. 65-77.

1935 Thin-section mechanical analysis of indurated sediments: Jour. Geology, v. 43, p. 482-496.

—— A time chart for mechanical analysis by the pipette method: Jour. Sed. Petrology, v. 5, p. 93-95.

—— (and Croneis, C. G.) Down to Earth, an introduction to geology: Chicago, Ill., Univ. Chicago Press, 501 p.

1936 Application of logarithmic moments to size-frequency distributions of sediments: Jour. Sed. Petrology, v. 6, p. 35-47.

—— The use of quartile measures in describing and comparing sediments: Am. Jour. Sci., v. 232, p. 98-111.

1937 (and Aberdeen, E. J.) The sediments of Barataria Bay (La.): Jour. Sed. Petrology, v. 7, p. 3-17.

—— Sediments and exponential curves: Jour. Geology, v. 45, p. 577-601.

—— Korngrösseneinteilungen und statische Analyse: Neues Jahrb. Mineralogie Beil.-Bd. 74, Abt. A., p. 137-150.

1938 (and Pettijohn, F. J.) Manual of sedimentary petrography: New York, D. Appleton-Century Co., 549 p.

—— (and Griffith, J. S.) Beach environment in Little Sister Bay, Wis.: Geol. Soc. America Bull., v. 49, p. 629-652.

—— Local areal variation of beach sands: Geol. Soc. America Bull., v. 49, p. 653-658.

—— Size frequency distributions of sediments and the normal phi curve: Jour. Sed. Petrology, v. 8, p. 84-90.

1939 (and Caldwell, L. T.) Areal variation of organic carbon content of Barataria Bay sediments, Louisiana: Am. Assoc. Petroleum Geologists Bull., v. 23, p. 582-594.

—— Tidal lagoon sediments on the Mississippi delta, *in* Trask, P. D., ed., Recent marine sediments: Tulsa, Am. Assoc. Petroleum Geologists, p. 178-194.

—— Graphic presentation and statistical analysis of sedimentary data, *in* Trask, P. D., ed., Recent marine sediments: Tulsa, Am. Assoc. Petroleum Geologists, p. 558-591.

—— Preferred orientation of pebbles in sedimentary deposits: Jour. Geology, v. 47, p. 673-706.

—— Application of photo-electric cell to the measurement of pebble axes for orientation analysis: Jour. Sed. Petrology, v. 9, p. 122-130.

—— Application of the photoelectric cell to the study of pebble size and shape: Proc. 6th Pacific Sci. Cong., v. 2, p. 769-777.

1940 (and Tisdel, F. W.) Size distribution of source rocks of sediments: Am. Jour. Sci., v. 238, p. 296-305.

—— Flood gravel of San Gabriel Canyon, Calif.: Geol. Soc. America Bull., v. 51, p. 639-676.

1941 (and Rasmussen, W. C.) The probable error of sampling beach sand for heavy mineral analysis: Jour. Sed. Petrology, v. 11, p. 10-20.

—— Influence of geophysics and geochemistry on the professional training of geologists: Am. Inst. Mining and Metall. Engineers, Tech. Pub. no. 1327 (Mining Technology, May 1941), 11 p.

—— The effect of abrasion on the size, shape, and roundness of rock fragments: Jour. Geology, v. 49, p. 482-520.

—— Measurement and geological significance of shape and roundness of sedimentary particles: Jour. Sed. Petrology, v. 11, p. 101.

—— Principles of sedimentation and the search for stratigraphic traps: Econ. Geology, v. 36, p. 786-810.

1942 Settling velocity and flume behavior of non-spherical particles: Am. Geophys. Union Trans., pt. II, p. 621-633.

—— Flood deposits of Arroyo Seco, Los Angeles County, California: Geol. Soc. America Bull., v. 53, p. 1335-1402.

—— (and Monk, G. D.) Permeability as a function of the size parameters of unconsolidated sand: Am. Inst. Mining and Metall. Engineers, Tech. Paper no. 1492 (Petroleum Technology, July 1942), 11 p.

—— Physical and chemical changes in sediments after deposition: Jour. Sed. Petrology, v. 12, p. 111-117.

—— Criteria for subsurface recognition of unconformities: Am. Assoc. Petroleum Geologists Bull., v. 26, p. 36-62.

1943 Fundamental attributes of sedimentary particles: Proc. 2d Hydraulic Conf., Univ. Iowa Studies in Engineering, p. 318-331.

1944 Shore processes and beach characteristics: Beach Erosion Board, U.S. Army Corps Engineers, Tech. Mem. no. 3, 35 p.

—— Shore currents and sand movement on a model beach: Beach Erosion Board, U.S. Army Corps Engineers, Tech. Mem. no. 7, 44 p.

1945 Sedimentary maps and oil exploration: N.Y. Acad. Sci. Trans., ser. II, v. 7, p. 159-166.

—— Recent sedimentation and the search for petroleum: Am. Assoc. Petroleum Geologists Bull., v. 29, p. 1233-1261.

1947 Analysis of sedimentation and diagenesis: Am. Assoc. Petroleum Geologists Bull., v. 31, p. 168-174.

—— Project 6, ancient sediments, comprehensive investigation of the sediments of a single limited stratigraphic section over a considerable area: Am. Assoc. Petroleum Geologists Research Comm., 1946-1947, Repts. on Proj. 1-12, p. 98-113.

—— Shales and their environmental significance: Jour. Sed. Petrology, v. 17, p. 101-108.

1948 Lithofacies maps and regional sedimentary-stratigraphic analysis: Am. Assoc. Petroleum Geologists Bull., v. 32, p. 1909-1923.

—— (and Dapples, E. C., and Sloss, L. L.) Tectonic control of lithologic associations: Am. Assoc. Petroleum Geologists Bull., v. 32, p. 1924-1947.

1949 (and Sloss, L. L., and Dapples, E. C.) Integrated facies analysis: Geol. Soc. America, Mem. 39, p. 91-122.

—— (and Sloss, L. L., and Dapples, E. C.) Sedimentary tectonics and sedimentary environments: Am. Assoc. Petroleum Geologists Bull., v. 33, p. 1859-1891.

—— Geology of beach engineering: Shore and Beach, Oct., p. 1-4.

—— (and Keulegan, G. H.) Stable configuration of bottom slope in a shallow sea and its bearing on geological processes: Am. Geophys. Union Trans., v. 30, p. 855-861.

1950 (and Dapples, E. C., and Sloss, L. L.) The organization of sedimentary rocks: Jour. Sed. Petrology, v. 20, p. 3-20.

—— (and Ohsiek, L. E.) Pulsational transport of sand by shore agents: Am. Geophys. Union Trans., v. 31, p. 216-220.

—— Grain-size measurements made in thin sections, comments: Jour. Geology, v. 58, p. 160.

—— Geological aspects of beach engineering, in Application of geology to engineering practice (Berkey Volume): Geol. Soc. America, p. 195-221.

1951 Littoral processes in lakes, in Johnson, J. W., ed., Coastal engineering: Proc. 1st Conf. on Coastal Engineering, 1950, Chap. 16, p. 155-160.

—— (and Sloss, L. L.) Stratigraphy and sedimentation: San Francisco, Calif., W. H. Freeman and Co., 497 p.

—— Regional stratigraphic analysis as a guide to geophysical exploration: World Oil, v. 132, p. 99-100.

—— Occurrence and lithologic associations of evaporites in the United States: Jour. Sed. Petrology, v. 21, p. 63-81.

—— Some relations among sedimentation, stratigraphy, and seismic exploration: Am. Assoc. Petroleum Geologists Bull., v. 35, p. 1505-1521.

1952 (and Garrels, R. M.) Origin and classification of chemical sediments in terms of pH and oxidation-reduction potentials: Jour. Geology, v. 60, p. 1-32.

1952 Principles of facies map interpretation: Jour. Sed. Petrology, v. 22, p. 200-211.
1953 (and Dapples, E. C., and Sloss, L. L.) Petrographic and lithologic attributes of sandstones: Jour. Geology, v. 61, p. 291-317.
—— (and Nagel, F. G.) Regional stratigraphic analysis of "Upper Cretaceous" rocks of Rocky Mountain region: Am. Assoc. Petroleum Geologists Bull., v. 37, p. 940-960.
—— (and Miller, R. L.) Design of experiments for statistical analysis of geological data: Jour. Geology, v. 61, p. 510-532.
—— Latin square experiments in sedimentary petrology: Jour. Sed. Petrology, v. 23, p. 280-283.
—— Statistical designs for sampling beach sands: Am. Geophys. Union Trans., v. 34, p. 857-868.
1954 The tetrahedron as a facies mapping device: Jour. Sed. Petrology, v. 24, p. 115-122.
—— Statistical significance of beach sampling methods: Beach Erosion Board, U.S. Army Corps Engineers, Tech. Mem. no. 50, p. 1-33.
—— Applications of statistical methods to sedimentary rocks: Jour. Am. Stat. Assoc., v. 49, p. 51-66.
—— (and Miller, R. L.) A note on transformation of data for analysis of variance, discussion: Jour. Geology, v. 62, p. 192-193.
—— Statistical problems of sample size and spacing on Lake Michigan (Ill.) beaches, *in* Johnson, J. W., ed., Coastal engineering: Proc. 4th Conf. on Coastal Engineering, Council on Wave Research, Berkeley, Chap. 9, p. 147-162.
1955 Experimental design in the earth sciences: Am. Geophys. Union Trans., v. 36, p. 1-11.
—— Composite end members in facies mapping: Jour. Sed. Petrology, v. 25, p. 115-122.
—— Statistical analysis of facies maps: Jour. Geology, v. 63, p. 452-470.
—— (and Slack, H. A.) Measurement and statistical evaluation of low-level radioactivity in rocks: Am. Geophys. Union Trans., v. 36, p. 460-464.
1956 (and Lieblein, J.) Geological application of extreme-value methods to interpretation of cobbles and boulders in gravel deposits: Am. Geophys. Union Trans., v. 37, p. 313-319.
—— (and Slack, H. A.) Relative efficiency of beach sampling methods: Beach Erosion Board, U.S. Army Corps Engineers, Tech. Mem. no. 90, 52 p.
—— (and Slack, H. A.) Statistical analysis of low-level radioactivity of Pennsylvanian black fissile shale in Illinois: Geol. Soc. America Bull., v. 67, p. 739-762.
—— Regional and local components in facies maps: Am. Assoc. Petroleum Geologists Bull., v. 40, p. 2163-2194.
—— (and Tukey, J. W.) Multivariate analysis of mineralogic, lithologic, and chemical composition of rock bodies: Jour. Sed. Petrology, v. 26, p. 322-337.
1957 (and Libby, W. G.) Application of moments to vertical variability maps of stratigraphic units: Am. Assoc. Petroleum Geologists Bull., v. 41, p. 197-211.
—— Comparison of percentage and ratio data in facies mapping: Jour. Sed. Petrology, v. 27, p. 293-297.
—— A method for specification of sand for beach fills: Beach Erosion Board, U.S. Army Corps Engineers, Tech. Mem. no. 102, 86 p.
1958 Measurement and error in regional stratigraphic analysis: Jour. Sed. Petrology, v. 28, p. 175-185.
—— (and Sloss, L. L.) High-speed digital computers in stratigraphic and facies analysis: Am. Assoc. Petroleum Geologists Bull., v. 42, p. 2650-2669.
1959 Trend surface analysis of contour-type maps with irregular control-point spacing: Jour. Geophys. Research, v. 64, p. 823-834.
—— The "sorting out" of geological variables illustrated by regression analysis of factors controlling beach firmness: Jour. Sed. Petrology, v. 29, p. 575-587.
1960 The "geological population" as a framework for analysing numerical data in geology: Liverpool and Manchester Geol. Jour., v. 2, p. 341-368.
—— Stratigraphic maps from data observed at outcrop: Proc. Yorkshire Geol. Soc., v. 32, p. 353-366.
—— Some problems in applying statistics to geology: Appl. Stat., v. 9, p. 82-91.

—— (and Dapples, E. C., and Sloss, L. L.) Lithofacies maps, an atlas of the United States and Southern Canada: New York, John Wiley & Sons, Inc., 126 p.
1961 The analysis of observational data from natural beaches: Beach Erosion Board, U.S. Army Corps Engineers, Tech. Mem. no. 130, p. 1-59.
1962 The computer in geology: Science, v. 136, p. 1087-1092.
—— Open and closed number systems in stratigraphic mapping: Am. Assoc. Petroleum Geologists Bull., v. 46, p. 2229-2245.
—— (and Allen, P.) Secondary trend components in the Top Ashdown Pebble Bed, a case history: Jour. Geology, v. 70, p. 507-538.
1963 (and Imbrie, J.) Stratigraphic factor maps: Am. Assoc. Petroleum Geologists Bull., v. 47, p. 698-701.
—— Confidence intervals on low-order polynomial trend surfaces: Jour. Geophys. Research, v. 68, p. 5869-5878.
—— (and Sloss, L. L.) Stratigraphy and sedimentation (2d ed.): San Francisco, Calif., W. H. Freeman and Co., 623 p.
1964 (and Harrison, W.) Interactions of the beach-ocean-atmosphere system at Virginia Beach, Virginia: Coastal Eng. Research Center, Tech. Memo. 7, 102 p.
—— A geological process-response model for analysis of beach phenomena: U.S. Office of Naval Research, Tech. Rept. 8, Task no. 389-135, 15 p.
—— Some remarks on the phi notation: Jour. Sed. Petrology, v. 34, p. 195-197.
—— (and Benson, B. T., and Hempkins, W. B.) *Whirlpool*, a computer program for sequential multiple regression: U.S. Office of Naval Research, Tech. Rept. 14, Task no. 389-135, 49 p.
—— (and Harrison, W., and Wilson, W. S.) Sedimentation at an inlet entrance (Rudee Inlet, Virginia Beach, Virginia): Coastal Eng. Research Center, Tech. Memo. 8, 42 p.
1965 (and Graybill, F. A.) An introduction to statistical models in geology: New York, McGraw-Hill Book Co., 475 p.
—— Sampling in paleontology, *in* Kummel, B. and Raup, D., eds., Handbook of paleontological techniques: San Francisco, Calif., W. H. Freeman and Co., p. 137-150.
—— (and Whitten, E. H. T., Waye, I., and Beckman, W. A.) A surface-fitting program for areally distributed data from the earth sciences and remote sensing: NASA Contractor Rept. CR-318, Washington, 146 p.
—— (and James, W. R.) A lognormal size distribution model for estimating stability of beach fill material: Coastal Eng. Research Center, Tech. Memo. 16, 17 p.
1966 The cyclothem as a response to sedimentary environment and tectonism, *in* Merriam, D. F., ed., Symposium on cyclic sedimentation: Kansas Geol. Survey Bull. 169, p. 239-247.
—— A comparison of polynomial and Fourier models in map analysis: U.S. Office of Naval Research, Tech. Rept. 2, Contract 1228(36), 54 p.
—— Classification of map surfaces based on the structure of polynomial and Fourier coefficient matrices: Kansas Geol. Survey Computer Contr. 7, p. 12-18.
—— Il calcolatore elettronico in geologia: Geologica Romana, v. 5, p. 335-338.
1967 The general linear model in map preparation and analysis: Kansas Geol. Survey Computer Contr. 12, p. 38-44.
—— Fortran IV computer programs for Markov chain experiments in geology: Kansas Geol. Survey Computer Contr. 13, 38 p.
—— (and LaMonica, G. B.) Classification and organization of quantitative data in geology: U.S. Office of Naval Research, Tech. Rept. 4, Task no. 388-078, Contract Nonr-1228(36), p. 339-354.
1968 Statistical models in sedimentology: Sedimentology, v. 10, p. 7-23.
—— Computer simulation of transgressive and regressive deposits with a discrete-state, continuous-time Markov model: Kansas Geol. Survey Computer Contr. 22, p. 11-18.
—— Fortran IV computer program for simulation of transgression and regression with continuous-time Markov models: Kansas Geol. Survey Computer Contr. 26, p. 1-37.
1969 The computer in geologic perspective, *in* Merriam, D. F., ed., Computer applications in the earth sciences: New York, Plenum Press, p. 251-275.

—— (and James, W. R.) Frequency distributions of stream link lengths: Jour. Geology, v. 77, p. 544-565.

—— (and Dacey, M. F.) Markov chains and embedded Markov chains in geology: Jour. Internat. Assoc. Math. Geology, v. 1, p. 79-96.

1970 (and Shreve, R. L.) Some statistical properties of dendritic channel networks: U.S. Office of Naval Research, Tech. Rept. 13, Task no. 389-150, Contract Nonr-1228(36), 117 p.

—— (and Frazee, C. J., and Fehrenbacher, J. B.) Loess distribution from a source: Soil Sci. Soc. America Proc., v. 34, p. 296-301.

—— (and Jones, T. A.) The influence of areal trends on correlations between sedimentary properties: Jour. Sed. Petrology, v. 40, p. 656-665.

—— (and Dacey, M. F.) Markovian models in stratigraphic analysis: Jour. Internat. Assoc. Math. Geology, v. 2, p. 175-191.

—— (and Scherer, W.) Structuring observational data for Markov and semi-Markov models in geology: U.S. Office of Naval Research, Tech. Rept. 15, Contract N00014-67-A0356-0018 (formerly Nonr-1228(36); NR 389-150), 59 p.

—— Geological models in transition, in Merriam, D. F., ed., Geostatistics: New York, Plenum Press, p. 143-161.

1971 (and Dacey, M. F.) Comments on spatial randomness in dendritic stream channel networks: U.S. Office of Naval Research, Tech. Rept. 17, Contract N00014-67-A0356-0018 (formerly Nonr-1228(36); NR 389-150), 18 p.

1972 Areal variation and statistical correlation, in Merriam, D. F., ed., Mathematical models in sedimentary processes: New York, Plenum Press, p. 167-173.

—— Areal variation and statistical correlation in open and closed number systems: Proc. 38th Sess. Internat. Stat. Inst., v. 2, p. 551-556.

—— (and Watson, G. S.) The effect of trends on correlation in open and closed three-component systems: Jour. Internat. Assoc. Math. Geology, v. 4, p. 317-330.

—— (and Orme, A. R.) Field mapping and computer simulation of braided stream networks: Geol. Soc. America Bull., v. 83, p. 3369-3380.

—— Probabilistic models and the quantification process in geology: Geol. Soc. America Spec. Paper 146, p. 1-10.

1973 (and Dacey, M. F.) Comments on randomness in spatial components of dendritic stream channel networks, in Recent researches in geology: Delhi, India, Hindustan Pub. Corp., p. 53-65.

1974 (and James, W. R.) Spatial and temporal variations in geometric and material properties of a natural beach: U.S. Army Coastal Eng. Research Center Tech. Rept. 44, Contracts DACW72-70-C-0011, DACW72-71-C-0007, DACW72-72-C-0002, DACW72-72-C-0010, 79 p.

—— Milestones along the highway of quantification in geology, in Merriam, D. F., ed., The impact of quantification on geology: Syracuse Univ. Geol. Contr. 2, New York, Springer-Verlag, p. 51-66.

1975 Markov models in the earth sciences, in McCammon, R. B., ed., Concepts in geostatistics: New York, Springer-Verlag (in press).

—— Probabilistic modelling in geology, in Merriam, D. F., ed., Random processes in geology: New York, Springer-Verlag (in press).

STRATIGRAPHIC, SEDIMENTARY, AND PALEONTOLOGICAL TOPICS

Phenetic Variation in Some Middle Ordovician Strophomenid Brachiopods

Peter W. Bretsky
and
Sara S. Bretsky

*Department of Earth and Space Sciences
State University of New York
Stony Brook, New York 11790*

ABSTRACT

In attempting to name species of *Oepikina* and *Strophomena* in a survey of Middle Ordovician faunas (Mifflin Member of the Platteville Formation) in Wisconsin, Illinois, Iowa, and Minnesota, we found that supposedly diagnostic characters of the various species, particularly characters describing the profile of the valves, appeared to intergrade, and that combinations of characters that were thought to have been consistently associated appeared sometimes to vary independently. We postulated that most of the Mifflin strophomenids represented species populations of either *Strophomena plattinensis* Fenton or of *Oepikina minnesotensis* (Winchell), and that each of these species displayed a certain amount of variability, probably environmentally controlled, in form. We investigated patterns of morphological variability within and among local populations in an attempt to map gradients of variability and to determine their probable relation to paleoecological parameters.

For pedicle and brachial valves of each species, a principal components analysis was performed on measurements of size, shape, inflation, and symmetry. The first principal component accounted for at least 40 percent of the variation in each of the four correlation matrices, and the first two together for more than 60 percent of the variation. Inflation characters loaded strongly on the first axis, whereas symmetry characters had the highest loading on the second axis.

Clinal maps, prepared for each valve type by plotting the mean values of one inflation character and one symmetry character in each of 16 sampling areas, showed isophene patterns that generally paralleled probable topographic features of the shallow sea floor. The inferred nearshore populations were generally flatter and less symmetrical than those from deeper water. Examination of patterns of intrapopulation variability showed that the level of variability, especially for the inflation character, was greater in the onshore populations than in the offshore ones. Greater morphological variability in shallow-water populations could reflect adaptation to increased levels of spatial heterogeneity, most likely resulting from nearshore substratum patchiness.

INTRODUCTION

One of the grand truisms in paleontology and, in fact, in all evolutionary studies is that each species incorporates a certain level of morphological variability. How else could natural selection act? But when systematic studies are based on scanty material, population polymorphism is often translated into nomenclatural proliferation. In a survey of the paleoecology of Middle Ordovician faunas of Wisconsin, Illinois, Iowa, and Minnesota, we discovered that this may have been the case in the taxonomy of brachiopods belonging to the genera *Strophomena* and *Oepikina*. Previous workers have identified three species of *Oepikina* and three species and four subspecies of *Strophomena* among samples from the Mifflin Member of the Platteville Formation. In attempting to apply these species names to our material, we found that supposedly diagnostic characters, particularly those describing the profile of the valves, appeared to intergrade among specimens from different localities, and that combinations of characters that were thought to have been consistently associated appeared to vary independently. Having at our disposal a considerably larger sample of specimens than had been available to earlier workers, and having the advantage of being able to use techniques of multivariate analysis to consider many characters simultaneously, we wished to define patterns of morphological variability and to determine their possible relation to paleoecological parameters.

McKerrow (1953), Ager (1965, 1967a, 1967b, 1971), Copper (1966, 1967), Mitra (1958), Bowen (1966), Temple (1968), Bassett (1970), and Winter (1971) have provided convincing illustrations of intraspecific variability in certain fossil brachiopods. DuBois (1916) and McCammon (1970) have recognized some of the factors that contribute to such variability in Holocene brachiopod species. We do not, however, rule out the possibility that subsequent studies may demonstrate the validity of some species-level separation of Mifflin specimens of *Strophomena* or of *Oepikina*. Originally, we had intended to deal with both qualitative and quantitative characters which some workers had regarded as diagnostic at the species level, but the generally poor preservation of the valve interiors caused us to limit the multivariate analysis to characters describing valve size, roundness, inflation, and symmetry. Also, some of the nominate species-group taxa to which other workers have assigned Mifflin specimens were originally described from areas outside the geographic scope of our study. Not having reviewed the original type material of these taxa, we have chosen not to present a formal synonymy of the Mifflin species at this time. We therefore present this discussion, less as a formal solution to a taxonomic problem than as an exemplification of some possible approaches to estimating the nature and degree of variability among fossil species populations. We expect that its results will help to provide answers to some of the following questions:

1. What is the degree of phenotypic variability within benthic marine invertebrate fossil species from epeiric seas of the North American midcontinent, and to what extent is this variability expressed in the form of gradients—*clines* in the sense of Huxley (1938; see Simpson, 1961, p. 179)?

2. When the same characters are measured in generically distinct groups of organisms, does each taxon show the same kinds of morphological changes over the same geographic range?

3. If taxonomic characters do define a clinal pattern of variation, what can then be said about changes in the degree of variability for any specified character within and among local populations? If indeed there are considerable differences in the level of variability within and among local populations, what sorts of environments tend to be inhabited by the more variable populations?

GEOLOGIC SETTING

Figure 1 shows the location of the study area. Samples that included strophomenid brachiopods were taken at 59 localities. At the scale of Figure 1, some of these localities would be indistinguishable on the map; the localities have thus been grouped into 16 major sampling areas (Fig. 1) averaging about 256 sq km (100 sq mi) each. The locations of these sampling areas and the number of individual localities comprising each are given in Table 1.

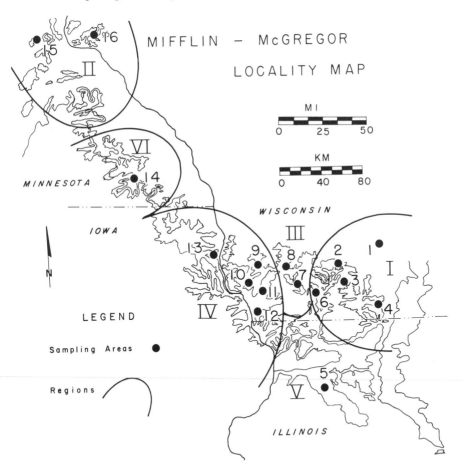

Figure 1. Location of the Mifflin-McGregor study area showing the outcrop pattern of the Platteville Limestone (geology compiled from various sources). Exposures of the Mifflin Member from which the strophomenid brachiopods have been collected are grouped into 16 major sampling areas (see also Table 1). In addition, these 16 areas define 6 major regions (I through VI) on the basis of overall lithologic-faunal similarity and geographic proximity.

TABLE 1. LOCATION OF SAMPLING AREAS

Area	Geographic locality	Number of local sections sampled	Specimens measured for principal components analysis				Total
			Oepikina pedicle	Oepikina brachial	Strophomena pedicle	Strophomena brachial	
1	Madison–Sun Prairie, Wisconsin	2	24	14	50	29	117
2	Mount Horeb, Wisconsin	5	13	14	8	5	40
3	New Glarus, Wisconsin	2	16	10	16	6	48
4	Janesville-Beloit, Wisconsin	4	3	1	7	1	12
5	Dixon-Oregon, Illinois	3	8	10	10	1	29
6	Darlington, Illinois	3	6	3	13	4	26
7	Mineral Point, Wisconsin	3	12	17	8	2	39
8	Dodgeville-Highland, Wisconsin	4	14	19	9	3	45
9	Fennimore, Wisconsin	3	33	27	20	6	86
10	Lancaster, Wisconsin	5	18	12	13	3	46
11	Platteville, Wisconsin	2	13	18	8	4	43
12	Dickeyville, Wisconsin	4	35	18	10	8	71
13	McGregor-Guttenberg, Iowa	7	7	5	4	1	17
14	Preston-Fountain-Chatfield, Minnesota	8	7	6	2	0	15
15	Minneapolis-St. Paul, Minnesota	2	39	6	23	11	79
16	Ellsworth, Wisconsin	2	8	1	10	7	26
	Total number of valves used in the study		256	181	211	91	739

The organisms and sediments of the midcontinental Ordovician strata were first described in the latter part of the 19th century by such eminent geologists as Owen, Meek, Worthen, Hall, Ulrich, Winchell, and Schuchert. Subsequent studies emphasized the setting up of a broad interregional temporal and paleogeographic framework. Correlations of the midcontinent Middle Ordovician with the Black River and lower Trenton beds in central New York and with various Middle Ordovician formations in Tennessee and Missouri are discussed and summarized by Kay (1935), Bays and Raasch (1935), and Cooper (1956). Descriptions of local stratigraphy, refinement of stratigraphic nomenclature, and understanding of facies relations were stimulated by interest in the possibility of major mineral discoveries. The findings of these studies are summarized in Agnew (1955), Agnew and others (1956), Weiss (1953, 1957), Weiss and Bell (1956), Heyl and others (1959), and in nine papers on the geology of quadrangles in Wisconsin, Iowa, and Illinois (U.S. Geological Survey Bulletins 1123A-1123I, 1961-1966). The recent increase in knowledge of modern carbonate environments has resulted in studies of patterns of lithofacies and biofacies and of the diagenetic history of the Middle Ordovician carbonates. Examples of these studies are those of Deninger (1964), Asquith (1967), Badiozamani (1972), and Sloan (1973).

There have been many taxonomic studies of individual groups of Middle Ordovician organisms. Some of these studies deal only with midcontinent faunas, but some incorporate data from a broader geographic area. Among these taxonomic works are the studies of Fenton (1928b), Salmon (1942), Weiss (1955), and Cooper (1956) on brachiopods; Thompson (1959), Webers (1966), and Sweet and others (1971) on conodonts; Karklins (1969) on cryptostome bryozoans; Kay (1934, 1935), Swain (1957), and Swain and others (1961) on ostracodes; and DeMott (1963) on trilobites.

The first important attempt at a synthesis of the "synecology" or "community ecology" of the midcontinent Middle Ordovician faunas was that of Scott (1962). Preliminary results of our own field studies over the past three years of the stratigraphy and paleoecology of the Platteville Formation and the overlying Decorah Formation are presented in Bretsky (1970) and Sulima (1972). Webers (1973) and Sloan (1973) have also carried out paleoecological investigations in the region.

The rocks from which the strophomenid samples come are carbonates of the Platteville Formation. The Mifflin Member consists of irregularly thin-bedded, nodular (wavy) limestone (normally a biopelmicrite) and dolomite; some sections also contain shale partings. In Wisconsin, Illinois, and Iowa, the Mifflin overlies the massive to thick-bedded Pecatonica Dolomite and underlies the more regularly bedded and thicker bedded dolomitic limestone of the Magnolia Member. The Mifflin is consistently about 4.6 m (15 ft) thick throughout this area (see Agnew and others, 1956, Fig. 34, p. 257, for a diagrammatic cross section of the Platteville, and Fig. 38, p. 268, for a stratigraphic column of the Platteville strata in the lead-zinc district). The precise location within the Mifflin section of the bedding planes sampled for the ecological survey has been recorded, but in this discussion of morphological variability in the strophomenid brachiopods, we do not attempt to deal with possible changes in form among stratigraphic levels. In southeastern Minnesota, the Platteville units tend to lose their identity, and our Minnesota samples come from the lower portions of what is termed the McGregor Member, a stratigraphic equivalent of the combined Mifflin and Magnolia Members farther to the south.

On the basis of comparison of the total benthic fossil population (Bretsky, 1974, in prep.), lithologic characteristics (Badiozamani, 1972), and geographic proximity, we have grouped the 16 areas into six major regions within the Mifflin-lower McGregor outcrop area (Fig. 1).

Region I

Dolomites and limy dolomites in the vicinity of Madison, Darlington, and Beloit, south-central Wisconsin (areas 1, 2, 3, 4, and 6; 16 collecting localities); characterized by a fauna of large mollusks, especially cyrtodontid bivalves and pleurotomariid gastropods.

Region II

Dolomites and limy dolomites from Minneapolis-St. Paul, Minnesota, and Ellsworth, Wisconsin (areas 15 and 16; 4 collecting localities); characterized by strophomenid brachiopods and cyrtodontid bivalves.

Region III

Sections where dolomite and limestone are mixed, in the vicinity of Mineral Point and Dodgeville, Wisconsin (areas 7 and 8; 7 collecting localities); comprising mainly assemblages of orthid brachiopods and ostracodes, or of murchisoniid gastropods.

Region IV

Interbedded limestones and limy shales in southwestern Wisconsin and eastern Iowa (areas 9, 10, 11, 12, and 13; 21 collecting localities); characterized by an orthid brachiopod-ostracode fauna, but with one or more beds of bellerophontid gastropods.

Region V

Limestones and limy dolomites in northern Illinois, where the total Platteville section thickens to about 26 m (85 ft), some 6.1 to 7.6 m (20 to 25 ft) of which are represented by the Mifflin Member (area 5; 3 collecting localities); again characterized by orthid brachiopod populations, with some pleurotomariid and murchisoniid gastropod-dominated assemblages; taxonomically a benthic fauna not unlike that of region III, but a comparison of numerical dominance points up major differences between the regions.

Region VI

Limestones composed of finely fragmental skeletal material, sparsely fossiliferous, in northeastern Iowa and southeastern Minnesota; grades northward into an argillaceous dolomite, presumably the Hidden Falls Member of Weiss and Bell (1956; area 14; 8 collecting localities); very sparse fauna, mainly atrypid-orthid brachiopod, cryptostome bryozoan, and crinoidal debris.

Previous interpretations of the paleoecology of the Middle Ordovician epeiric seas have regarded them as shallow seas probably located in a tropical or subtropical belt (Williams, 1969; Whittington and Hughes, 1972). The extremely high diversities of benthic organisms in the Platteville Formation (Bretsky, 1970, and unpub. data) tend to confirm these reconstructions. The entire Platteville appears to have been deposited under quiet subtidal conditions; there is no petrographic evidence of supratidal environments, and there are only sparse indications of intermittent high-energy bottom conditions (compare Badiozamani, 1972). The Mifflin sediments were presumably originally lime muds. Variations in the petrology of the Mifflin rocks seem to be less a result of differences in the original composition of the sedimentary materials than of differences in the composition of the fauna and in the degree and kind of biogenic reworking. As an example, a few large burrows (1.0 to 1.5 cm in diameter) are characteristic of localities in portions of regions I, II, and III, where they are commonly associated with a fauna of large mollusks. In region IV the burrows are normally finer (about 1 mm in diameter), numerous, and the sediment contains a high percentage of pelletal material; the dominant members of the shelly benthos are typically nonmolluscan (orthid brachiopods, cryptostome bryozoans, and ostracodes), and all the animals are relatively small sized (not greater than 1 cm in largest dimension).

ORGANISMS

The two taxa of strophomenid brachiopods involved in this study of population variability and clines, to which we refer as *Oepikina minnesotensis* (Winchell) and *Strophomena plattinensis* Fenton, each account for about 3 percent by number of the total Mifflin fossil fauna (Bretsky, unpub. data). They were widespread throughout the upper Mississippi valley and lived in a concave-up position (see Richards, 1972, for discussion of living position) on the predominantly lime-mud substratum. Pope (1966) and Williams and others (1965) have reviewed the morphology and systematics of the strophomenoids. Pope's study is particularly significant because he has recounted in detail the evolutionary history of the shield-shaped, concavo-convex Oepikinidae and the convexi-concave Strophomenidae. The Oepikinidae, a primitive family ancestral to the Rafinesquinidae, are characterized by a large diductor muscle field completely surrounding the adductor scar and by prominent brachial septa (Pls. 1 and 2). The Strophomenidae are convexi-concave in profile and resupinate; Pope (1966, p. 99–103) reviewed the taxonomy and phylogeny of the genus *Strophomena*. Thus, the midcontinent Middle Ordovician seas were populated not only by a representative of the ancestral strophomenoid stock, *Oepikina minnesotensis*, but also by the phylogenetically more advanced *Strophomena plattinensis* (Pls. 3 and 4).

Species-level taxonomy in both these Middle Ordovician genera has tended toward increasing the number of species names to express what appear to have been only slight morphological differences among contiguous populations. In most cases, the definitions of species have been almost monothetic; that is, species of *Oepikina* and *Strophomena* were defined on the degree of valve inflation or convexity, or else on the size and placement of the adductor and diductor muscle scars in the pedicle valve.

In a review of Middle Ordovician species from North America previously assigned to the genus *Rafinesquina*, Salmon (1942) introduced the genus *Oepikina* (type species *O. septata* Salmon, 1942, p. 591). Of seven species which she referred to the genus, three—*O. minnesotensis* (Winchell, 1881), *O. transitionalis* (Okulitch, 1935), and *O. inquassa* (Sardeson, 1892)—were thought to be represented in the Platteville Formation. In her key to the species of *Oepikina*, Salmon employed the position of the point of greatest curvature of the shell (that is, a measure of the symmetry of the valve profile) as the basis for species identification. The degree of inflation of the shell was also used to discriminate among species that were inflected at the same position relative to the umbo. She did mention having had some difficulty in applying these definitions in individual cases, mainly because

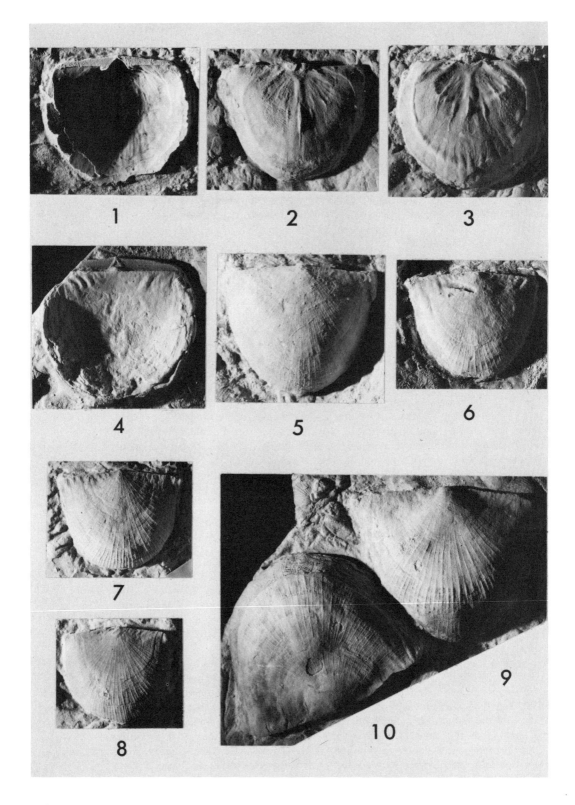

PLATE 1

OEPIKINA MINNESOTENSIS (WINCHELL), 1881

Figure 1. N.Y.S.M. no. 12877, loc. 2, area 12, pedicle internal.

Figure 2. N.Y.S.M. no. 12878, loc. 2, area 12, brachial internal.

Figure 3. N.Y.S.M. no. 12879, loc. 206, area 9, brachial internal.

Figure 4. N.Y.S.M. no. 12880, loc. 2, area 12, brachial external, articulated valves; upper left lost through careless preparation of a petrographic thin section.

Figure 5. N.Y.S.M. no. 12881, loc. 15, area 9, pedicle external.

Figure 6. N.Y.S.M. no. 12882, loc. 52, area 11, pedicle external.

Figure 7. N.Y.S.M. no. 12883, loc. 14, area 9, pedicle external.

Figure 8. N.Y.S.M. no. 12884, loc. 206, area 9, pedicle external.

Figure 9. N.Y.S.M. no. 12885, loc. 209, area 4, pedicle external.

Figure 10. N.Y.S.M. no. 12886, loc. 209, area 4, pedicle external.

All figures are × 2.

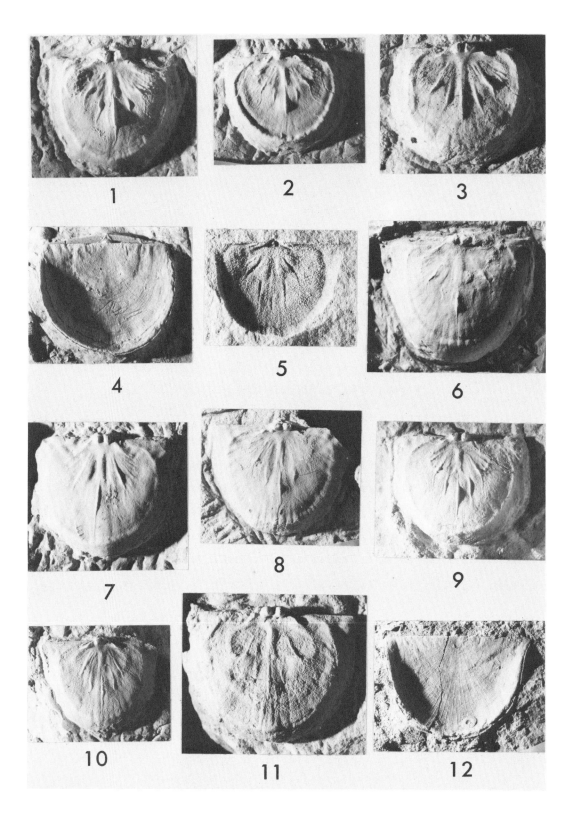

PLATE 2

OEPIKINA MINNESOTENSIS (WINCHELL), 1881

Figure 1. N.Y.S.M. no. 12887, loc. 2, area 12, brachial internal.

Figure 2. N.Y.S.M. no. 12888, loc. 2, area 12, brachial internal.

Figure 3. N.Y.S.M. no. 12889, loc. 2, area 12, brachial internal.

Figure 4. N.Y.S.M. no. 13185, loc. 14, area 9, brachial internal.

Figure 5. N.Y.S.M. no. 13186, loc. 42, area 1, brachial internal mold.

Figure 6. N.Y.S.M. no. 13187, loc. 203, area 12, brachial internal.

Figure 7. N.Y.S.M. no. 13188, loc. 2, area 12, brachial internal.

Figure 8. N.Y.S.M. no. 13189, loc. 65, area 7, brachial internal.

Figure 9. N.Y.S.M. no. 13190, loc. 305, area 2, brachial internal.

Figure 10. N.Y.S.M. no. 13191, loc. 206, area 9, brachial internal.

Figure 11. N.Y.S.M. no. 13192, loc. 211, area 4, brachial internal.

Figure 12. N.Y.S.M. no. 13193, loc. 54, area 7, brachial external, valves articulated.

All figures are × 2.

PLATE 3

STROPHOMENA WINCHELLI HALL AND CLARKE, 1892

Figure 1. N.Y.S.M. no. 13194, loc. 203, area 12, brachial external.

Figure 2. N.Y.S.M. no. 13195, loc. 14, area 9, pedicle internal.

STROPHOMENA PLATTINENSIS FENTON, 1928

Figure 3. N.Y.S.M. no. 13196, loc. 206, area 9, pedicle internal.

Figure 4. N.Y.S.M. no. 13197, loc. 205, area 10, brachial internal.

Figure 5. N.Y.S.M. no. 13198, loc. 15, area 9, pedicle internal.

Figure 6. N.Y.S.M. no. 13199, loc. 42, area 1, pedicle external.

Figure 7. N.Y.S.M. no. 13200, loc. 2, area 12, brachial external.

Figure 8. N.Y.S.M. no. 13201, loc. 55, area 7, pedicle internal.

Figure 9. N.Y.S.M. no. 13202, loc. 205, area 10, pedicle internal.

Figure 10. N.Y.S.M. no. 13203, loc. 2, area 12, brachial internal.

All figures are $\times$ 2.

PLATE 4

STROPHOMENA PLATTINENSIS FENTON, 1928

Figure 1. N.Y.S.M. no. 13204, loc. 228, area 4, pedicle internal.

Figure 2. N.Y.S.M. no. 13205, loc. 209, area 4, pedicle internal, upper left partial *Thaleops* pygidium.

Figure 3. N.Y.S.M. no. 13206, loc. 1, area 3, pedicle internal mold.

Figure 5. N.Y.S.M. no. 13208, loc. 61, area 16, pedicle internal mold.

Figure 6. N.Y.S.M. no. 13209, loc. 65, area 7, pedicle internal.

STROPHOMENA PLATTINENSIS?

Figure 4. N.Y.S.M. no. 13207, loc. 311, area 8, pedicle internal.

All figures are × 2.

of the possibility of subtle gradations in symmetry and inflation among various species. No specific collecting localities were mentioned for the various Platteville species; the total Platteville sample studied by Salmon consisted of 23 specimens, which she assigned as follows:

Oepikina minnesotensis (7 specimens). Highly inflated; asymmetrical, with point of inflection of shell located near the middle or slightly toward the anterior portion of the valve (point of maximum convexity occurring about 10 mm from beak on a specimen about 17 mm long).

Oepikina inquassa (4 specimens; see also Winchell and Schuchert, 1895, p. 401, Pl. 31). Inflation low to moderately high; shell profile symmetrical and gently curved or asymmetrical and broadly angular; larger than *O. minnesotensis*, with point of inflection located near shell anterior, about 18 mm from beak; considered a variety of *O. minnesotensis* by Winchell and Schuchert (1895) and Shrock and Raasch (1937).

Oepikina transitionalis (12 specimens). Inflation low; point of maximum inflation located about 16 mm from beak on a specimen about 26 mm long; flatter and more broadly rounded than either of the two preceding species.

In our sample of over 1,500 *Oepikina* specimens (437 of which were measured for the multivariate analysis) from the Mifflin and lower McGregor Members of the Platteville Formation, we have found these earlier species concepts impossible to apply in practice. Cooper (1956), although agreeing with Salmon about the taxonomic significance of the degree of convexity and the position of the point of inflection or geniculation, differed with her about the identification of Platteville specimens. He listed only *O. minnesotensis* as occurring in the Platteville-McGregor strata of the upper Mississippi valley. He thought *O. inquassa* to be present in the overlying Decorah Formation (Cooper, 1956, p. 912) and could not clearly identify *O. transitionalis*. He also concurred with Fenton (1928a) and Shrock and Raasch (1937) in regarding *O. minnesotensis* as being extremely wide-ranging in the upper Mississippi valley. We therefore view the *Oepikina* populations sampled from the various Mifflin and lower McGregor local sections as belonging to a single polymorphic species that ranged over a wide expanse of the Middle Ordovician epeiric sea. Although the characters of valve profile, which the earlier workers regarded as diagnostic of species, appear to intergrade completely, we realize that it is quite possible that closely related species could have responded to the same environmental influences in such a way as to produce similar valve profiles and that future studies using material in which the shell interiors are better preserved may well justify the recognition of more than one Platteville species of *Oepikina*.

With the other common Mifflin strophomenoid genus, *Strophomena*, problems of species-level taxonomy also abound and are complicated by the fact that previous definitions of *Strophomena* species in the upper Mississippi valley have tended to be inspired more by stratigraphic considerations than by a careful study of the limits of intraspecific variability. Furthermore, there has been a controversy over the taxonomic usefulness of various morphological criteria in the definition of species. In the original definition of *Strophomena plattinensis*, the species to which we tentatively assign most of our more than 1,500 Mifflin-McGregor specimens of *Strophomena*, Fenton (1928b) contended that shell shape had little significance in discriminating among closely related (presumably congeneric) brachiopod species (compare Foerste, 1912). Rather, species differentiation, according to Fenton, should be based on internal features, especially muscle scars and medial and lateral ridges. Cooper (1956, p. 926), while believing that all morphological features should initially be taken into account, considered that musculature, especially that of the pedicle valve, is variable. As in defining species of *Oepikina*, he regarded valve shape as the most useful discriminating character.

Previous workers (Shrock and Raasch, 1937; Cooper, 1956) have identified the following species and subspecies of *Strophomena* among material from the Platteville Formation:

Strophomena plattinensis Fenton, 1928b (see Pls. 3 and 4). Valves large, flat; considered characteristic of the Mifflin-McGregor as well as the underlying Pecatonica Dolomite; sometimes misidentified as *S. filitexta* Hall, 1847, a New York and Ontario species (Fenton, 1929; Cooper, 1956).

S. plattinensis crassa Raasch, 1937 (in Shrock and Raasch, 1937). Valves flat; convexity of pedicle valve extremely low; differentiation from *S. plattinensis* vague.

S. auburnensis Fenton, 1928b. Similar in size to *S. plattinensis* (length about 25 to 35 mm) but more strongly inflated; brachial valve asymmetrical, with inflection point located about two-thirds of the way from the posterior margin; posterior margin flattened; pedicle valve regularly concave; also previously confused with *S. filitexta*.

S. auburnensis impressa Raasch, 1937. Supposedly more convex than *S. a. auburnensis*; vascular sinuses present (perhaps their absence in other "species" is only a preservational phenomenon; see Pl. 4, fig. 3).

S. winchelli Hall and Clarke, 1892 (Pl. 3, figs. 1 and 2). Valves very highly inflated; convexity twice that of *S. auburnensis*; anterior margin lobate rather than rounded, producing a triangular shell shape; assigned to *Trigrammaria* Wilson, 1945, by Cooper (1956).

Specimens readily assignable to the distinctive *S. winchelli* are present, but rare, in our material. We have measured 302 individuals for the multivariate analysis from the remaining collections. We agree with Cooper (1956, p. 926-927) that "*Strophomena* must some day be studied as a whole to make the group useful, but the study will require enormous collections." Since we have a substantial collection, but one which comes only from a restricted stratigraphic interval and geographic region, we have again chosen to treat all members of what appears, at least from consideration of characters of the valve profile, to be a single unit with a good deal of intrataxon variability as a single widespread species population.

Richards (1969, 1972) and Alexander (1972) have suggested that intraspecific variability in inflation and in symmetry of the valves of concavo-convex and convexi-concave brachiopods could have resulted from variations in the rate of sedimentation or in the degree of firmness of the substratum (for example, soft versus firm lime muds or shell beds). Both authors indicated that these brachiopods may have possessed a high degree of phenotypic plasticity—presumably an ability of any individual organism to adapt itself readily to the particular microenvironmental conditions in which it occurred, rather than a high level of genetic polymorphism within the population as a whole. Rudwick (1965) and Richards (1972) have speculated that highly inflated specimens may have been those that sank rather deeply into a soft substratum and thus had to attain a curved profile to keep the anterior margin clear of the substratum. Rudwick believed that these brachiopods may have thus been essentially infaunal with the concave valve filled with sediment and only the commissure projecting above the sediment. Richards added that rapid growth to keep up with a rapid influx of sediment, rather than sinking into a soft substratum, could result in a similar valve shape. Hence, it seems that the profile of the valves may have been adapted rather closely to the immediate environmental setting.

Most of the Mifflin brachiopods occur as disarticulated valves which are nearly all convex up; however, the relatively few articulated specimens are found in the concave-up position. It appears reasonable to suppose that the concave-up occurrences represent life positions for both *Strophomena* and *Oepikina*. The disarticulated valves are rarely broken and never abraded; it thus seems likely that postmortem reorientation was accomplished either by gentle water movements or by the activities of other organisms. We believe that the variations in inflation and symmetry which some workers have regarded as species-specific characters are, in fact, as Rudwick and Richards have suggested, more likely to be the results of adaptation to slightly different environmental conditions, perhaps largely the fluidity of the substratum or the rate of sedimentation, or both. Whether the variation within a species was the result of a high potential for adaptability of individual organisms or of a stringent selection for characteristics suitable to a given microenvironment so that only those organisms that were suited to a particular set of conditions survived to become adults, we cannot at this point be certain.

MEASUREMENTS AND ANALYSIS

Preparation of the specimens to be measured consisted of a preliminary identification, including inspection for breakage or signs of obvious compression or flattening, followed by the complete removal of any sediment that was encrusting the valve surface or any portion of the valve margins. A second identification was made on the cleaned valve, and specimens were then coated with latex (Permamold, Polymer Chemical Co., Cincinnati, Ohio). When hardened, the latex impression was removed. The location of the medial-sagittal axis was carefully inscribed on the mold, and the latex impression was then carefully cut along this plane (essentially bisecting the "shell" along the medial anterior-posterior axis). The two halves

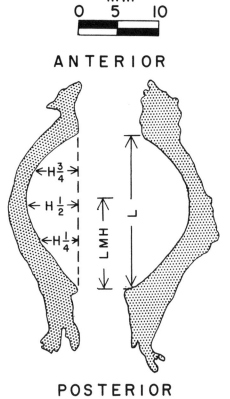

Figure 2. Line drawing of medially (anterior-posterior) bisected, mounted latex impression showing the symmetrical profiles. Taken from photographed specimen 10, location 2, area 12; *Oepikina* pedicle external. Parameters measured along the medial-sagittal plane are discussed in text. Shape: Log L; HW/L. Inflation: $H^1/_4/L$; $H^1/_2/L$; $H^3/_4/L$. Symmetry LMH/L; $H^1/_4/H^3/_4$.

were mounted in a fine sand mold, and the cut planar surfaces were coated with India ink. The mounted latex impressions were photographed, and the measurements of valve inflation and symmetry were made with a vernier caliper from the photographs. There were 256 *Oepikina* pedicle valves, 181 *Oepikina* brachials, 211 *Strophomena* pedicle valves, and 91 *Strophomena* brachials (see Table 1 for geographic distribution of the samples of each valve type).

Figure 2 illustrates the measurements that were made to define valve size, shape, inflation, and symmetry. The value of shell length (L) was converted to its base-10 logarithm (log L) in order to make this measurement more nearly commensurate in magnitude with the other characters, which were expressed as ratios. The ratio of hinge-line width to length (HW/L) measures the general shape of the shell. The height (H) of the shell above the horizontal plane was measured at points taken at $1/4$, $1/2$, and $3/4$ the distance from the posterior to the anterior margin of the shell. Ratios of these heights to valve length ($H^{1/4}/L$, $H^{1/2}/L$, and $H^{3/4}/L$, respectively) measure the inflation. The distance from the posterior margin to the highest point on the valve profile (that is, the point of inflection or of maximum convexity of the valve) was recorded. The ratio of this value to total valve length (LMH/L) and the ratio $H^{1/4}/H^{3/4}$ (which compares the degree of inflation at a point near the hinge line to that at a point near the anterior margin of the valve) measure the symmetry of the shell; a perfectly symmetrical valve profile would have $LMH/L = 0.50$ and $H^{1/4}/H^{3/4} = 1.00$.

We used the Numerical Taxonomy System (NTSYS) collection of programs for multivariate analysis, developed for Stony Brook's IBM-360 computer by F. J. Rohlf, J. Kishpaugh, and D. Kirk of the Department of Ecology and Evolution, Stony Brook, to carry out a principal components analysis on each of the four valve types—*Oepikina* brachial valves, *Oepikina* pedicles, *Strophomena* brachials, and *Strophomena* pedicles. Principal components analysis is a technique of multivariate morphometrics (Blackith and Reyment, 1971) which can accomplish several important aims in a study of the sort reported here. Some of the several characters measured are likely to be more or less highly correlated, thus representing, in effect, only one independent dimension of variation. The technique of transforming the correlated original measurements to uncorrelated principal component axes also reduces the complexity of the data matrix, permitting one to concentrate on explaining patterns of variability of those characters that contribute most to the total variation within the group of characters being studied. Similarly, determination of character correlations and the uncorrelated principal component axes permits separate analysis of those characters (or character complexes) that are relatively independent of other characters or groups of characters. Also, by plotting the position of individual specimens relative to the principal component axes (compare Rohlf, 1968; Temple, 1968), one can attempt to define distinct groups of specimens—or to convince oneself that intergradation is complete.

Using the NTSYS programs of Sokal and Sneath (1963, p. 293-295), we standardized the character measurements for each of the four 7-character × N-specimen matrices. The 7×7 matrix of correlations among all possible pairs of characters was computed, and the first three principal components were extracted from the correlation matrix. For all four valve types, the first principal component accounts for at least 40 percent of the variation within the correlation matrix, and the first two together for more than 60 percent of the variation (Tables 2B, 3B, 4B, and 5B). In all four cases, the three inflation characters ($H^{1/4}/L$, $H^{1/2}/L$, $H^{3/4}/L$) loaded strongly on the first principal component axis (Tables 2A, 3A, 4A, and 5A); the absolute magnitudes of the loadings always exceeded 0.8 and frequently were greater than 0.95. This axis, therefore, can be considered primarily a measure

TABLE 2. RESULTS OF PRINCIPAL COMPONENTS ANALYSIS OF *OEPIKINA* PEDICLE VALVES

A. Loadings of characters on first three principal component axes

Character	Axis 1	Axis 2	Axis 3
Log L	0.693	−0.075	0.336
HW/L	−0.253	0.189	−0.898
$H^{1/4}/L$	0.876	−0.189	−0.246
$H^{1/2}/L$	0.952	−0.204	−0.125
$H^{3/4}/L$	0.949	0.034	−0.176
LMH/L	0.315	0.813	0.068
$H^{1/4}/H^{3/4}$	−0.389	−0.754	−0.053

B. Proportion of variability accounted for by first three principal component axes

Axis	Eigenvalue	Percent of variability	Cumulative percentage
1	3.370	48.1	48.1
2	1.423	20.3	68.4
3	1.035	14.8	83.2

C. Matrix of correlations among characters

	Log L	HW/L	$H^{1/4}/L$	$H^{1/2}/L$	$H^{3/4}/L$	LMH/L	$H^{1/4}/H^{3/4}$
Log L	1.000						
HW/L	−0.299	1.000					
$H^{1/4}/L$	0.481	−0.099	1.000				
$H^{1/2}/L$	0.568	−0.195	0.922	1.000			
$H^{3/4}/L$	0.502	−0.120	0.845	0.929	1.000		
LMH/L	0.163	0.002	0.006	0.122	0.339	1.000	
$H^{1/4}/H^{3/4}$	−0.198	0.016	−0.093	−0.220	−0.354	−0.492	1.000

TABLE 3. RESULTS OF PRINCIPAL COMPONENTS ANALYSIS OF *OEPIKINA* BRACHIAL VALVES

A. Loadings of characters on first three principal component axes

Character	Axis 1	Axis 2	Axis 3
Log L	0.522	−0.025	0.580
HW/L	−0.286	−0.064	−0.806
$H^{1/4}/L$	0.819	−0.292	−0.315
$H^{1/2}/L$	0.960	−0.175	−0.099
$H^{3/4}/L$	0.940	0.000	−0.166
LMH/L	0.140	0.860	−0.166
$H^{1/4}/H^{3/4}$	−0.359	−0.785	0.005

B. Proportion of variability accounted for by first three principal component axes

Axis	Eigenvalue	Percent of variability	Cumulative percentage
1	2.977	42.5	42.5
2	1.477	21.1	63.6
3	1.151	16.4	80.0

C. Matrix of correlations among characters

	Log L	HW/L	$H^{1/4}/L$	$H^{1/2}/L$	$H^{3/4}/L$	LMH/L	$H^{1/4}/H^{3/4}$
Log L	1.000						
HW/L	−0.309	1.000					
$H^{1/4}/L$	0.202	−0.040	1.000				
$H^{1/2}/L$	0.412	−0.195	0.822	1.000			
$H^{3/4}/L$	0.355	−0.156	0.742	0.933	1.000		
LMH/L	−0.014	0.003	−0.049	−0.024	0.179	1.000	
$H^{1/4}/H^{3/4}$	−0.141	0.111	−0.063	−0.223	−0.285	−0.498	1.000

TABLE 4. RESULTS OF PRINCIPAL COMPONENTS ANALYSIS OF *STROPHOMENA* PEDICLE VALVES

A. Loadings of characters on first three principal component axes

Character	Axis 1	Axis 2	Axis 3
Log L	0.141	−0.558	0.231
HW/L	0.054	−0.561	0.587
$H^{1/4}/L$	−0.926	−0.236	−0.089
$H^{1/2}/L$	−0.984	−0.067	0.035
$H^{3/4}/L$	−0.960	0.121	0.056
LMH/L	−0.045	0.765	0.157
$H^{1/4}/H^{3/4}$	0.065	−0.394	−0.794

B. Proportion of variability accounted for by first three principal component axes

Axis	Eigenvalue	Percent of variability	Cumulative percentage
1	2.777	39.6	39.6
2	1.441	20.6	60.2
3	1.065	15.2	75.4

C. Matrix of correlations among characters

	Log L	HW/L	$H^{1/4}/L$	$H^{1/2}/L$	$H^{3/4}/L$	LMH/L	$H^{1/4}/H^{3/4}$
Log L	1.000						
HW/L	0.173	1.000					
$H^{1/4}/L$	−0.047	0.014	1.000				
$H^{1/2}/L$	−0.078	0.006	0.896	1.000			
$H^{3/4}/L$	−0.113	−0.087	0.802	0.942	1.000		
LMH/L	−0.162	−0.202	−0.174	0.009	0.916	1.000	
$H^{1/4}/H^{3/4}$	0.071	−0.045	0.065	−0.053	−0.122	−0.176	1.000

TABLE 5. RESULTS OF PRINCIPAL COMPONENTS ANALYSIS OF *STROPHOMENA* BRACHIAL VALVES

A. Loadings of characters on first three principal component axes

Character	Axis 1	Axis 2	Axis 3
Log L	0.290	−0.617	0.514
HW/L	−0.304	0.339	0.800
$H^{1/4}/L$	0.965	−0.088	0.027
$H^{1/2}/L$	0.975	0.082	0.004
$H^{3/4}/L$	0.913	0.328	−0.013
LMH/L	−0.916	0.750	−0.160
$H^{1/4}/H^{3/4}$	−0.188	−0.714	−0.241

B. Proportion of variability accounted for by first three principal component axes

Axis	Eigenvalue	Percent of variability	Cumulative percentage
1	2.967	42.4	42.4
2	1.691	24.2	66.6
3	0.990	14.1	80.7

C. Matrix of correlations among characters

	Log L	HW/L	$H^{1/4}/L$	$H^{1/2}/L$	$H^{3/4}/L$	LMH/L	$H^{1/4}/H^{3/4}$
Log L	1.000						
HW/L	−0.034	1.000					
$H^{1/4}/L$	0.296	−0.251	1.000				
$H^{1/2}/L$	0.204	−0.217	0.940	1.000			
$H^{3/4}/L$	0.043	−0.145	0.829	0.914	1.000		
LMH/L	−0.376	0.175	−0.252	−0.105	0.080	1.000	
$H^{1/4}/H^{3/4}$	0.193	−0.176	−0.067	−0.163	−0.337	−0.262	1.000

of inflation. Similarly, the highest loading on the second principal component axis for all four valve types was that of the symmetry character LMH/L. For *Oepikina* pedicle and brachial valves and *Strophomena* brachial valves, the second symmetry character, $H^{1/4}/H^{3/4}$, had a loading almost as high as that of LMH/L; however, its loading on this axis for *Strophomena* pedicle valves was lower (Tables 2A, 3A, 4A, and 5A). The principal components analysis suggests, therefore, that the most important independent dimensions of variability in these brachiopods are those of inflation and symmetry (the latter being best defined in terms of the position of the point of inflection of the valve profile). We have accordingly analyzed trends in variability in both these features.

MAPPING OF CLINES

When only relatively few individual specimens are considered in a multivariate study, it is frequently possible to delimit clusters or to recognize marked trends by plotting the position of each individual relative to pairs of principal component axes. But because of the large number of specimens used in this study, such plots of the position of individual strophomenid specimens of a given valve type relative to the first two principal components gave pictures that were difficult to interpret. We therefore concentrated on examining possible changes from place to place within the study area in the mean values of characters that loaded strongly on the first two. We did so by constructing clinal maps based on the mean value of the selected character for all the specimens from a given sampling area (Fig. 1; Table 1); we should emphasize that the values plotted on these maps are means of the original character data rather than those of character loadings on principal component axes. The contours were drawn by hand, as we did not have access to a mechanical plotter. K. W. Flessa independently, without knowledge of the inferred paleogeography of the area, contoured the data and produced maps similar to ours.

As the correlation matrices (Tables 2C, 3C, 4C, and 5C) show, all three inflation characters are highly correlated. In making the clinal maps, the value $H^{3/4}/L$ is used to represent inflation. The "symmetry factor" is represented by LMH/L, which consistently had a high loading on the second principal component axis. Thus, mean values of $H^{3/4}/L$ and LMH/L for each valve type at each of the 16 sampling areas were computed (Tables 6, 7, 8, and 9), plotted on base maps, and contoured (Figs. 3 through 10). The trends of the contours (isophenes) outline clinal patterns which, though not identical in all valve types or for both characters, show a striking correspondence in several features. R. Cowen, in his review of the manuscript, remarked that area 5 (region V) seems to dominate the isophene patterns, and he submitted alternate reconstructions in which area 5 is viewed as an "outlier," hence negating the development of the south-trending salient in some of the figures. This approach tends to produce a bilaterally symmetrical contour pattern with gradients trending north-to-south, southeast, and southwest, but one which over the bulk of the study area is not unlike patterns constructed with the inclusion of the salient. We also considered fitting trend surfaces to the data but were advised (J. Grover, 1973, oral commun.) that the points appeared to be too irregularly spaced for the assumptions of the technique to be valid. The spacing of data points, of course, has resulted from the location of accessible exposures of the Mifflin Member.

With the exception of *Strophomena* brachial valves (Fig. 9), specimens became more inflated in a generally south, southeast, and southwest direction (Figs. 3, 5, and 7); the localities with the lowest inflation values predominate along a northern

TABLE 6. WITHIN-AREA MEANS AND STANDARD DEVIATIONS OF INFLATION AND SYMMETRY CHARACTERS, *OEPIKINA* PEDICLE VALVES

Area	Inflation ($H^{3}/_{4}/L$) mean (S.D.)	Symmetry (LMH/L) mean (S.D.)	Number of specimens
1	0.207 (0.055)	0.580 (0.097)	24
2	0.191 (0.050)	0.568 (0.084)	13
3	0.202 (0.050)	0.552 (0.074)	16
4	0.223 (0.042)	0.513 (0.159)	3
5	0.197 (0.036)	0.545 (0.094)	8
6	0.207 (0.046)	0.552 (0.079)	6
7	0.157 (0.067)	0.517 (0.065)	12
8	0.161 (0.067)	0.554 (0.102)	14
9	0.205 (0.081)	0.568 (0.120)	33
10	0.202 (0.082)	0.565 (0.079)	18
11	0.188 (0.060)	0.564 (0.087)	13
12	0.233 (0.081)	0.535 (0.094)	35
13	0.233 (0.074)	0.584 (0.056)	7
14	0.230 (0.039)	0.586 (0.092)	7
15	0.176 (0.055)	0.595 (0.098)	39
16	0.129 (0.052)	0.591 (0.100)	8
Total			256

periphery of the exposures, generally in regions II, III, and portions of I (Fig. 1). In each case, there also appears to be a south-trending salient or "outlier" of lower inflation values; as constructed, the salient extends into northern Illinois. If the general pattern of morphological variability reflects systematic variation in some environmental parameter, the correlation may have been with topography of the sea floor. The isophene pattern corresponds broadly to the contours of the buried Precambrian surface that defines the Wisconsin Dome (Ostrom and others, 1970, Fig. 3, p. 6). It is also in general agreement with the isopach map of Kistler (in Sloss and others, 1960; Fig. 11 herein) for the St. Peter–Platteville interval, in which one sees the development of a well-defined basin, the Illinois Basin, to the south and southeast of the Wisconsin Dome, as well as a gradual

TABLE 7. WITHIN-AREA MEANS AND STANDARD DEVIATIONS OF INFLATION AND SYMMETRY CHARACTERS, *OEPIKINA* BRACHIAL VALVES

Area	Inflation ($H^{3}/_{4}/L$) mean (S.D.)	Symmetry (LMH/L) mean (S.D.)	Number of specimens
1	0.159 (0.041)	0.659 (0.101)	14
2	0.136 (0.045)	0.676 (0.075)	14
3	0.096 (0.052)	0.607 (0.193)	10
4	omitted	omitted	(1)
5	0.134 (0.038)	0.635 (0.055)	10
6	0.130 (0.050)	0.630 (0.200)	3
7	0.160 (0.052)	0.652 (0.084)	17
8	0.124 (0.041)	0.685 (0.099)	19
9	0.157 (0.055)	0.647 (0.115)	27
10	0.170 (0.060)	0.650 (0.064)	12
11	0.166 (0.060)	0.609 (0.122)	18
12	0.173 (0.070)	0.634 (0.088)	18
13	0.140 (0.012)	0.654 (0.074)	5
14	0.177 (0.032)	0.708 (0.052)	6
15	0.118 (0.049)	0.713 (0.066)	6
16	omitted	omitted	(1)
Total			181

TABLE 8. WITHIN-AREA MEANS AND STANDARD DEVIATIONS OF INFLATION AND SYMMETRY CHARACTERS, *STROPHOMENA* PEDICLE VALVES

Area	Inflation ($H^{3/4}/L$) mean (S.D.)	Symmetry (LMH/L) mean (S.D.)	Number of specimens
1	0.086 (0.042)	0.602 (0.090)	50
2	0.094 (0.072)	0.522 (0.213)	8
3	0.054 (0.045)	0.494 (0.191)	16
4	0.117 (0.053)	0.490 (0.124)	7
5	0.084 (0.055)	0.579 (0.085)	10
6	0.081 (0.061)	0.565 (0.122)	13
7	0.117 (0.059)	0.632 (0.162)	8
8	0.114 (0.046)	0.627 (0.109)	9
9	0.120 (0.057)	0.522 (0.076)	20
10	0.118 (0.077)	0.505 (0.096)	13
11	0.112 (0.073)	0.589 (0.090)	8
12	0.148 (0.044)	0.684 (0.045)	10
13	0.165 (0.017)	0.592 (0.065)	4
14	omitted	omitted	(2)
15	0.085 (0.055)	0.635 (0.152)	23
16	0.051 (0.031)	0.699 (0.202)	10
Total			211

thickening of sediments toward the southwest into what may be a northeastward extension of the Forest City Basin. Detailed mapping of facies in the Platteville and Decorah (Agnew and others, 1956, Fig. 34, p. 257; Badiozamani, 1972; Sulima, 1974, in prep.) has demonstrated an intermingling of limestone and dolomite facies near the Wisconsin-Illinois state line and in northwestern Illinois. These facies patterns might suggest the existence of some topographic irregularities along the southern margin of the dome, perhaps extending into northwestern Illinois as an expression of the Wisconsin Arch. If the inflation gradients generally parallel bathymetric gradients, the increase in inflation accompanying increased water depth represents a correlation with, but not necessarily an immediate causal factor for, the clines.

TABLE 9. WITHIN-AREA MEANS AND STANDARD DEVIATIONS OF INFLATION AND SYMMETRY CHARACTERS, *STROPHOMENA* BRACHIAL VALVES

Area	Inflation ($H^{3/4}/L$) mean (S.D.)	Symmetry (LMH/L) mean (S.D.)	Number of specimens
1	0.133 (0.058)	0.522 (0.072)	29
2	0.120 (0.019)	0.478 (0.057)	5
3	0.135 (0.046)	0.507 (0.016)	6
4	omitted	omitted	(1)
5	omitted	omitted	(1)
6	0.110 (0.036)	0.442 (0.044)	4
7	omitted	omitted	(2)
8	0.190 (0.062)	0.447 (0.092)	3
9	0.162 (0.050)	0.475 (0.032)	6
10	0.107 (0.045)	0.483 (0.055)	3
11	0.107 (0.124)	0.440 (0.064)	4
12	0.135 (0.051)	0.541 (0.072)	8
13	omitted	omitted	(1)
14	omitted	omitted	0
15	0.133 (0.027)	0.511 (0.063)	11
16	0.123 (0.048)	0.523 (0.061)	7
Total			91

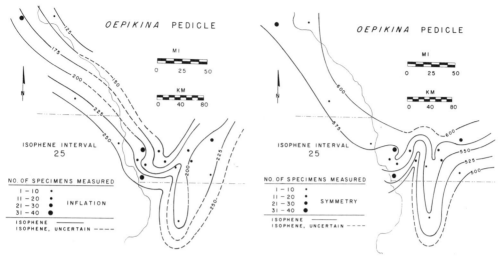

Figure 3. Clinal map showing isophenes based on the mean value ($\times 10^3$) of $H^{3/4}/L$ for *Oepikina* pedicle valves at each of the sampling areas (see also Table 6).

Figure 4. Clinal map showing isophenes based on the mean value ($\times 10^3$) of LMH/L for *Oepikina* pedicle valves at each of the sampling areas (see also Table 6).

The maps of variation in valve symmetry again exhibit the same general north-to-south, southeast, and southwest trends (Figs. 4, 6, and 8). In every case, asymmetrical specimens are more common along the northern and northwestern perimeter of the study area, and the south-trending salient again appears. Although a comparison of the details of the various clinal maps shows numerous local dissimilarities, especially with regard to the local steepness of the clines and to the precise position of the south-trending salient, the overall patterns of variability are strikingly similar. Whatever environmental factors are involved in selecting for the phenetic characteristics of the local populations, quite similar clinal expressions of two characters have resulted even though the characters are *not* highly correlated (Tables 2, 3, 4, and 5).

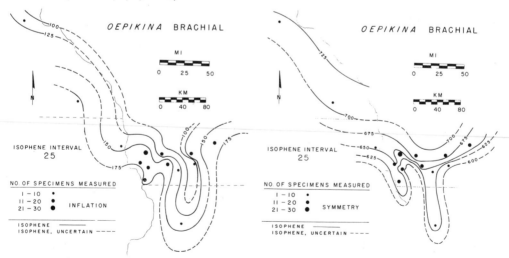

Figure 5. Clinal map showing isophenes based on the mean value ($\times 10^3$) of $H^{3/4}/L$ for *Oepikina* brachial valves at each of the sampling areas (see also Table 7).

Figure 6. Clinal map showing isophenes based on the mean value ($\times 10^3$) of LMH/L for *Oepikina* brachial valves at each of the sampling areas (see also Table 7).

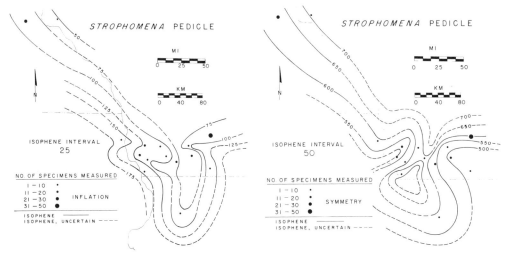

Figure 7. Clinal map showing isophenes based on the mean value ($\times\ 10^3$) of $H^{3/4}/L$ for *Strophomena* pedicle valves at each of the sampling areas (see also Table 8).

Figure 8. Clinal map showing isophenes based on the mean value ($\times\ 10^3$) of LMH/L for *Strophomena* pedicle valves at each of the sampling areas (see also Table 8).

Although inflation and symmetry account for much of the morphological variability in both *Oepikina* and *Strophomena* populations, the other characters also may show a moderate to high degree of correlation with either inflation or symmetry, as shown by the values of their loadings on the first or second principal component axis. A more detailed look at these character correlations may give further insight into the geographic expression of a suite of integrated but intraspecifically variable characters.

Briefly, for the *Oepikina* pedicle and brachial valves, valve size (log L) loads fairly heavily on the first principal component axis (Tables 2A and 3A), being correlated at about the 0.4 to 0.5 level with each of the inflation characters (Tables 2C and 3C). It would thus appear that smaller valves should generally be less highly inflated than larger ones, and that the population along the northern rim should, on the whole, be smaller than those to the south, southeast, and southwest.

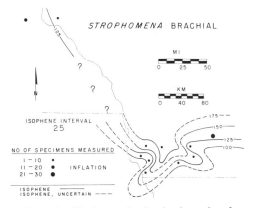

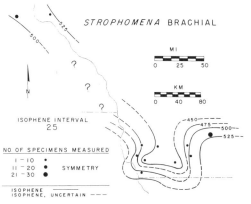

Figure 9. Clinal map showing isophenes based on the mean value ($\times\ 10^3$) of $H^{3/4}/L$ for *Strophomena* brachial valves at each of the sampling areas (see also Table 9).

Figure 10. Clinal map showing isophenes based on the mean value ($\times\ 10^3$) of LMH/L for *Strophomena* brachial valves at each of the sampling areas (see also Table 9).

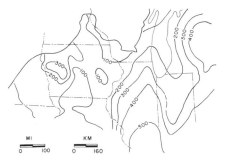

Figure 11. Simplified isopach map of the St. Peter-Simpson-Platteville interval taken from Sloss and others (1960). Contours are in feet and indicate the development of the Illinois and Forest City Basins, respectively, to the southeast and southwest of the Wisconsin Dome.

It appears that the maximum size attained by specimens from Minnesota (regions II and VI, see Fig. 1), but *not* Wisconsin (region III), populations tends to be somewhat less than to the south; however, the overall inflation of the Minnesota specimens remains well within the range of values attained by those populations from the more southeasterly localities. R. P. Richards (June 1973, written commun.) has suggested, as an explanation of the correlation between size and degree of inflation, that any initial inflation in the upper valve would act as an effective sediment trap; thus inflation would subsequently proceed at an increasingly greater rate in order to keep the shell margin above the accumulating sediment cover. One could also envision a gradual decline in overall shell growth rate, whereas accumulation of sediment, proceeding at a fairly regular rate, dictated an increasing inflation throughout ontogeny to maintain the plane of commissure above the sediment-water interface. In order to test any of these hypotheses, it would be necessary to develop an independent criterion of age for any given brachiopod fossil shell.

VARIATIONS WITHIN AND AMONG POPULATIONS

In the preceding section, we have characterized each of the 16 major sampling areas semitypologically by the mean value of its inflation or symmetry character for each of the four valve types. We now consider how variable these characters are within each of the local populations, and we should especially like to determine whether there is any systematic change in the level of variability that can be related to the previously reported clines. Perhaps it is not only our "residual typological bias" (Gould, 1969, p. 409), but also the fact that most familiar techniques of statistical analysis theoretically require equality of variances or homogeneity of variance-covariance matrices, which leads us to skim lightly over the existence of intrasample variability. That the maintenance of a high level of variability may be as much an adaptation to the environment as is the possession of any given allele or morphological character has recently been postulated by Slobodkin and Sanders (1969) and Bretsky and Lorenz (1970). Evidence has been presented that there is a strong correlation between the degree of genetic and morphological variability, even though the genetic and morphological characters on which measurements of variability are made may be responding to different factors of the environment (Soulé, 1972; Soulé and others, 1973). [For a summary of some of the varied arguments on the relation between environmental, genetic, and morphological variability, see also Valentine and others (1973), Bretsky (1973), and Levinton (1973).]

Table 10 records means and coefficients of variation of $H^{3/4}/L$ and LMH/L for *Oepikina* pedicle and brachial valves and *Strophomena* pedicle valves from each of the six major regions (Fig. 1). [C_v = (standard deviation/mean) × 100;

TABLE 10. WITHIN-REGION MEANS AND COEFFICIENTS OF
VARIATION FOR INFLATION ($H^{3/4}/L$) AND SYMMETRY (LMH/L)

Region	Inflation		Symmetry	
	Mean	C_v	Mean	C_v
A.		Oepikina pedicle		
I	0.206	24.8	0.553	18.0
II	0.153	35.8	0.593	16.7
III	0.159	42.2	0.536	15.6
IV	0.212	35.7	0.563	15.5
V	0.197	18.3	0.545	17.2
VI	0.230	17.0	0.586	15.7
B.		Oepikina brachial		
I	0.130	37.9	0.643	22.5
II	0.118	41.5	0.713	9.3
III	0.142	33.5	0.669	13.7
IV	0.161	36.7	0.639	14.3
V	0.134	28.4	0.635	8.7
VI	0.177	18.1	0.708	7.3
C.		Strophomena pedicle		
I	0.086	65.9	0.535	28.3
II	0.068	62.8	0.667	26.4
III	0.116	45.4	0.630	21.5
IV	0.133	43.6	0.578	13.3
V	0.084	65.5	0.579	14.7
VI		omitted		omitted

Strophomena brachial valves were omitted from this analysis because of the relatively small sample size of this valve type.] In Figures 12, 13, and 14, values of the within-region means for the inflation and symmetry characters of a given valve type have been plotted against the values for C_v for that region. In all three valve types, there is a tendency for C_v to increase sharply with decrease in inflation. Hence, those populations, mostly along the northern periphery of the Mifflin exposures, that have low mean values of inflation also have a high level of variability in inflation (for contouring of C_v of inflation for Oepikina and Strophomena pedicle valves, see Figs. 15 and 16). Conversely, variation in symmetry is low and shows no regular trends for the Oepikina valves (Figs. 12 and 13). With the exception of region I, C_v for Strophomena pedicle valves increases as the magnitude of LMH/L (symmetry) decreases (Fig. 14). Regions II and III, in the north-northwest, have high C_v values for symmetry in Strophomena pedicle valves (for contouring, see Fig. 17), as they did for inflation in Oepikina pedicle and brachial valves and Strophomena pedicle valves.

For both Strophomena pedicle and Oepikina brachial valves, region I appears anomalous (compare Figs. 13 and 14) in that variation is rather high although the valves are fairly symmetrical, whereas other regions characterized by more or less symmetrical valves tend to have the lowest values of C_v for a given valve type. On the whole, the localities along the northern periphery of exposures tend to contain the most highly variable populations, especially with regard to inflation (Figs. 15 and 16). In one case, this is also true of symmetry (Fig. 17), but the magnitude of variation in symmetry appears to be less than that of variation in inflation.

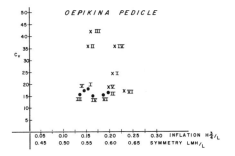

Figure 12. Plots of values of the within-region means for the inflation (×) and symmetry (●) characters of *Oepikina* pedicle valves and the corresponding values of coefficient of variation (see also Table 10A).

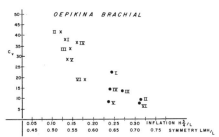

Figure 13. Plots of values of the within-region means for inflation (×) and symmetry (●) characters of *Oepikina* brachial valves and the corresponding values of coefficient of variation (see also Table 10B).

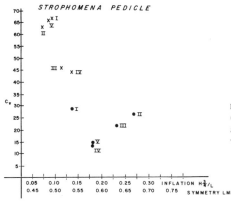

Figure 14. Plots of values of the within-region means for the inflation (×) and symmetry (●) characters of *Strophomena* pedicle valves and the corresponding values of coefficient of variation (see also Table 10C).

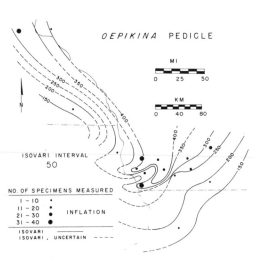

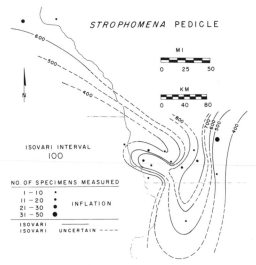

Figure 15. Clinal map showing isovaris based on the coefficient of variation (× 10) for inflation ($H^{3/4}/L$) for *Oepikina* pedicle valves at each of the sampling areas (see also Table 10A; compare Fig. 3).

Figure 16. Clinal map showing isovaris based on the coefficient of variation (× 10) for inflation ($H^{3/4}/L$) for *Strophomena* pedicle valves at each of the sampling areas (see also Table 10C; compare Fig. 7).

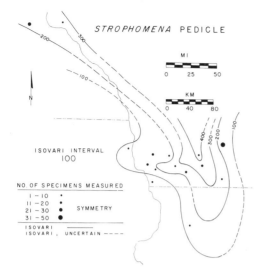

Figure 17. Clinal map showing isovaris based on the coefficient of variation ($\times$ 10) for symmetry (LMH/L) for *Strophomena* pedicle valves at each of the sampling areas (see also Table 10C; compare Fig. 8).

POSSIBLE INTERPRETATIONS OF THE C_v DATA

One way of interpreting the patterns of variability indicated by Figures 15, 16, and 17 is to view the position of a population relative to the abscissa on Figures 12, 13, and 14 (that is, the regional mean of inflation or of symmetry) as an indication of the "average" type of valve which was favored in a particular environment, and to regard the ordinate C_v as an axis that represents the relative strength of selective pressures for stabilization around that mean value. We noted earlier that the inflation and symmetry isophenes correspond broadly to contours on isopach and lithofacies maps that portray a northerly positive or dome region flanked on the south, east, and west by basins, the best developed of which lies immediately to the south-southwest of the dome. Furthermore, the isovari maps of *Oepikina* and *Strophomena* pedicle valves correspond to these isophene patterns. As mentioned above, some authors (Rudwick, 1965; Richards, 1972) have attempted to link the shapes of brachiopod valves to the firmness of the substratum, and it is possible that the greater degree of within-region variability in the northern populations (see Figs. 15, 16, and 17) may record the existence of a more heterogeneous substratum with a greater number of microhabitats, as is characteristic of shallow-water regions. For example, the shallow-water environments may have contained hard patches where flatter valves were favored and soft patches where more inflated valves were produced; whereas in the deeper water regimes, the substratum may have been uniformly soft, leading both to a higher magnitude of inflation and to less variation in inflation within the deeper water populations. It is a temptation, though less likely to be a valid explanation in this specific example, to regard the higher level of variability in the shallow-water strophomenids as confirmation of the hypotheses of Slobodkin and Sanders (1969) and Bretsky and Lorenz (1970), who have interpreted shallower water regions as areas of generally greater environmental unpredictability, in which populations tend to maintain a high level of genetic variability, reflected also in greater within-region variability of phenetic characters.

SUMMARY AND CONCLUSIONS

This study of morphological variability in Middle Ordovician strophomenid brachiopods from midcontinent epeiric seas shows that inflation and symmetry characters accounted for most of the variability within samples of pedicle and brachial valves of *Oepikina minnesotensis* and *Strophomena plattinensis*. Clinal maps, prepared for each valve type by plotting the mean values of one inflation character and one symmetry character at each major sampling area, show isophene patterns that may have generally paralleled topographic features of the sea floor. Nearshore populations, at least of *Strophomena* pedicle valves and of both types of *Oepikina* valves, were usually flatter and less symmetrical than those from deeper water. Examination of patterns of intrapopulation variability showed that the level of variability at a given locality may have been related to the environmental setting. Especially for the inflation character, inferred onshore populations appear to be more variable than offshore ones. There is a difference among species in the degree of variation in symmetry, with *Strophomena* pedicle valves showing more pronounced variability among populations than either type of *Oepikina* valves. Perhaps there may be slight differences in the underlying genetic mechanisms that control variation in symmetry in taxa which are distinct at the generic level. Again, the nearshore populations of *Strophomena* pedicle and *Oepikina* brachial valves, both dorsal valves, were more variable in symmetry than were the corresponding offshore populations. It is difficult to determine whether the variations in symmetry and inflation within and among populations result from direct phenotypic response to different environments or from differential survival of individuals having the genetic potential to develop shells of a particular shape; some independent criterion of age determination is needed. It is more probable that the greater variability in shallow-water populations reflects adaptation to a more complex mosaic of environmental "patches" in the nearshore region, rather than having resulted from a general tendency toward selection for a higher level of genetic variability as a result of greater overall environmental unpredictability in shallow waters.

ACKNOWLEDGMENTS

We thank B. M. Bell, Z. P. Bowen, R. Cowen, K. W. Flessa, J. S. Levinton, D. M. Lorenz, H. M. McCammon, R. P. Richards, L. L. Sloss, and J. W. Valentine for reading various drafts of this paper and offering many helpful suggestions. John Kishpaugh and David Schwimmer taught us how to use the NTSYS package of programs. For assistance in preparing and photographing specimens, making measurements, and keypunching data, we are indebted to Gail Houart, Osa Kendrick, Pauline Mohr, Dorothy Brown, Gayle Lohmann, and Kathleen Simmons. The field collection of over 5 tons of Middle Ordovician limestones and dolomites from which the strophomenid specimens were gleaned was carried out with the assistance of undergraduate volunteers from two introductory geology courses at Northwestern University (fall 1969 through spring 1970), and without their help the present study, as well as subsequent paleoecological surveys of these Ordovician epicontinental seas, would still be in the planning stage. All illustrated and measured specimens are deposited in the New York State Museum (N.Y.S.M.), Albany, New York 12224. The present study is supported by grants from the Biological Oceanography Program (Grant GA-25340) and the Systematic Biology Program (Grant GB-27441) of the National Science Foundation.

REFERENCES CITED

Ager, D. V., 1965, The adaptation of Mesozoic brachiopods to different environments: Palaeogeography, Palaeoclimatology, Palaeoecology, v. 1, p. 143-172.

——1967a, Some Mesozoic brachiopods in the Tethys region, in Adams, C. G., and Ager, D. V., eds., Aspects of Tethyan biogeography: Systematics Assoc. Pub., no. 7, p. 135-151.

——1967b, Brachiopod paleoecology: Earth-Sci. Rev., v. 3, p. 157-179.

——1971, Space and time in brachiopod history, in Middlemiss, F. A., Rawson, P. F., and Newell, G., eds., Faunal provinces in space and time: Liverpool, Seel House Press, p. 95-110.

Agnew, A. F., 1955, Facies of Middle and Upper Ordovician rocks of Iowa: Am. Assoc. Petroleum Geologists Bull., v. 39, p. 1703-1752.

Agnew, A. F., Heyl, A. V., Behre, C. H., Jr., and Lyons, E. J., 1956, Stratigraphy of Middle Ordovician rocks in the lead-zinc district of Wisconsin, Illinois, and Iowa [Mississippi valley]: U.S. Geol. Survey Prof. Paper 274-K, p. 251-312.

Alexander, R. R., 1972, The autecology of the Cincinnatian (Upper Ordovician) brachiopod *Rafinesquina*: A biometric analysis: Geol. Soc. America, Abs. with Programs for 1972 (North-Central Sec.), v. 4, no. 5, p. 305-306.

Asquith, G. B., 1967, The marine dolomitization of the Mifflin Member Platteville Limestone in southeast Wisconsin: Jour. Sed. Petrology, v. 37, p. 311-326.

Badiozamani, K., 1972, The dorag dolomitization model—application to the Middle Ordovician of Wisconsin [Ph.D. thesis]: Northwestern Univ., Evanston, Ill., 94 p.

Bassett, M. G., 1970, Variation in the cardinalia of the brachiopod *Ptychopleurella bouchardi* (Davidson) from the Wenlock Limestone of Wenlock Edge, Shropshire: Palaeontology, v. 13, p. 297-302.

Bays, C. A., 1937, The stratigraphy of the Platteville Formation [Ph.D. thesis]: Univ. Wisconsin, Madison, 294 p.

Bays, C. A., and Raasch, G. O., 1935, Mohawkian relations in Wisconsin, in Kansas Geol. Survey Guidebook 9th Ann. Field Conf.: p. 296-301.

Blackith, R. E., and Reyment, R. A., 1971, Multivariate morphometrics: London, Academic Press, Inc., 412 p.

Bowen, Z. P., 1966, Intraspecific variation in the brachial cardinalia of *Atrypa reticularis*: Jour. Paleontology, v. 40, p. 1017-1022.

Bretsky, P. W., 1970, Benthic community structure of Middle Ordovician epeiric seas, upper Mississippi valley: Geol. Soc. America, Abs. with Programs for 1970 (Ann. Mtg.), v. 2, no. 7, p. 501.

——1973, A reflection on genetics, extinction, and the "killer clam": Geology, v. 1, p. 157.

Bretsky, P. W., and Lorenz, D. M., 1970, Adaptive response to environmental stability: A unifying concept in paleoecology: North American Paleontol. Conv., Chicago, 1969, Proc. E: Lawrence, Kans., Allen Press, p. 522-550.

Cooper, G. A., 1956, Chazyan and related brachiopods: Smithsonian Misc. Coll., v. 127, pt. 1 (text), pt. 2 (plates), 1024 p., 269 pls.

Copper, P., 1966, Ecological distribution of Devonian atrypid brachiopods: Palaeogeography, Palaeoclimatology, Palaeoecology, v. 2, p. 245-266.

——1967, Adaptations and life habits of Devonian atrypid brachiopods: Palaeogeography, Palaeoclimatology, Palaeoecology, v. 3, p. 363-379.

DeMott, L. L., 1963, Middle Ordovician trilobites of the upper Mississippi valley [Ph.D. thesis]: Harvard Univ., Cambridge, Mass., 214 p.

Deninger, R. W., 1964, Limestone-dolomite transition in the Ordovician Platteville Formation in Wisconsin: Jour. Sed. Petrology, v. 34, p. 281-288.

DuBois, H. M., 1916, Variation induced in brachiopods by environmental conditions: Puget Sound Marine Sta. Pub. 1, no. 16, p. 177-183.

Fenton, C. L., 1928a, The stratigraphy and larger fossils of the Plattin Formation in Ste. Genevieve County, Missouri: Am. Midland Naturalist, v. 11, p. 125-143.

——1928b, Forms of *Strophomena* from Black River and Richmond formations: Am. Midland Naturalist, v. 11, p. 144-158.

——1929, *Strophomena filitexta* Hall, a valid species: Am. Midland Naturalist, v. 11, p. 500-502.

Foerste, A. F., 1912, *Strophomena* and other fossils from the Cincinnatian and Mohawkian horizons, chiefly in Ohio, Indiana, and Kentucky: Denison Univ. Sci. Lab. Bull., v. 17, p. 17-173.

Gould, S. J., 1969, An evolutionary microcosm: Pleistocene and Recent history of the land snail *P.* (*Poecilozonites*) in Bermuda: Mus. Comp. Zoology Bull., v. 138, p. 407-531.

Hall, J., and Clarke, J. M., 1892, An introduction to the study of the genera of Paleozoic Brachiopoda: New York Geol. Survey, Palaeontology, v. VIII, pt. 1, 367 p.

Heyl, A. V., Agnew, A. F., Lyons, E. J., and Behre, C. H., Jr., 1959, The geology of the upper Mississippi valley lead-zinc district: U.S. Geol. Survey Prof. Paper 309, 310 p.

Huxley, J. S., 1938, Clines: An auxiliary taxonomic principle: Nature, v. 142, p. 219.

Karklins, O. L., 1969, The cryptostome Bryozoa from the Middle Ordovician Decorah Shale, Minnesota: Minnesota Geol. Survey, SP-6, 121 p.

Kay, G. M., 1934, Mohawkian Ostracoda: Species common to Trenton faunules from the Hull and Decorah Formations: Jour. Paleontology, v. 8, p. 328-343.

———1935, Ordovician system in the upper Mississippi valley, *in* Kansas Geol. Survey Guidebook 9th Ann. Field Conf.: p. 281-295.

Levinton, J. S., 1973, Genetic extinction hypothesis and its critics: Geology, v. 1, p. 157-158.

McCammon, H. M., 1970, Variation in Recent brachiopod populations: Uppsala Univ. Geol. Inst. Bull., N.S., v. 2, no. 5, p. 41-48.

McKerrow, W. S., 1953, Variation in the Terebratulacea of the Fuller's Earth rock: Geol. Soc. London Quart. Jour., v. 109, p. 97-124.

Mitra, K. C., 1958, Variation in *Goniorhynchia boueti* from Normandy and Dorset: Jour. Paleontology, v. 32, p. 992-1006.

Ostrom, M. E., Davis, R. A., Jr., and Cline, L. M., 1970, Field trip guidebook for the Cambrian-Ordovician geology of western Wisconsin: Wisconsin Geol. and Nat. History Survey Inf. Circ., no. 11, 131 p.

Pope, J. K., 1966, Comparative morphology and shell histology of the Ordovician Strophomenacea (Brachiopoda) [Ph.D. thesis]: Univ. Cincinnati, Cincinnati, Ohio, 270 p.

Richards, R. P., 1969, Biology and ecology of *Rafinesquina alternata* (Emmons): Geol. Soc. America, Abs. with Programs for 1969, Pt. 6 (North-Central Sec.), p. 41-42.

———1972, Autecology of Richmondian brachiopods (Late Ordovician of Indiana and Ohio): Jour. Paleontology, v. 46, p. 386-405.

Rohlf, F. J., 1968, Stereograms in numerical taxonomy: Systematic Zoology, v. 17, p. 246-255.

Rudwick, M.J.S., 1965, Ecology and paleoecology, *in* Moore, R. C., ed., Treatise on invertebrate paleontology, Pt. H, Brachiopoda: Boulder, Colo., Geol. Soc. America, v. 1, p. H199-H214.

Salmon, E. S., 1942, Mohawkian Rafinesquinae: Jour. Paleontology, v. 16, p. 564-603.

Scott, R. J., 1962, Lithologic control of faunas in the Middle Ordovician Platteville Formation of Wisconsin [M.S. thesis]: Univ. Wisconsin, Madison, 34 p.

Shrock, R. R., and Raasch, G. O., 1937, Paleontology of the disturbed Ordovician rocks near Kentland, Indiana: Am. Midland Naturalist, v. 18, p. 532-607.

Simpson, G. G., 1961, Principles of animal taxonomy: New York, Columbia Univ. Press, 247 p.

Sloan, R. E., 1973, Notes on the Platteville Formation, southeastern Minnesota, *in* Webers, G. F., and Austin, G. S., eds., Field trip guidebook for Paleozoic and Mesozoic rocks of southeastern Minnesota: Minnesota Geol. Survey Guidebook Ser. no. 4, p. 43-53.

Slobodkin, L. B., and Sanders, H. L., 1969, On the contribution of environmental predictability to species diversity, *in* Woodwell, G. M., and Smith, H. H., eds., Diversity and stability in ecological systems: Clearinghouse for Federal Scientific and Technical Information, U.S. Natl. Bur. Standards, Springfield, Va., Brookhaven Symposium Biology, no. 22, p. 82-95.

Sloss, L. L., Dapples, E. C., and Krumbein, W. C., 1960, Lithofacies maps: New York, John Wiley & Sons, Inc., 108 p.

Sokal, R. R., and Sneath, P.H.A., 1963, Principles of numerical taxonomy: San Francisco, W. H. Freeman and Co., 359 p.

Soulé, M. E., 1972, Phenetics of natural populations. III. Variation in insular populations of a lizard: Am. Naturalist, v. 106, p. 429-446.

Soulé, M. E., Yang, S. Y., and Weiler, M.G.W., 1973, Island lizards: The genetic-phenetic variation correlation: Nature, v. 242, p. 191-193.

Sulima, J. H., 1972, Community ecology of a Middle Ordovician epeiric sea (Decorah Formation, upper Mississippi valley): Geol. Soc. America, Abs. with Programs for 1972 (Ann. Mtg.), v. 4, no. 7, p. 681.
Swain, F. M., 1957, Early Middle Ordovician Ostracoda of the eastern United States: Jour. Paleontology, v. 31, p. 528-570.
Swain, F. M., Cornell, J. R., and Hansen, D. L., 1961, Ostracoda of the families Aparchitidae, Aechminidae, Leperditellidae, Drepanellidae, Eurychilinidae and Punctaparchitidae from the Decorah shale of Minnesota: Jour. Paleontology, v. 35, p. 345-372.
Sweet, W. C., Ethington, R. L., and Barnes, C. R., 1971, North American Middle and Upper Ordovician conodont faunas, in Sweet, W. C., and Bergström, S. M., eds., Symposium on conodont biostratigraphy: Geol. Soc. America Mem. 127, p. 163-206.
Temple, J. T., 1968, The lower Llandovery (Silurian) brachiopods from Keisley, Westmorland: Palaeontographical Soc. Mon., v. 122, no. 521, 58 p.
Thompson, W. H., 1959, The conodonts of the Platteville Formation of southeastern Minnesota [M.S. thesis]: Univ. Minnesota, Minneapolis, 73 p.
U.S. Geological Survey Bulletins 1123A-1123I, 1961-1966, Geology of parts of the upper Mississippi valley zinc-lead district: 571 p.
Valentine, J. W., Hedgecock, D., Zumwalt, G. S., and Ayala, F. J., 1973, Mass extinctions and genetic polymorphism in the "killer clam," Tridacna: Geol. Soc. America Bull., v. 84, p. 3411-3414.
Webers, G. F., 1966, The Middle and Upper Ordovician conodont faunas of Minnesota: Minnesota Geol. Survey, SP-4, 123 p.
——1973, Paleoecology of the Ordovician strata of southeastern Minnesota, in Webers, G. F., and Austin, G. S., eds., Field trip guidebook for Paleozoic and Mesozoic rocks of southeastern Minnesota: Minnesota Geol. Survey Guidebook Ser. no. 4, p. 25-41.
Weiss, M. P., 1953, The stratigraphy and stratigraphic paleontology of the upper Middle Ordovician rocks of Fillmore County, Minnesota [Ph.D. thesis]: Univ. Minnesota, Minneapolis, 615 p.
——1955, Some Ordovician brachiopods from Minnesota and their stratigraphic relations: Jour. Paleontology, v. 29, p. 759-774.
——1957, Upper Middle Ordovician stratigraphy of Fillmore County, Minnesota: Geol. Soc. America Bull., v. 68, p. 1027-1062.
Weiss, M. P., and Bell, W. C., 1956, Middle Ordovician rocks of Minnesota and their lateral relations, in Schwartz, G. M., ed., Lower Paleozoic of the upper Mississippi valley: Geol. Soc. America Guidebook, 1956 Ann. Mtg. (Minneapolis), p. 55-73.
Whittington, H. B., and Hughes, C. P., 1972, Ordovician geography and faunal provinces deduced from trilobite distribution: Royal Soc. London Philos. Trans. B, v. 263, no. 850, p. 235-278.
Williams, A., 1969, Ordovician faunal provinces with reference to brachiopod distribution, in Wood, A., ed., The Pre-Cambrian and Lower Paleozoic rocks of Wales: Cardiff, Univ. Wales Press, p. 117-154.
Williams, A., and others, 1965, Brachiopoda, in Moore, R. C., ed., Treatise on invertebrate paleontology, Pt. H: Boulder, Colo., Geol. Soc. America, 2 vols., 927 p.
Winchell, N. H., and Schuchert, C., 1895, The Lower Silurian Brachiopoda of Minnesota: The geology of Minnesota: Minneapolis, Geol. and Nat. History Survey of Minnesota, v. III, pt. I, p. 333-474.
Winter, J., 1971, Brachiopoden-Morphologie und Biotop—ein Vergleich quantitativer Brachiopoden-Spektren aus Ahrdorf—Schichten (Eifelium) der Eifel: Neues Jahrb. Geologie u. Paläontologie Monatsh., h. 2, p. 102-132.

Manuscript Received by the Society September 10, 1973
Revised Manuscript Received November 29, 1973

Laws of Distribution Applied to Sand Sizes

E. C. DAPPLES

Department of Geological Sciences
Northwestern University
Evanston, Illinois 60201

ABSTRACT

The concept that with continued current transport all sand-grain distributions, when logarithmically transformed, become symmetrical (normal) has been taken to indicate a progressive approach toward ideal textural maturity. Whereas residual soil appears to display a distribution of sizes that obeys Rosin's law of crushing, streams that carry such debris are considered to modify the distribution until it becomes log normal. Rosin's law appears to be well founded on crushing theory, but there is no established basis for the development of a log-normal size distribution as currents shift sediment from one locality to another. Although Rosin's and log-normal distributions do not stem from the same general equation and are not mathematically related, they yield graphic similarities for certain special sand populations. For such examples, the distributions appear to have the same degree of symmetry. Sand transported by long streams, for example, develops a steady-state size distribution, apparently related to the crushing law.

A steady-state, log-normal size distribution of pronounced symmetry is produced along shorelines where "winnowing" is the current process. Ideal cratonic shoreline sands (pure quartz and laminated quartz-glauconite sandstones) illustrate that symmetric, log-normal size distributions are developed when laminar flow dominates the current regime. Conversely, turbulent flow introduces asymmetry to size distributions.

A series of special cases characterizes sand-size distributions in streams and along shorelines; hence, individual distributions should be compared with known standards of reference. The standard for streams should be taken from a very long stream such as the Mississippi, and for shorelines from ideal, stable-cratonic sandstones such as those in western Wisconsin.

INTRODUCTION

Investigations into the processes controlling deposition of sedimentary particles have characterized size distributions with increasing accuracy. Hence, there have been many modern studies to distinguish between sands accumulated as river bars, dunes, beaches, offshore bars, and along shorelines in general (for example, Keller, 1945; Doeglas, 1946; Inman, 1949; Passega, 1957; Fuller, 1961; Friedman, 1961, 1967; Shepard and Young, 1961; Moss, 1962a, 1962b; Koldijk, 1968; Moiola and Weiser, 1968).

In the 19th century, the existence of size distributions was demonstrated by sedimentary geologists; in the USA, analyses culminated in the classic investigations and concepts of Udden (1914). He distinguished "drifting" and "blowing" as processes moving the traction load in one direction, "silting" and "dusting" to be mechanisms of settling from suspension involving both low and high velocity currents, and "washing" or "winnowing" to result from discontinuous currents of a dominant direction primarily along the sediment-"fluid" interface. Each of these depositional processes was conceived as reflecting complex physical laws of sediment transport which produced a distribution of particles characterized by a "chief-ingredient size." If the sizes are separated into groups, with the average diameter of each group bearing a constant ratio to the diameter of the next group, then the mass in each succeeding group diminishes by a fixed ratio from the mass of the chief-ingredient size. Udden considered this ratio an index of sorting to be used empirically to distinguish between environments of deposition. Indirectly, such concepts of size distribution led to the application by Wentworth (1922, 1929) of more rigorous statistical methods.

A major advance in the application of statistical methods to size distribution was the concept of the phi-diameter scale proposed by Krumbein (1934, 1936). The purpose of the phi scale is to symmetrize the frequency distribution of grain-diameter measurements; strongly skewed distributions based on millimeters are "normalized" by this logarithmic transformation. The application of logarithmic moments advocated by Krumbein (1936) gave more meaning to the computed parameters, because all parts of the distribution are involved. Previously, parameters derived from the methods of Trask (1932) were based on the central half of the distribution only. Krumbein's (1938) concept facilitated the use of known statistical parameters by simplification of the graphic representation of grain-size distributions.

The advantages of involving all of the grain sizes present and the phi transformation have become clear, and various procedures have been proposed to determine significant estimators of the distribution (McCammon, 1962; Jones, 1970). Among the most generally accepted are those of Inman (1952), Folk and Ward (1957), Friedman (1962), Middleton (1962), and Folk (1966), all of which are suitable for graphic techniques. However, an important practical feature (demonstrated by Inman, 1952) involves the limited significance of values in the extreme sizes. Inman (p. 128-130) showed that in the ranges coarser than 5 percent and finer than 95 percent of a distribution, errors due to fluctuation in sampling, splitting, and sieving are measurable in amounts that question the obtained numerical values. Other significant errors, such as "smoothing" curves and interdependencies of parameters inherent in selected methods, have been demonstrated by Jones (1970). In consequence, similarities or dissimilarities between many distributions are commonly more apparent than real.

Despite the exceptionally large errors involved in determining the distribution of sizes in most sediments, Rogers (1965) showed that sand samples analyzed in three separate laboratories produced essentially uniform values of size distributions; previously, Otto (1937) demonstrated the error in splitting and sieving sands to be minor. Hence, errors tend to be reduced to a minimum when the sediment is entirely in the sand range. Nevertheless, because the finest fractions are extremely important in defining skewness and kurtosis values, the distribution of silt and clay in a mixture of sand must be determined very accurately (Folk, 1966, p. 79). In such a mixture, silt and clay sizes are smaller than the feasible sieving limits, and their proportions must be obtained by a method of separation which is largely incompatible with sieving. The resulting values of percentile sizes can be grossly in error.

Samples involving fragments larger than granules pose exceptional difficulties. Measurement of the abundance of sizes of such particles cannot be compared with corresponding values of particles in the clay range. Hence, for gravelly sands, glacial tills, or residual soils, the errors are so large in sampling, splitting, and obtaining the size distribution that the pebble and clay fractions are known only by very approximate values. The amounts in these two sizes commonly exceed the coarsest 5 percent and the finest 5 percent of the sample; the extent to which the "tail" fractions represent the true random sample depends on how closely their sizes approach the sand diameters. The spread of sizes about the mean diameter and the skewness and kurtosis values of the distribution can be only rough approximations. For example, differences in values of sizes obtained by repeated analysis of samples of the same crushed rock are illustrated in a later section.

CRUSHING AND NATURAL PHYSICAL DISINTEGRATION

Concepts Relating Crushing and Transportation Laws

Krumbein and Tisdel (1940) examined the distribution of fragment sizes of exfoliation shells of certain granite boulders, residual micaceous soils, glacial tills, and pyroclastic debris. They determined that these distributions were predictable by an equation later identified as Rosin's law of crushing (Rosin and Rammler, 1933). Bennett (1936, App.) developed a theory based on random fracturing along planes oriented in three dimensions to account for the empirical relations observed by Rosin. Since then, other naturally fragmented rock materials have been reported to show size distributions of this type (Geer and Yancey, 1938; Landers and Reid, 1946; Pettijohn, 1957; McEwen and others, 1959; Dapples, 1959; Kittleman, 1964). In the initial broken rubble, correspondence with the ideal Rosin's distribution is frequently poor; only after repeated crushing do the observed and theoretical size distributions tend to coincide. This strongly implies that Bennett's assumption that breakage occurs along randomly oriented planes does not apply to all rocks. Both Rosin and Bennett were interested in the distribution of sizes in crushed bituminous coal, which tends to rupture along bedding planes and small joints randomly inclined and oriented. Other rocks rarely have closely spaced, randomly oriented, predetermined fracture planes, and fractures caused by crushing follow irregular surfaces. Fracture surfaces of most rocks are therefore unlike those in the theoretical assumptions, and close approximation to the general law should not be anticipated.

Epstein (1947) noted that data obtained as a consequence of breakage of certain solids showed distributions of sizes that could be considered log normal. He developed a theoretical breakage model that could produce distributions of types based on the following assumptions: (a) the breakage process occurs in discrete steps, and for each step, many fragments must be produced; (b) the probability of breakage of any piece is a constant that is independent of size, presence of other pieces, and number of breakage steps prior to a given step; and (c) the distribution of particle sizes obtained by a single breakage step is independent of initial particle size. Although assumption (a) appears to be reasonable, there is no established evidence for (b) and (c). In fact, the probability of breakage of large pieces in a crusher is different from that of small particles. Among granular rocks, the distribution of particles resulting from a single breakage step does not appear to be independent of the dimension of an individual fragment. Hence, although the theoretical analysis is adequate, the observed size distribution of broken rock debris is not approximated closely by a log-normal density function.

Interest in Rosin's law was renewed by Kittleman (1964), who concluded that mechanically derived sediments should be examined for departures from the ideal Rosin's distribution. Moreover, he compared various theoretical distributions employing specified values of Trask's sorting coefficient (Trask, 1932) in order to determine the closeness of fit predicted by Rosin's and log-normal equations. Interpretation of Kittleman's work leads to the conclusion that, as the number of size grades is increased, the Rosin's and log-normal density functions become increasingly graphically similar.

Krumbein and Tisdel (1940) inferred that an important but unknown natural law was the theoretical basis of Rosin's density function. They reasoned that the observed distribution of particle sizes is the product of rock disintegration in cold, wet climates. The processes involved dislodge mineral grains and fracture the rock into pieces similar to those produced in a crushing process. As long as such material remains in a closed system, natural processes bring about repeated crushing of fragments and produce, ideally, a size distribution predicted by Rosin's law. As erosion removes the products from the weathering site, the closed system no longer exists. Currents also modify the size distribution. In the ultimate stages of transport, the sediment's size distribution was believed by Krumbein (1938) to be represented by a log-normal distribution of pronounced symmetry.

The degree of relation between the two theoretical distribution laws is of considerable importance. If disintegration products in the closed system develop a size distribution predictable by Rosin's law, but slowly approach a log-normal distribution with increasing transport, some measure of textural maturity (Folk, 1951) is indicated by the departure from Rosin's distribution. On the other hand, under certain natural conditions, if the breakage processes produce a log-normal size distribution, the concept of textural maturity as commonly defined is not justifiable.

Middleton (1970) recognized that few attempts have been made to provide basic theory to account for the observed log-normal size distributions, despite the common acceptance that they characterize many sands. Whereas Rosin's distribution appears to be supported by adequate theory, this is not the case for a log-normal distribution of sizes assumed to be derived by some current transport mechanism. Middleton pointed out that the central-limit theorem permits generation of either normal or log-normal distributions by small changes in the basic assumptions of the models. For example, Epstein's (1947) derivation of the log-normal size distribution from breakage is based on the multiplication model (Middleton, 1970).

Rogers and others (1963) considered Epstein's theoretical analysis to provide an adequate basis for the size distribution resulting from current transport. That is, during transportation a multiple stage breakage and abrasion process leads to the progressive removal of random proportions of specific grain sizes. Ultimately a distribution of sizes is produced that is asymptotically log normal. A more conservative viewpoint was advocated by Middleton (1970) and other investigators, who reasoned that the locally existing current conditions at the site of final deposition exert more control on the size distribution than does breakage resulting from weathering and transportation. According to Middleton, sand deposited by a stream or by waves on a beach can be interpreted as a deposit of final size distribution that arises from the influence of individual sorting events on an initial size distribution. A sorting event is regarded as an independent current surge that both deposits and removes some sand. The weight of any grain size remaining as a deposit after a sorting event depends only on the proportion of that size originally present in the source. This proportionate-effects model is based on somewhat different assumptions than Epstein's multiplication model, and it leads to a log-normal distribution of particle sizes.

Factors Influencing the Crushed Sample

The products of single-stage laboratory crushing differ considerably from those of natural disintegration, because quarried rock is broken to pass between crusher jaws or rollers set to some predetermined opening, and the coarse sizes are truncated at this dimension. When the material has passed through the jaws, its size distribution (as determined by sieving through screens of known aperture) is skewed toward the fine sizes and is typically expressed by a crushing law. Theoretically, by this single laboratory procedure, the maximum "diameter" of the largest particle is known. In fact, an arbitrary, intermediate particle length that cannot exceed the crusher opening is determined. There is a range in maximum size that depends on the tendency of the rock to break into platelike fragments or to remain in more equidimensional form. If the rock is finely crystalline and also homogeneous (that is, the tendency to fracture at a certain stress is about the same in all directions), the true particle diameters in the coarse sizes are closely predictable by the openings of the measuring sieves. Conversely, if the rock is coarsely granular, fractures into clusters of grains, or tends toward schistosity, the diameter of the coarse particles is determined with low levels of precision by screen openings. Observed size distributions do not coincide with theoretical ones in such instances. If the size of the crusher opening is reduced to smaller than the grain size, the accuracy of measuring the diameter of particles by the use of sieves is increased distinctly (Gaudin, 1926).

Glacial breaking and grinding and explosive volcanism produce particle size distributions that approach those produced by laboratory crushing. Breakage due to weathering, however, does not proceed without some chemical breakdown; the generation of clay minerals increases the amount of very small particles. Hence, the size distribution of any soil in situ contains some excess of clay sizes greater than that predicted by crushing laws.

With large fragments it is difficult to determine what constitutes the limiting dimension of the largest particles included in the sample needed to represent the entire volume of the crushed product (sediment) at a locality. For example, the dislocation of blocks from bed rock and the size of the largest fragments are controlled by the joint pattern, the intensity of jointing, and the separation of bedding intervals or sheeting. Hence, the maximum size of the disrupted particles cannot be determined by the investigator, and the true size distribution of the fragmented rock is more symmetrical about a mean size than the gathered sample. The investigator selects a maximum size according to some limit that he imposes on the sample volume; this uppermost limit of the selected size represents a truncation of the actual distribution, similar to that of laboratory crushing.

A serious biasing effect is introduced by the amount of coarse fraction included in the sample. Because standard methods employ the weight-percent frequency in each size grade, the weight of the largest fragments disproportionately influences the entire sample and correspondingly modifies the frequency distribution. Similarly, how much of the fine fraction should be included or excluded from the total sample in order to represent the true proportion relative to the amount of coarse material selected? How much of the fine fraction should be excluded because it is a product of chemical change from feldspar to clay mineral, rather than fragmentation of the feldspar to clay size? The investigator is almost forced to select the amount of coarse material in proportion to fine material according to some predetermined concept. For example, if he considers the distribution is predictable by Rosin's law, he would select the proper weight of each size fraction to make up the total weight of the sample. Clearly, this is not an acceptable procedure, although the spectrum of sizes that constitutes the average disintegrated rock at the outcrop precludes recognition of the true population (Krumbein, 1960). For such debris, the values obtained for sizes in the lower and higher percentiles of the distribution are so much in error that the moment parameters of skewness and kurtosis lack significance, and the graphic representation is deceptive. As yet, no procedure is known that avoids considerable error (Krumbein and Pettijohn, 1938, p. 20–22; Blatt and others, 1972, p. 67; Krumbein, 1960), although use of the Edgeworth series advocated by Jones (1970) merits attention.

Obviously, as the size range decreases in magnitude, the accuracy of representing the true distribution increases (Folk and Ward, 1957). Sampling sands from a sand bed or silt from a silt bed becomes much more reproducible from sample to sample and by individual investigators. For these reasons, this study is confined to sediment almost exclusively in the sand-size range in order to compare size distributions with greatest equivalency.

RELATIONSHIPS OF ROSIN'S AND THE LOG-NORMAL LAWS

General Equations

Otto (1939) showed that certain sediments from environments dominated by a single regime possess size distributions that appear to be modified normal distributions when expressed in phi sizes. The distributions in question were ideally represented by sands where the washing and winnowing processes of Udden prevailed over extended time. The distribution of sizes is skewed, but the phi transformation produces pronounced symmetry that closely approximates the normal distribution. The inference has been accepted, however, that the disaggregated rock at the outcrop (crushed according to Rosin's law) is, by transportation, modified gradually into a more symmetrical distribution about the mean size. Eventually, sands from along shorelines develop frequency distributions with exceptional symmetry, if plotted in phi diameters. There is reason to suspect, therefore, that Rosin's law and the log-normal distribution are related in some fundamental way and could be special examples of a more fundamental equation. This would resolve uncertainties raised earlier by Middleton (1970) regarding the basic assumptions underlying breakage and transportation processes that yield size distributions derived from crushing laws.

Among statisticians, Rosin's equation is known as the Weibull probability law (Johnson and Kotz, 1971). As yet, no derivation of the law is known, but it has appeal because it is readily fitted to various scattered data. For application to the distribution of particle sizes, let $F(D)$ represent the proportion of particles having sizes less than D. These sizes obey Rosin's law if

$$F(D) = \int_0^D f(x)\, dx,$$

$$= 1 - e^{-(D/b)^c}, \quad D > 0,$$

where the density function

$$f(x) = cb^{-c} x^{c-1} e^{-(x/b)^c},$$

in which b and c are parameters dependent on the distribution of particle size; c determines the shape of the probability functions, namely, skewness and kurtosis, and b is an arbitrary scaling factor of particle size.

Kolmogoroff (1941; see Middleton, 1970) derived a crushing law using the multiplicative form of the central-limit theorem. He showed that particle sizes obey the log-normal probability law, so that

$$F(D) = \int_0^D f(x)\, dx, \quad D > 0,$$

where the function under the integral sign is

$$f(x) = (x\sigma\sqrt{2\pi})^{-1} \exp[-(\log x)^2 / 2\sigma^2],$$

which is the density function of the log-normal probability law.

If y (the cumulated weight percent) $= \log x$, then y obeys the normal probability law with a mean of zero and a variance of σ^2. The density functions of Rosin's and log-normal equations are not the same. Transformation of particle size distributions from Rosin's to a normal type is not regarded as theoretically reasonable, and the two equations are not related as descendents of a common equation of distribution.

Particular conditions can be demonstrated, however, in which the parameters c and σ^2 cause log-normal and Weibull distributions to show similarity when plotted graphically (Atchison and Brown, 1957; Johnson and Kotz, 1971). This is illustrated in Figure 1 in which three log-normal distributions of increasing values of σ^2 ranging from 0.1 to 2.0 are plotted. When σ^2 has a very low value, the distribution tends to be concentrated in a narrow range and is more symmetrically distributed about the value of the mean. Increase in the value of σ^2 shows a corresponding decrease in symmetry and the strong development of positive skewness. The same illustration shows certain Weibull distributions that differ in the value of c superposed on the log-normal distributions for comparison of shape. Low values of c show distributions that are skewed, whereas increasing values of c show distributions more symmetrical around the mean. When $c = 1$, the Weibull distribution is skewed graphically so that it tends to coincide with a log-normal distribution of $\sigma^2 = 2$ over much of the distribution, except the extremes. Similarly, as c increases to

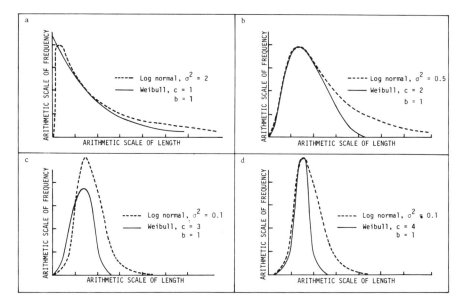

Figure 1. Selected distributions of Weibull functions of four values of parameter c. Corresponding log-normal equations for different values of σ^2 have been superposed to illustrate similarities in the two distributions. Note tendency for development of symmetry with decreasing values of σ^2 and increase in c. Individual log-normal size distributions can simulate unrelated Weibull distributions, particularly if extremes of the distribution are poorly known as a result of errors in determination.

values approaching 4, the distribution tends to become symmetrical and approaches the shape of the log-normal distribution of $\sigma^2 = 0.1$. The inference to be drawn is that, despite an absence of relation in the fundamental equations, the distributions show graphic similarity as they become increasingly skewed and also when they are symmetrical around a mean value. Hence, certain sediments should fit Rosin's and log-normal distributions equally well, within the limits of accuracy of the data. There is no fundamental reason, however, to suggest that a crushed rock obeying Rosin's law should be transformed into one with a log-normal distribution by continuation of the same breakage processes. As the crushed sediment is transported downstream from its source, it could be transformed by processes of continued breakage or current action into a size distribution that approaches either a Rosin's or log-normal distribution.

Similarity in Rosin's and Log-Normal Distributions

A comparison of Rosin's and log-normal distributions for particular sediments is best understood when examined visually. Figure 2 illustrates such relations using samples representing artificially and naturally crushed rock. Selected homogeneous rocks (chert, gabbro, and fossiliferous limestone) were passed through a jaw crusher once, the product was split and sieved repeatedly, and the mean weight-percent frequencies were plotted on Rosin's and log-normal scales. The gabbro, a relatively homogeneous rock, was selected to illustrate the difference in size distribution resulting from a single passage through the jaw crusher and repeated passage through an identical opening in the jaw crusher. Samples of disaggregated arkose, desert granitic soil, and humid-climate-weathered granite "soil" represent natural disintegration products. Each sediment was split from a kilogram bulk sample and sieved. Five separate splits and sievings were made for each sample, after which the mean weight-percent frequency for each one-half phi size was plotted on log-normal and Rosin's scales. Each scale has the same overall length from 0 to 100 percent cumulated weight frequency (inset, Fig. 2) but differs in spacing of percent intervals. Values of Δy represent the cumulated differences (on an arbitrary length scale) between the plotted positions at each half-phi size of a single analysis when plotted on Rosin's and log-normal scales. Departures between the two curves (Δy) represent the cumulative effect of Δy with progressively increasing positive phi values. Certain of the artificially crushed homogeneous rocks (chert, limestone, shale, and repeatedly crushed gabbro) yield data that tend to simulate departures shown by the ideal Rosin's distribution. Other disintegration sediments, particularly the granite soil weathered in a humid climate and the once-crushed gabbro, tend to show considerably lower values of Δy. Such low values show that in the coarser fraction the distribution tends to obey a log-normal law, whereas in finer sizes Rosin's law expresses the distribution. This tendency appears to be typical of single crushings, as shown by the bimodal shapes of other crushed rocks (Figs. 3, 4, and 5). Repeated crushing, illustrated by the gabbro, reduced the bimodality and resulted in a close approximation to Rosin's distribution.

Crushed Products and Rosin's Distribution

Figures 3, 4, and 5 illustrate selected size distributions of samples crushed from 17 different rock types ranging from homogeneous to somewhat inhomogeneous and, for comparison, one residual soil derived from one of the granites (see Marathon Granite, Fig. 5). Except for this soil, each represents a large specimen (1 kg)

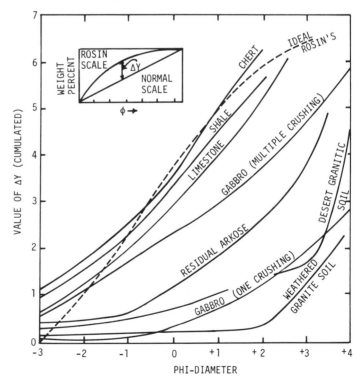

Figure 2. Departure of Rosin's and log-normal scales as indicated by values of Δy measured with increasing phi sizes. Δy is the relative cumulated distance of departure at each phi size from 0.1 to 99.9 cumulated weight percent. Inset illustrates that ideal log-normal distribution plots as a straight line, and corresponding Rosin's distribution is curved. Large fragments of nodular chert, poorly bedded shale, fossiliferous limestone, and finely crystalline gabbro were crushed and plotted on log-normal and Rosin's scales, from which values of Δy were determined. Curves of natural disintegration products—arkose, desert granite "soil," and humid-climate granite "soil" (leucogranite, Spruce Pine, North Carolina)—do not parallel curves of artificially crushed rocks, nor do the distributions parallel the ideal Rosin's law (dashed line).

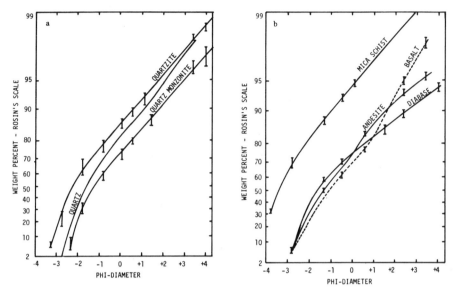

Figure 3. Distributions of particle sizes plotted on Rosin's scale after single passage through a jaw crusher. a. Structurally homogeneous rocks. b. Inhomogeneous rocks showing slight cleavage or schistosity. Barred ordinates show ranges in values for repeated splits and sievings of original crushed sample. Line represents distribution of sizes in a single split and is not a mean of values. Distributions of sizes in each split show the same shape of curve and vary in slope only. Ideal Rosin's distribution plots as a straight line. Departure from this line shows a tendency toward two distinct distributions changing in the granule size. Curve for quartz in (a) represents data from Gaudin (1926).

broken into coarse fragments and passed once through a jaw crusher that opened to a maximum of 4 mm and closed to a 1-mm minimum. Each crushed product was split to a sample of about 250 g and sieved through openings of $\sqrt{2}$ intervals. The sieved portion was returned to the original bulk-crushed rock and a new sample was split and sieved. Following this procedure, the size distribution of each bulk-crushed rock was determined for seven separate splits and plotted on Rosin's scale to indicate the nature of the distribution. The range in cumulated weight percent for each phi size is indicated by a vertical bar, and each line represents a single analysis. Other splits for each sample tend to approximate the same curve, but to plot them all would merely cause superposition of lines. A single bulk-crushed rock produces a family of curves of similar shape that lie within the bracketed ordinates.

Replicate splitting and sieving of the same crushed product can result in a wide range of weight percentage in the coarse sizes; for an individual grade size, this range may be as much as 10 percent of the total sample weight. In the sand-size range, differences due to replication are much smaller, but they are large enough to indicate that a single analysis has a low level of precision, even in the central part of the distribution; values in the extremes lack significance. The size distribution of the quartz sample (Fig. 3a) is from Gaudin's (1926) data, which result from somewhat different crushing and sieving procedures, but the shape of the distribution parallels that of the quartzite and quartz monzonite crushed in our laboratory. The similarities suggest that comparable results are obtained by slightly different crushing procedures, particularly when the particles have been reduced to sand sizes.

Figure 3b represents the size distributions that result from crushing rocks considered less homogeneous than those represented in Figure 3a. With the exception of the mica schist, these distributions tend to show the anticipated significant departures from Rosin's law, particularly the basalt which showed a slight cleavage. Figures 4 and 5 represent the distribution of crushed rocks regarded as homogeneous, ranging from chert and fossiliferous limestone to gabbro, rhyodacite, and various granites. Except for vitrophyric granite, Figure 4b illustrates rocks with a slight cleavage revealed only in thin section, and these crush as two populations. For comparison, Figure 5b illustrates the distribution of the artificially crushed Marathon Granite and a single split of its naturally disintegrated soil. Of all the distributions in Figures 3, 4, and 5, the chert and quartz conglomerate (Fig. 4a) and the granite soil (Fig. 5b) most closely approximate Rosin's distribution. The conglomerate is extremely well cemented with quartz and behaves as a homogeneous rock. Other single crushings of this rock produced distributions that differ one from another, but each closely simulated Rosin's distribution.

The soil in Figure 5b is interpreted as the product of repeated crushing in the natural state by the cycle of freezing and thawing in a region of humid climate. Its parent rock tends to break into a similar distribution after a single laboratory crushing. In both cases, fracture tends to deliver two populations, each constituting approximately one-half of the sample. The overlap in size distribution between the natural and artificially crushed product lends confidence to the assumption that the disintegration tendency of a homogeneous rock in the natural state parallels artificial crushing. In the example of the biotite granite, a single crushing revealed its inhomogeneity; the fracturing process yielded a bimodal distribution, which

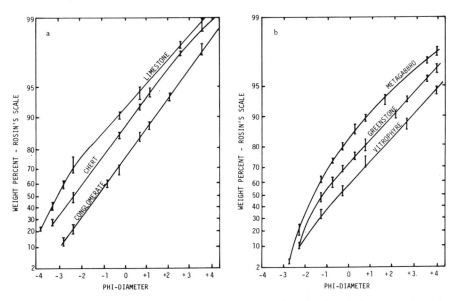

Figure 4. Distributions of particle sizes plotted on Rosin's scale after single passage through a jaw crusher of selected, apparently homogeneous rocks. The solid lines indicate distributions of single splits; barred ordinate lines indicate ranges of splits and sievings from a single crushing. Ranges in weight percent shown by barred ordinates illustrate degree of variability to be expected in splitting and sieving such sizes. The most reliable values are means of repeated splits and sievings of the same large sample. a. Initial crushing of the conglomerate and chert follows the ideal Rosin's distribution. (*Note:* this chert is a different sample from that in Fig. 2.) The fossiliferous limestone breaks as two distinct distributions regarded as typical of single crushing. b. Slight inhomogeneity causes initial crushing to depart from Rosin's law. Note tendency for two distributions in the lines for greenstone and vitrophyre.

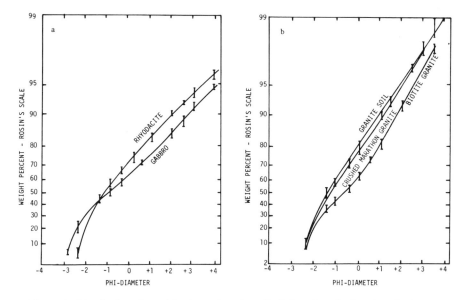

Figure 5. Distribution of particle sizes, plotted on Rosin's scale, of selected rocks based on a single crushing, but with repeated splits and sievings. a. Note tendencies shown by other homogeneous rocks to crush initially as two distributions changing at the granule size. Fine sizes tend to follow Rosin's law. b. Biotite granite illustrates influence of platy, hard, and soft minerals on the initial crushed distribution. The crushed non-mica Marathon Granite, Wisconsin, is closely paralleled by its "soil" (humid climate) disintegration product. Similarity in the two distributions indicates that natural soils approach Rosin's law. Size distribution of the soil is based on a single split only.

is regarded as a commonly occurring feature (Figs. 3a and 5a). The results of all the crushings suggest that the appearance of a rock is not a reliable criterion to predict whether it will break with a distribution closely approximated by Rosin's law. Rather, there is reason to suspect that, at least in the initial stages of crushing, some other law is important.

The influence of inhomogeneity of a rock is substantially reduced, as predicted, when the dimensions of the coarser sizes of the crushed product are reduced to sand sizes and finer. In almost all of the rocks crushed, this tendency is made clear by the approximation of the sand and finer fraction to Rosin's distribution. Except for rocks having pronounced schistosity, this tendency should prevail as mineral grains or aggregates of more uniform diameters are produced by crushing.

The effect of prolonged crushing and some grinding in modifying the size distribution is illustrated by Figure 6 on the basis of Gaudin's (1926) data. This plot shows different size distributions of quartz grains after selected periods of crushing. Gaudin's data have been transformed and plotted to show the approach to Rosin's distribution with the progress of crushing (Fig. 6a). As crushing was prolonged, beginning with 4 min and ending with 4 hr, the distribution decreased in median size and progressively approximated Rosin's distribution. The approach to the ideal Rosin's distribution is regarded as a function of the length of time that repeated crushing is in progress and of the reduction in particle diameters so that all grains are in the sand-silt size range. A similar interpretation was implied by McEwen and others (1959, p. 484), who noted that the disintegration products of granites do not follow Rosin's distribution unless a considerable amount of fine debris is produced.

The data of Figure 6a are plotted in Figure 6b on a log-normal scale. Although with increasing crushing time there is a distinct tendency for the curves to become straight lines, they are not phi normal (that is, symmetrical about a mean diameter). Presumably, additional time would produce no significant change in curve shape, but repeated or prolonged crushing tends to eliminate the bimodality of the distributions which is typical of initial stages of crushing.

DISTRIBUTIONS APPROACHING EITHER ROSIN'S OR LOG-NORMAL TYPES

River Sands

McEwen and others (1959, p. 491) observed that size distributions of disintegrated granitic soil continue to obey Rosin's equation when they become stream-deposited sands a short distance from the weathering outcrop. Should further studies prove this to be characteristic of the general condition, an investigator can assume that debris deposited in the headwaters of streams eroding homogeneous rocks initially has a Rosin's distribution. Such distributions are modified downstream, although Friedman (1961, p. 820) considered the unidirectional flow of rivers to be responsible for maintaining the general positive skewness in the deposited sand.

On beaches and dunes, waves and currents wash or winnow the sand, and by removing the finer fraction modify the size frequency as anticipated. Although such sands are approximately symmetrically log normal (that is, the phi-transformed distribution is normal), their size distributions commonly tend to be negatively skewed (Friedman, 1961, p. 521). Medium- to fine-grained beach sands tend to

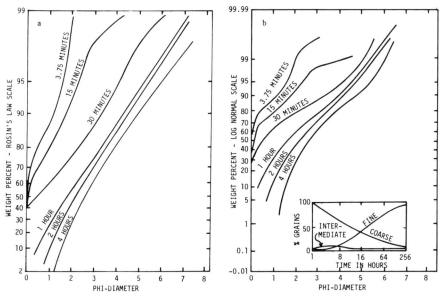

Figure 6. Influence of continued crushing of quartz on distribution of particle sizes, based on data from Gaudin's (1926) experiments. a. Distributions after selected intervals of crushing time plotted on Rosin's scale. Short-time crushing results in distributions that differ strongly from Rosin's equation, but a steady-state, relative-size distribution approaching Rosin's law is attained after 1 to 2 hr of continued crushing. b. Same data as in (a) plotted on log-normal scale. Continued crushing introduces symmetry into the distribution, but symmetrical log-normal equation is not attained. Inset shows the small change in relative size distribution during 4 hr of crushing.

be better sorted than river sands of comparable grain size (Friedman, 1961, p. 524). The skewness of the distributions changes gradually from markedly positive in a stream system to nonskewed or slightly negative on the beaches and, locally, to slightly positively skewed in adjacent dunes.

Size distributions of two river sands with a minimum content of silt and clay are shown in Figures 7 and 8. The sand represented in Figure 7 was collected from a natural levee sloping toward the channel of the lower Mississippi River at Plaquemines, Louisiana. The sand shown in Figure 8 is from a point bar on which very small dunes (<50 cm in height) have developed adjacent to the major channel of the Platte River at Morse Bend, Nebraska. Note that the natural-levee sand (Fig. 7) is so well sorted that approximately 90 percent of the distribution is included between two phi sizes. Sand from the point bar, also partially winnowed by dune activity, is more poorly sorted and some silt and clay fraction remains. Neither river sand shows a size distribution that, when plotted in phi sizes, is normally distributed, and they do not fit Rosin's equation because deficiencies occur in the fine-sand sizes. However, the natural-levee sand fits Rosin's distribution better than the phi-normal condition (Fig. 7). The Platte River point-bar sand (Fig. 8) does not fit one distribution any better than the other.

If the distributions of Figures 7 and 8 are typical of sands deposited by low-velocity streams, the tendency to depart from Rosin's distribution and to approach a phi-normal condition is not a gradual process developed by long-stream transport. Rather, the distributions suggest that they are predicted by a Weibull distribution (probability law) that has values of c in the 3 to 4 range (Fig. 1); that is, the sediment is concentrated in a few size classes and there is a corresponding reduction in skewness. Moreover, this distribution can be expected to continue as long as

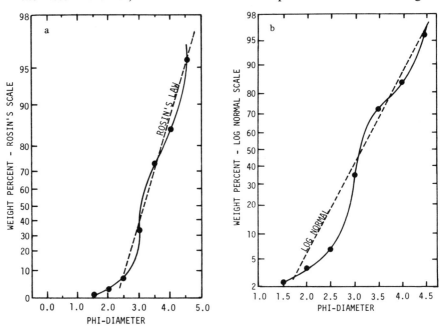

Figure 7. Distribution of sand sizes from river-channel slope of a natural levee of the Mississippi River, Plaquemines, Louisiana. a. Plotted on Rosin's scale. b. Plotted on log-normal scale. Solid lines connect points in distribution; dashed lines indicate ideal Rosin's or log-normal distributions. Note that the distribution approximates Rosin's law somewhat more closely than the log normal.

the influence of the river environment persists. The currents in the river system modify the distribution of particle sizes so that they fit Rosin's equation with high values of c. The underlying assumptions of the breakage process of Rosin's law need not be applied in stream transport, because the Weibull distribution is not necessarily based on such assumptions. For the present, an empirical standard of reference could be established for stream-deposited sand; it would be based on a Weibull distribution selected to fit the size distribution of a well-sampled long river such as the Mississippi. The intent would be to avoid the present overlap in estimators of distribution that are based on similarity to a phi-normal distribution.

Beach Sand of Glacial Origin

Sands along the southwest shore of Lake Michigan are primarily derived from erosion of glacial till and, hence, begin movement along the strand with a size distribution closely approximated by Rosin's law (Krumbein and Tisdel, 1940, p. 304). Figure 9a illustrates numerous similar size distributions of beach and nearshore sands sampled about 30 years ago from naturally accumulated deposits. Two samples, one an offshore sand and another a beach sand from the same transect, are plotted on Rosin's and phi-normal scales to show the shapes of the distributions. Except for relatively small differences in the coarse and fine fractions, the distributions tend to be approximated more closely by the phi-normal equation than by Rosin's equation, although the departures are small.

Modification of the initial Rosin's distribution by longshore transport follows a progression different from the behavior of the stream-transported sand. The

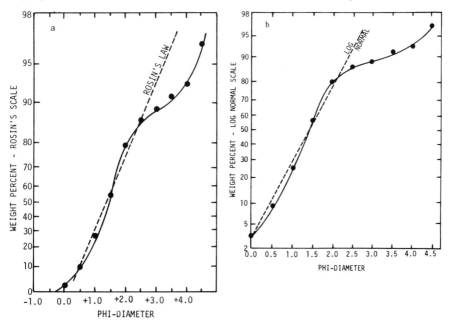

Figure 8. Distribution of sand sizes from point bar in complex channel of Platte River at Morse Bend, Nebraska. a. Plotted on Rosin's scale. b. Plotted on log-normal scale. Solid lines connect points of actual distribution, dashed lines represent ideal Rosin's law or log-normal distributions. Note that the distribution does not fit either ideal condition, with departures primarily in the fine sizes.

influence of washing action by the shoreline currents rapidly imposes different laws of physical behavior and produces a log-normal frequency distribution of pronounced symmetry in the sand sizes. As the sediment is delivered to a new environment where different physical laws prevail, there is an abrupt change in size distribution, and the geographic site that contains sand transitional between the two types of distribution is narrowly defined.

SAND SIZE DISTRIBUTIONS ALONG MARINE TRANSGRESSIVE SHORELINES

Cratonic Sands

Wherever longshore transport has shifted sands over long intervals of time, their size distribution should be normal when converted into phi sizes. Excellent examples occur among certain cratonic sandstones in western Wisconsin where

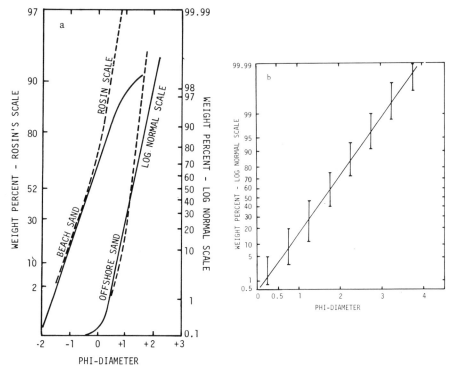

Figure 9. Distribution of sizes in selected sands deposited in the nearshore environment. a. Sand deposited from erosion of glacial-till bluffs, western shore of Lake Michigan (Evanston, Illinois), plotted on Rosin's scale (dashed line) and log-normal scale (solid line). Ideal distributions are straight lines. Beach and offshore sands more closely approximate a phi-normal distribution than Rosin's law. Note that the largest departures from the ideal distribution are in fine sizes of beach sand which skew the distribution positively. b. Distribution of sizes of Cambrian pure quartz sandstones outcropping in the Tomah Quadrangle, Wisconsin. Solid line illustrates nearly ideal log-normal distribution of one example. Other distributions are similar and closely approximate the one shown, even in general slope of line. Ordinates with bars indicate ranges in cumulated weight percent for each screen size to include all samples analyzed. Samples include both coarse and fine sands interpreted as beach and offshore deposits of slow marine transgression.

lower Paleozoic strata are an assemblage of dolomites and sandstones representing deposits of cycles of marine transgression and regression (Twenhofel and others, 1935). Certain parts of the Cambrian sandstones are characterized by an abundance of glauconite in the form of pellets. For the most part, these pellets were transported with quartz sand to accumulate as glauconite sandstone currently interpreted as "tidal" flat and offshore-bar sediments. Shoreward, the glauconite-bearing beds pass into pure quartz sandstones which show local occurrences of large-scale crossbedding. This bedding is interpreted as a dune section accumulated on a beach-complex system. Seaward from the old shoreline positions, the strata become shallow-water dolomites.

The areal extent of some of the sandstones is exceptionally widespread, but the general conditions of deposition appear to have been similar for all. Each contains abundant evidence that it was laid down primarily along a nearshore and beach complex. Except for a few beds of intraformational conglomerate, the remainder of the sandstones contains particles no larger than small granules; in the fine sizes, there is relatively little silt and clay. Throughout the entire geographic area examined, the sandstones are unusually friable and disaggregate completely. Also, the restriction in the size range permits the observer to have confidence that he has obtained the coarsest and finest fractions in any suitable sample. Repetitive sampling of outcrops by several investigators indicates that very small variability is introduced by different methods of sampling and that a true random sample can be gathered with relatively good assurance. Indeed, lateral uniformity of particle sizes has been recognized for many years for such beds as the Galesville and Jordan Sandstones, among others. Experienced observers have emphasized the similarity in the distribution of the sand sizes, although overall regional differences exist (Ostrom, 1970; Dapples, 1955). Hence, the stratum represented by an individual sample cannot be identified with complete assurance.

Approximately 100 individual kilogram samples, each from a selected sedimentation unit (Otto, 1938), were taken from widely scattered exposures in western Wisconsin in order to obtain a spectrum of the ranges of size and distribution. Figures 9b and 10 represent the data for these samples; each figure represents the ranges of size and distribution in a geographic area and not from a single formation. The size distribution of a single sample is drawn as a straight line that almost precisely connects plotted points on an ideal phi-normal scale. Each figure shows much the same slope of line, indicating only small differences in the standard deviation between the samples selected from widely spaced geographic areas. As a visual aid, each bracketed vertical line indicates the range in percent for that size and is used in preference to plotting a bundle of overlapping and crossing lines representing the assemblage of individual distributions. The range in values of weight-percent frequency is as much as 40 percent of the total in the central parts of the distribution; hence, the mean size and kurtosis of each distribution differs substantially among individual samples and formations. The reader should understand, however, that no attempt has been made to single out coarse sands from fine sands, nor to separate individual formations. Each figure contains samples from the same individual formations, and the only distinction is geographic separation. An envelope connecting the horizontal bars at the tops and bottoms of the ordinates includes the size distributions of all the samples taken in each geographic area. With few exceptions, a line indicating the size distribution of one of the samples is quite straight, although not necessarily parallel to the representative line drawn on the figure. By far, the largest number of the distributions are phi normal and the standard deviations are clustered about the same values (0.6 phi unit).

The spectrum of size distributions includes sands accumulated in a beach-shoreline complex including backshore dunes, offshore bars, and spits. Locally, spits isolated quietwater areas, which are thought to have been "tidal" flats. In such areas marginal to the strand, deposition of very fine sands and silts occurred.

Except for the presence of glauconite in certain of the tidal flat sandstones described below, quartz is the remaining ingredient in the samples selected. The quartz appears to have been moved from one environment to another, so that the same debris was modified only by the winnowing properties of the specific environment in which the sand was eventually deposited. The size distributions (Figs. 9b and 10) indicate that, if characteristic shapes are produced by different environments, they tend to be subtle and are represented principally by differences in mean size and kurtosis. An individual sample from one environment cannot be distinguished with confidence from that of another. Indeed, it appears that a steady-state condition was developed during the slow marine transgressions moving sand from one localized environment into an adjacent one. Sand was lost and gained in each environment as in the model proposed by Middleton (1970). Moreover, this condition prevailed for such long episodes that resulting size distributions represent a culminating equilibrium state for each environment. Even with longer time, no further changes in the size distributions could be produced, except for reductions in mean size brought about by abrasion or by localized lower current energies.

Assuming that the size distributions have attained their culminating equilibrium states, they represent the ultimate shape that the nearshore environments can impose on them. The distribution is clearly symmetrically log normal (phi normal), but the variability is concentrated largely in the kurtosis of each distribution. The shapes of such size distributions could be accepted as the standards of ultimate textural maturity from which departures could be compared; this is particularly true for standard deviation and kurtosis. Other nearshore sands should show somewhat different size distributions, and the departures in shape from the standards would represent a measure of approach to the condition of maximum textural maturity for that environment (Fig. 9a).

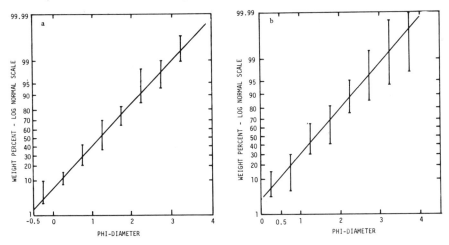

Figure 10. Distribution of sizes in selected pure quartz Cambrian sandstones of western Wisconsin. a. From Mauston Quadrangle. b. From Elk Mound Quadrangle. Samples are from a variety of stratigraphic positions to illustrate ranges in distribution of grain sizes. A selected sample in each case shows an ideal symmetrical log-normal distribution. With few exceptions, most other examples are similar but differ in mean size and sorting. All fall within the limits of the barred ordinates.

Quartz-Glauconite Sandstones

Sandstones that are mixtures of quartz grains and spheroidal pellets of glauconite are commonplace in the Lone Rock (Franconia) Formation (Cambrian). They are intimately intertongued with beds interpreted as parts of the nearshore-beach complex (Berg, 1954). Strata rich in glauconite do not occur as beach or dune deposits, nor as offshore-bar deposits, the sand of which is derived from the dunes. Also, glauconite is usually absent in sandy units that pass laterally into the carbonates. Abundant glauconite occurs in strata that were marginal to the strand lines and locally interfingered with the seaward portions of the wave-protected parts of the tidal flats.

During an episode of time, the quartz-glauconite sands constituted a relatively narrow band between the strand and the outer margins of the infralittoral region. Inside this belt the water was exceptionally shallow, and the currents, which could move glauconite pellets, ranged from velocities capable of developing small-scale cross-bedding to deposition of very thin laminae. Such laminae consist of alternate concentrations of glauconite and quartz, each ranging to 5 mm thick. Individual layers are distinct because each consists of high concentrations of glauconite or of quartz. Hence, dark green layers are separated from colorless layers as glauconite predominates over quartz. Some laminae are only a few grains thick, but those of quartz tend to be thicker. Invididual thin layers tend to be separated by films of mud-cracked shale which permit tracing single beds for a distance of several meters. Such features indicate former water depths of only a few centimeters.

Alternate laminae of preponderant glauconite or quartz are typical also of the small-scale cross-bedded strata; individual foreset layers are often no more than a few grains thick, consisting primarily of one mineral. Such concentrations of quartz and glauconite indicate that the locally prevailing currents were finely adjusted to gather selectively, move, and deposit quartz and glauconite in some general repetitive alternation. Periodically, a stronger current produced local erosion of the previously alternating laminae of quartz and glauconite and deposited a local rubble of partially cemented sandstone. Later, conditions returned to deposition of laminae marked by concentrations of either quartz or glauconite.

Except for the scattered intraformational conglomerates, the quartz-glauconite sandstones are interpreted as deposits of equilibrium conditions in which an alternation of currents preferentially moved and deposited glauconite and quartz in a repetitive sequence. Such currents represent the ideal conditions of the washing process described by Udden (1914; Emery and Stevenson, 1950), and the size distribution should be predicted as phi normal. Figure 11a illustrates the size distribution of ten samples of the quartz-glauconite sandstones. The line has been drawn as the best fit to the plotted points for one sample and illustrates a close approximation to an ideal log-normal distribution symmetrical about mean size. Other samples show size distributions of equally good symmetry but vary somewhat in mean size, standard deviation, and kurtosis. The entire spectrum, however, falls within the limits of the bracketed ordinates as shown.

Five samples of very friable, laminated glauconite sandstone (from those of Fig. 11a) were selected from widely separate geographic localities, disaggregated, and repeatedly passed through a magnetic separator to concentrate the quartz and glauconite into individual fractions. The size distribution of each (Figs. 11b and 12a) was determined by sieving. In each illustration, the line represents the distribution of one example, and the bracketed ordinates are the limiting ranges of percentage values for each size. All of the distributions are phi normal (even in the extreme sizes) and are better examples of the ideal symmetry than the pure quartz sandstones lacking glauconite (Figs. 9b and 10).

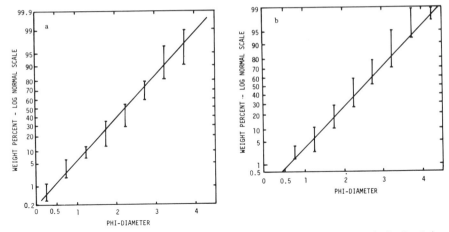

Figure 11. Ranges in distribution of grain sizes in quartz-glauconite sandstone in the Cambrian Lone Rock (Franconia) Formation taken from scattered Wisconsin localities. a. Line represents distribution of a single bulk sample of sandstone consisting of alternating thin laminae of quartz and glauconite. Other similar samples fall within the barred ordinates and are also symmetrically log normal. b. Line indicates distribution of sizes of glauconite grains only in the same sample shown in (a). Barred ordinates show ranges in values in other samples. These are also nearly ideal phi-normal distributions excluding the extreme sizes where errors are significant due to aggregates of grains and crushed grains.

Examples of the shapes of the frequency distributions of the quartz and glauconite separated from the same bulk sample are shown in Figure 12b. The limiting sizes and modal diameters are virtually the same, but the concentration of quartz is greater in its modal size than is the corresponding glauconite fraction.

Mean values of the specific gravities of the glauconite in several samples range between 2.45 and 2.77. The most common value is slightly less than 2.7, which is approximately the same as the associated quartz. The two minerals differ in that the glauconite pellets may be slightly heavier and more spherical. Such differences are small, but the currents were capable of concentrating one mineral over the other in alternating rhythms.

Deposition of the laminated quartz-glauconite sandstones appears to have been dominated by traction motion as indicated by the restricted spread of the sizes of individual grains, the small-scale foreset cross-beds, scattered ripple marks with internal bedding, and very thin layers separated by films of mud-cracked shale. Nowhere do the sands show characteristics of suspension settling. Quartz and glauconite grains responding to minor differences in current properties were moved and segregated repetitively. The thickness of a few grains piled one upon another to constitute an individual lamina of glauconite indicates that the grains were moved by currents at the lowermost part of the traction load only. The grains were rolled into position as a monomineralic layer at the very boundary of the sediment-water interface. In this part of the current regime, where an individual grain was rolled into final position, the driving force must have been dominated by laminar flow which moved either quartz or glauconite as separate grains, but not in equal proportions. Very subtle differences in the laminar-flow regime dictated whether quartz or glauconite was to dominate a lamina.

As stated earlier, the selected quartz-glauconite sandstones show size distributions that are more ideally phi normal than certain of the associated pure quartz units. Of the latter, some show local occurrence of laminated, gently sloping, large-scale cross-bedding which is interpreted as part of offshore-bar or spit deposition. Samples from such strata show log-normal distributions with slight skewness. This skewness is sufficiently persistent to be regarded as genuine and is attributed to differences in the flow regimes. Among the quartz-glauconite sandstones, laminar flow is regarded as the unique condition that produced the rhythmic lamination and the symmetrical phi-size distribution. Similarly, in the laminated beds of pure quartz sandstone, which appear to be deposits of the lowermost part of the traction load, the phi-size distribution is symmetrical. Elsewhere in the same strata, where depositional conditions suggest a strong influence of turbulent flow (for example, show pronounced scour and fill), the size distribution is obviously skewed. Accordingly, phi-normal distributions were produced under conditions in which washing or winnowing was accomplished by dominant laminar flow. Wherever turbulent flow dominated the regime, a skewed log-normal distribution is to be expected. Presumably, as the influence of turbulence increased, the skewness of the distribution and the tendency toward concentration into a few grain sizes increased.

CONCLUSION

Arguments presented above have been directed toward recognition of certain items pertaining to size distributions of sandy sediments. Among other things, such items concern the magnitude of errors inherent in the estimation of particle sizes, special cases that simulate size distributions of rocks crushed in the laboratory, and examples of size distributions that approximate both log-normal and Weibull functions. Rosin's distribution is not to be regarded as representing an extraordinary condition among sediments, but rather to represent the expected size frequencies of well-disintegrated "soil." This is the distribution of the relative sizes of particles fed to the headwaters of streams, where the primary sediment is the product of disintegration.

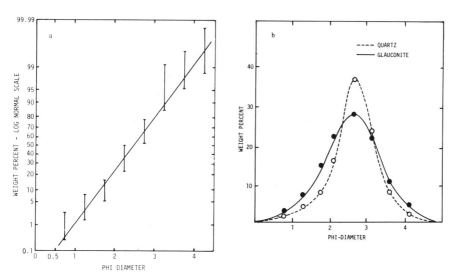

Figure 12. Grain sizes in quartz-glauconite sandstones in the Lone Rock (Franconia) Formation. a. Ranges and distribution of sizes of quartz grains only from samples plotted in Figure 11a. Line is one example of ideal phi-normal type; other samples yield similar lines. b. Frequency distributions of mean values from separated quartz and glauconite grains in samples used in Figure 11a. Note the strong correspondence in mean sizes and extreme limits of the distributions; the distinction is reflected by differences in concentrations of grain sizes.

Sampling of residual soil cannot be very precise, particularly because of the quantity and sizes of coarse particles included in a sample. Repeated crushing in the laboratory develops a size distribution that approaches Rosin's law far more closely than either the naturally or artificially partially crushed aggregate. In the natural state, repeated "crushing" is accomplished by repeated freezing and thawing, and the disintegrated rubble is reduced slowly to a much finer soil. Hence, a sample truncated at the small-pebble size is likely to provide a suitable representation of the general shape of the size distribution entering a stream's headwaters. Moreover, the observed size distribution can be compared with Rosin's distribution as a measure of departure from the ideal disintegration condition.

Sand debris that is deposited by a stream of low velocity maintains unique relative size distribution throughout its length. This distribution is regarded as a special case developing from Rosin's distribution in which the c parameter attains certain high values. For this reason, the size distribution of river sands should be compared with their own natural representatives that have been established as the standards of reference. Broad streams of low velocity best represent the drifting process (unidirectional flow) as defined by Udden. The special size distribution must, in the long run, be developed from known theory of hydraulic conditions, that is, relationships between traction-load and suspension-load deposition.

The process identified by Udden as washing or winnowing provides the physical condition suited to produce the phi-normal frequency distribution, particularly where laminar flow controls the movement of grains. Laminar flow is regarded as extraordinarily sensitive in selective movement of grains of approximately the same size and specific gravity, but which differ somewhat in sphericity as illustrated by the Cambrian quartz-glauconite sandstones. Where turbulent flow is important, despite deposition of sand from the traction load, departures from the ideal symmetrical log-normal distribution are primarily in skewness and kurtosis. For these distributions, selected natural examples established as standards of reference provide a better basis of comparison than the use of the ideal normal model.

The concept of a progression in textural maturity from an immature stage of size distribution represented by Rosin's law to an ultimate stage of maturity when the distribution becomes phi normal appears not to represent the general case. Although a size distribution represented by Rosin's law indicates an unstable condition, the modifications brought about by long-stream transport can be considered to culminate in a steady-state situation. Sand debris that is progressively deposited, eroded, and remixed, later to be redeposited farther downstream, achieves a size distribution that culminates in a distribution represented by a Weibull equation of certain parameters, the values of which are undetermined as yet. Such a distribution should be regarded as the ultimate condition of maturity for a stream-sand deposit. Dominance of turbulent flow in the stream precludes the development of symmetry in the log-normal distribution, even in that part deposited from the traction load.

To produce the symmetrical log-normal size distribution, a restraining condition of laminar flow must be placed on the current regime. The most commonplace environments in which laminar flow prevails are the shallow water, tidal flat, and special environments of the shoreline complex; hence, sands such as those of the Cambrian-Ordovician strata described above are ideal representatives of the expected size distribution. Elsewhere, in the same strata where the environment is not so strongly influenced by laminar flow, corresponding sands of a different lithofacies will have size distributions that are not symmetrically log normal; yet, in a sense, they are just as texturally mature. Their size distribution is a response to a culminating condition in which a turbulent-current regime moves and deposits the traction load.

Obviously, in order to separate the opposing influences of turbulent- and laminar-flow regimes on the culminating size distributions, many available distributions need to be re-examined in the light of their best fit to a known distribution. The purpose is to establish the values of parameters of the selected equations and to provide the necessary reference standards characteristic of specific current-depositional environments. Perhaps the results of such research will remove the uncertainty that currently prevails from overlap in sorting, skewness, and kurtosis when distributions are compared on the basis of their departure from the ideal normal distribution. Perhaps such a different approach to an old problem will assist in re-evaluation of the concept of textural maturity.

ACKNOWLEDGMENTS

This investigation is an outgrowth of a more extensive study of the Cambrian sandstones of Wisconsin which has been supported by National Science Foundation Grant GA-16524. I thank my colleague Michael Dacey for his counsel regarding frequency distributions.

REFERENCES CITED

Atchison, J., and Brown, J.A.C., 1957, The lognormal distribution: New York, Cambridge Univ. Press, 176 p.

Bennett, J. G., 1936, Broken coal: Jour. Inst. Fuel, v. 10, p. 22-39.

Berg, R. R., 1954, Franconia Formation of Minnesota and Wisconsin: Geol. Soc. America Bull., v. 65, p. 857-882.

Blatt, H., Middleton, G. V., and Murray, R., 1972, Origin of sedimentary rocks: Englewood Cliffs, N.J., Prentice-Hall, Inc., 634 p.

Dapples, E. C., 1955, General lithofacies relationships of St. Peter sandstone and Simpson group: Am. Assoc. Petroleum Geologists Bull., v. 39, p. 444-467.

——1959, Basic geology in science and engineering: New York, John Wiley & Sons, Inc., 609 p.

Doeglas, D. J., 1946, Interpretation of the results of mechanical analyses: Jour. Sed. Petrology, v. 16, p. 19-40.

Emery, K. O., and Stevenson, R. E., 1950, Laminated beach sand: Jour. Sed. Petrology, v. 20, p. 220-223.

Epstein, B., 1947, The mathematical description of certain breakage mechanisms leading to the logarithmico-normal distribution: Jour. Franklin Inst., v. 244, p. 471-477.

Folk, R. L., 1951, Stages of textural maturity in sedimentary rocks: Jour. Sed. Petrology, v. 21, p. 127-130.

——1966, A review of grain-size parameters: Sedimentology, v. 6, p. 73-93.

Folk, R. L., and Ward, W. C., 1957, Brazos River bar: A study in the significance of grain-size parameters: Jour. Sed. Petrology, v. 27, p. 3-26.

Friedman, G. M., 1961, Distinction between dune, beach and river sands from their textural characteristics: Jour. Sed. Petrology, v. 31, p. 524-529.

——1962, On sorting, sorting-coefficients, and the log-normality of the grain-size distribution of sandstones: Jour. Geology, v. 70, p. 737-754.

——1967, Dynamic processes and statistical parameters compared for size-frequency distribution of beach and river sands: Jour. Sed. Petrology, v. 37, p. 327-354.

Fuller, A. O., 1961, Size-distribution characteristics of shallow marine sands from the Cape of Good Hope, South Africa: Jour. Sed. Petrology, v. 31, p. 256-261.

Gaudin, A. M., 1926, An investigation of crushing phenomena: Am. Inst. Mining and Metallurgy Eng. Trans., v. 73, p. 253-316.

Geer, M. R., and Yancey, H. F., 1938, Expression and interpretation of the size composition of coal: Am. Inst. Mining and Metallurgy Eng. Trans. Coal Div., v. 130, p. 250-269.

Inman, D. L., 1949, Sorting of sediments in the light of fluid mechanics: Jour. Sed. Petrology, v. 19, p. 51-70.

——1952, Measures for describing the size distribution of sediments: Jour. Sed. Petrology, v. 22, p. 125-145.

Johnson, N. I., and Kotz, S., 1971, Continuous univariate distributions: Boston, Houghton Mifflin Co., 300 p.

Jones, T. A., 1970, Comparison of the descriptors of sediment grain-size distributions: Jour. Sed. Petrology, v. 40, p. 1204-1215.

Keller, W. D., 1945, Size distribution of sand in some dunes, beaches, and sandstones: Am. Assoc. Petroleum Geologists Bull., v. 29, p. 215-221.

Kittleman, L. R., 1964, Application of Rosin's distribution in size-frequency analysis of clastic rocks: Jour. Sed. Petrology, v. 34, p. 483-502.

Koldijk, W. S., 1968, On environment-sensitive grain-size parameters: Sedimentology, v. 10, p. 57-69.

Kolmogoroff, A. N., 1941, Über das logarithmisch normale Verteilungsgesetz der Dimensionen der Teilchen bei Zerstückelung: Dokl. Acad. Nauk SSSR, v. 31, p. 99.

Krumbein, W. C., 1934, Size-frequency distribution of sediments: Jour. Sed. Petrology, v. 4, p. 65-77.

——1936, Application of logarithmic moments to size-frequency distribution of sediments: Jour. Sed. Petrology, v. 6, p. 35-47.

——1938, Size frequency distribution of sediments and the normal phi curve: Jour. Sed. Petrology, v. 8, p. 84-90.

——1960, The "geological population" as a framework for analyzing numerical data: Liverpool and Manchester Geol. Jour., v. 2, p. 341-368.

Krumbein, W. C., and Pettijohn, F. J., 1938, Manual of sedimentary petrography: New York, D. Appleton-Century Co., 549 p.

Krumbein, W. C., and Tisdel, F. W., 1940, Size distribution of source rocks of sediments: Am. Jour. Sci., v. 238, p. 296-305.

Landers, W. S., and Reid, W. T., 1946, A graphical form for applying the Rosin and Rammler equation to the size distribution of broken coal: U.S. Bur. Mines Inf. Circ. 7346.

McCammon, R. B., 1962, Efficiencies of percentile measures for describing the mean size and sorting of sedimentary particles: Jour. Geology, v. 70, p. 453-466.

McEwen, M. C., Fessenden, F. W., and Rogers, J.J.W., 1959, Texture and composition of some weathered granites and slightly transported arkosic sands: Jour. Sed. Petrology, v. 29, p. 477-492.

Middleton, G. V., 1962, On sorting, sorting coefficients, and log normality of the grain-size distribution of sandstones—A discussion: Jour. Geology, v. 70, p. 754-756.

——1970, Generation of the log-normal frequency distribution in sediments, in Romanova, M. A. and Sarmanov, O. V., eds., Topics in mathematical geology: New York, Consultants Bureau, p. 34-42.

Moiola, R. J., and Weiser, D., 1968, Textural parameters: An evaluation: Jour. Sed. Petrology, v. 38, p. 45-53.

Moss, A. J., 1962a, The physical nature of common sandy and pebbly deposits, Pt. 1: Am. Jour. Sci., v. 260, p. 337-373.

——1962b, The physical nature of common sandy and pebbly deposits, Pt. 2: Am. Jour. Sci., v. 261, p. 297-343.

Ostrom, M. E., 1970, Lithologic cycles in lower Paleozoic rocks of western Wisconsin; field trip guidebook for Lower Cambrian and Ordovician geology of western Wisconsin: Univ. Wisconsin Inf. Circ. 11, 131 p.

Otto, G. H., 1937, The use of statistical methods in effecting improvements on a Jones sample splitter: Jour. Sed. Petrology, v. 7, p. 110-132.

——1938, The sedimentation unit and its use in field sampling: Jour. Geology, v. 46, p. 529-582.

——1939, A modified logarithmic probability graph for the interpretation of mechanical analyses of sediments: Jour. Sed. Petrology, v. 9, p. 62-76.

Passega, R., 1957, Texture as characteristic of clastic deposition: Am. Assoc. Petroleum Geologists Bull., v. 41, p. 1952-1984.

Pettijohn, F. J., 1957, Sedimentary rocks (2d ed.): New York, Harper & Bros., 719 p.
Rogers, J.J.W., 1965, Reproducibility and significance of measurements of sedimentary size distributions: Jour. Sed. Petrology, v. 35, p. 722-732.
Rogers, J.J.W., Krueger, W. C., and Krog, M., 1963, Sizes of naturally abraded materials: Jour. Sed. Petrology, v. 33, p. 628-632.
Rosin, P., and Rammler, E., 1933, The laws governing the fineness of powdered coal: Jour. Inst. Fuel, v. 7, p. 29-36.
Shepard, F. P., and Young, R., 1961, Distinguishing between beach and dune sands: Jour. Sed. Petrology, v. 31, p. 196-214.
Trask, P. D., 1932, Origin and environment of source beds of petroleum: Houston, Gulf Pub. Co., 323 p.
Twenhofel, W. H., Raasch, G. O., and Thwaites, F. T., 1935, Cambrian strata of Wisconsin: Geol. Soc. America Bull., v. 46, p. 1687-1743.
Udden, J., 1914, Mechanical composition of clastic sediments: Geol. Soc. America Bull., v. 25, p. 655-744.
Wentworth, C. K., 1922, A scale of grade and class terms for clastic sediments: Jour. Geology, v. 30, p. 377-392.
——1929, Methods of computing mechanical composition types in sediments: Geol. Soc. America Bull., v. 40, p. 771-790.

MANUSCRIPT RECEIVED BY THE SOCIETY OCTOBER 1, 1973
REVISED MANUSCRIPT RECEIVED MAY 10, 1974

Burrowing Depths of Some Neogene Heterodont Bivalves

David E. Ogren

Department of Geology
Georgia State University
Atlanta, Georgia 30303

Keewhan Choi

Department of Mathematics
Georgia State University
Atlanta, Georgia 30303

Martin D. Fraser

Department of Mathematics
Georgia State University
Atlanta, Georgia 30303

ABSTRACT

Using a classification procedure based on the principal components method and the Mahalanobis distance, we classified 11 species of fossil bivalves into three burrowing depth groups: shallow, medium, and deep. The classification is consistent with data from the literature concerning burrowing depths of modern examples of these species.

INTRODUCTION

The purpose of this study is to develop a statistical method for classifying burrowing habits in a selected group of fossil bivalves and, from that, to be able to predict the burrowing habits from certain morphological attributes of the shell. It has been generally assumed that the morphological characteristics of an organism are related to its life style; that is, that morphology is related to function in the environment.

To test the possibility of predicting the burrowing habits, 20 single, mature shells for each of the 11 species of bivalve collected from the Neogene Caloosahatchee Formation in Florida were examined. These collections were obtained from dredgings from the Caloosahatchee Formation along the Caloosahatchee River near Moore Haven, Florida, and from lime pits near Belle Glade and South Bay, Florida. The member of the Caloosahatchee Formation from which the specimens were obtained is unknown. Judgments as to which measurements to make were based on a knowledge of the burrowing behavior of several species of modern bivalves as summarized by Trueman (1968, p. 167-186).

Features of the shell that were chosen for measurement were those which would seem to be most directly related to the burrowing process. The shell features selected are (A) shell length measured approximately parallel to the hinge line; (B) shell height measured approximately perpendicular to the hinge line; (C) the A/B ratio; (D) shell inflation/A ratio (inflation is the distance from the valve pair symmetry plane to the outermost point on the shell); (E) shell thickness/A ratio (thickness is the distance from the inside of the shell to the outside, measured at the approximate midpoint of the shell); (F) length of pallial sinus; (G) anterior muscle scar length/B ratio; (H) posterior muscle scar length/B ratio; (I) central cardinal tooth length/B ratio; and (J) penetrability measured by pushing the shell with a pocket penetrometer into loose, dry sand until just buried.

Ratios are used to subdue differences in absolute size between species. The species selected for study are given in Table 1, along with their code numbers, burrowing depths of modern examples from the literature, and references. All forms selected for use in the study are either represented in the modern fauna or, at least, very similar species are alive today.

STATISTICAL METHOD

The problem is that of classifying these 11 species of bivalves into groups according to their burrowing depth. The classification is based on the previously defined measurements taken from a sample of 20 shells for each of the 11 species, a total of 10 measurements on 220 shells. Table 2 lists the sample means and standard deviations of the 10 measurements for each of the 11 species. Burrowing depth is not a directly observable measurement on fossil shells. Rather, the burrowing depth must be inferred from features of the shell.

Sample correlation coefficients between measurements are given in Table 3. All correlation coefficients are sufficiently small to warrant retention of all 10 measurements in the calculations.

Our classification procedure is based on the principal components method (Nie and others, 1970, p. 208-244; App. 1) and the Mahalanobis distance (Rao, 1970, p. 351-378; App. 1). In the first stage of this procedure, a tentative classification into three groups was arrived at by principal components analysis. Because the principal components analysis used only the first three most important components, this tentative classification was refined in the second stage using a classification procedure based on the Mahalanobis distance.

Stage 1

The first three most important principal components, which account for 85 percent of the total variation, were used to divide tentatively the 11 species into three groups. Figure 1 is a plot of the first two principal components for each of the 11 species. Because the same clustering is obtained using the first three principal

TABLE 1. BIVALVE SPECIES WITH REFERENCES TO BURROWING DEPTHS

No.	Species	Burrowing depth	References
1	*Eucrassatella gibbesii* (Toumey and Holmes) 1856, Olsson and Harbison, 1953	Shallow(?)	
2	*Lucina pennsylvanica* (Linne) 1758, Olsson and Harbison, 1953	Shallow	Allen, 1958, p. 426; Kauffman, 1969, p. 164; Stanley, 1970, p. 152
3	*Anondontia alba* Link, 1807	Shallow	Kauffman, 1969, p. 164; Stanley, 1970, p. 149
4	*Trachycardium emmonsi* (Conrad) 1867, Gardner, 1943	Shallow(?)	
5	*Macrocallista nimbosa* (Solander) 1786, Dall, 1903	Medium	Kauffman, 1969, p. 164
6	*Macrocallista maculata* (Linne) 1758, Perry, 1940	Medium to deep	Kauffman, 1969, p. 164; Aurelia, 1970, p. 44
7	*Chione cancellata* (Linne) 1758, Palmer, 1927	Shallow	Perry and Schwengel, 1955, p. 69; Stanley, 1970, p. 161
8	*Chione latilirata athleta* (Conrad) 1863, Mansfield, 1932	Shallow	Perry and Schwengel, 1955, p. 69
9	*Dosina elegans* (Conrad) 1844, Dall, 1903	Medium	Norton, 1947, p. 203; Stanley, 1970, p. 166
10	*Tellina tayloriana* Sowerby, 1867	Deep	Stanley, 1969, p. 634; Aurelia, 1970, p. 44
11	*Tagelus divisus* (Spengler) 1794, Dall, 1900	Deep	Frey, 1967, p. 573; Stanley, 1970, p. 175

components as with the first two principal components, only the first two principal components are plotted for ease of graphic representation. From the properties of principal components analysis, the distance between two species represented by their principal components indicates the degree of similarity in their morphology; that is, species that are near to each other in Figure 1 are morphologically similar. In Figure 1, above the horizontal axis, there are two clusters and two species not in a cluster. Below the horizontal axis, such tight clustering is not apparent. However, species 5 and 9 are separated from species 6, 10, and 11 by the vertical axis.

At this point, external information was used to form three tentative groups. Table 2, column F (length of pallial sinus), shows that all species above the horizontal axis (1, 2, 3, 4, 7, and 8) can be distinguished by their having no or very short pallial sinuses. It seems reasonable that species which have short or no pallial sinuses have short siphons and therefore would burrow shallowly. Long siphons would imply nonshallow burrowing. From Table 2, column D (inflation), a ranking of the sample mean values served to distinguish species 1, 2, 3, 4, 7, and 8 from 5, 6, 9, 10, and 11. This same ranking also distinguishes species 5 and 9 from 10 and 11. This is consistent with the observation that shallow burrowers generally have inflated shells and that the inflation value generally decreases with increasing burrowing depth. Species 6 could not be definitely assigned to any group on the basis of pallial sinus length or inflation. However, because of the proximity of species 6 to species 10 and 11 in Figure 1, it was assigned to that group. In summary, the tentative groupings are 1, 2, 3, 4, 7, and 8, shallow-depth forms; 5 and 9, medium-depth forms; and 6, 10, and 11, deep forms.

As used here, the terms "shallow," "medium," and "deep" are purely qualitative. The statistical procedure only classifies; it does not identify the meaning of the groups. The meaning of the groups was established by reference to statements

TABLE 2. SAMPLE MEANS AND STANDARD DEVIATIONS (IN PARENTHESES) OF
TEN MEASUREMENTS ON ELEVEN SPECIES OF BIVALVES

Species	Measurement									
	A	B	C	D	E	F	G	H	I	J
1	66.079	46.263	0.700	0.237	0.046	0.0	0.177	1.229	0.268	0.288
	(3.030)	(2.648)	(0.020)	(0.017)	(0.013)	(0.0)	(0.024)	(0.200)	(0.021)	(0.018)
2	37.150	35.675	0.970	0.346	0.064	0.0	0.051	1.570	0.410	0.324
	(6.111)	(2.667)	(0.067)	(0.039)	(0.018)	(0.0)	(0.025)	(0.192)	(0.043)	(0.033)
3	45.125	41.275	0.916	0.310	0.033	0.0	0.187	2.245	0.460	0.275
	(3.316)	(2.398)	(0.026)	(0.036)	(0.017)	(0.0)	(0.038)	(0.399)	(0.031)	(0.022)
4	38.825	47.125	0.820	0.493	0.032	0.0	0.058	2.770	0.275	0.318
	(2.858)	(3.437)	(0.040)	(0.043)	(0.011)	(0.0)	(0.028)	(0.592)	(0.056)	(0.036)
5	104.050	55.100	0.532	0.174	0.019	23.000	0.066	2.745	0.278	0.340
	(16.560)	(3.291)	(0.074)	(0.085)	(0.006)	(5.947)	(0.008)	(0.474)	(0.028)	(0.036)
6	48.100	35.025	0.741	0.223	0.025	17.550	0.078	0.857	0.287	0.322
	(11.201)	(3.582)	(0.059)	(0.018)	(0.005)	(1.677)	(0.008)	(0.402)	(0.014)	(0.028)
7	33.225	28.775	0.870	0.301	0.063	3.925	0.078	0.897	0.256	0.226
	(4.089)	(2.479)	(0.046)	(0.034)	(0.016)	(3.341)	(0.010)	(0.109)	(0.023)	(0.028)
8	30.675	27.000	0.881	0.371	0.104	3.025	0.105	0.905	0.208	0.223
	(1.712)	(1.395)	(0.031)	(0.034)	(0.021)	(0.413)	(0.015)	(0.122)	(0.021)	(0.025)
9	59.333	51.929	0.864	0.190	0.015	20.976	0.073	2.681	0.362	0.350
	(7.070)	(9.993)	(0.122)	(0.023)	(0.003)	(2.442)	(0.008)	(0.782)	(0.025)	(0.030)
10	44.600	29.650	0.669	0.149	0.018	28.425	0.062	0.735	0.321	0.209
	(5.355)	(2.550)	(0.055)	(0.062)	(0.023)	(6.477)	(0.014)	(0.139)	(0.030)	(0.021)
11	37.075	12.825	0.345	0.090	0.012	12.825	0.061	0.572	0.480	0.435
	(2.483)	(3.310)	(0.076)	(0.013)	(0.004)	(3.334)	(0.015)	(0.050)	(0.062)	(0.056)

in the literature about observed depths of burrowing. Because this information is not only sparse (Stanley, 1970, p. 3) but also generally qualitative, it should be understood that our use of the terms is purely relative.

Stage 2

One species was selected from each tentative group from stage 1 as the initial "seed" species. Each of the remaining eight species was assigned to the nearest seed, thus forming three groups. To determine the seed nearest to a species, the Mahalanobis distance between the two species was calculated, using all 10 measurement values. The Mahalanobis distance is a numerical measure of the similarity between two species that takes into account any possible correlation among the 10 measurements. In each group, a new seed was computed to be the set of 10 averaged measurement values of all the species assigned to the group. The steps above were repeated, using a new seed as the initial seed, until no change occurred in the assignment of species to groups.

TABLE 3. SAMPLE CORRELATION COEFFICIENTS BETWEEN MEASUREMENTS

	Measurement									
	A	B	C	D	E	F	G	H	I	J
A	1.000	0.688	−0.370	−0.415	−0.404	0.472	0.083	0.478	−0.122	0.136
B	0.688	1.000	0.291	0.192	−0.211	0.101	0.160	0.822	−0.246	−0.063
C	−0.370	0.291	1.000	0.690	0.515	−0.461	0.172	0.246	−0.197	−0.443
D	−0.415	0.192	0.690	1.000	0.584	−0.717	0.066	0.310	−0.333	−0.265
E	−0.404	−0.211	0.515	0.584	1.000	−0.577	0.146	−0.220	−0.409	−0.422
F	0.472	0.101	−0.461	−0.717	−0.577	1.000	−0.414	−0.017	0.002	0.154
G	0.083	0.160	0.172	0.066	0.146	−0.414	1.000	0.023	0.060	−0.362
H	0.478	0.822	0.246	0.310	−0.220	−0.017	0.023	1.000	0.038	0.215
I	−0.122	−0.246	−0.197	−0.333	−0.409	0.002	0.060	0.038	1.000	0.540
J	0.136	−0.063	−0.443	−0.265	−0.422	0.154	−0.362	0.215	0.540	1.000

This procedure was repeated for the 36 possible combinations of initial seed species. The results are shown in Table 4, which gives the percentage of times each species was assigned to various burrow-depth groups. For example, species 1 was assigned to the shallow group 50 percent of the time, to the medium group 28 percent, and to the deep group 22 percent. Therefore, species 1 was finally assigned to the shallow group. Each species, except 2 and 8, was finally assigned to the same group as in the tentative grouping of stage 1 a majority of times. Species 2 was assigned to the shallow group more frequently than to the medium group. Species 8 was assigned to the shallow group and the medium group with equal frequency.

CONCLUSIONS

The final results of the classification procedure produced this classification: Species 1, 2, 3, 4, 7, and 8 are shallow burrowers; species 5 and 9 are medium-depth burrowers; and species 6, 10, and 11 are deep burrowers. These results may be compared with Table 1, which contains a consensus from the literature of burrowing depths for the species considered in this paper. It should be noted that there is not always complete agreement among workers about depth of burrowing in modern bivalves. For example, species 2 was reported by two workers to be a shallow burrower and by another to be a deep burrower. Our classification assigns species 2 to the shallow group.

Table 5 presents sample means and standard deviations (in parentheses) of the 10 measurements for the species classified into three depth groups. These sample means may be taken as representative values of measurements for each depth group. As such, these average measurements are quantitative morphologic attributes that distinguish burrowing habit. Moreover, these sample means of the three depth groups can be used to predict a depth group for a new species. For a new species, calculate the Mahalanobis distance to each depth group and assign the new species to the nearest group. At present, the estimate of the covariance matrix used in the Mahalanobis distance is based on 220 samples on 11 species. When more data on a larger number of species are available, the estimate of the covariance

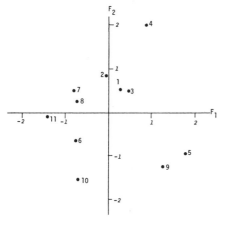

Figure 1. Plot of averaged principal components of 11 species of bivalves. Each species is plotted in the F_1-F_2 plane, where F_1 is the first principal component axis and F_2 is the second principal component axis.

TABLE 4. PERCENTAGES OF GROUP ASSIGNMENTS BASED ON MAHALANOBIS DISTANCES

Species	Burrowing depth groupings		
	Shallow (%)	Medium (%)	Deep (%)
1	50.0	27.8	22.2
2	44.4	41.7	13.9
3	52.8	36.1	11.1
4	55.6	22.2	22.2
5	8.3	91.7	0
6	16.7	30.6	52.8
7	52.7	2.8	44.4
8	38.9	38.9	22.2
9	16.7	66.7	16.7
10	19.4	19.4	61.1
11	25.0	11.1	63.9

TABLE 5. SAMPLE MEANS AND STANDARD DEVIATIONS (IN PARENTHESES) OF
TEN MEASUREMENTS FOR THREE DEPTH GROUPS

Group	Measurement									
	A	B	C	D	E	F	G	H	I	J
Shallow	42.013	37.686	0.860	0.343	0.126	1.158	0.109	1.603	0.313	0.274
	(3.771)	(2.575)	(0.041)	(0.035)	(0.016)	(1.374)	(0.024)	(0.320)	(0.035)	(0.028)
Medium	81.692	53.514	0.698	0.182	0.017	21.988	0.070	2.713	0.320	0.345
	(12.732)	(7.439)	(0.101)	(0.062)	(0.005)	(4.546)	(0.008)	(0.647)	(0.027)	(0.033)
Deep	43.258	77.500	0.585	0.154	0.018	19.600	0.067	0.721	0.363	0.322
	(7.310)	(3.177)	(0.064)	(0.038)	(0.014)	(4.316)	(0.013)	(0.247)	(0.041)	(0.038)

matrix will improve, thereby permitting increasing accuracy of the Mahalanobis distance used for prediction.

ACKNOWLEDGMENT

We thank Miriam and Howard Schriner, Jr., of LaBelle, Florida, for lending some of the material used in this study.

APPENDIX 1. STATISTICAL METHODS

Principal Component Analysis

Let $A, B, C, D, \ldots$, and J be 10 measurements. Each shell has 10 measurement values. The first principal component of the 10 measurements is that linear compound

$$Y_1 = a_1 A + a_2 B + a_3 C + \ldots + a_{10} J,$$

where $a_1, a_2, \ldots, a_{10}$ are constants. This accounts for the largest percentage of the total variance of our multivariate data.

The second principal component of the 10 measurements is that linear compound

$$Y_2 = b_1 A + b_2 B + b_3 C + \ldots + b_{10} J,$$

where $b_1, b_2, \ldots, b_{10}$ are constants. This accounts for the second largest percentage of the total variance.

The third and all the other principal components up to the tenth are defined in the same manner.

The first principal component is the most important linear compound, the second principal component is the second most important linear compound, and so on, because the first (second) principal component explains the largest (second largest) amount of the total variance, and so on. Each shell has 10 principal component values that are computed from the original measurement values.

One important benefit of the principal components analysis is that it enables us to summarize most of the variation in a multivariate system in fewer *new* variables, which are called principal components. The percentage of the total variance explained by each principal component is given in Appendix Table 1. Also, the average values of principal components for each species are given in Appendix Table 2. From Appendix Table 1, we see that by using only the first three principal components, we can explain 85 percent of the total variance. Each species can be plotted in a three-dimensional space whose three axes are the first three principal components when each species is represented by the average values of the first three principal components.

APPENDIX TABLE 1. PERCENT OF VARIATION ACCOUNTED FOR BY EACH OF THE
TEN PRINCIPAL COMPONENTS

Principal component	Variation (%)	Principal component	Variation (%)
1st	34.8	6th	3.5
2d	24.8	7th	1.8
3d	14.7	8th	1.4
4th	11.8	9th	0.6
5th	6.3	10th	0.2

One technical advantage of the principal components analysis is that the analysis does not put any restrictions on kinds of measurements. Measurements could be absolute values of some physical measurements or their ratios (see Rao, 1970, for further mathematical details of principal components analysis).

Mahalanobis Distance

Let $P_1 = (A_1, B_1, \ldots, J_1)$ and $P_2 = (A_2, B_2, \ldots, J_2)$ be row matrices of average values of the 10 measurements for any two species, and the matrix S be the covariance matrix of the 10 measurements. The Mahalanobis distance between the species is defined to be the positive square root of the quantity D^2, where

$$D^2 = (P_1 - P_2) S^{-1} (P_1 - P_2)^T,$$

where the superscripts -1 and T denote the matrix inverse and transpose (Rao, 1970, p. 355).

One important property of D is that it incorporates the dependent structure of the measurements into the definition in terms of the covariance matrix S.

APPENDIX TABLE 2. AVERAGE VALUES OF PRINCIPAL COMPONENTS FOR
ELEVEN SPECIES OF BIVALVES

Species	Principal components									
	1st	2d	3d	4th	5th	6th	7th	8th	9th	10th
1	0.273	0.505	−0.646	−0.832	1.649	−0.614	−0.104	0.431	−1.191	−1.534
2	−0.065	0.822	1.314	1.083	−1.178	0.199	0.971	1.166	−0.434	−0.428
3	0.452	0.488	1.542	0.686	1.933	−0.356	−0.427	−0.056	0.714	1.008
4	0.883	1.992	−0.696	−0.303	−1.005	0.213	−1.036	−0.986	0.993	−0.343
5	1.749	−0.964	−0.397	−1.427	−0.422	0.155	0.454	1.445	0.278	0.846
6	−0.788	−0.646	−0.954	0.742	0.112	0.903	−0.804	0.704	0.709	−0.495
7	−0.809	0.479	−0.647	0.442	−0.366	−0.764	0.129	0.289	−1.064	0.997
8	−0.768	0.216	−1.107	0.235	0.349	−0.199	2.009	−0.638	0.578	0.494
9	1.224	−1.247	0.179	1.329	−0.193	0.777	−0.231	−1.158	−0.901	−0.486
10	−0.727	−1.535	0.203	−0.121	−0.673	−1.724	−0.719	−0.570	0.801	−0.259
11	−1.424	−0.111	1.210	−1.815	−0.205	1.411	−0.243	−0.627	−0.482	0.201

REFERENCES CITED

Allen, J. A., 1958, On the basic form and adaptations to habitat in the Lucinacea (Eulamellibranchia): Royal Soc. London Philos. Trans. B, v. 241, p. 421-484.

Aurelia, M., 1970, The habitats of some subtidal pelecypods in Harrington Sound, Bermuda: Bermuda Biol. Research Sta. Pub. 6, p. 39-52.

Frey, R. W., 1967, The lebensspuren of some common marine invertebrates near Beaufort, North Carolina. I. Pelecypod burrows: Jour. Paleontology, v. 42, p. 570–574.

Kauffman, E. G., 1969, Form, function and evolution, *in* Moore, R. C., and Teichert, C., eds., Treatise on invertebrate paleontology, Pt. N: Geol. Soc. America and Univ. Kansas, v. 1, 489 p.

Nie, N., Bent, D. H., and Hull, C. H., 1970, Statistical package for the social sciences: New York, McGraw-Hill Book Co., 343 p.

Norton, O. A., 1947, Some ecological observations on *Donsinia discus* Reeve at Beaufort, North Carolina: Ecology, v. 28, p. 199–204.

Perry, L. M., and Schwengel, J. S., 1955, Marine shells of the western coast of Florida: Ithaca, N.Y., Paleont. Research Inst., 318 p.

Rao, C. R., 1970, Advanced statistical methods in biometric research: Darien, Conn., Hafner Pub. Co., Inc., 390 p.

Stanley, S. M., 1969, Bivalve mollusk burrowing aided by discordant shell ornamentation: Science, v. 166, no. 3905, p. 634–635.

——1970, Relation of shell form to life habits in the Bivalvia (Mollusca): Geol. Soc. America Mem. 125, 296 p.

Trueman, E. R., 1968, The burrowing activity of bivalves, *in* Fretter, V., ed., Studies in the structure, physiology and ecology of molluscs: Symposia of the Zool. Soc. London and Malacological Soc. London, no. 22, 377 p.

MANUSCRIPT RECEIVED BY THE SOCIETY SEPTEMBER 12, 1973
REVISED MANUSCRIPT RECEIVED DECEMBER 12, 1973

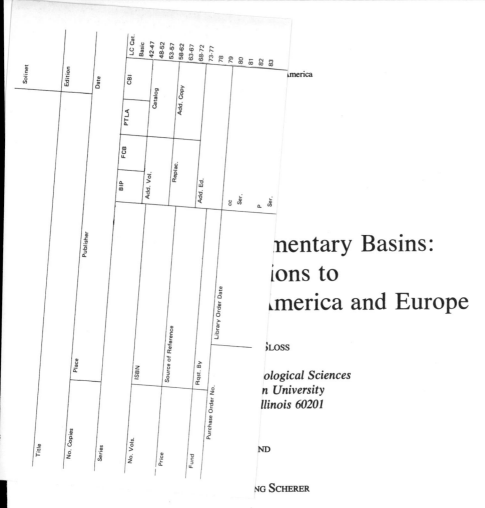

...nentary Basins: ...ons to ...merica and Europe

...LOSS

...ological Sciences
...n University
...llinois 60201

...ND

...NG SCHERER

ITS Servicios Tecnicos, S.A.
Apartado 60982
Caracas, Venezuela

ABSTRACT

The three-dimensional geometry of a time-stratigraphic unit, representing deposition in a craton–interior basin during a discrete increment of time, reflects the degree and mode of subsidence of the basin during accumulation. Although it is clear that subsidence rates and geometric forms of basins are variable over the course of major depositional cycles (±50 m.y.), it has not been possible to express these variations quantitatively so that the geometric evolution of individual basins or comparisons between basins could be analyzed rigorously. Application of a surface-fitting expression, a modification of the bivariate-normal distribution function, to thickness data on successive time-stratigraphic units permits the derivation of a number of geometric parameters that are significant basin descriptors. Of these descriptors, the most readily applicable to basin analysis are (1) position of the hinge line or locus of maximum rate of change of thickness, (2) basin slope at the hinge line, and (3) basin dimensions and shape as defined by the length of principal elliptical axes at the hinge line. Given independently derived geochronologic data, each parameter may be monitored with respect to time and expressed as values per unit time.

Analysis of Devonian units of the Michigan, Elk Point, and Moscow Basins in terms of slope and dimensional parameters, plus additional data on volume and areal distribution, indicates a substantial degree of similarity and synchrony in the development of the three basins and a particularly close kinship between the widely separated Elk Point and Moscow Basins. If parallelism and synchrony in the evolution of basins in the interiors of separate cratons are confirmed by further study, there is a strong implication that globally effective marine transgressions of continents are accompanied by epeirogenic tectonism within continents.

INTRODUCTION

The shapes of the sediment fills that occupy basins of craton interiors reflect a complex history of downwarping and uplifting. In the course of a major cycle of deposition and erosion (perhaps 50 to 100 m.y.), the cross-sectional shape of a typical basin may evolve as shown in Figure 1. At T_0 (time at the initiation of sedimentation), the basinal site is a peneplane near base level. Between T_0 and T_1 (time at the end of deposition of an identifiable time-stratigraphic unit), subsidence of the basin interior is initiated; this depresses the interior below base level so that sediment, confined to the center region of the basin, accumulates. Typically, as in the interval between T_1 and T_2, the entire region subsides below base level, and deposition spreads from the basin axis to the flanks and beyond to surrounding shelf areas; at the same time, the rate of subsidence (and of concomitant deposition) is slightly greater in the interior. Differential subsidence of the basin interior relative to its flanks and to the surrounding shelf reaches a climax in the time span T_2 to T_3. During this time depositional environments, influenced by rates of subsidence, are strongly differentiated, and the lithofacies patterns of the resulting accumulation conform closely to isopach lines defining the basinal geometry. For example, such episodes are commonly marked by euxinic dark shales in the interior of the basin, surrounded by concentric belts of carbonates (attributable to banks and reefs) and peripheral evaporites of lagoonal and supratidal origins. Between T_3 and T_4 the basin interior continues to subside more rapidly than surrounding areas but with a decreasing rate of change in subsidence rate from the margins to the interior. The product is a broader, less well-defined basinal geometry commonly lacking a clear correlation with the distribution of environments and resulting facies.

In the next increment of time (not illustrated), differential subsidence of the basin is progressively reduced until deposition in the area is essentially that of stable-shelf conditions. Broad epeirogenic uplifting of the area ensues (through stages T_6, T_7 or, in exteme cases, T_8, Fig. 1) carrying the basin and its margins above base level and into an erosional regime. Erosion is often halted at stages equivalent to T_6 or T_7 by the initiation of a succeeding depositional cycle; nevertheless, erosion significantly alters the geometry of time-stratigraphic units, with the effects increasing from the basin center to the margins and surrounding shelves and from the oldest to the youngest unit deposited.

The above is a qualitative and oversimplified description of a scenario that is repeated with countless variations by cratonic basins on all continents. In spite of the obliteration of parts of depositional history by erosion and regardless of the lack of quantitative rigor, it is clear that the geometry of a particular sedimentary basin is not fixed or static. Moreover, although individual basins may differ from adjacent or distant counterparts in many geometric parameters, there is a degree of craton-wide, and perhaps world-wide, synchrony in the major stages of basin

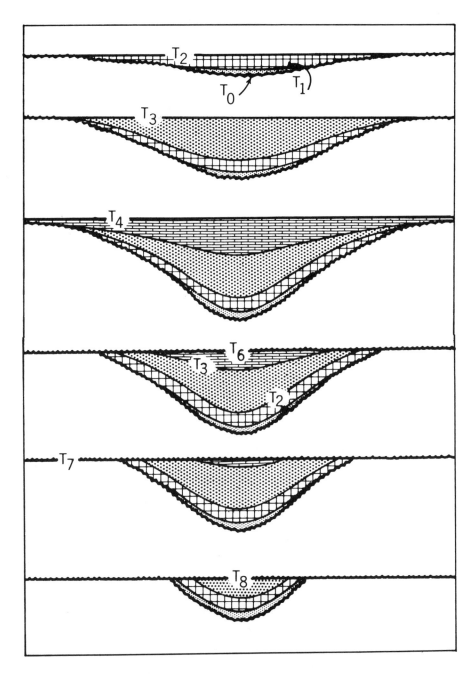

Figure 1. Stages in the subsidence, filling, and erosion of a sedimentary basin. At time of initiation of a depositional cycle (T_0), basin area is an essentially planar surface marking the conclusion of a preceding erosional episode. In initial depositional stage (T_0 to T_1), deposition is confined to basin-center area, spreading over a greater area in the next stage (T_1 to T_2). The interval T_2 to T_3 is marked by differential subsidence of basin interior relative to margins. Differential subsidence, and concomitant deformation of earlier, originally planar surfaces, continues to time T_4 but with less difference between subsidence rates of basin margins and interior. After time T_4 differential subsidence ceases, and at some time between T_4 and T_6 regional uplift is initiated and continues to T_8 with increasing erosional effect on the distribution and geometry of the units preserved. Width of the basin illustrated is measured in hundreds of kilometers; total thickness accumulated is measured in kilometers.

evolution (Sloss, 1972). This latter possibility suggests an evaluation of cratonic history in terms of global tectonic concepts (Sloss and Speed, 1974).

Development of an acceptable theory for time-dependent changes in basin form and for relating such changes to the mechanics of crustal deformation is not possible in the absence of objectively derived quantitative parameters capable of yielding rigorous geometric descriptions of basins in space and time. Further, useful descriptors are limited to those that (1) are readily derivable from currently available data and (2) lead to geologically meaningful analyses of the evolution of individual basins and comparisons between basins. Our aim in this paper is to identify descriptive parameters that meet these standards, present a methodology for quantitative derivation of certain of these parameters, and review some of the results obtained.

BASIN PARAMETERS

Spatial Measures

Basin analysis is dependent on an adequate array of three-dimensional data such as those produced by widespread drilling accompanied by electrical, radioactive, sonic, and other mechanical logs, supported by detailed observations of cores and samples. These data may be amplified by observations of outcrops and may be supplemented or substituted by seismic profiling with close velocity control. If an adequate data array is available and if significant time-stratigraphic markers can be correlated by paleontologic or lithostratigraphic techniques, successive isochronous units may be identified for analysis. In the absence of erosional effects, the depositional geometry at the close of each time increment and, hence, the areal variation in rates of subsidence prevailing during each time increment are shown by isopach maps. Thus, the fundamental geometric display is the isopach that is prepared, manually or by computer, by interpolation between discrete thickness values obtained from well bores, outcrop measurements, or seismic profiles. Indeed, all of the essential spatial (as opposed to temporal) information is contained in the isopach map, and the problem becomes one of abstracting from the map useful geometric parameters, including the following: geographic coordinates of the position of maximum thickness, basin length, basin width, orientation of long and short axes, position of hinge line, length of axes at intersection with hinge line, maximum thickness, mean thickness, and maximum rate of change of thickness.

Of the parameters listed above, only the location of the basin center (the position of maximum thickness) and the maximum thickness can be read directly from the isopach map without subjective evaluation. Measurement of basin dimensions such as length and width requires definition of the basin margins and, except where erosion has removed the unit in question from surrounding shelf areas, margins cannot be defined readily by simple map inspection. Very few basins assume uncomplicated subcircular or elliptical shapes; therefore, the orientation of principal basin axes is also a matter of gross approximation from inspection. The hinge line (the position of the circumbasin line of maximum rate of change of thickness) can be observed on some maps within the belt of the most closely spaced isopach contours; many maps, however, fail to display this feature unequivocally. Without the ability to define the hinge line, the dimensions of the major basin axes at the hinge line are similarly unattainable. Such dimensions are significant descriptors of the geometry of basin interiors. The mean thickness of any unit analyzed is a measure of the average net subsidence of the basin area during the time span considered, less the effects of subsequent erosion. Mean thickness is not directly derivable from isopach maps, nor can it be derived by simple averaging of thickness at data points. The value of the basin slope, the maximum rate of change of thickness values, is an extremely important parameter, because it is independent of the size of the basin and permits comparisons among basins of any dimensions. Again, experience has shown that attempts to determine basin slope by direct measurement on isopach maps, even where hinge lines are obvious, commonly produce an unacceptable spread of values.

Temporal Measures

There are means for achieving quantitative expressions for the spatial parameters described above. Time, treated on an interval scale, is also an important dimension in basin analysis. While it is useful to note that the basinal form of the increment from T_2 to T_3 (Fig. 1) evolved to a different shape in the younger increment from T_3 to T_4, significantly more information is added by expressing spatial parameters as rates of change relative to time. When measures such as maximum thickness, mean thickness, and basin slope are expressed as values per unit time, meaningful comparisons between stages in the same basin and between equivalent stages in separate basins become possible.

Geochronology, in its present state, has sufficient accuracy to resolve many Phanerozoic time-stratigraphic units in terms of absolute age at the series level (10- to 20-m.y. increments). In exceptional circumstances, such as those provided by closely spaced and accurately dated Late Cretaceous bentonites, resolution at the stage level (2 to 10 m.y.) is possible. Thus, while rates of change in basin geometry relative to time may commonly be calculated with a degree of accuracy for relatively long time increments involving tens of millions of years, rates operating during spans of time an order of magnitude less cannot often be attained at a high level of confidence.

Our experience suggests that some of the most significant events and episodes of basin evolution occur at intervals of a few million years and that critical changes are revealed only if units shorter than those resolved by radiometric methods are analyzed. In order to study these high-frequency changes, approximations must be based on two assumptions: (1) the rates of organic evolution and, therefore, the numbers of identified biostratigraphic zones per unit time approach a constant over periods of several million years; and (2) major areas of cratons (as opposed to individual basins) subside and accumulate sediments at relatively constant rates for similar periods of time. Time-stratigraphic stages are delimited by range zones, and in the growth of stratigraphic nomenclature, stages tend to have been erected at approximately equal intervals in terms of the number of zones recognized in each. In the absence of conflicting data, we have therefore divided time spans between dated points by assigning equal intervals to stages. When geologically unreasonable time values are produced by this process, the values are modified by applying regional mean thicknesses according to assumption 2. The same premise is applied to subdivisions within stages or to units that are undefined by biostratigraphic zones.

In view of the importance attributed to the time dimension in geometric analysis, there is a natural tendency to question results that involve first approximations such as those described. Fortunately, changes in geometric parameters relative to time are commonly sufficiently large to transcend time-assignment errors by a factor of at least two.

QUANTIFICATION OF BASIN PARAMETERS

Map Digitization

As noted above, most of the significant spatial parameters of basins are not explicit in the raw data nor can they be derived by inspection of an isopach integration. All of the necessary information, however, is inherent in the isopach map, and some of this information is retrievable by simple data-processing methods. Our first methodologic step has been digitization of the isopach map, by hand or automatic digitizer, to produce equally spaced, orthogonally gridded thickness values by interpolation between isopach contours. The resulting array of thickness values readily yields a variety of measures such as the statistical moments of the distribution. We attempted (Sloss, 1972) to use the relations among certain moments for analytical purposes, but with the exception of mean thickness, their geologic significance remains obscure or, at best, subject to multiple interpretations. It is evident that higher levels of analysis are required.

Trend-Surface Analysis

The difficulties in statistical treatment of thickness values are those commonly encountered in analysis of areally distributed data by methods that remove the influence of geographic variations of the data. One approach to the resolution of such difficulties is through the fitting of mathematical surfaces to the areally distributed data, followed by the derivation of meaningful coefficients of the most suitable ("best-fitting") mathematical surface.

Trend-surface analysis is an attractive approach to the present problem because a real, rather than abstract, surface is involved. One of the premises of basin analysis is that basins remain filled to depositional base level by rates of sediment supply equal to or in excess of rates of subsidence; departures from this rule (other than relatively rare "starved basin" interludes) are considered trivial relative to the thickness of sediment considered. Thus, at any time during a depositional episode, such as the initiation of a time interval to be analyzed, the topography of a basin floor is an essentially planar surface that is progressively deformed by differential subsidence during deposition. The degree of deformation and the geometry of the surface assumed by the close of the interval studied are represented by areal variations in thickness values. The coefficients of a trend surface fitted to the thickness values provide mathematical descriptors of the surface, while distracting local perturbations are filtered out. The gridded data produced by digitization of isopach maps are readily amenable to standard trend-surface methods; in a majority of basin examples tested, third-degree polynomial and low-order double Fourier functions provided acceptable fits. However, in our present state of knowledge, we are unable to identify the geological significance of the multitude of coefficients produced by these standard methods, and we are left with elegant expressions that do not aid our interpretations. Once more, other approaches are indicated.

Bivariate-Normal Model

In discussion of the problems of selection of an appropriate surface-fitting function from which useful basin descriptors might be derived, it was pointed out by W. R. James (1971, oral commun.) that certain properties of basin profiles (inflection points symmetrically disposed on either side of a central maximum thickness value; negligible change in rates of thickness beyond basin perimeters) resemble the properties of inverted Gaussian or normal distributions. By extension, bivariate (three-dimensional) normal-distribution functions would generate surfaces closely approximating the shapes of sedimentary basins, as shown in Figure 2.

Our application of a normal-distribution function to the shapes of sedimentary basins bears no probability-theory implications. Rather, it is an admittedly simplistic selection of a well-understood expression that yields a number of measures of basin properties, including geographic coordinates of basin center, position of hinge line at the principal axes, orientation of principal axes, basin length and width at the hinge line, and maximum rate of thickness change at principal axes. Each of these may be monitored through time, as previously described. Mathematical development of the model is given in the appendix.

ANALYSIS OF DEVONIAN BASINS

The sedimentary basin fills that were actively subsiding in Devonian time are appropriate examples for testing an experimental geometric model. In the majority of mid-Paleozoic craton–interior basins, Devonian deposition begins (as at T_0 of Fig. 1) with the transgression of an essentially planar erosion surface, the sub-Kaskaskia unconformity, in Siegenian or Emsian (late Early Devonian) time. Further, Devonian basinal strata are commonly protected from radical erosional effects by a cover of lower Carboniferous beds so that preservation tends to approach the condition shown at T_6 (Fig. 1). Perhaps most importantly, primary data on several Devonian basins have been documented and integrated by previous workers. For the Michigan Basin we have drawn on an unpublished dissertation by Gardner (1971), supplemented by graduate-student class projects on Upper Devonian units not studied by Gardner. The Elk Point Basin data are from maps of western Canada and Northern Great Plains States presented by Grayston and others (1964), Belyea (1964), and Baars (1972), with amplification by students of detail on certain units near the base of the section. Data on the Moscow Basin were derived from the maps of Domrachev and Tikhii (1961) in the Russian Platform atlas.

Depositional Areas and Rates

Four geometric parameters for each of the basins analyzed are graphically presented in Figure 3. The points and connecting lines of the graphs on the right show the preserved areal extent of each unit; note that the Elk Point and Moscow Basins are an order of magnitude greater in area than the Michigan Basin. The difference is exaggerated by limitation of data to the southern peninsula of Michigan; Devonian strata preserved beneath Lakes Michigan and Huron and on the Niagara Peninsula of Ontario are not taken into account, but even if these areas were included, the Michigan Basin would remain significantly smaller than the other two.

The bars of the same column of graphs at the right show the volumes of preserved sediment that accumulated per unit time. Effective subsidence below base level, which permitted the deposition and preservation of a sedimentary record, began in the Riga area of the Russian Platform in Siegenian time and spread across the Moscow Basin proper in Eifelian and Givetian times to reach a climax in Frasnian time. Sedimentation in the Elk Point Basin began slightly later but followed a similar pattern, being confined northwest of the Meadow Lake escarpment until Eifelian time, and then spreading to maximum in Frasnian time. Rates of deposition were roughly proportionate with areas of subsidence in both basinal areas.

Our data indicate a somewhat independent course for the Michigan Basin. A large part of its area subsided below base level early in the Kaskaskia cycle so that, by the close of Early Devonian time, the areas receiving sediment in the three basins approached equality. Maximum areas and maximum rates of deposition for the Michigan Basin appear to have been achieved in Middle Devonian time, with a decline of both parameters setting in while the other basins were continuing to expand in size and accelerate in subsidence. The degree to which these contrasts reflect the limitation placed on the Michigan Basin data by the undrilled areas of Lakes Michigan and Huron is indeterminate without consideration of other parameters.

Basin Widths and Slopes

The left and central bar diagrams of Figure 3 show the variations through time of two parameters derived from bivariate-normal functions applied to the three basins investigated. The two derived parameters, basin width (z) and basin slope (α), are illustrated in Figure 4, which shows basin profiles drawn along one of the principal axes and represents two time-stratigraphic intervals. In each case the unit is confined between synchronous surfaces: the upper surface is essentially horizontal at the close of the time interval considered; the lower line represents an originally horizontal surface (at the close of deposition of the preceding time-stratigraphic unit), deformed during accumulation of the unit analyzed by differential subsidence of the basin interior relative to the margins. The measure of basin width (z) is defined as the distance from the basin center to the inflection point of the profile calculated along one of the principal basin axes. The value graphed on Figure 3 is the mean ($\bar{z}$) of the intercepts calculated on the long and short axes.

Basin slope (α) is defined as the angle between the upper and lower surfaces that bound the time-stratigraphic unit calculated at the inflection point along the principal axes. Thus, basin slope is measured at the point of maximum rate of change of thickness of each unit analyzed. The value graphed is the larger of

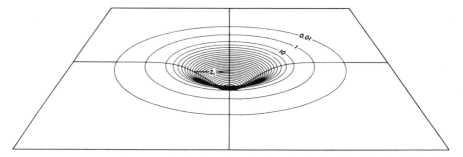

Figure 2. Perspective view of a sedimentary basin simulated by the bivariate-normal function; view is normal to the long principal axis. Vertical exaggeration ×200; maximum subsidence during accumulation of unit analyzed (maximum depression at basin center) = 250 m; distance from basin center to inflection point on long axis (z_1) = 25 km; distance from basin center to inflection point on short axis = 15 km; depths below horizontal plane shown by contours in meters; contour interval 10 m (except outer two contours).

two angles calculated at the principal axes, expressed as meters per kilometer per million years. The same relative variations of basin slope as a function of time would be shown if average or minimum values were presented.

The degree to which the bivariate-normal model represents the actual geometries of the three basins at stages in their evolution is measured by the "goodness of fit" of the model to observed thickness values and their areal distribution. Table 1 indicates the degree to which the data on the 28 units analyzed are approximated by the model. In the method employed, a 100 percent reduction in the corrected sums of squares would represent a perfect fit; the acceptability of lesser percentage reductions in the sums of squares is dependent, in part, on the volume of data employed. The present study involves an average of more than 100 data points per map for the Michigan Basin, more than 130 points for the Elk Point Basin, and more than 220 points for the Moscow Basin area.

As might be expected, the bowl-like and uncomplicated geometry of the Michigan Basin is well suited to approximation by the bivariate-normal expression. All units analyzed in the Elk Point Basin area, except the early Frasnian Beaverhill Lake Formation, also yield good fits, although the Givetian Upper Elk Point unit requires treatment as two subbasins, northwest and southeast of the Meadow Lake escarpment. Summations and averages of the parameters derived from analysis of both subbasins are used to represent the Upper Elk Point unit in Figure 3. The larger the basin area analyzed, the greater the probability that second-order tectonic features will complicate the geometry of a stratigraphic unit and reduce the degree to which that geometry can be approximated by a simple mathematical expression. This principle is exemplified by the results of analysis of the very broad Moscow Basin, where the goodness of fit of the bivariate-normal model is markedly below that

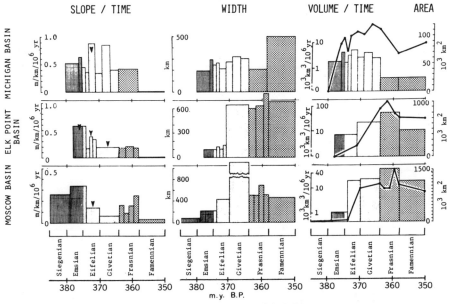

Figure 3. Variations of selected geometric parameters of three basins relative to Devonian time. Devonian stages, except Gedinnian (earliest), and their approximate absolute-time limits are shown. Preserved areal extent of each unit analyzed is indicated by points and connecting lines of the graphs on the right, superimposed on a logarithmic plot of volumes preserved per million years. Arrowheads indicate positions of basin-center salt deposits. Predicted increase of basin slope associated with Eifelian salt in Moscow Basin and Givetian salt in Elk Point Basin is not resolved by analysis of available isopach maps covering long time increments.

achieved in the other two areas included in the present study. Nevertheless, with the exception of the Givetian unit, products of analysis of the Moscow Basin Devonian units approach or exceed acceptable levels. Subjectively considered, the relative values of the parameters derived by analysis of Moscow Basin units agree with relative values predicted by visual map inspection. In the current state of the art, we are unable to define the geologic confidence level attributable to parameters derived from mathematical surfaces that involve a large number of data points and reduce the sums of squares to a range between 40 and 50 percent. The following discussion is predicated on the premise that slope and width parameters derived for the two Moscow Basin units in this range are provisionally acceptable for our interpretative purposes and that reductions higher than 60 percent are clearly significant.

One would anticipate a strong negative correlation between basin slope and width; that is, a broad basin (Fig. 4, top) should have a relatively gentle slope, and a narrower basin (Fig. 4, bottom) would have steeper slopes. If an inverse correlation exists, one of the measures is redundant since both contain the same information. In the oldest Devonian units of the Moscow Basin area (Fig. 3, bottom), however, both slope and width (Fig. 3, left and center) increase from Siegenian into Emsian time; the slope then decreases as width continues to increase during the Eifelian. Consideration of concomitant changes in areal and volumetric data (Fig. 3, right) clearly indicates that understanding and description of basin geometry require integration of several parameters.

Although the Moscow Basin is characterized by greater widths and gentler slopes than the Elk Point Basin, there are interesting similarities. Slopes per unit time achieve maximums in Emsian to early Eifelian times and decline irregularly to minimums in Famennian time. Basin widths increase until the end of Givetian time in both basins and are relatively constant in the Late Devonian, although at a decreased level in the Moscow Basin. The parameters of the Michigan Basin again appear to follow a partially independent history, perhaps because of real differences or as previously noted, because of geographic constraints on the data.

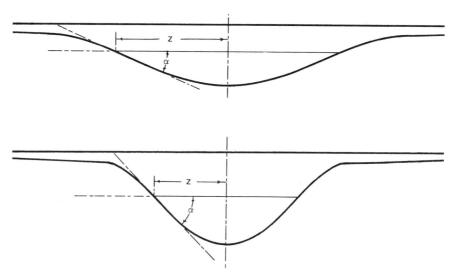

Figure 4. Basin width (z) and basin slope (α) showing relation of these parameters to inflection point of basin profile. Diagram illustrates basin geometry at stages of moderate (top) and extreme (bottom) differential subsidence.

TABLE 1. PERCENTAGES OF CORRECTED SUMS OF SQUARES ASSOCIATED WITH BIVARIATE-NORMAL MODEL APPLIED TO INDIVIDUAL STRATIGRAPHIC UNITS IN THREE DEVONIAN BASINS

	Michigan Basin	Elk Point Basin	Moscow Basin
Famennian	97	99	45
Frasnian	98	77 75 23	57 61 71 42
Givetian	94 91 94	81 (NW), 90 (SE)*	29
Eifelian	92 80 90	84 95 74 82	79 43
Emsian	95 95		
Siegenian			84

Note: Stratigraphic units listed here in chronological sequence from youngest (Famennian) to oldest (Siegenian).
*Upper Elk Point unit analyzed as occupying two basins, northwest and southeast of the Meadow Lake escarpment. Table gives sums-of-squares reductions for both subbasins.

Nevertheless, the Michigan Basin shares with the other basins analyzed the late Middle Devonian decline and Famennian minimum in slope values.

The presence of a significant volume of salt in cratonic basins is an indicator of rapid subsidence, although there is dispute as to whether subsidence precedes or accompanies salt accumulation. In either case, one would expect episodes of salt deposition to be reflected in basin slopes and widths. Such episodes, however, tend to be extremely brief, and their analysis demands detailed, fine-scale correlation leading to the development of stratigraphic maps covering short time increments. Late Eifelian salt of the Michigan Basin ("Horner salt member" of Gardner, 1971) and the Elk Point Basin (Cold Lake Salt) are partially resolved by maps of brief time increments in these areas; they appear as steep-slope, narrow-width perturbations in Figure 2. An equivalent salt is present in the Moscow Basin, but here the greater part of Eifelian time is represented by a single map (Domrachev and Tikhii, 1961, map 17), and the record of the salt episode is lost by smoothing over a long time interval. We confidently predict that mapping and analysis of finer-scale, time-stratigraphic subdivisions of the Moscow Basin Eifelian will reveal a perturbation of basin geometry at the time of salt deposition. It is equally predictable that the Givetian Prairie salt unit of the Elk Point Basin will appear as a spike on the slope graph and a depression on the width plot when detailed maps are available for analysis. These would match the perturbation of Michigan Basin geometry revealed by slope analysis of Givetian subdivisions in the Michigan Basin, although the latter, deposited under open-circulation environments, lack salt.

CONCLUSIONS

Quantitative analysis of the parameters of basin geometry offers opportunities for more rigorous descriptions of sedimentary basins in space and time in order to investigate the evolution of individual basins, to compare and contrast separate basins, to relate basin geometry to facies responses, and to provide a geometric framework for the formulation of sound tectonic theory of basin subsidence. We cannot claim that our method makes it possible to realize all of the potential opportunities; rather, we hope to have demonstrated that, given a sufficient mass of three-dimensionally distributed stratigraphic data, there are objective means for the reduction of such data to geologically useful forms.

One example of a "geologically useful" product is demonstration that the dimensions and shapes of sedimentary basins change as functions of relatively short (1 to 5 m.y.) time increments and that some changes are synchronous in widely separated basins. Variations in slope and width parameters support an older contention (Sloss and others, 1949) that the transgressions represented by major stratigraphic cycles or sequences are accompanied by accelerations in the rates of basin subsidence. This thesis was developed further (Sloss, 1964, 1966) to suggest that rates of basin subsidence during the deposition of "epeirogenic" sequences, such as the Kaskaskia, are marked by two stages of rapid subsidence separated by a mid-sequence stage of slower relative subsidence. The data presented here confirm the first (Middle Devonian) stage of basin subsidence during Kaskaskia deposition and the mid-sequence lull near the close of the Devonian. It is predictable that analysis of early Carboniferous units will reveal the second basin-subsidence stage of Kaskaskia deposition.

Although it is heartening to find support for old, qualitatively derived postulates, our newer data make it obvious that the evolution of sedimentary basins is too complex to be accommodated by a simplistic model. A more nearly complete model to represent tectonic evolution during Kaskaskia deposition has been proposed by Johnson (1971). His Kaskaskia depophases I and II fit well with the Emsian to Frasnian increases in submerged areas, and rates of deposition documented by us in the Elk Point and Moscow Basins and Famennian decreases in both parameters agree with Johnson's minor regressive episode preceding his depophase III.

Demonstration of concomitant marine transgression and basin subsidence leads to questioning of the theory that marine transgression of continental interiors is a wholly eustatic effect of elevated mid-ocean ridges and the resulting decrease in volume capacity of ocean basins. Our data are very difficult to reconcile with the latter assertion, which has been often expressed (for example, Armstrong, 1969) and widely accepted.

Demonstration of apparent synchrony and parallelism in the geometric evolution of basins so widely separated as central Russia and the Prairie Provinces of Canada demands confirmation by analysis of other sedimentary basins. Such expanded investigation, however, is not possible without detailed stratigraphic data permitting correlation and basin-wide isopach mapping of time-stratigraphic units at (and preferably below) the level of stages. Unfortunately, although raw data from drilling and logging have been acquired for many basins in many parts of the world, these data have not been integrated in a published form suitable for analysis.

If the apparent similarity of distant basins that our study suggests is real and pervasive, then basin models have wide applicability as predictive devices for assessment of the thickness, geometry, and facies reponses of little-known and undrilled areas. Attention is naturally drawn to the submerged continental shelves. Qualitative evaluation of the narrow belts of subsurface data derived from onshore and nearshore drilling on the Atlantic and Gulf coasts of North America indicates that these regions subsided and accumulated sediments during Mesozoic time in harmony with basins of the continental interior. The latter are represented by a wealth of subsurface data, but few have been adequately integrated in isopach form to permit geometric analysis and the development of models potentially applicable to the outer continental shelf.

If, as we believe, some of the geometric differences apparently expressed in the Michigan Basin (as compared to the other two examples in this study) are not artifacts created by geographic limitations on the data, such differences offer opportunities for further study. If global controls on tectonic patterns of cratonic basins exist, are these controls locally overridden by regional factors and to what degree and extent? That is, when the Devonian stratigraphy of the Illinois Basin has been resolved, will geometric analysis indicate a kinship with the Michigan Basin and thus suggest that eastern and western North America represent slightly different Devonian evolutionary models, with western North America following the pattern of the Russian Platform in relatively fine detail?

Finally, we do not wish to leave the impression that our experimental geometric model is so successful as to discourage further experimentation. Any surface-fitting approach that involves a relatively simple function is not uniformly suitable in application to all sedimentary basins. Some large basins are compound and display more than two inflection points on basin profiles; these must, in our method, be subdivided with consequent loss of information on the basin area as a whole. We have encountered a number of stratigraphic units in certain basins that cannot be fitted by our expressions. This occurs where, for the time span under analysis, a basin has a broad, flat floor and thus departs from the parabolic shape required by the bivariate-normal model. Other problems emerge where data limitations (as in the Michigan Basin) or map boundaries prevent analysis of an entire basin and its margins. The effects of erosion are largely ignored in our model and yet these effects must ultimately be accounted for in any complete basin model. The problem of erosional factors in the preserved thickness of a stratigraphic unit has been approached statistically by Kolmogorov (1951) in terms of conditional probability distributions, but his method assumes an unattainable degree of knowledge of "original" (that is, pre-erosion) thicknesses. Other aspects of basin models and basin analysis have been studied by a number of workers (for example, Potter and Pettijohn, 1963; Harbaugh and Bonham-Carter, 1970), but it remains evident that there are abundant opportunities for more investigators and investigations.

ACKNOWLEDGMENTS

We have previously acknowledged the contribution of William R. James to our investigation; others who aided in the development and implementation of the mathematical model include Michael F. Dacey, William C. Krumbein, and graduate-student collaborators Douglas Lorenz, Niel Plummer, and Richard Wetterauer. Much of the map digitization, data collation, and computer processing was done by Nancy Millea. Stratigraphic gaps in our coverage were filled by the efforts of Bruce Brown, Robert Langan, Paul Smith, and Charles Spirakis.

The original manuscript was reviewed by E. H. Timothy Whitten and Michael F. Dacey whose helpful comments on both form and substance are most gratefully received. Special thanks are extended to Colin Stearn who identified inadequacies in our treatment and suggested avenues toward clarification.

Our list of references is short because we have made no attempt to record the hundreds of publications from which our stratigraphic data were drawn. We trust that our neglect of direct acknowledgment will not offend the many workers who have produced the primary information required for our syntheses.

Research was supported by National Science Foundation Grant GA-22844, by a fellowship provided by the American Oil Company, and by departmental funds.

APPENDIX 1. MATHEMATICAL FORMULATION OF THE BASIN MODEL

James (1968) noted that there are a number of geologic investigations involving measurable responses to phenomena for which there is no present basis in fundamental physical theory; nevertheless, interrelations between the measurable variables can be stated as mathematical expressions. James has proposed recognition of a special class of substantive statistical models for such expressions, and it is clear that basin models fall within this class.

Given an isopach map of a stratigraphic unit deposited in a sedimentary basin and an orthogonal array of thickness values derived by digital interpolation of thickness contours, the problem becomes one of selecting the most appropriate statistical model. A number of algebraic and trigonometric functions are suitable; of these, as discussed in the body of this paper, we have applied a variation of the bivariate-normal expression. Our choice is dictated by the obvious similarity between basin profiles and inverted Gaussian or normal-distribution curves, by the fact that the properties of normal distributions have been exhaustively investigated, and because relatively simple manipulation of such expressions is capable of yielding quantitative measures of fundamental basin parameters. Our model for the geometry of sedimentary basins is defined for very large areas, theoretically for the Euclidean plane. The model is continuous in space and time and is given by the bivariate, negative exponential function

$$w = f(U, V, t)$$
$$= w_0 t \exp\left\{-\left[\left(\frac{U-c_1}{s_1}\right)^2 - \beta\left(\frac{U-c_1}{s_1}\right)\left(\frac{V-c_2}{s_2}\right) + \left(\frac{V-c_2}{s_2}\right)^2\right]\right\} \quad (1)$$

where w is thickness at the location (U, V) in cartesian coordinates; t is geologic time in years; point (c_1, c_2) is the position of maximum thickness; s_1 and s_2 are related to the magnitude of the basin in the U and V directions, respectively; w_0 is maximum thickness; and β is a cross-product term that expresses the angle between the principal basin axes and the geographic coordinate system.

Estimates of fundamental basin parameters are obtained by solution of equation (1) using a surface-fitting method applied to orthogonally arrayed thickness data derived from digital interpolation of the isopach map. James (1968) developed program MODFIT for iterative least-squares estimation of the coefficients of normal distributions. For our purposes we have slightly modified James's program to produce program BASIN (this and other FORTRAN programs mentioned in this paper were presented, with complete listings and sample problems for input, by Scherer, 1973).

The surface-fitting program requires entry of reasonable initial estimates of the variables by inspection of the isopach map. The maximum thickness is the initial estimate for w_0, and the U, V coordinate readings at this point are the initial estimates for c_1 and c_2. Estimation of subsidence parameters s_1 and s_2 is more difficult because it involves (1) visually finding the inflection point of the subsidence surface and (2) multiplying the distance from center to inflection point along a line parallel to the coordinate axis by $\sqrt{2}$, in the absence of any rotation. Fortunately, the parameters s_1 and s_2, as well as the other parameters of the subsidence model, are relatively insensitive to poor initial estimates. Specifically, s_1, s_2, and β may be underestimated or overestimated by a factor of two to three, and the model will still stabilize at the correct parameters.

The parameter β indicates the rotation of the principal axes of the basin from the map coordinate system; its limiting values are $+2$ and -2 for valid representations of the subsidence model, that is, for parabolic surfaces. In most practical cases, it is safe to give an initial estimate of $\beta = 0$.

The initial estimates for w_0, c_1, c_2, s_1, s_2, and β, together with the U, V coordinates and thickness measurements from map digitization, constitute input for program BASIN. The program is normally run for a maximum of 30 iterations, which are generally sufficient for convergence on the correct set of parameters. Signs of a good fit are that the final parameters for w_0, c_1, and c_2 should be close to the initial estimates, β should be within the range $-2 < \beta < +2$, and the sum of squares reduction (the percent of the total variability accounted for by the model) should be reasonably high (>70%).

Solution of equation (1) yields quantitative values for a number of fundamental geometric basin parameters. To obtain other significant measures—the positions and values of the maximum rate of change of thickness along the principal elliptical axes of the basin, the lengths of the axes at these positions, and the angle of rotation of the axes relative to the coordinate system used—it is useful to relate equation (1) to the bivariate-normal distribution for which there are well-known procedures for the derivation of auxiliary parameters. The density function of the bivariate-normal distribution was written by Yule and Kendall (1948) in the form

$$g(U, V) = [2\pi\sigma_1\sigma_2(1-\rho^2)^{1/2}]$$
$$\cdot \exp\left\{\frac{-1}{2(1-\rho^2)}\left[\left(\frac{U-c_1}{\sigma_1}\right)^2 - 2\rho\left(\frac{U-c_1}{\sigma_1}\right)\left(\frac{V-c_2}{\sigma_2}\right) + \left(\frac{V-c_2}{\sigma_2}\right)^2\right]\right\}, \quad (2)$$

so that $f(U, V, t)$ differs from $g(U, V)$ only by change of notation and a normalizing constant. To relate the two expressions, the metric of measurement is selected so that

$$wt = 1/2\pi\sigma_1\sigma_2(1-\rho^2)^{1/2},$$

where

$$\sigma_1 = s_1 \bigg/ \left[2\left(1-\frac{1}{2}\beta\right)^2\right]^{1/2}, \quad (3a)$$

$$\sigma_2 = s_2 \bigg/ \left[2\left(1-\frac{1}{2}\beta\right)^2\right]^{1/2}, \quad (3b)$$

and

$$\rho = \frac{1}{2}\beta. \quad (3c)$$

By establishing this relation between equations (1) and (2), properties given by Yule and Kendall (1948) of the bivariate-normal distribution are readily related to the basin model of equation (1). In particular, the angle of rotation θ of the principal axes of the basin is given by

$$\theta = 1/2 \arctan \beta s_1 s_2 / (s_1^2 - s_2^2).$$

The angle of rotation θ is positive if the U-axis is rotated counterclockwise from its original position in order to coincide with the major principal axis.

The two standard deviations z_1 and z_2 along the principal axes of the ellipse are also obtained from the relation between equations (1) and (2). Yule and Kendall (1948, p. 232) gave a system of two simultaneous equations that relate z_1 and z_2 to the parameters of the bivariate-normal density as follows:

$$z_1^2 + z_2^2 = \sigma_1^2 + \sigma_2^2$$

$$z_1 z_2 = \sigma_1 \sigma_2 (1-\rho^2)^{1/2}.$$

Put

$$A = \sigma_1^2 + \sigma_2^2$$

and

$$B = \sigma_1 \sigma_2 (1 - \rho^2)^{1/2}.$$

Then, a few algebraic manipulations give

$$z_1 = [(A - 2B)^{1/2} + (A + 2B)^{1/2}]/2$$

$$z_2 = [(A + 2B)^{1/2} - (A - 2B)^{1/2}]/2.$$

Replacing A and B by the expressions (3a), (3b), and (3c) gives the required result.

Continuing with the analogy between the geometric properties of basins and the properties of surfaces generated by bivariate-normal distribution functions, it is obvious that a basin profile along either of the principal elliptical axes has the properties of a normal-distribution curve. Therefore, at either axis the positions of maximum rate of change of thickness are analogous to the inflection points of the normal curve, and these are one standard deviation on either side of the mean. Thus, determination of basin slope becomes a problem of calculating the slope of the equivalent normal curve at its inflection points.

The thickness at any point along a principal basin axis is given by the equation

$$w = f(x) = w_0 \exp\left\{-\frac{1}{2}(x/z_i)^2\right\}, \quad i = 1, 2,$$

where w = thickness, w_0 = maximum thickness, x is the distance from the center point (c_1, c_2), and z_1 and z_2 are the standard deviations.

Taking the derivative of w relative to x and evaluating at $x = z_i$ gives inflection points at

$$a/z_i e^{1/2}, \quad -a/z_i e^{1/2}, \quad i = 1, 2.$$

For each axis there are two slopes of the same magnitude and opposite sign at the inflection points.

These slopes are now defined for basin analysis purposes as

$$\alpha_1 = \pm w_0 / z_1 e^{1/2}$$

$$\alpha_2 = \pm w_0 / z_2 e^{1/2}.$$

Computer program PARAM (Scherer, 1973) calculates the above coefficients from the parameters estimated by program BASIN.

Basin shape is one of a number of other measures that can be derived directly from the fundamental parameters. An easily obtained shape measure is given by the ratio of the lengths of the principal axes at the inflection points and is the maximum of z_1/z_2 and z_2/z_1.

As noted in the body of this paper, measurement of time-dependent changes in basin geometry permits analysis of the evolution of individual basins and comparisons between basins. Given an independent determination of geologic time, the rate of subsidence of a basin during a particular time span, for example, may be obtained by differentiating the maximum thickness (w_0) relative to time (t):

$$\text{rate of subsidence} = dw_0/dt.$$

The calculation assumes that the loss of stratigraphic thickness by compaction and erosion is trivial and that the basin is filled to depositional base level at the close of the episode analyzed. Considering the relatively long time spans involved, the comparative rarity of severe erosion at basin centers, and the speed of compaction, these assumptions do not commonly introduce significant errors. Rates of change of other geometric variables relative to time may be similarly obtained; analysis of time-related changes in basin slopes (α_{max} or $\bar{\alpha}$) has proved particularly valuable in studies of basin evolution and in comparisons and contrasts between separate basins.

REFERENCES CITED

Armstrong, R. L., 1969, Control of sea level relative to the continents: Nature, v. 221, p. 1042-1043.

Baars, D. L., 1972, Devonian System, *in* Mallory, W. W., ed., Geologic atlas of the Rocky Mountain region: Denver, Colorado, Rocky Mtn. Assoc. Geologists, p. 90-99.

Belyea, H. R., 1964, Upper Devonian, *in* McCrossan, R. G., and Glaister, R. P., eds., Geologic history of western Canada: Calgary, Alberta Soc. Petroleum Geologists, p. 60-85.

Domrachev, S. M., and Tikhii, V. N., 1961, Maps 15 through 23, *in* Vinogradov, A. P., and Nalivkin, V. D., eds., Atlas of litho-paleogeographical maps of the Russian Platform and its geosynclinal framing, Pt. I, Late Precambrian and Paleozoic: Moscow-Leningrad, Acad. Sci. USSR.

Gardner, W. C., 1971, Environmental analysis of the Middle Devonian of the Michigan Basin [Ph.D. thesis]: Evanston, Ill., Northwestern Univ., (Univ. Microfilms 72-7783), 145 p.

Grayston, L. D., Sherwin, D. F., and Allan, J. F., 1964, Middle Devonian, *in* McCrossan, R. G., and Glaister, R. P., eds., Geologic history of western Canada: Calgary, Alberta Soc. Petroleum Geologists, p. 49-56.

Harbaugh, J. W., and Bonham-Carter, G., 1970, Computer simulation in geology: New York, Interscience Pubs., Inc., 575 p.

James, W. R., 1968, Development and application of non-linear regression models in geology [Ph.D. thesis]: Evanston, Ill., Northwestern Univ. (Univ. Microfilms 69-6947).

Johnson, J. G., 1971, Timing and coordination of orogenic, epeirogenic, and eustatic events: Geol. Soc. America Bull., v. 82, p. 3263-3298.

Kolmogorov, A. N., 1951, Solution of a problem in probability theory connected with a problem of the mechanism of stratification: Am. Math. Soc. Trans. no. 53, p. 3-8.

Potter, P. E., and Pettijohn, F. J., 1963, Paleocurrents and basin analysis: New York, Academic Press, Inc., 296 p.

Scherer, W., 1973, A mathematical model for the differential subsidence of intra-cratonic basins [Ph.D. thesis]: Evanston, Ill., Northwestern Univ., 60 p.

Sloss, L. L., 1964, Tectonic cycles of the North American craton, *in* Merriam, D. F., ed., Symposium on cyclic sedimentation: Kansas Geol. Survey Bull. 169, v. 2, p. 449-460.

——1966, Orogeny and epeirogeny—The view from the craton: New York Acad. Sci. Trans., ser. 2, v. 28, p. 579-587.

——1972, Synchrony of Phanerozoic sedimentary-tectonic events of the North American craton and the Russian Platform: Internat. Geol. Cong., 24th, Montreal 1972, sec. 6, p. 24-32.

Sloss, L. L., and Speed, R. C., 1974, Relationships of cratonic and continental-margin tectonic episodes, *in* Dickinson, W. R., ed., Tectonics and sedimentation: Soc. Econ. Paleontologists and Mineralogists Spec. Pub. 22, p. 98-119.

Sloss, L. L., Krumbein, W. C., and Dapples, E. C., 1949, Integrated facies analysis, *in* Longwell, C. R., chm., Sedimentary facies in geologic history: Geol. Soc. America Mem. 39, p. 91-123.

Yule, G. U., and Kendall, M. G., 1948, An introduction to the theory of statistics (13th ed.): New York, Hafner Pub. Co., Inc., 570 p.

MANUSCRIPT RECEIVED BY THE SOCIETY NOVEMBER 2, 1973
REVISED MANUSCRIPT RECEIVED MARCH 22, 1974

Cross-Bed Variability in a Single Sand Body

R. STEINMETZ

*Amoco Production Company
Research Center, P.O. Box 591
Tulsa, Oklahoma 74102*

ABSTRACT

In order to determine detailed variations in cross-bed orientation, 600 measurements were taken in a nested sampling design from one continuously exposed Wasatch (Eocene) channel sandstone in southwest Wyoming. The sandstone is fine-grained, well-sorted quartz arenite and crops out for a distance of 1,830 m. The outcrop nearly parallels the overall flow direction of N. 70° E. Twenty-four sample sites spaced 60 m apart were established along the outcrop. At every site, 25 cross-bed measurements (true dip direction and angle), arranged in a 5 × 5 sample grid with 50-cm spacing, were recorded.

Results show that vector average dip directions for the 24 sample sites fall in the 275° sector between 319° and 234° (0° = North), with a grand average of 070°. Arithmetic mean dip angles for 24 sites range from 9° to 31°, with a grand mean of 18°. Sand-body thickness (5.0 to 15.5 m) is directly proportional to the amount of variability in cross-bed orientations within each sample site.

Analysis of variance indicates that a minimum of 125 and a maximum of 375 cross-bed measurements are required to estimate adequately the overall flow direction in this sand body. Because local cross-bed variability may be large and the distribution of variation probably differs among sand bodies, it is suggested that future studies of paleostreamflow directions may require more detailed sampling than has been performed in the past.

INTRODUCTION

In a review paper on sand bodies and sedimentary environments, Potter (1967, p. 361) stated, "The principal difficulty . . . is the lack of systematic, quantitative data on the petrology, texture, sedimentary structures, and internal organization of sand bodies." The present study is a detailed statistical investigation of cross-bed orientation and variability in a single channel sand body with continuous exposure.

The purpose of this study is to estimate the minimum number of cross-bed measurements necessary to determine the average direction of cross-beds and paleostreamflow.

Previous workers in other areas have investigated the problem of how many cross-beds to measure. For example, Potter and Olson (1954) concluded that four measurements per outcrop with one outcrop per section and two sections per township or range would have sufficed in their study of basal Pennsylvanian sandstones in the Illinois Basin. Potter and Siever (1956a) determined that only two foreset measurements per outcrop with two outcrops per section are adequate. In the final paper of their trilogy, Potter and Siever (1956b, p. 449) stated, ". . . the relative order of variability of cross-bedding is as follows: within outcrops << between outcrops > between arbitrary 6- or 12-mile intervals along the outcrop belt." Looff and Hubert (1964) concluded that 10 to 20 cross-bed azimuths are sufficient in a Pennsylvanian channel sandstone in Missouri because of considerable homogeneity in orientation along the 275-m outcrop.

In marked contrast to the above studies, Wurster (1958) measured 2,500 cross-beds in a sand pit near Stuttgart. These measurements were taken along the pit faces over a horizontal distance of 250 m and in a vertical distance of 6 m. Wurster did not state what number of cross-bed measurements would have sufficed.

Other statistical studies of orientation data indicate that local variability can be very large. Pincus (1951), in a study of fracture orientation, concluded that 80 joints should be measured at each locality. Larsson (1952), in a petrofabric study of gneiss, indicated that 165 biotite poles per thin section should be measured.

The present study uses the statistics proposed by Fisher (1953) for rock magnetism. The statistical tests follow those of Watson and Irving (1957) and Watson (1966); the computational and graphical procedures are after Steinmetz (1962). Recent papers concerned with the statistical analysis of directional data were by Watson (1970), Schuenemeyer and others (1972), and Rao and Sengupta (1972).

Location of the sand body studied is shown in Figure 1. It is located in T. 25 N., R. 103 W., Sweetwater County, Wyoming. The channel sand body is in the Wasatch Formation, Eocene in age, and in the north-central part of the Green River Basin.

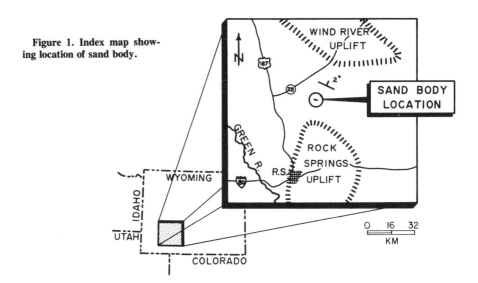

Figure 1. Index map showing location of sand body.

SAMPLING PROCEDURE

Figure 2 illustrates the nature of the sand-body outcrop. The channel sandstone is exposed continuously in cliffs and rounded outcrops for almost 1,830 m along the hillside. It has a maximum thickness of 15.5 m. The sandstone is a fine-grained, well-sorted, friable quartz arenite.

The outcrop pattern of this sand body is plotted on a topographic map in Figure 3A. The sinuous outcrop pattern represents the exposed portion of the sand body viewed from above. This outcrop pattern is not the original depositional configuration of the channel sandstone.

Figure 3B shows the location of sample sites along the outcrop and the arbitrarily chosen division boundaries that were used to establish a nested sampling design. There are three divisions and eight sample sites within each division for a total of 24 sample sites. At each sample site, 25 cross-bed measurements were taken for a total of 600 cross-bed measurements.

Field measurements started at the southeast end of the sand body at sample site 11. Sample sites are spaced 60 m apart along the outcrop and are located in the lower one-third of the sand body (Fig. 3B).

At each sample site, measurement locations were arranged in a 5 × 5 grid pattern. Figure 4 is a sketch of the sampling pattern at site 17. Grid spacing is 50 cm vertically and horizontally. This means that each sample site covers an area of 4 m² (2 m × 2 m).

The grid pattern was placed arbitrarily over each sample site, and the 25 cross-bed measurements were taken wherever the grid intersections fell. At each grid point, a hole was excavated in the friable sandstone, and true dip direction and true dip angle were measured. Measurements were taken with a Brunton compass and the aid of a dip-direction indicator (Pryor, 1958).

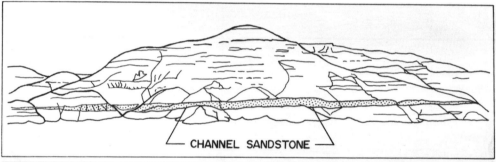

Figure 2. Photograph of hillside with several sand bodies interbedded with variegated shales. Hill is approximately 150 m high. View toward north. Line drawing shows the channel sandstone studied.

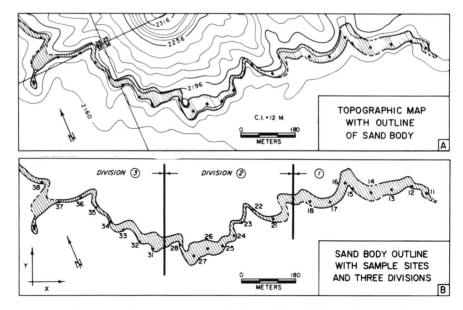

Figure 3. A, topographic map of the south flank of the hillside in Figure 2. Outcrop of exposed portion of sand body indicated by ruled area. Located in T. 25 N., R. 103 W., Sweetwater County, Wyoming. B, map of sand-body outcrop with location of sample sites and divisions indicated.

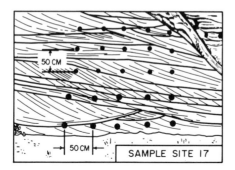

Figure 4. Field sketch of 5 × 5 sampling grid at site 17. Cross-bed measurements were taken at each of the 25 grid intersections which are spaced 50 cm apart.

RESULTS

The summary cross-bed data for each sample site are presented in Table 1. Data on sand-body thickness at each sample site are listed also. The cross-bed statistics follow the methods of Fisher (1953) and Steinmetz (1962).

GRAPHICAL ANALYSIS OF DATA

The first level of cross-bed variability is that found among 25 sets of measurements taken at each sample site. Figure 5 shows the range of variations encountered. Each circle represents a stereographic projection of the lower hemisphere on which 25 cross-bed measurements are plotted. The points represent true dip direction and dip angle of each cross-bed measured in the field. The points are not poles

TABLE 1. SUMMARY DATA

	1	2	3	4	5	6	7	8
Sample site	$\bar{A}_{25}$	$\bar{A}_2$	$\bar{A}_2 - \bar{A}_{25}$	$\bar{D}_A$	$\bar{D}_V$	R	$\theta_{0.05}$	T
Division 1								
11	136°	150°	14°	20.8°	31°	17.09*	20°	4.9+
12	79°	98°	19°	15.9°	20°	20.16*	14°	5.8+
13	86°	94°	8°	18.6°	31°	15.19*	24°	9.8
14	71°	81°	10°	22.3°	29°	19.47*	16°	9.8+
15	131°	96°	−35°	31.2°	47°	16.98*	20°	11.0
16	327°	13°	46°	20.8°	36°	15.00*	24°	10.7+
17	48°	60°	12°	28.3°	32°	22.42*	10°	8.5
18	33°	8°	−25°	14.6°	25°	14.81*	25°	9.4+
Division 2								
21	50°	37°	−13°	27.3°	33°	21.11*	13°	5.8+
22	53°	293°	−120°	21.2°	37°	14.75*	25°	5.5
23	45°	56°	11°	19.9°	40°	13.01*	29°	7.3
24	44°	348°	−56°	18.4°	75°	8.13†	44°	11.6
25	319°	253°	−66°	22.0°	61°	10.58*	35°	12.5
26	336°	271°	−65°	17.4°	62°	8.40†	42°	11.6
27	347°	312°	−35°	15.2°	35°	11.53*	32°	14.9
28	234°	284°	50°	12.7°	57°	6.48N.S.	52°	13.7
Division 3								
31	163°	108°	−55°	10.7°	32°	8.80†	41°	15.5
32	197°	89°	−108°	9.4°	30°	8.00N.S.	44°	13.7
33	357°	9°	12°	15.4°	24°	16.10*	22°	14.6
34	104°	110°	6°	13.3°	18°	18.55*	18°	6.7
35	116°	127°	11°	12.6°	19°	16.85*	21°	8.2
36	112°	56°	−56°	11.6°	63°	5.62N.S.	56°	8.5
37	100°	96°	−4°	18.0°	21°	21.58*	12°	6.7
38	39°	325°	−74°	14.2°	41°	9.19†	40°	5.5

Column 1. Resultant vector direction of cross-bed dip measured in degrees clockwise from North = 0°. Each value is based on 25 measurements of cross-bed dip direction and is computed by using direction cosines. No correction had to be made for structural dip.

Column 2. Vector mean dip direction in degrees of the two steepest cross-beds at each sample site.

Column 3. Arithmetic difference in degrees between columns 2 and 1. Negative values mean that $\bar{A}_{25}$ is larger than $\bar{A}_2$ in a clockwise direction. Positive values mean that $\bar{A}_{25}$ is displaced counterclockwise from $\bar{A}_2$.

Column 4. Arithmetic mean angle of cross-bed dip measured in degrees below the horizontal.

Column 5. Resultant vector angle of cross-bed dip. $\bar{D}_V$ is calculated from direction cosines and only used to plot the resultant vector on the lower hemisphere of a Schmidt net (Fig. 5).

Column 6. Magnitude of the resultant cross-bed vector. R is dimensionless with a maximum value of 25, the number of measurements at each sample site. R values are compared with a table (Watson, 1956) of significance points to test the observed values for random orientation. N.S. stands for nonsignificant and means that the 25 cross-beds at sample sites 28, 32, and 36 have random orientation. All other sample sites differ significantly from randomness at either the 1% level (*) or the 5% level (†).

Column 7. Radius of the circle of confidence around the resultant vector defined by $\bar{A}_{25}$ (col. 1) and $\bar{D}_V$ (col. 5). Although $\theta_{0.05}$ is inversely proportional to R, both parameters are a measure of cross-bed variability.

Column 8. True stratigraphic thickness in meters of the sand body at each sample site. A plus sign means that contacts with the bounding shales are covered, so the sandstone is somewhat thicker.

of cross-beds, but represent the intersection of the cross-bed vectors with the lower hemisphere.

The measurements from sample site 17 are tightly clustered (Fig. 5) in the northeast quadrant and have fairly steep dip angles. The magnitude (R) of the resultant vector ($\bar{A}$) approaches the maximum value 25 and is significant at the 1 percent level. The $\theta_{0.05}$ value is the smallest one found in this investigation (Table 1). As sketched in Figure 4, the foreset-type cross-beds are very uniform at sample site 17.

As the data points spread (Fig. 5), the magnitude (R) decreases and $\theta_{0.05}$ increases. Sample site 31 exhibits a spread of cross-bed measurements with a cluster in the south-southeast. Here, R is barely significant at the 5 percent level, and the radius of confidence is large. Site 36 shows low dip angles spread uniformly around the compass. The resultant vector is nonsignificant, and the $\theta_{0.05}$ value is the largest calculated for any sample site (Table 1).

The important points thus far are (1) there is a wide range in cross-bed variability at each sample site, and (2) this variability exists within an area of 2 m^2.

The next level of variability to examine is that found among the 24 sample sites along the outcrop. This variation is illustrated in Figure 6. Here, the vector average dip directions ($\bar{A}_{25}$) and the significance level of R are plotted for each site. Even though the spread of sample-site vector directions about the grand average vector of 070° seems large, an overall pattern seems to exist. Average dip directions in divisions 2 and 1 shift from north through northeast to east, progressively along the outcrop from west to east.

The variability that exists at the division level is shown in Figure 7. Here, the resultant vectors are all significant at the 1 percent level, and the $\theta_{0.05}$ values are small. This is due to the large number ($N = 200$) of measurements that enter the computation.

The contour diagram for division 1 (Fig. 7) shows a strong mode of 068° which essentially equals the vector average of 072°. The contour diagram for division 2 exhibits a disperse mode in the northeast quadrant; the diagram for division 3 has a bimodal distribution.

The last level of cross-bed variability is that found among all 600 measurements.

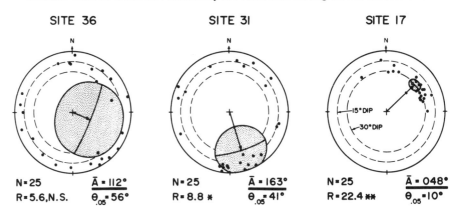

Figure 5. Stereographic projections of cross-bed measurements plotted by using true dip direction and dip angle. North at top with an azimuth of 0°. Number of measurements = $N = 25$. $\bar{A}$ is the resultant vector dip direction. R is the magnitude of the resultant cross-bed vector. $\theta_{0.05}$ is the radius of the circle of confidence about R.

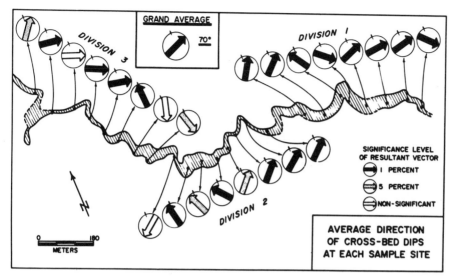

Figure 6. Vector average dip directions ($\bar{A}_{25}$) for 24 sample sites. Significance level of resultant cross-bed vector (R) indicated by pattern inside arrow. Note orientation of North arrow.

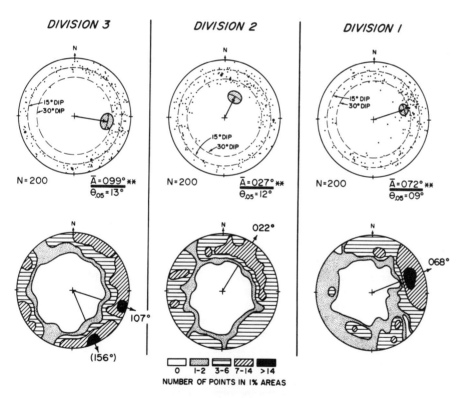

POINT AND CONTOUR DIAGRAMS OF TRUE DIP ANGLES AND DIRECTIONS

Figure 7. Cross-bed dip directions and angles plotted as points on a stereographic projection. $N = 200$ points in each division. $\bar{A}$ is resultant dip direction, and $\theta_{0.05}$ is radius of circle of confidence. Contour diagrams constructed from respective point diagrams.

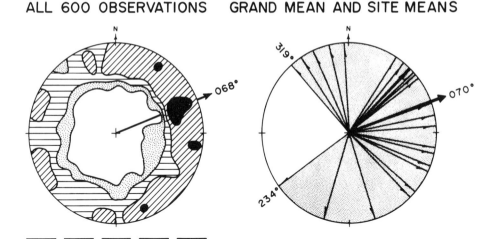

Figure 8. Summary of cross-bed data from entire sand body. Left, contour diagram of all 600 cross-bed measurements. Azimuth of strongest mode is 068°. Right, circular plot of vector averages for 24 sample sites and grand vector mean of 070°.

This is shown in the contour diagram of Figure 8. There are four modes scattered over the east half of the diagram, with the strongest mode having an azimuth of 068°. For comparison, the grand vector mean (070°) and the 24 site means are also plotted in Figure 8. These two graphs demonstrate that a high degree of cross-bed variability exists throughout the entire sand body and among the sample sites.

STATISTICAL ANALYSIS OF DATA

In order to evaluate all the cross-bed measurements statistically, analyses of variance were run. The analytical design used follows that of Watson and Irving (1957) and Watson (1966).

Table 2 is a summary of the analyses of variance calculated for each of the three divisions (Fig. 3B). All the F-ratios are significant at the 1 percent level. This means that variation among the eight sample sites is significantly different when compared with variation present at all sample sites within each division.

Results of the analysis of variance calculated for the entire sand body are presented in Table 3. Here, the conclusion is the same; there is a significant difference among all 24 sample sites. Examination of the sum-of-squares values shows that 79 percent of the total variation exists at the sample site level and only 21 percent of the total variation exists between the sites. Identical results were obtained using the angular variance method of Larochelle (1967).

The mean square values in the analyses of variance (Tables 2 and 3) can be analyzed further by calculation of expected values. To make these calculations, one must assume that the number of cross-beds measured at all levels is sufficient to estimate the population variance accurately.

TABLE 2. SUMMARY OF ANALYSES OF VARIANCE CALCULATED BY DIVISIONS

Source of variation	Degrees of freedom (d.f.)	Mean square values			F-ratios		
		Div. 1	Div. 2	Div. 3	Div. 1	Div. 2	Div. 3
Between sites ($B = 8$)	7	4.16	1.93	4.42	13.5*	3.50*	8.91*
Within sites ($W = 25$)	192	0.307	0.552	0.496	..	..	..
Total ($N = 200$)	199	0.442	0.601	0.634			

Note: For d.f. = 7/200, the F-ratio = 2.05 at 5% and 2.73 at 1% (*) level of significance.

The following equations (Watson, 1966) were used:

$$\text{M.S.}_W = 1/\hat{\omega}, \quad (1)$$

and

$$\text{M.S.}_B = 1/\hat{\omega} + \overline{W}/\hat{\beta}, \quad (2)$$

where M.S.$_W$ is the mean square value within sample sites, and M.S.$_B$ is the mean square value between sample sites. $\overline{W}$ equals 25, the number of measurements taken at each sample site. Equation (1) yields an estimate of variance ($\hat{\omega}$) within sample sites. Equation (2) gives an estimate of variance ($\hat{\beta}$) between sample sites.

The numerical results of equations (1) and (2) for each of the three divisions and the entire sand body are presented in Table 4. In all cases, between-site variations ($\hat{\beta}$) are larger than within-site variations ($\hat{\omega}$). The ratio $\hat{\beta}/\hat{\omega}$ affords a quantitative comparison of between-site variations by setting $\hat{\omega}$ equal to unity.

Using the rounded-off values of this ratio, estimates of the minimum number of cross-bed measurements that could have been recorded are calculated in Table 5. The assumption here is that 25 measurements at each sample site are the minimum number that can be recorded, because 79 percent of the variation exists at this level.

This assumption is supported by two other considerations. First, the $\theta_{0.05}$ values in Table 1 show that the radius of the circle of confidence for most of the sample sites exceeds 20°. Second, the differences between resultant vector directions calculated for all 25 measurements and for the two steepest dipping cross-beds are very great (Table 1, col. 3). These differences are shown in a graph (Fig. 9) where 10 sample sites have $\bar{A}_2$ minus $\bar{A}_{25}$ values greater than 45°.

TABLE 3. ANALYSIS OF VARIANCE CALCULATED FOR ENTIRE SAND BODY

Source of variation	Degrees of freedom (d.f.)	Sum of squares	(%)	Mean square	F-ratio
Between sites ($B = 24$)	23	68.1	21	2.96	6.54*
Within sites ($W = 25$)	576	260.2	79	0.452	..
Total ($N = 600$)	599	328.3	100	0.548	

Note: For d.f. = 24/400, the F-ratio = 1.54 at 5% and 1.84 at 1% (*) level of significance.

TABLE 4. EXPECTED VALUES OF MEAN SQUARE

Source of variation	*	Divisions† 1	2	3	Entire sand body§
Between sites	$\hat{\beta}$	6.49	18.10	6.37	9.98
Within sites	$\hat{\omega}$	3.26	1.81	2.02	2.21
	$\hat{\beta}/\hat{\omega}$	1.99	10.00	3.15	4.51
		(≈ 2)	($=10$)	(≈ 3)	(≈ 5)

* See text for equations and explanation of terms.
† Calculated from mean square values in Table 2.
§ Calculated from mean square values in Table 3.

The estimates of the minimum number of cross-bed measurements (Table 5) show that, for the entire sand body, a total of 125 measurements would be adequate to obtain an average dip direction with the same precision that all 600 measurements have. In other words, the grand vector mean of 070° (Fig. 8) could have been obtained by measuring only 25 cross-beds at five sample sites.

In order to take into accounnt the variation in cross-bed directions among divisions (Fig. 7), estimates of the minimum number of measurements were made for each division. The results in Table 5 indicate that 250 measurements are required in division 2 alone and that the total number would be 375. This greater number is due largely to the nature of cross-bed variation in division 2. Table 2 shows that the largest mean square value within sample sites occurs in division 2, and Figure 6 illustrates the marked bimodality of the average cross-bed directions of the eight sites in this division.

In contrast, divisions 1 and 3 have smaller mean square values within sample sites (Table 2), and the sample-site averages are more uniformly distributed (Fig. 6). Figure 7 illustrates that these two divisions have more distinct clustering of cross-bed measurements than division 2, with division 1 having a stronger mode

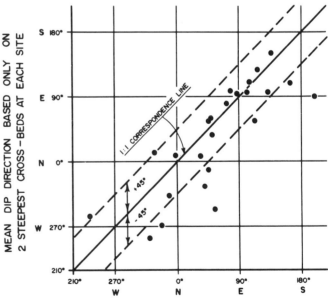

Figure 9. Cross plot of mean dip directions based on 25 measurements versus the two steepest cross-beds at each sample site. See Table 1 for data; see text for details.

TABLE 5. ESTIMATES OF MINIMUM NUMBER OF CROSS-BED MEASUREMENTS

Division	Number of sites*	Number per site	Total
1	2	25	50
2	10	25	250
3	3	25	75
Entire sand body	5	25	125

*Based on $\hat{\beta}/\hat{\omega}$ ratios in Table 4.

than division 3. The estimates of the minimum number of sites (Table 5) reflect these observations.

The last data to be analyzed are the sand-body thickness values. These are recorded in Table 1 and contoured on a map in Figure 10A. The average cross-bed dip angles for each sample site (Table 1, col. 4) are contoured in Figure 10B, and an isopleth map of $\theta_{0.05}$ values is drawn in Figure 10C.

Comparison of the three maps in Figure 10 suggests that the thickest part of this sand body has the lowest average cross-bed dips. In turn, the lowest dip

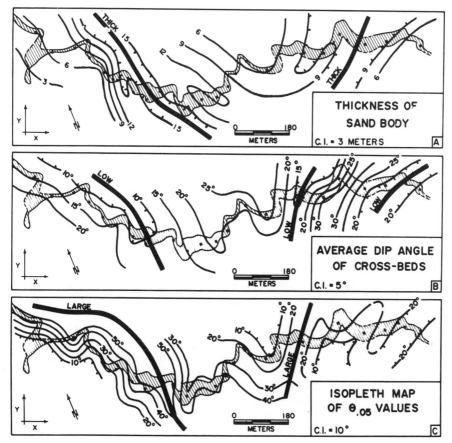

Figure 10. A, isopach map of sand-body thickness. B, contour map of average cross-bed dip angles for each sample site. C, isopleth map of $\theta_{0.05}$ values. Data for these maps are listed in Table 1.

angles seem to coincide with the zone of greatest cross-bed variability (largest $\theta_{0.05}$ values).

These observations are verified by the linear correlation coefficients in Table 6. There is a significant inverse relation between average dip angle (Fig. 10B) and radius of the circle of confidence (Fig. 10C). The positive correlation between sand-body thickness and circle of confidence is also significant. However, even though there is an inverse relation between thickness and dip angle, it is nonsignificant.

Finally, it is instructive to compare the maps of Figure 10 with the pattern of average cross-bed dip directions in Figure 6. Such comparison indicates that current flow direction was more variable in the thick sand axes and locally more uniform in the areas of thin sand.

TABLE 6. LINEAR CORRELATION COEFFICIENTS

	Sand-body thickness (T)	Average dip angle (D_A)	Radius of circle of confidence ($\theta_{0.05}$)
Thickness (T)	1.00	−0.28	+0.48†
Average dip ($\bar{D}_A$)		1.00	−0.56*
Radius ($\theta_{0.05}$)			1.00

* Significant at 1% level.
† Significant at 5% level.

INTERPRETATION OF DATA

From the quantitative data presented here, it is not possible to determine the origin of this sand body. However, based on the nature of the trough cross-beds, the presence of clay chips, and the stratigraphic relations, it is possible to recognize that this sand body was deposited by channel processes in shallow water in a continental environment.

The grand vector mean of cross-bed azimuths is 070° and indicates that overall flow direction was toward the east-northeast. Vector means at each of the 24 sample sites (Fig. 6) indicate a change in flow direction along the sand-body outcrop. This variation seemingly can be explained by the sand-body isopach map (Fig. 10A). It appears that mean cross-bed directions either point toward the thick channel axis, or parallel re-entrants in the contours, which may represent sand bars in the channel.

In addition, the maps of Figures 10B and 10C, along with the r-values of Table 6, demonstrate that the sample sites with the steepest average cross-bed dips have the least variability (smallest $\theta_{0.05}$ values). In turn, the most variable sets of cross-bed data (largest $\theta_{0.05}$ values) are found along the thick channel axis. These relations suggest more uniform flow directions along the shallower flanks of the channel.

Finally, within the 2 m² area of each sample site, variation among the 25 cross-bed azimuths and dip angles can be large. Figure 5 portrays the range of variations that exists, and the $\theta_{0.05}$ values (Table 1, col. 7) exceed 20° at 16 sample sites. This means that cross-bed variability at the most detailed level of this study can be almost as large as that found along the entire 1,830-m outcrop.

CONCLUSIONS

Unlike other statistical studies, such as Potter and Siever (1956a, 1956b), this study shows that cross-bed directions in a single channel sand body can be highly variable. Large variation exists within each sample site and along the entire outcrop of this Wasatch sand body. In fact, 79 percent of the total variability for the entire sand body exists at the sample-site level. The result is that a minimum of 125, and possibly a maximum of 375, cross-bed measurements are required to estimate adequately the overall flow direction.

This conclusion leads to the following suggestions: Not all channel sand bodies are alike, and to obtain meaningful paleostreamflow directions from cross-bed measurements, it is necessary to run a preliminary test to determine the variability at all sampling levels. The results of such preliminary tests will vary according to the type of channel sandstone. Potter and Siever (1956a, 1956b) found very little variation at the outcrop level; this study found large variation at all levels. Such different results are inherent in the rocks and will dictate the sampling plan to be used and the number of cross-bed measurements to be recorded. It is possible that future investigators may find the number of cross-bed measurements required are quite sizable and considerably larger than expected.

ACKNOWLEDGMENT

This study is an outgrowth of part of my Ph.D. dissertation from Northwestern University, and I thank W. C. Krumbein for his guidance during the initial stages of the study.

REFERENCES CITED

Fisher, R. A., 1953, Dispersion on a sphere: Royal Soc. [London] Proc., ser. A, v. 217, p. 295-306.
Larochelle, A., 1967, A re-examination of certain statistical methods in palaeomagnetism: Canada Geol. Survey Paper no. 67-18, 19 p.
Larsson, I., 1952, A graphic testing procedure for point diagrams: Am. Jour. Sci., v. 250, p. 586-593.
Looff, K. M., and Hubert, J. F., 1964, Sampling variability of paleocurrent cross-bed data in the post-Myrick Station channel sandstone (Pennsylvanian), Missouri: Jour. Sed. Petrology, v. 34, p. 774-776.
Pincus, H. J., 1951, Statistical methods applied to the study of rock fractures: Geol. Soc. America Bull., v. 62, p. 81-130.
Potter, P. E., 1967, Sand bodies and sedimentary environments: A review: Am. Assoc. Petroleum Geologists Bull., v. 51, p. 337-365.
Potter, P. E., and Olson, J. S., 1954, Variance components in some basal Pennsylvanian sandstones, Pt. 2, Geological application: Jour. Geology, v. 62, p. 50-73.
Potter, P. E., and Siever, R., 1956a, Sources of basal Pennsylvanian sediments in the Eastern Interior Basin, Pt. 1, Cross-bedding: Jour. Geology, v. 64, p. 225-244.
———1956b, Sources of basal Pennsylvanian sediments in the Eastern Interior Basin, Pt. 3, Some methodological implications: Jour. Geology, v. 64, p. 447-455.
Pryor, W. A., 1958, Dip direction indicator: Jour. Sed. Petrology, v. 28, p. 230.
Rao, J. S., and Sengupta, S., 1972, Mathematical techniques for paleocurrent analysis: Treatment of directional data: Internat. Assoc. Math. Geol. Jour., v. 4, p. 235-248.
Schuenemeyer, J. H., Koch, G. S., Jr., and Link, R. F., 1972, Computer program to analyze directional data, based on the methods of Fisher and Watson: Internat. Assoc. Math. Geol. Jour., v. 4, p. 177-202.

Steinmetz, R., 1962, Analysis of vectorial data: Jour. Sed. Petrology: v. 32, p. 801-812.

Watson, G. S., 1956, A test for randomness of directions: Royal Astron. Soc. Monthly Notices, Geophysical Supplement, v. 7, p. 160-161.

——1966, The statistics of orientation data: Jour. Geology, v. 74, p. 786-797.

——1970, The statistical treatment of orientation data, *in* Merriam, D. F., ed., Geostatistics: A colloquium: New York, Plenum Press, p. 1-9.

Watson, G. S., and Irving, E., 1957, Statistical methods in rock magnetism: Royal Astron. Soc. Monthly Notices, Geophysical Supplement, v. 7, p. 289-300.

Wurster, P., 1958, Geometrie und Geologie von Kreuzschichtungskörpern: Geol. Rundschau, v. 47, p. 322-359.

MANUSCRIPT RECEIVED BY THE SOCIETY SEPTEMBER 12, 1973
REVISED MANUSCRIPT RECEIVED FEBRUARY 27, 1974

SEDIMENT TRANSPORT

Model of Recurring Random Walks for Sediment Transport

MICHAEL F. DACEY

Department of Geological Sciences
Northwestern University
Evanston, Illinois 60201

ABSTRACT

Several studies have modeled the movement of sediment particles as random walks in which the movement space is completely surrounded by reflecting barriers. As a consequence, a particle never escapes from the movement space. For some applications, this is an undesirable attribute in that it artificially counteracts the natural movement (downstream, downslope, or settling) of a particle. In the recurring random-walk model, the movement of a particle is also described by a random walk but in a space that is bounded partly by reflecting barriers and also by an absorbing barrier through which the particle escapes. To compensate for this removal, new particles are added to the movement space at periodic intervals. The recurring random-walk model may be formulated as a sequence of independent random walks that differ only in the times at which the movement of a particle is initiated. It is shown that, under general conditions, a steady-state distribution of particle movement is attained. Measures of the sediment concentration profile at steady state are obtained.

INTRODUCTION

The behavior of particles whose movement is characterized by erratic, unpredictable motions is frequently modeled as a random walk. The classical application of the random-walk model is to Brownian motion. Under certain conditions, the movement of sediment particles is irregular and apparently random; as one example, Strahler (1952, p. 933–934) suggested that soil creep is the result of random movements of grains. Random-walk models for the transfer and transport of sediment particles have been investigated in several studies, including those of Culling (1963), Scheidegger and Langbein (1966), Conover and Matalas (1967), and Heiskanen (1972). The approach of these studies was to model as a random walk the movement behavior of a single particle, which is treated as a representative member of a large ensemble of sediment particles. Though the short-run movement of a particle may be haphazard, there typically is a long-run drift, such as downstream, downslope, or settling. It is frequently difficult to encompass this long-run drift in a random-walk model; if the space for movement is bounded, the particle passes through the space in a finite amount of time and there is then no further movement to describe, but the condition of an unbounded space (such as an unending slope or bottomless ocean) is unrealistic. A consequence is that, to analyze the long-term, steady-state behavior of particles, some rule must be introduced that returns the escaping particle to the movement space. For example, in the sediment-transport model of Conover and Matalas, a particle does not settle upon reaching the bottom surface of a stream but instead is reflected back into the stream flow. Some type of reflecting principle is evidently a component of all conventional random-walk models for sediment movement. For many modeling objectives, it is an unsatisfactory component in that it artificially counteracts the natural movement of a particle.

An alternative to the single-particle, random-walk model is a model that treats the movement of each particle in an ensemble of particles as an independent random walk. The model of Conover and Matalas (1969) was of this type, but it incorporated a reflecting principle so that particles do not escape from the movement space. This study describes a multiple-particle, random-walk model that does not incorporate a reflecting principle. It allows particles to escape from the movement space, but compensates for this removal by the addition of new particles at periodic intervals. One direct application of the model is to movement of sediment particles in turbulent water. Particles enter the top of the water at a constant rate and drift through the water until they are submerged at the bottom surface. The vertical movements of all drifting particles are described by independent random walks. Regardless of the rates of entry and submergence, the concentration of drifting particles is shown to have an equilibrium, and the density concentration profile of sediments in equilibrium is obtained. This use of the model for settling sediments in a water column was described by Lerman and others (1974). Other interpretations of the model will be indicated.

PRELIMINARIES

This section provides a verbal description of the structure that, in the next section, is formulated as a mathematical model for movement of sediment particles.

Interpretation and Objectives of the Model

It is convenient to formulate the model for movement over time of sediment particles in a linear space T. Both time and space are discrete; let $t_0, t_1, t_2, ...$ be the points of time, and let $E_1, E_2, ..., E_J$ be the finite number J of positions of the space T. The positions are ordered so that E_J is between E_{J-1} and E_{J+1}. Particles enter the space T through E_1, depart through E_J, and between entrance and departure, move in a prescribed, though possibly erratic, manner through the positions of T. Once a particle leaves T, it does not return. It is convenient to identify a position E_{J+1} that does not belong to T and serves only to account for particles that have passed through and permanently left T. Figure 1a illustrates the positions of T and E_{J+1}; movement is allowed between adjacent positions that are separated by dashed lines, whereas solid lines indicate movement is restricted to one direction.

A particle is said to be in the space T if it is located in any of the positions

E_1 through E_J. Particles are entered into T in the following way: At the initial time t_0, T is empty of particles. At each of the times t_1, t_2, ..., a particle enters T by being placed in position E_1. A particle leaves T when it reaches E_{J+1}, and once in E_{J+1}, it remains there. The rules for movement of a particle that is in the space T constitute a random walk. From one time point to the next, a particle moves from its present position to an adjacent position or remains at the same position. The choices of move one position lower, move one position higher, and no change depend upon chance. The statement of the model identifies the probability of occurrence for each of the three alternatives, and there is wide latitude in specifying these probabilities, but the same probability structure governs the movement of all particles. Whereas the movement of each particle describes a random walk with one reflecting barrier at E_1, and one absorbing barrier at E_{J+1}, the model incorporates the movement of numerous particles that move independently of each other. Because a new particle having an independent random walk enters T at each time point, the model is aptly described as a sequence of recurring random walks that operate independently in the same space and time span. This system is analyzed to obtain the concentration of particles in each of the positions E_1, E_2, ..., E_J after the system has been operating for a sufficient length of time to approach its steady-state behavior.

One possible interpretation of the space T would be as a vertical profile of a water column of depth X where E_1 corresponds to the interval from the surface to depth X/J and subsequent positions occur at equal intervals of X/J, so that E_J corresponds to the interval $(J-1)X/J$, and the bottom is at depth X. Sediments enter at the top surface, position E_1, and drift through the water until buried at the bottom surface, position E_{J+1}, and thereby leave the space. Another possible interpretation of the space T is as intervals on a hillside, with the movement of sediments corresponding to the creep of soil from hilltop to river valley. These and other possible applications are encompassed within a single model by varying the specification both of the positions of T, which need not be of equal length, and of the probability structure that governs the passage of particles through T.

The space of movement is readily extended to two or three dimensions. Figure 1b shows one possible arrangement of J positions in a planar space S. Again, movement is allowed between adjacent positions separated by dashed lines, whereas solid lines restrict movement to one direction. A simple modification of the model is to add a position E_{J+2} so as to distinguish particles that leave S by downward movement from those that leave by forward movement. A possible interpretation of the space S is as a cross section of a stream link or other open channel in which particulate matter enters the stream link at the source position E_1 and leaves either by being carried downstream to the junction E_{J+1} with another stream link or by being buried at the bottom surface E_{J+2} of the stream. This interpretation is similar to that used by Conover and Matalas (1967, 1969) for a random-walk model on S. Their model had similar rules for particle movement but differed from the recurring random-walk model in that particles did enter or leave the movement space.

The regularity in size and shape of positions suggested by Figure 1 is convenient for notation but is not a requirement of the model. Also, there is no loss of generality by restricting the formulation of the model to the linear space T, but the notation is much simpler. The model requires that all particles enter the movement space through the same position. This condition of a "point source" may be overly restrictive for many types of sediments, but it may be appropriate for discharge of sewage or pollutants into a lake that is represented by a three-dimensional movement space.

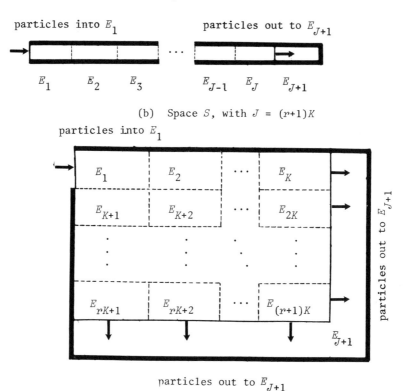

Figure 1. Examples of positions of the movement spaces *T* and *S*. A particle may pass both ways between positions separated by a dashed line but only one way between positions separated by a solid line; movement through a thick, solid line is prohibited.

Structure and Notation

Let $\Omega = \{\omega_m : m \geq 1\}$ denote a collection of particles. The particle ω_n enters the space T at time t_n. The location of ω_m at t_n, $n \geq m$, is denoted by $X_n(m)$ and is an integer from the collection $\mathbf{E} = \{1, 2, ..., J + 1\}$, where J is finite. The elements of $\mathbf{E}$ are called *states*. The states 1 through J correspond to the positions E_1 through E_J of a particle within the linear space T, while state $J + 1$ represents the position of a particle that has left the space T. Accordingly, $X_n(m)$ is not defined for $m < n$, and for $n \geq m$, $X_n(m)$ has a value from $\mathbf{E}$.

The movement in T of a particle is described by a structure of probabilities and conditional probabilities. The event that ω_m has position E_j at time t_n is denoted by $\{X_n(m) = j\}$, and the probability of this event is $P\{X_n(m) = j\}$. The conditional probability that ω_m has position E_j at time t_{n+1}, given its position at t_n, is denoted by $P\{X_{n+1}(m) = j \mid X_n(m)\}$.

Two properties of movement of particles are that the particle ω_n enters T through position E_1 at time t_n and that a particle that leaves T never returns. The properties are treated in the model as events that occur with probability 1. So, two conditions of the model are made explicit by

$$P\{X_n(n) = 1\} = 1 \tag{1}$$

and

$$P\{X_{n+1}(m) = J + 1 | X_n(m) = J + 1\} = 1. \tag{2}$$

The probabilities that govern the movement of a particle within T depend neither upon the time of the system nor the time at which the particle first enters the system. Thus, the system is time homogeneous, and this condition is made explicit by noting that for all positive integers n, n', m, and m' ($n \geq m$ and $n' \geq m'$),

$$P\{X_{n+1}(m) = j | X_n(m) = i\} = P\{X_{n'+1}(m') = j | X_{n'}(m') = i\} \tag{3}$$

for all i and j in $\mathbf{E}$. Because the conditional probability of $X_{n+1}(m)$ does not depend upon either n or m, it is convenient to use the notation

$$P(i, j) = P\{X_{n+1}(m) = j | X_n(m) = i\}, \tag{4}$$

which is defined for all i and j in $\mathbf{E}$.

Since $P(i, j)$ is a conditional probability, it is required that all

$$P(i, j) \geq 0. \tag{5}$$

Moreover, a particle that has a position at time t_n also has a position at t_{n+1}. So, for each $i \in \mathbf{E}$

$$\sum_{j \in \mathbf{E}} P(i, j) = 1. \tag{6}$$

The modified random-walk principles that control movement of a particle require that, in successive time periods, the position of a particle is unchanged or is changed to an adjacent position. These restrictions on movement of a particle are made explicit by

$$P(1, 1) + P(1, 2) = 1 \tag{7}$$

and

$$P(j, j-1) + P(j, j) + P(j, j+1) = 1, \quad j = 2, 3, ..., J. \tag{8}$$

Equation (2) implies that

$$P(J + 1, J + 1) = 1. \tag{9}$$

To insure that movement from E_1 to E_{J+1} is possible, it is also specified that

$$P(j, j+1) > 0, \quad j = 1, 2, ..., J. \tag{10}$$

The *initial condition* (1) and the *transition probabilities* (5) through (10) describe completely the behavior of one particle through T from the time it enters the system at E_1 until it departs at E_{J+1}. A succinct way to summarize this movement is by a *transition matrix* **P** having entries $P(i, j)$ that give the probability of a move or transition in one step (or time unit) from state i to state j. Since **E** has $J + 1$ states, the matrix **P** has $J + 1$ rows and $J + 1$ columns. Equations (5) through (9) indicate that the matrix **P** has the following typical form:

$$\mathbf{P} = \begin{Vmatrix} P(1,1) & P(1,2) & 0 & 0 & \cdots & 0 & 0 & 0 \\ P(2,1) & P(2,2) & P(2,3) & 0 & \cdots & 0 & 0 & 0 \\ 0 & P(3,2) & P(3,3) & P(3,4) & \cdots & 0 & 0 & 0 \\ \cdot & \cdot & \cdot & \cdot & & \cdot & \cdot & \cdot \\ \cdot & \cdot & \cdot & \cdot & & \cdot & \cdot & \cdot \\ \cdot & \cdot & \cdot & \cdot & & \cdot & \cdot & \cdot \\ 0 & 0 & 0 & 0 & \cdots & P(J,J-1) & P(J,J) & P(J,J+1) \\ 0 & 0 & 0 & 0 & \cdots & 0 & 0 & 1 \end{Vmatrix} \quad (11)$$

Alternatively, **P** is specified by $\{P(i, j) : i, j \in \mathbf{E}\}$.

This matrix completely describes the movement of one particle through the space T. It can be shown that with probability 1 a particle leaves T and thereby is assigned position E_{J+1} in a finite number of moves. The space does not, however, become empty of particles because at each point of time a new particle enters T, and at any time t_n, $n \geq J$, there will be at least J particles in T. The number of these particles in each of the J positions yields a particle concentration profile for the discrete space T at time t_n. Because the movement of particles is subject to the laws of chance, the number of particles in a position is a random variable.

Let $N_n(j)$ be the number of particles in position E_j at time t_n. This random variable may be described by its probability terms or, more succinctly, by its moments. Of particular interest is its expected value $\mathscr{E}N_n(j)$, since a concentration profile of particles in T at time t_n may be constructed from the J expected values $\mathscr{E}N_n(1), \mathscr{E}N_n(2), \ldots, \mathscr{E}N_n(J)$. If a steady state exists, in the sense that after a sufficiently long time particles leave T from E_J at the same rate that they enter T through E_1, then the limiting values of the $\mathscr{E}N_n(j)$ describe the concentration of particles when the system attains its steady-state behavior.

RECURRING RANDOM-WALK MODEL

This model incorporates the structure identified in the preceding section for movement of particles through the discrete, linear space T. The movement of each particle through T is in accordance with the assumptions expressed by equations (1) through (10). Time is also discrete, and at each unit of time a new particle enters T.

Statement of the Model

DEFINITION 1. The *state space* is $\mathbf{E} = \{1, 2, \ldots, J + 1\}$, where $J < \infty$. The *transition probabilities* $P(i, j)$ for a move from state $i \in \mathbf{E}$ to state $j \in \mathbf{E}$ are given by the matrix $\mathbf{P} = \{P(i, j) : i, j \in \mathbf{E}\}$. The $X_n(m)$ is an integer-valued *random variable* that is defined only for $m \geq 1$ and $n \geq m$.

ASSUMPTION 1. The matrix $\mathbf{P}$ satisfies the following conditions:

(a) $P(i, j) \geq 0, \quad i, j \in \mathbf{E}$,

(b) $\sum_{j=1}^{J+1} P(i, j) = 1, \quad i \in \mathbf{E}$,

(c) $P(1, 1) + P(1, 2) = 1$,

(d) $P(j, j - 1) + P(j, j) + P(j, j + 1) = 1, \quad j = 2, 3, \ldots, J$,

(e) $P(j, j + 1) > 0, \quad j = 1, 2, \ldots, J$,

(f) $P(J + 1, J + 1) = 1$.

In particular, $\mathbf{P}$ has the structure of (11).

ASSUMPTION 2. Each random variable $X_n(m)$ of the collection $\{X_n(m) : n \geq m, m \geq 1\}$ obeys the following conditions:

(a) $P\{X_n(m) \in \mathbf{E}\} = 1$,

(b) $P\{X_n(n) = 1\} = 1$,

and for each $j \in \mathbf{E}$

(c) $P\{X_{n+1}(m) = j \mid X_n(m)\} = P(X_n(m), j)$.

Remark. The random variable $X_n(m)$ is not defined for $n < m$.

Remark. Notice that the collection $\{X_n(m) : n \geq m \geq 1\}$ contains a double sequence. For a fixed value of m, it contains the sequence $X_m(m), X_{m+1}(m), \ldots, X_{m+k}(m), \ldots$, and there is in the collection a similar sequence for each positive integer m.

DEFINITION 2. The sequence of random variables $\{X_n(m) : n \geq m, m \geq 1\}$ that satisfies assumptions 1 and 2 is called a *recurring random walk* with state space $\mathbf{E}$ and transition matrix $\mathbf{P}$.

DEFINITION 3. The particle ω_m is said to be in state j at time t_n if, and only if, the event $\{X_n(m) = j\}$ occurs. Put

$$I_n(m, j) = 1, \quad \text{if } n \geq m \text{ and } X_n(m) = j,$$

$$= 0, \quad \text{otherwise.}$$

For each $j \in \mathbf{E}$, put

$$N_n(j) = \sum_{m=1}^{n-j+1} I_n(m, j).$$

The quantity $N_n(j)$ is called the *counting function* for state j at time t_n.
Property 1. The number of particles in state j at time t_n is $N_n(j)$.

DEFINITION 4. Put $\mathbf{E}^* = \{1, 2, ..., J\}$ and $N(j) = \lim_{n\to\infty} N_n(j)$. The collection of J numbers $\{\mathscr{E}N(j) : J \in \mathbf{E}^*\}$ is called the *density concentration profile*.
The derivation of this profile makes use of the model of definition 5.

DEFINITION 5. The sequence of random variables $\{X_n : n \geq 0\}$ is a Markov chain with state space $\mathbf{E} = \{1, 2, ..., J + 1\}$, transition matrix $\mathbf{P}$, as described by assumption 1, and starting vector $\pi = \{1, 0, 0, ..., 0\}$, a vector of $J + 1$ elements of which the last J are equal to 0.

Remark. The sequence $\{X_n : n \geq 0\}$ is a finite state, *absorbing Markov chain*, and the single absorbing state is $J + 1$. Kemeny and Snell (1960, chap. 3) provided an elementary but reasonably complete discussion of this type of Markov chain. The sequence is also called a *random walk* with one reflecting barrier and one absorbing barrier.

Property 2. Let m have a fixed value and put $Y_{n-m}(m) = X_n(m)$. For any n, the random variable $Y_n(m)$ of the sequence $\{Y_n(m) : n \geq 0\}$ is identically distributed as the random variable X_n of the sequence $\{X_n : n \geq 0\}$.

Property 3. For fixed value of m, the sequence $\{X_{n-m}(m) : n \geq m\}$ is a Markov chain with state space E, transition matrix $\mathbf{P}$, and starting vector π.

More specifically, property 3 implies property 4.
Property 4. For any $n \geq m$ and $j \in \mathbf{E}$,

$$P\{X_{n-m}(m) = j\} = P\{X_n = j\}.$$

A consequence of these properties is that the sequence $\{X_n(m) : n \geq m, m$ fixed$\}$ differs from $\{X_n : n \geq 0\}$ only by shift of the time scale forward m units. Since the collection $\{X_n(m) : n \geq m, m = 1, 2, ...\}$ contains a sequence $\{X_n(m) : n \geq m, m$ fixed$\}$ for each positive integer m, it contains a countably infinite number of Markov chains, each of which differs from $\{X_n : n \geq 0\}$ only by shift of the time scale. The n Markov chains that are initiated by time t_n have the same state space $\mathbf{E}$, the same transition matrix $\mathbf{P}$, the same starting vector π, and they are mutually independent and differ only in the starting times. The important consequence of these remarks is that properties of $N_n(j)$ may be obtained by analysis of n independent, finite state, absorbing Markov chains that differ only in the times at which they are initiated.

Comparison with the Conventional Random Walk

A feature of the recurring random-walk model is that each particle eventually reaches state $J + 1$ in the sense that for any m

$$\lim_{n\to\infty} P\{X_n(m) = J + 1\} = 1.$$

It is, accordingly, uninteresting to analyze the long-run behavior of any single particle. It is for this reason that models for movement behavior of a single particle in a state space $\mathbf{E}$ do not use a transition matrix similar to $\mathbf{P}$. In particular, transport of sediments is not adequately modeled by a conventional random walk having transition matrix $\mathbf{P}$. To clarify this statement, as well as for subsequent analysis, some properties of the random walk are identified.

DEFINITION 6. For the random walk $\{X_n : n \geq 0\}$, put

$$I_n(j) = 1, \quad \text{if } X_n = j,$$

$$= 0, \quad \text{otherwise.}$$

Remark. The number of particles of the random walk $\{X_n : n \geq 0\}$ in state j at time t_n is $I_n(j)$.

Well-known properties of the random walk $\{X_n : n \geq 0\}$ are given by property 5.

Property 5. For each $j \in \mathbf{E}^*$, as $n \to \infty$,

$$I_n(j) \to 0, \text{ and } P\{X_n = j\} \to 0.$$

These properties indicate that the particle escapes the space T in a finite time with probability 1. As a consequence, there is no suitable measure, such as $\mathscr{E} N_n(j)$, to construct the density concentration profile for a process that has been operating for a long time. To preclude the particle escaping T and being absorbed by a single state, we must construct a different type of transition matrix by redefining T to include E_{J+1} and by invoking some type of reflection principle. One form of this reflection principle is to replace the condition $P(J + 1, J + 1) = 1$ by $P(J + 1, J) > 0$, which, in effect, was the strategy used by Conover and Matalas (1967, 1969). An alternative form continually repeats the random walk by replacing $P(J + 1, J + 1) = 1$ by $P(J + 1, 1) = 1$. This reflecting principle has the effect of replacing the absorbing barrier, position E_{J+1}, of the random walk of definition 5 by a reflecting barrier so that the particle remains forever within T. For this random-walk model with two reflecting barriers, the density concentration profile of a steady-state system is given by the limiting values of the $\mathscr{E} I_n(j)$ which, incidentally, are the limiting probabilities that the particle is in each of the $J + 1$ positions.

Basic Properties of the Model

This section displays the measures $\{\mathscr{E} N(j) : j \in \mathbf{E}^*\}$ for the density concentration profile and other properties of the recurring random-walk model that are pertinent to sediment transport. Some of these properties were used by Lerman and others (1974). The following section provides the formal analysis.

DEFINITION 7. For $k \geq 1$, the kth power of $\mathbf{P}$ is $\mathbf{P}^k$, and its entries are denoted by $P^k(i, j)$. The matrix $\mathbf{P}^0$ has 1's on the main diagonal and 0's elsewhere. The $\mathbf{P}^k$ is called the *k-step transition probability matrix*.

Remark. The entry $P^k(i, j)$ is the probability that a particle in state i moves from i to state j in exactly k steps; that is, $P\{X_{n+k}(m) = j \mid X_n(m) = i\} = P^k(i, j)$.

DEFINITION 8. Let $\mathbf{I}$ be the $(J \times J)$ identity matrix that has 1's on the main diagonal and 0's elsewhere. Let $\mathbf{Q} = \{Q(i, j) : i, j \in \mathbf{E}^*\}$ be a $J \times J$ matrix with entries given by $Q(i, j) = P(i, j)$. Also, put $\mathbf{R} = (\mathbf{I} - \mathbf{Q})^{-1}$.

Property 6. For $n \geq m$ and $j \in \mathbf{E}$,

$$P\{X_n(m) = j\} = P^{n-m}(1, j).$$

Proof. Since $P\{X_m(m) = 1\} = 1$, then

$$P\{X_n(m) = j\} = P\{X_n(m) = j \mid X_m(m) = 1\}$$

$$= P\{X_{n-m} = j \mid X_0 = 1\}, \quad \text{(property 4)}$$

which is the probability that the Markov chain of definition 5 moves from state 1 to state j in exactly $n - m$ steps. Hence, it is the $n - m$ step transition probability from state 1 to state j. By the Chapman-Kolmogorov theorem (Kemeny and Snell, 1960, p. 33), this probability is $P^{n-m}(i, j) \cdot P\{X_m(m) = 1\} = P^{n-m}(i, j)$.

Property 7. For $j \in \mathbf{E}$,

$$\lim_{n \to \infty} P^n(1, j) = 0, \quad j \in \mathbf{E}^*,$$

$$= 1, \quad j = J + 1.$$

This result is a consequence of property 4 and the fact that $\{X_n : n \geq 0\}$ is a finite state, absorbing Markov chain. By a theorem in Kemeny and Snell (1960, p. 43), the chain enters an absorbing state in a finite number of steps with probability 1.

Property 8. For each $j \in \mathbf{E}$,

$$\mathscr{E} I_n(m, j) = P^{n-m}(1, j), \quad n \geq m + j - 1,$$

$$= 0, \quad \text{otherwise.}$$

Property 9. For each $j \in \mathbf{E}$ and $n \geq j$,

$$\mathscr{E} N_n(j) = \sum_{k=j-1}^{n-1} P^k(1, j).$$

Proof. Definition 3 implies

$$\mathscr{E} N_n(j) = \mathscr{E} \sum_{m=1}^{n-j+1} I_n(m, j)$$

$$= \sum_{m=1}^{n-j+1} \mathscr{E} I_n(m, j)$$

$$= \sum_{m=1}^{n-j+1} P^{n-m}(1, j) \quad \text{(property 8)}$$

$$= \sum_{k=j-1}^{n-1} P^k(1, j).$$

Property 10. For each $j \in \mathbf{E}$,

$$\mathscr{E} N_{n+1}(j) = \mathscr{E} N_n(j) + P^n(1, j).$$

Property 11. For each $j \in \mathbf{E}$,

$$\mathscr{E} N(j) = \sum_{k=0}^{\infty} P^k(1, j).$$

Proof.

$$\mathscr{E} N(j) = \lim_{n \to \infty} \mathscr{E} N_n(j)$$

$$= \sum_{k=j-1}^{\infty} P^k(1, j) \qquad \text{(property 9)}$$

$$= \sum_{k=0}^{\infty} P^k(1, j),$$

since $P^k(1, j) = 0$ for $k < j - 1$.

THEOREM 1. For each $j \in \mathbf{E}^*$,

$$\mathscr{E} N(j) = R(1, j),$$

which is finite valued.

The proof of this result is given in the next section.
Properties 7 and 11 imply property 12.
Property 12. $\mathscr{E} N_n(J + 1) \to \infty$ as $n \to \infty$.

Theorem 1 identifies the values $\mathscr{E} N(1)$, $\mathscr{E} N(2)$, ..., $\mathscr{E} N(J)$ used to construct the density concentration profile, which are obtained by calculating the inverse of $\mathbf{I} - \mathbf{Q}$. It also establishes that a steady-state profile exists, in the sense that the limiting values of $\mathscr{E} N_n(j)$ are finite for all $j \in \mathbf{E}^*$. This result implies that, no matter how close to 0 are the $P(j, j + 1)$, after a sufficient length of time, particles enter state $J + 1$ at the same rate at which they are introduced into state 1.

Relation to Random Walks

It having been verified that the recurring random walk of definition 2 and the random walk of definition 5 are different models of particle movement, a fundamental relation between random variables for the two models is now established. The random variable for the random walk is given by definition 9.

DEFINITION 9. For each $j \in \mathbf{E}$, put

$$V_n(j) = \sum_{m=0}^{n-1} I_m(j), \qquad n \geq 1,$$

and

$$V(j) = \lim_{n \to \infty} V_n(j).$$

The $V_n(j)$ is called the *occupation time* of state j prior to time t_n.

This occupation time for the conventional random walk with transition matrix $\mathbf{P}$ and starting vector π is compared with the number $N_m(j)$ of particles in state j at time t_n for the recurring random walk with the same transition matrix $\mathbf{P}$.

THEOREM 2. For any $n \geq 1$ and $j \in E$, $N_n(j) = V_n(j)$.
Proof. First consider

$$P\{I_n(m, j) = 1\} = P\{X_n(m) = j\} \qquad \text{(definition 3)}$$

$$= P\{X_{n-m} = j\} \qquad \text{(property 4)}$$

$$= P\{I_{n-m}(j) = 1\}. \qquad \text{(definition 6)}$$

Since $I_n(m, j)$ and $I_{n-m}(j)$, for $n \geq m \geq 1$ and $j \in E$, take on only values 0 and 1, they are identically distributed as are the partial sums

$$\sum_{m=0}^{r} I_n(m, j), \quad \sum_{m=0}^{r} I_m(j).$$

Since $I_m(j) = 0$ for $m < j - 1$,

$$\sum_{m=0}^{r} I_m(j) = \sum_{m=j-1}^{r} I_m(j).$$

Now consider

$$P\{N_n(j) = k\} = P\left\{\sum_{m=1}^{n-j+1} I_n(m, j) = k\right\}$$

$$= P\left\{\sum_{m=1}^{n-j+1} I_{n-m}(j) = k\right\}$$

$$= P\left\{\sum_{m=j-1}^{n-1} I_m(j) = k\right\}$$

$$= P\left\{\sum_{m=0}^{n-1} I_m(j) = k\right\}$$

$$= P\{V_n(j) = k\}.$$

The identity of theorem 2 holds for any positive integer n, which implies theorem 3.

THEOREM 3. For $j \in E$, $N(j) = V(j)$.

The important implication of these results is that the particle counts $N_n(j)$ and $N(j)$ for the recurring random walk may be analyzed as the occupation times $V_n(j)$ and $V(j)$ for a conventional random walk. This identity facilitates analysis because the occupation time of absorbing Markov chains has been studied extensively and many of its properties are known. As one example, this relation yields the sediment concentration profile. Specifically, theorem 1 is obtained directly from a theorem in Kemeny and Snell (1960, p. 52). The same theorem verifies theorem 4.

THEOREM 4. Let $\mathscr{V} N_n(j)$ be the variance of $N_n(j)$. Then for $j \in \mathbf{E}^*$,

$$\mathscr{V} N(j) = R(1, j) [2R(j, j) - R(1, j) - 1].$$

It is worth noting that the relation between $N_n(j)$ and $V_n(j)$ is not restricted to the basic, simple model treated in this study. This relation can be effectively exploited for analysis of extensions and generalizations of the recurrent random-walk model, some of which may reflect the subtleties and intricacies of sediment transport with greater fidelity than does the basic model.

Example. The general conditions of assumption 1 are replaced by the specific values of the following assumption.

ASSUMPTION 1'. The p and q are non-negative, real numbers such that $q < 1$, $p > 0$, and $q + 2p = 1$. The matrix $\mathbf{P}$ is given by the following conditions:

$$P(1, 1) = q + p,$$

$$P(j, j) = q, \quad j = 2, 3, ..., J,$$

$$P(J + 1, J + 1) = 1,$$

$$P(j, j - 1) = p, \quad j = 2, 3, ..., J,$$

$$P(j, j + 1) = p, \quad j = 1, 2, ..., J,$$

and all other $P(i, j)$ entries in $\mathbf{P}$ are 0.

In particular, $\mathbf{P}$ has the following typical form:

	1	2	3	4		$J-1$	J	$J+1$
1	$p+q$	p	0	0	...	0	0	0
2	p	q	p	0	...	0	0	0
3	0	p	q	p	...	0	0	0
4	0	0	p	q	...	0	0	0
.	.	.	.	.	.	.	.	.
.	.	.	.	.	.	.	.	.
.	.	.	.	.	.	.	.	.
$J-1$	0	0	0	0	...	q	p	0
J	0	0	0	0	...	p	q	p
$J+1$	0	0	0	0	...	0	0	1

Since the probabilities of moving to the adjacent lower and higher valued positions are equal, it is appropriate to say that assumptions 1' and 2 define the *symmetric recurring random walk*.

Property 13. For the symmetric recurring random walk,

$$\mathscr{E} N(j) = \frac{J-j+1}{p} \quad \text{and} \quad \mathscr{V} N(j) = \frac{J-j+1}{p} \left[\frac{J-j+1}{p} - 1 \right], \quad j \in \mathbf{E}^*.$$

Proof. It is readily verified by direct calculation that $(\mathbf{I} - \mathbf{Q})\mathbf{R} = \mathbf{I}$ when

$$R(i, j) = (J - j + 1)/p, \quad i \le j,$$
$$= (J - i + 1)/p, \quad i \ge j.$$

These values, along with theorems 1 and 4, yield the expected value and variance of $N(j)$.

For this example, the variance of $N(j)$ is large relative to its expected value. Relatively large variances are evidently a characteristic feature of the recurring random-walk model so that the density concentration profile for a realization of the model may vary greatly from its expected shape. The expected shape declines linearly from J/p for position E_1 to $1/p$ for position E_J. Assume T has length J. Then it is plausible to represent the density of particles at distance x by

$$D(x) = \frac{J - x + 1/2}{p}, \quad 0 \le x \le J.$$

Justification for this representation is provided by property 14.

Property 14.

$$\int_0^J D(x) \, dx = \int_{j=1}^J \mathscr{E} N(j),$$

$$\int_{j-1}^j D(x) \, dx = \mathscr{E} N(j), \quad j \in \mathbf{E}^*.$$

This example is interesting in that it indicates that variation in the density of sediments results even when particles move in a completely random fashion. It does not imply that concentration profiles are necessarily linear decay functions. Lerman and others (1974) illustrated a transition matrix for which the shape of the concentration profile is an inverted V.

ACKNOWLEDGMENT

The support of the National Science Foundation, Grant GS-2967, is gratefully acknowledged.

REFERENCES CITED

Conover, W. J., and Matalas, N. C., 1967, A statistical model of sediment transport: U.S. Geol. Survey Prof. Paper 575-B, p. 60-61.
——1969, Statistical model of turbulence in sediment-laden streams: Am. Soc. Civil Engineers Proc., Jour. Eng. Mechanics Div., v. 95, p. 1063-1081.
Culling, W.E.H., 1963, Soil creep and the development of hillside slopes: Jour. Geology, v. 71, p. 125-137.
Heiskanen, K. I., 1972, Diffusion model of sedimentation from turbulent flow, in Merriam, D. F., ed., Mathematical models of sedimentary processes: New York, Plenum Press, p. 125-137.
Kemeny, J. G., and Snell, J. L., 1960, Finite Markov chains: Princeton, N.J., Van Nostrand, 210 p.
Lerman, A., Lal, D., and Dacey, M. F., 1974, Settling and chemical reactivity of suspended particles in natural waters, in Gibbs, R. J., ed., Suspended solids in water: New York, Plenum Press, p. 17-48.
Scheidegger, A. E., and Langbein, W. B., 1966, Probability concepts in geomorphology: U.S. Geol. Survey Prof. Paper 500-C., p. 1-14.
Strahler, A. N., 1952, Dynamic basis of geomorphology: Geol. Soc. America Bull., v. 63, p. 923-938.

Manuscript Received by the Society July 2, 1973
Revised Manuscript Received October 25, 1973

Printed in the U.S.A.

Multilayer Markov Mixing Models for Studies of Coastal Contamination

WILLIAM R. JAMES

U.S. Army Coastal Engineering Research Center
Fort Belvoir, Virginia 22060

ABSTRACT

Results of sediment tracer studies in nearshore and beach environments indicate that mixing of sediment occurs within a sediment bed through thicknesses in excess of the maximum amplitude of the bed forms. This phenomenon may be important in problems related to coastal contamination by particulate pollutants. Hazardous concentrations may be established deep within a bed where mixing occurs very slowly. This may have the effect of storing pollutant within the bed for long time periods following an initial "spill"-type accident.

A tridiagonal Markov chain is adopted as a general model for describing mixing processes between adjacent layers within a bed and across the bed-fluid interface. For this model the statistical expectation of particle residence time within a bed depends only on the sediment exchange rate between the bed and the fluid, and the thickness of the bed through which mixing is possible. However, the variance of bed residence time also depends on the number of layers considered in the model and the manner in which transition probabilities vary among the layers. Several models are explored; for certain initial conditions, the predicted duration of contamination at a particular locality may exceed that predicted by the simple one-layer Einstein model by one or two orders of magnitude.

INTRODUCTION

Aside from biological mechanisms, mixing processes within sediment beds have not received much attention from sedimentologists and hydraulic engineers. The very term "bed" implies absolute inertness of the particles within it. Historically, sediment motion has been considered in terms of movement within the confines of a cross-sectional area of a fluid, either as lateral and vertical mixing within the fluid mass or as drift in bed load and suspended load. Sediment particles in a bed have been considered, by definition, to be at absolute rest. Accordingly, the only particles susceptible to motion are those within a thin layer at the bed surface. The mechanisms for initiating motion of these particles are normally limited to hydraulic lift forces created by shear stress over the grain boundary and impact by saltating grains. Particles buried below the surface layer have not been considered available for exchange with particles in the fluid until the overlying particles are eroded. The material within a bed was considered to be in a rigid framework where intergranular boundaries are in a fixed relation. Given these concepts, the only physical mechanisms available for causing short-term vertical dispersion of particles within the bed itself is the migration of bed forms. With these mechanisms the vertical scale of mixing is controlled by the amplitude of the bed forms, and the temporal scale is controlled by the bed-form migration rates.

The above concepts provide reasonable explanations for vertical mixing of sediment tracers under conditions of uniform flow in laboratory flumes (Crickmore and Lean, 1962; Hubbell and Sayre, 1964). Consequently, one might expect these concepts to apply in natural counterparts such as rivers, estuaries, and tidal inlets. However, observations of tracer mixing within a bed in the littoral zone indicate tracer dispersion to depths far greater than can be reasonably explained by fluctuations in bed-surface elevations. An early study by King (1951) showed disturbance over a one-day interval of the positions of particles at several centimeters below the bed surface on the beach face, in the breaker zone, and just seaward of the breaker zone. These experiments also show an apparent correlation of the maximum depth of disturbance with the grain-size characteristics of the material—the coarser, more permeable material exhibiting grain disturbance at greater depth. Courtois and Monaco (1969) reported results from core samples collected during a tracer experiment in the nearshore zone. Their experiment was conducted over a time interval of many days and shows tracer dispersion to depths exceeding 20 cm below the bed surface. Komar (1969) reported vertical tracer dispersion within sediment on the beach face to depths exceeding 10 cm in a period of a few hours.

From the above studies it is not clear what role bed-form migration or longer term cut-and-fill processes played in accomplishing tracer burial. A tracer experiment conducted in the nearshore zone by the Coastal Engineering Research Center showed significant tracer dispersion to depths within the bed exceeding 14 cm over a 23-hr period (James and Duane, unpub. data). During the experiment, a bottom elevation monitor, placed near the coring site, continuously recorded bed-surface elevations to the nearest 1/4 cm. The relief of this record showed a maximum change of less than 3 cm. These results and those of the experiments described above suggest that in nearshore marine environments, sedimentary material does not always rest in a rigid structure within the bed. During the passage of waves, particle locations within the bed may be disturbed at depths far exceeding the amplitudes of the bed forms at the surface.

It is not difficult to imagine how this might arise under oscillatory flow conditions. Sleath (1970) and Putman (1949) described laboratory measurements of oscillating pressures within sand beds under wave action and derived a theory relating amplitude of the pressure and fluid velocity variations to depth in the bed, bed permeability, and wave characteristics. These studies show an approximately exponential decay of fluid velocity amplitudes dependent on depth in the bed (for thick beds) with the decay constant dependent on wave period. Martin (1970) described the effects of vertical flow through porous media as insignificant under normal wave action, but his concern was with development of incipient motion in particles at the bed surface, and he restricted his laboratory studies to uniformly graded, highly spherical sand. Madsen (1974) has developed a theory that predicts a much stronger effect accompanying the horizontal flow components within the bed when waves become very asymmetrical, as near the breakpoint. According to Madsen's calculations, bed fluidization is possible under breaking or near-breaking waves, through bed thicknesses approximately 15 percent of the wave height. The dispersion of bed particles under these circumstances would probably be of very short duration and result in a minor jostling effect rather than total failure of the bed, owing to the short time in which sufficient flow conditions exist preceding passage of a steep wave crest.

It is conceivable that wave-induced flow through a porous bed could result in minor jostling of particles and slight readjustment of grain positions deep within the bed, particularly if grain boundaries are irregular in shape and the grains are of variable size. If slight adjustments in grain positions are possible during the passage of a single wave, then mixing of particles within the bed will occur and vertical dispersion of tracer will be noticeable after a sufficient number of waves has passed.

This within-bed mixing has potentially serious implications for beach contamination by particulate pollutants. The depths within a bed to which material may disperse in hazardous concentrations is not limited by the range of bed-surface elevations. Moreover, the temporal scale of the mixing processes deep within the bed may be quite large. This could have the effect of "storing" pollutant in hazardous concentrations deep within a bed under normal conditions and releasing it to the environment or exposing it at the bed surface at a later date under conditions of severe erosion, such as storms or during seasonal profile adjustments. At the present time it is not possible to evaluate the significance of such phenomena because so little is known about the character of the mixing processes involved. However, it is possible to look at a variety of models for the mixing processes and to investigate the implications these models have for problems involving beach contamination.

If an accident occurs in which a pollutant is suddenly introduced into a nearshore marine environment in ecologically hazardous concentrations, wave and current action will usually transport this material along the shoreline in a fairly confined band relative to the width of the surf zone (Inman and Brush, 1973). If this pollutant is chemically, physically, and biologically stable so that the only mechanisms available for reducing its concentration are the hydrodynamic mixing processes present in its environment and it consists of, or is securely attached to, negatively buoyant particles so that its behavior is similar to sediment naturally present in the environment, then under normal conditions, this material will move quickly along the shoreline borne by longshore currents, mixing (exchanging positions) with bed material as it moves. The process of exchange between bed material and material in transit will cause a depletion of the initial concentration of the material in transit, gradually distributing the pollutant over a length of beach and ultimately reducing the concentrations of material in transit to nonhazardous levels. In addition, subsequent dispersion of material within the bed, exchange of material between bed and fluid, and offshore motion of material in rip currents will ultimately reduce pollutant concentrations below the hazardous level everywhere. The length of beach initially contaminated and the duration of contamination at specific locations obviously depend on the amount of pollutant initially released and on the spatial and temporal scales of the advective and diffusive processes present in the environment.

This paper presents various models for describing the vertical mixing of particles within a sediment bed and investigates the implications of these models for the period of time that might be required for natural dispersion processes to reduce concentrations of pollutants below hazardous levels. The models are developed within the general framework of finite Markov processes. The choice of a stochastic framework, as opposed to a deterministic one, is primarily for convenience (as suggested by Krumbein, 1967). Although all results presented here could have been derived from strictly deterministic considerations, the stochastic approach is conceptually simpler because it tends to unify Lagrangian and Eulerian frames of reference. The location of an individual particle and the duration of time a particle spends in a particular location may be considered random variables that are essentially Lagrangian in nature. By integrating (or summing in the finite case) the probabilities of occurrence of specific events for a large number of particles, one obtains the Eulerian counterpart in the form of vertical distribution of concentration as a function of time.

MULTILAYER MIXING MODELS AND MARKOV CHAINS

Einstein (1937) adopted a simple stochastic process model to describe interaction between material traveling as bed load in streams and material at rest in a sediment bed, which was treated as a static layer of uniform thickness. In this model, a particle in the bed layer at time t has a fixed probability of being eroded and entering the fluid bed-load layer during a subsequent time interval of duration dt. In order to maintain a mass balance, the probability for a particle entering the bed from the bed-load layer in a unit surface area must equal the probability for a particle exiting the bed into the bed-load layer through this surface area during the same time increment. Under these conditions erosion of particles from a bed is a simple Poisson process (described by Feller, 1957, p. 400), and the distribution of particle rest periods within the bed is exponential. Einstein showed that if exponential hop lengths are assumed for the distribution of distances of particle transit in the bed load before redeposition in the bed, and the average time a particle spends in transit is negligible compared to the average time it spends at rest in the bed, then the Bessel probability density function describes the temporal and lateral distribution of particles following initial deposition. A recent description of the properties of the Bessel density function can be found in Feller (1966, p. 59).

The Einstein model is a convenient starting point for the problems considered in this paper. In his model, particle motion out of the bed layer in the downward direction is prohibited. This simplifies the mathematical problem because a reasonably tractable analytic solution is available if one has the capabilities for rapid evaluation of Bessel functions. However, it complicates the conceptual problem because the sediment-exchange rate between the bed and the fluid layers cannot be separated from the total sediment-transport rate. If the initial concentration of pollutant is very high, then the concentration level, which can be defined as "safe," is much lower. In the Einstein model initial passage of the pollutant in the fluid layer cannot be considered independently from pollutant that has entered the bed at least once and been rapidly reintroduced into the environment. The rise and fall of pollutant concentration in the bed layer closely accompanies the rise and fall of pollutant concentration in the overlying fluid. This prediction arises from the absence of possibilities for sediment mixing downward into substrata less available to immediate erosion following initial deposition.

This paper is not concerned with time scales involved in the initial passage of a pollutant cloud over a region; rather, it is an attempt to identify problems that might arise long after the initial damaging effects of the accident have ceased. In the nearshore environment, where very slow long-term mixing may occur within the bed itself, contributions of pollutant to the bed following the initial passage of the contaminated fluid layers can probably be neglected. Hence, all considerations in this paper proceed from an initial condition in which a specific concentration of pollutant is "instantaneously" introduced into the surface bed layer, and only negligible quantities of pollutant enter the bed from the fluid subsequent to this initial pulse.

Under these simplified conditions, the one-layer model predicts an exponential decay in concentration at a particular site in a bed following establishment of the initial concentration. If c_s represents a limit below which concentrations may be considered nonhazardous, and c_0 represents the initial concentration, then the time required for vertical dispersion to reduce concentrations to safe levels is proportional to the logarithm of the ratio of the initial concentration to the safety limit,

$$t_s = \lambda \log(c_0/c_s), \tag{1}$$

where λ is the expected value of particle residence time within the bed.

An obvious extension of this model to account for mixing processes within the bed itself is to allow passage downward from the surface bed layer into one or more underlying layers. Crickmore and Lean (1962) considered two- and three-layer bed models to explain retardation of tracer transport greater than that predicted by the Einstein model. Their experiments, however, dealt with uniform flow conditions, and they did not consider the possibility of mixing within the bed itself. They subdivided the bed within the range spanned by the amplitudes of the bed forms and were concerned with the statistical distribution of times between re-exposures of a given elevation to fluid action, given a random configuration of the migrating bed forms.

Extension of Einstein's model in this paper involves exchange of particles within the bed itself among layers below the surface layer. Such models can be conveniently explored using finite Markov chains. A general description of the connection between Markov chains, random walks, and diffusion processes can be found in Feller (1957). Krumbein and Dacey (1969) described a variety of geological applications.

The general approach adopted in this paper is as follows. The bed is considered a vertical succession of a finite number of horizontal layers of equal thickness. During the passage of a wave surge, a particle in a specific layer may remain in that layer or may exchange positions with particles in an adjacent layer. The probability for a particle moving through two layer boundaries in the same time increment is assumed to be negligibly small. In order to conserve mass within layers, the probability for a particle moving downward across a layer boundary must equal the probability for a particle in the layer below to move upward through the same boundary in the same time increment. In the bottom layer, passage through the lower boundary is not permitted; in random-walk terminology, this boundary represents a reflecting barrier.

In contrast, particles exiting the surface layer into the fluid layer are removed from the system and replaced with new particles deposited after transport from an updrift source. The initial condition considered in this paper is that immediately following passage of the pollutant-laden cloud in the overlying fluid, the subsequent contribution of pollutant particles to the bed at any specific locality can be considered negligibly small. Hence, for the purposes of this study, the bed-fluid interface can be considered mathematically equivalent to an absorbing barrier.

The general system can be described in terms of a vector of transition probabilities. Each probability is associated with particle transition through the upper boundary of each layer during the passage of a wave surge with period Δt. If the transition probabilities do not change with time, the system can be classified as a Markov chain (Feller, 1957, p. 340). The vector of particle transition probabilities can be used to construct a tridiagonal transition-probability matrix as follows. Let p_i represent the probability that a particle in the ith layer at time t makes a transition downward through the lower boundary of the ith layer in the subsequent time interval of duration Δt. As discussed above, p_i must equal the probability that a particle in layer $i + 1$ makes a transition upward into the ith layer during the time interval Δt. Inasmuch as transitions through more than one layer boundary are not permitted in a single time increment, the probability associated with a particle remaining in the ith layer (not passing through any layer boundaries) is $1 - p_i - p_{i-1}$. Furthermore, because downward passage through the lowermost layer boundary is not permitted, the probability for remaining in the lowest layer is $1 - p_{k-1}$ for a k-layer model. In contrast, because the overlying fluid is an absorbing barrier for the system, the probability for particle return from the fluid layer to the surface bed layer at the same locality is zero. Table 1 shows the structure of the transition-probability matrix, which is a common means of representing Markov chains, for a four-layer model. Various mathematical properties associated with transition matrices were described by Feller (1957, Chaps. XV and XVI), and computational procedures that have been used in other geological studies were given by Krumbein (1967).

This general Markov model may be used to explore the implications of a variety of more specific mixing models that result from specification of the number of layers and various ways in which the transition probabilities vary with depth in the bed.

SPECIFIC MODELS

Little is known about wave-induced mixing processes within a sediment bed. Three different possibilities are considered in this paper with regard to the variation of transition probabilities with depth in the bed. For each of these possibilities, several different choices for the number of layers are explored. The simplest model for transition probabilities is that they are all equal so that no gradation of mixing properties with depth in the bed in the mixing zone exists. This model, the "constant-p model," is probably unrealistic because of the known decrease in the magnitude of pressure and fluid velocity fields with depth in a porous bed. It is included, however, for comparative purposes. The simplest model that admits decreasing

TABLE 1. STRUCTURE OF THE TRIDIAGONAL PROBABILITY TRANSITION MATRIX FOR A FOUR-LAYER MODEL

		Absorbing state	Bed layer 1	Bed layer 2	3	4
	Absorbing state	1	0	0	0	0
Bed layer	1	p_0	$1 - p_0 - p_1$	p_1	0	0
	2	0	p_1	$1 - p_1 - p_2$	p_2	0
	3	0	0	p_2	$1 - p_2 - p_3$	p_3
	4	0	0	0	p_3	$1 - p_3$

magnitude of mixing processes with depth in the bed, the "linear-p model," is one in which transition probabilities decrease linearly from the bed-surface boundary to a zero value at the lower boundary of the bottom layer. Finally, a convex decrease in transition probabilities is considered in the form of a quadratic relation in which the transition probability for crossing the lower boundary from the bottom layer is defined as zero (as required by the general model), and transition probabilities increase as the square of distance above this point toward the bed-surface boundary; this is the "quadratic-p model." For a k-layer model, the transition probabilities associated with particle passage downward from the ith layer to layer $i + 1$ can be expressed as

$$p_i = p_0 \left(\frac{k - i}{k}\right)^\alpha, \qquad (2)$$

where p_0 is the probability that a particle in the surface layer exits the system by passing into the overlying fluid layers from the surface layer; k is the number of bed layers considered in the model; and α can be either 0, 1, or 2 corresponding to the constant, linear, or quadratic models, respectively. These transition probabilities may be used to generate the full transition-probability matrix as described in the previous section. Hence, the system may be fully defined by specification of the number of layers (k), the value of α (0, 1, or 2), the system exit probability (p_0), and the initial concentration of pollutant in the surface bed layer (c_0).

The number of layers considered and the value selected for α are the important parameters for distinguishing among implications of the various models, whereas the value of p_0 controls the time scale for all of the models. In this paper, no reference is made to absolute time scales because these will vary among different localities and within single localities depending upon wave climate, water depth, bulk-sediment properties, and other factors. It is therefore convenient to adopt a dimensionless time scale for the purpose of comparing alternative models. A reasonable choice for this parameter is related to a variable defined by Einstein (1937, 1950) as the sediment exchange rate (ϵ), which is the mass of sediment exchanging positions between a bed and the fluid per unit surface area of the bed, per unit time. If the bed layers are of thickness Δz and the bulk density of bed sediment is ρ_b, then the sediment exchange rate is related to the parameters of the models described in this paper by

$$\epsilon = p_0 \left(\frac{\rho_b \Delta z}{\Delta t}\right). \qquad (3)$$

If it is assumed that details regarding the form of the diminution of mixing processes within the bed below the surface layer do not affect the rate at which sediment is exchanged between the bed and the fluid, then all models can be investigated with reference to a fixed exchange rate. If the dimensions Δz and Δt are considered constant for the system within which alternative models are being explored, then the appropriate dimensionless time scale for comparing models is given by

$$\tau = t/(\Delta t/p_0), \qquad (4)$$

where t is total elapsed time, Δt is the average wave period, and p_0 is the transition probability for a particle in the surface layer to enter the fluid. The connection of this time scale with particle residence time distributions is discussed below, and it is shown that for the Einstein model, the expected residence time for a particle to remain in the bed is unity on this time scale. The dimensionless time variate selected can therefore be regarded, for a one-layer-bed model, as the ratio of absolute elapsed time to the expected particle residence time in the bed.

RESIDENCE TIME DISTRIBUTIONS

For the single-layer Einstein model, a particle in a bed layer can exit the bed after a time increment Δt with probability p_0, or it can remain in the bed with probability $1 - p_0$. Particle exit is possible after each time increment, and because p_0 does not change with time, the probability that a particle remains in the bed for exactly $j - 1$ time increments and exits after the jth time increment is given by

$$P_j = (1 - p_0)^{j-1} p_0. \tag{5}$$

This is the well-known geometric distribution, with an expected value of $1/p_0$ and variance $1/p_0^2 - 1/p_0$. The geometric distribution is the discrete analogue to the continuous exponential distribution used by Einstein. The expected value described above is for the number of time increments a particle rests in the bed before its exit. To obtain an actual time dimension this must be multiplied by the time increment Δt. This product, $\Delta t/p_0$, appears as the denominator in equation (4), which defines the dimensionless time scale τ, as discussed above. On the dimensionless time scale, therefore, the expected bed residence time for a one-layer model is unity.

When two bed layers are considered, a particle originally in the surface layer may exit either upward or downward with relative probabilites $p_0/(p_1 + p_0)$ and $p_1/(p_0 + p_1)$, respectively. If it exits downward, it is not lost from the bed but spends a period of residence or "sojourn" in the lower layer, after which it returns to the surface layer where it has an additional sojourn before it may exit the bed. Its next transition may again be either upward or downward. It is readily seen that the particle may have an indefinite number of transitions between layers, as well as having indefinite sojourns within each layer before transition. Although the particle's sojourns within individual layers have a geometric distribution, the total bed-residence-time distribution is exceedingly complex, and no tractable mathematical representation has been found for it. However, it is possible to arrive at simple expressions for the expected value and variance of bed residence times for the general case of the k-layer model. These variables are useful in arriving at generalities regarding the comparative behavior of particles in the specific models described in this paper and elsewhere. Derivations of these relations are in the appendix.

Expected Values of Bed Residence Times

Surprisingly, the expected number of time increments a particle spends within a bed in a k-layer model is simply k/p_0. In terms of the dimensionless time scale (τ), this expectation is

$$E_k(\tau) = k. \tag{6}$$

The expected bed residence time does not depend at all upon the manner in which transition probabilities, and hence mixing processes, vary with depth in the bed. The expected value of bed residence time is the same for constant-, linear-, and quadratic-p models, or any other tridiagonal model that contains the same number of layers. If sediment mixing is at all possible between the surface layer and underlying layers, the expected residence time will be greatly increased regardless of the probability for transition into lower layers. If a value, Δz, is assigned to layer thickness and the value $D = k \Delta z$, is assigned to total bed thickness, then equation (6) implies that the average particle residence time within the bed is $(\rho_b D)/\epsilon$. If mixing can occur at indefinite depths within the bed, then regardless of the temporal or length scales of the processes involved, the average bed residence time for particles will also be indefinite. This has serious implications for problems of beach contamination, because it is not yet possible to measure an absolute limit to the depth at which any mixing can occur. It is not possible to say therefore what the average particle residence time at a specific locality will be.

Variance of Bed Residence Times

The general expression for the variance of the number of time increments (r) a particle spends in a bed for the k-layer model is given below:

$$\mathrm{Var}_k(r) = \left(\frac{k^2}{p_0^2} - \frac{k}{p_0}\right) + \frac{2}{p_0} \sum_{j=1}^{k-1} \frac{(k-j)^2}{p_j}. \tag{7}$$

A derivation of this expression is presented in the appendix.

The first term in this expression is similar to the expression of variance for the geometric distribution, with the exception that the number of layers (k) appears in the numerator. The first term does not contain transition probabilities for any of the lower layer boundaries, although the second term contains all of them. Unlike the expected value, the variance of bed residence time depends on the manner in which transition probabilities change with depth in the bed as well as on the number of layers.

The differences in variances predicted by the three models for transition probabilities are found by substituting equation (3) into equation (7) and using values 0, 1, and 2 for the exponent of α. Neglecting insignificant terms and transforming the results to the dimensionless time scale yields the three expressions:

$$\mathrm{Var}_k(\tau) = 2k^3/3 + k/3 \quad \text{constant-}p \text{ model,}$$

$$\mathrm{Var}_k(\tau) = k^3 \quad \text{linear-}p \text{ model,} \tag{8}$$

and

$$\mathrm{Var}_k(\tau) = 2k^3 - k^2 \quad \text{quadratic-}p \text{ model.}$$

Surprisingly, these expressions show that for all three models, the variance of bed residence times rises as the cube of the number of layers. This is significant because the variance, as opposed to the expected value, is a much more significant indicator of the distribution of probabilities for the long-duration residence times in the "tail" of the residence-time distributions. Assuming a large number of layers, the model used for transition probabilities results in constant factors of 2, 1, and

$^2\!/_3$ for the residence-time variances of the quadratic, linear, and constant-p models, respectively. These results do not depend in any way upon the dimensions of the system. However, if a transform is made to fixed system dimensions, in a fashion similar to that made in the discussion of expected values of bed residence time, the number of layers considered no longer vanishes as a parameter.

Again consider some fixed thickness, D, through which mixing is possible, and let the individual layer thickness, Δz, be defined as D/k. Using equations (4) and (5), the resulting expressions for variance of the residence-time distributions become

$$\text{Var}_k(t) = (D\rho_b/\epsilon)^2 2k/3 + \tfrac{1}{3}k \qquad \text{constant-}p \text{ model,}$$

$$\text{Var}_k(t) = (D\rho_b/\epsilon)^2 k \qquad \text{linear-}p \text{ model,} \tag{9}$$

and

$$\text{Var}_k(t) = (D\rho_b/\epsilon)^2 (2k - 1) \qquad \text{quadratic-}p \text{ model.}$$

From these expressions it is apparent that k is the most important parameter in the model, which implies that for almost any application of the general model, the choice of the number of layers considered will have the most significant influence on the accuracy of the prediction of long-term effects.

The effects of choice of parameters are indicated more clearly in Figure 1, in which each curve represents a trace of the percentage of original pollutant that remains in the bed (has not yet exited through the bed-fluid interface) after the mixing process has operated for time τ after initial contamination of the surface bed layer. Both vertical and horizontal axes are logarithmic scales; each individual

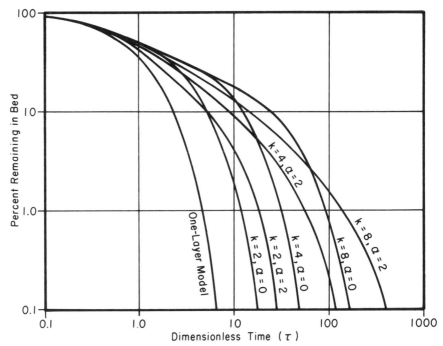

Figure 1. Depletion curves for various models.

curve represents the predicted depletion of pollutant for a specific choice of the parameters α and k. The initial rate of depletion is about the same for all of the models considered. The differences become apparent only after $\tau = 1$, at which time the one-layer model predicts a continuing rapid depletion and diverges greatly from the other models. The depletion curves for all of the multilayer models exhibit a region of linearity on this log-log scale, following their divergence from the depletion curve for the one-layer model. This linear region indicates the span of time during which the location of concentration maximums within the bed is no longer in the bed surface layer and is progressively moving downward from layer to layer. This linear region is generally longer and has a gentler slope for models with a large number of layers. The effect of α can also be seen here. In all cases shown, the quadratic-p model exhibits a steeper linear region than the constant-p model, which indicates that during the transfer of the concentration maximum from the surface layer to the bottom layer, the actual rate of loss of pollutant from the bed is more rapid for the quadratic-p model than for the constant-p model. However, the linear region in all quadratic-p cases is much longer in duration than in the corresponding constant-p model, and the curvature of the quadratic-p depletion curves in the end regions is less extreme where the concentration maximum is in the lowest layer. Thus, the curves always cross, and the slowest depletion rates ultimately become associated with the quadratic-p model.

DURATION OF CONTAMINATION

If at a specific locality, a polluting accident establishes in a bed surface layer an initial concentration, c_0, of pollutant that exceeds some tolerable concentration limit, c_s, by definition the bed will be contaminated as long as the concentration anywhere within the bed remains above the value c_s. Inasmuch as the maximum concentration within the bed decreases monotonically with time, the duration of the period of contamination will be the amount of time intervening between the initial establishment of the concentration c_0 and the time when the maximum concentration within the bed falls below c_s. The duration of contamination obviously depends upon the ratio c_0/c_s, the hazard ratio, and on the temporal scale of the mixing processes within the bed.

The effects of selection among the various models on the duration of contamination may be investigated in the framework of the dimensionless time scale defined previously. If τ_s represents the dimensionless duration of contamination, its value will depend upon the hazard ratio and the model used for prediction. Figure 2 shows graphs on a logarithmic scale of τ_s versus c_0/c_s for four models, the constant and quadratic-p models for both two- and eight-layer beds. The corresponding curve for the one-layer model is also plotted for comparative purposes.

Figure 2 shows that there is little discrimination among models for low values of the hazard ratio. This corresponds to the similar initial behaviors observed in the depletion curves, as discussed above. Thus, for low initial concentrations of pollutant in the surface layer, the mixing processes within the bed have little substantial effect on the duration of contamination. In fact, for these low initial concentrations, mixing processes reduce the duration of contamination because mixing downward from the surface bed layer accelerates the initial depletion of concentration from this layer. For hazard ratios exceeding ten, the one-layer model shows rapid divergence from all of the multilayer models, predicting much shorter contamination periods than the other models.

All of the multilayer models exhibit an irregular "scalloped" feature in the curves that begins at about $\tau_s = 1$. This reflects the period in which a downward transition of concentration maximums occurs through the layers. For the two-layer models there is only one such transition, and this is reflected by a single first-order discontinuity in the curves for these models. Among the two-layer models, the constant-p model exhibits this transition first because it has a higher rate of dispersion into the lower layer. However, the duration of contamination for hazard ratios exceeding ten is longer for the two-layer quadratic-p model.

For the eight-layer model, transition of the maximums through the individual layers gives to the curves an irregular shape that occurs over a broad range of the dimensionless time scale. Surprisingly, the two eight-layer models behave quite similarly for hazard ratios less than 100, even though the transition of maximal concentration into the eighth layer for the constant-p model occurs at a hazard ratio near 40, and the same transition does not occur for the quadratic-p model for hazard ratios less than 300. During this transition period both curves are remarkably linear on this log-log scale, suggesting a general power-law approximation for the transition phase. For hazard ratios exceeding 300, the two eight-layer models indicate rapid divergence in the duration of contamination.

Finally, Figure 2 also shows that consideration of more than one layer for the bed has profound effects on the duration of contamination for hazard ratios exceeding 100. In the most extreme case plotted, with a hazard ratio of 1,000, the eight-layer quadratic-p model predicts a duration of contamination 314 times greater than the expected residence time for a one-layer model with the same sediment-exchange rate. For the same hazard ratio, the one-layer model predicts a duration of 6.9 time units, a factor of 45 less than the eight-layer model. The eight-layer models predict durations greater than ten times the corresponding prediction for the one-layer

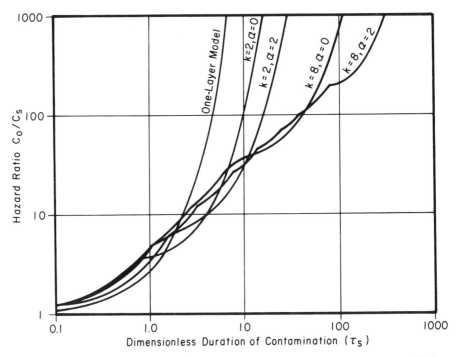

Figure 2. Duration of the period of contamination as a function of the hazard ratio considering two- and eight-bed layers, for both constant- and quadratic-p models.

Einstein model for hazard ratios exceeding 100. These curves also display a consistent trend toward increasing values of this duration ratio for hazard ratios in excess of those plotted.

It is difficult to assess the significance of the above results in a practical framework because typical numerical values for sediment exchange rates and particle bed residence times are not available. However, a rough approximation of a typical value of average bed residence time can be obtained utilizing some of Komar's (1969, p. 53) data on sand advection rates determined from tracer experiments in the surf zone in southern California. If the amount of time a particle spends in transit is small compared to the amount of time it rests in the bed, then the average velocity of the particle is closely approximated by the ratio of its average alongshore hop length to the average bed residence time. Komar's data indicate average surf-zone particle advection rates ranging between 0.03 and 0.65 cm/sec under differing surf conditions. Assuming that the duration of a single particle hop is approximately equivalent to a typical wave period of 12 sec, and that while in motion the particle moves with an alongshore velocity of 30 cm/sec, then an approximate average hop length is 360 cm. This implies values of average bed residence times that range between 10 min and $3^1/_2$ hr.

The one-layer Einstein model predicts that following an initial accident that establishes a hazard ratio of 1,000, the bed will be contaminated for a period ranging from approximately 70 min to 24 hr. If the same average bed residence times are applied to an eight-layer quadratic-p model, the corresponding predicted duration of contamination ranges from 32 hr to 45 days. The difference in the implications of these two predictions for the ecology of the environment could be highly significant.

CONCLUSION

Very little physical theory is available that can be used to describe sediment mixing processes within sedimentary beds. However, limited data are available that strongly suggest such processes do occur, at least in nearshore coastal environments. The tridiagonal Markov chain represents a convenient general framework for exploring a variety of models for vertical mixing that might apply to this process. Analysis of these models indicates some unexpected implications, particularly for the duration of contamination within a bed following a "spill"-type accident. If mixing processes occur deep within a bed, even on an apparently insignificant scale, some surface particles can work their way down to these depths. Given sufficiently high initial concentrations in the surface layer, hazardous concentrations of pollutants can therefore occur at any depth where mixing is possible.

The expected value of bed residence time is independent of the number of layers in the model or the manner in which transition probabilities vary from layer to layer. It depends only on the sediment-exchange rate between the fluid and the bed and on the total bed thickness through which any mixing occurs. However, the variance of bed residence time depends upon both the number of layers considered and the manner in which transition probabilities vary among the layers. For many purposes, particularly those that only involve use of the average bed residence time, it is not necessary therefore to have detailed knowledge of the mixing processes within the bed itself. For long term, low probability events, however, these details become increasingly important.

The results of this study present an important problem: In the absence of physically identifiable boundaries within a bed, there is no obvious way to choose an appropriate number of layers to consider in the model, although for many applications this choice is important. The analysis in this paper indicates that this problem probably cannot be solved by standard sediment-tracer techniques. Differences in predictions based on the various models occur in the "tails" of the residence-time distributions when tracer concentrations will normally be quite low and change slowly. Hence, the sampling and measurement problems encountered in standard work with sediment tracers will be compounded in studies related to this problem. More sensitive experimental tracer techniques must be developed before this question can be resolved.

The inadequacy of available sediment-tracer techniques reinforces the idea described by Krumbein and Dacey (1969, p. 94) that the development of mathematical models for geological processes is often sequential, with alternation between data collection and model building. Examination of the initial model will often point out inadequacies in the observations rather than support or reject the model. In this study limited observations suggest that mixing does occur within the nearshore sediment bed. Development of various models for the mixing processes suggests that important differences exist in the implications of the various models for problems involving bed contamination. New measurement techniques must be developed to provide the kind of data needed for discrimination among the models.

ACKNOWLEDGMENTS

Many of the ideas expressed in this paper arose from observations of sediment-tracer behavior during several field experiments in the littoral zone of southern California. D. B. Duane, E. Acree, and H. Brashear were chiefly responsible for designing and conducting these experiments, and they provided many insights and stimulating discussions of problems related to tracer burial. J. H. Balsillie assisted in preparing the figures. The research was conducted jointly by the U.S. Army Coastal Engineering Research Center under the Civil Works Research and Development program of the U.S. Army Corps of Engineers and the Atomic Energy Commission through the Oak Ridge National Laboratories. Permission of the Chief of Engineers to publish this information is appreciated. The findings of this paper are not to be construed as official Department of the Army position unless so designated by other authorized documents.

APPENDIX 1. MOMENTS OF BED RESIDENCE TIME DISTRIBUTIONS

Given a multilayer-bed model, consider the surface layer distinct from the set of lower layers. During time Δt, a particle in the surface layer can either exit the system (with probability p_0), enter the lower set of layers (with probability p_1), or remain in the surface layer with probability $1 - p_0 - p_1$. Thus, the distribution of residence time for a single sojourn within the surface layer is geometric, with expected value and variance given by

$$E_1(j) = 1/(p_0 + p_1),$$
$$\mathrm{Var}_1(j) = [1/(p_0 + p_1)]^2 - [1/(p_0 + p_1)], \tag{A1}$$

where j is the number of time increments constituting a single sojourn within a layer.

When a particle leaves the surface layer, the conditional probability for system exit, as opposed to entry into the set of lower layers, is $p_0/(p_0 + p_1)$. The conditional probability for entering the lower layers is $p_1/(p_0 + p_1)$, so that the sum of these two conditional probabilities is unity. Inasmuch as the particle must pass through the surface layer before leaving the system, the number of residence times (n) a particle spends in the surface layer prior to exit from the system is also geometrically distributed, according to

$$v_n = p_1^{n-1} p_0/(p_1 + p_0)^n, \quad n = 1, 2, \ldots, \infty. \tag{A2}$$

It also follows that the number of times a particle rests in the set of layers below the surface layer is one less than the number of times it spends within the surface layer. Separate layer-sojourn times are independent, random variables, and the total residence time is their sum. Thus, the conditional expected value and variance of residence time, given the number of separate sojourns in the surface layer, obey the relations

$$E_k(j|n) = n E_1(j) + (n - 1) E_{k-1}^*(j), \tag{A3a}$$

$$\text{Var}_k(j|n) = n \text{Var}_1(j) + (n - 1) \text{Var}_{k-1}^*(j), \tag{A3b}$$

where the notation $|n$ refers to values conditional on the number of sojourns in the surface layer, and the star refers to values for the set of $k - 1$ layers below the surface layer.

The total expected value of residence time is found from

$$E_k(j) = \sum_{n=1}^{\infty} E_k(j|n) v_n. \tag{A4}$$

Combining equations (A1) through (A4) yields the expression

$$E_k(j) = (1/p_0) + (p_1/p_0) E_{k-1}^*(j). \tag{A5}$$

The above relation may be used for an inductive proof of the general relation $E_k(j) = k/p_0$, as shown below.

Consider the effect of adding a new surface layer to the k-layer system so that subscripts on all the transition probabilities are increased by one and a new value appears for p_0. Adopting the proposed expression for expected residence time within the set of k lower layers implies $E_k^*(j) = k/p_1$. Substituting this relation into equation (A5) yields the result $E_{k+1}(j) = (k + 1)/p_0$, which proves the general relation for expected value of bed residence time.

The derivation of the general expression for the variance of bed residence time is not quite so straightforward. From the definition of variance, it follows that

$$E_k(j^2|n) = \text{Var}_k(j|n) + E^2(j|n). \tag{A6}$$

Using the previously derived relation for expected value in equation (A3), it follows that

$$E_k(j|n) = \frac{n}{p_1 + p_0} + \frac{(n-1)(k-1)}{p_1}. \tag{A7}$$

As in equation (A4) it also holds that

$$E_k(j^2) = \sum_{n=1}^{\infty} E_k(j^2|n) v_n. \tag{A8}$$

Finally, the variance of residence time is defined by

$$\text{Var}_k(j) = E_k(j^2) - E_k^2(j), \tag{A9}$$

where the last term is already known to be k^2/p_0^2. Then, using equations (A3b), (A6), (A7), and (A8) in equation (A9), the end result is the recursive expression

$$\text{Var}_k(j) = \frac{k^2}{p_0^2} - \frac{1}{p_0} + \frac{(k-1)^2}{p_0 p_1} + \frac{p_1}{p_0} \text{Var}^*_{k-1}(j). \tag{A10}$$

Mathematical induction may again be used to prove that this recursive relation is satisfied by

$$\text{Var}_k(j) = \frac{k^2}{p_0^2} - \frac{k}{p_0} + \frac{2}{p_0} \sum_{j=1}^{k-1} \frac{(k-j)^2}{p_j}. \tag{A11}$$

REFERENCES CITED

Courtois, G., and Monaco, A., 1969, Radioactive methods for the quantitative determination of coastal drift rate: Marine Geology, v. 7, p. 183-206.

Crickmore, M. J., and Lean, G. H., 1962, The measurement of sand transport by the time integration method with radioactive tracers: Royal Soc. [London] Proc., ser. A., v. 270, p. 27-47.

Einstein, H. A., 1937, Der Geschiebetrieb als Wahrscheinlichkeitsproblem: Zurich, Verlag Rascher and Co.

——1950, The bed-load function for sediment transportation in open channel flows: U.S. Dept. Agriculture Tech. Bull., no. 1026, 71 p.

Feller, W., 1957, An introduction to probability theory and its applications, Vol. 1: New York, John Wiley & Sons, Inc., 461 p.

——1966, An introduction to probability theory and its applications, Vol. 2: New York, John Wiley & Sons, Inc., 626 p.

Hubbell, D. W., and Sayre, W. W., 1964, Sand transport studies with radioactive tracers: Am. Soc. Civil Engineers Proc., Jour. Hydraulics Div., v. 90, no. HY3, p. 39-68.

Inman, D. L., and Brush, B. M., 1973, The coastal challenge: Science, v. 181, p. 20-32.

King, C. A. M., 1951, Depth of disturbance of sand on sea beaches by waves: Jour. Sed. Petrology, v. 21, p. 131-140.

Komar, P. D., 1969, The longshore transport of sand on beaches [Ph.D. dissert.]: San Diego, Univ. California, San Diego.

Krumbein, W. C., 1967, FORTRAN IV computer programs for Markov chain experiments in geology: Kansas Geol. Survey Computer Contr. 13, 38 p.

Krumbein, W. C., and Dacey, M. F., 1969, Markov chains and embedded Markov chains in geology: Jour. Internat. Assoc. Math. Geol., v. 1, p. 79-96.

Madsen, O. S., 1974, The stability of a sand bed under the action of breaking waves: MIT R. M. Parsons Lab. Tech. Rept., no. 182, 75 p.

Martin, C. S., 1970, Effects of a porous sand bed on incipient sediment motion: Water Resources Research, v. 6, no. 4, p. 1162-1174.

Putman, J. A., 1949, Loss of wave energy due to percolation in a permeable sea bottom: Am. Geophys. Union Trans., v. 30, no. 3, p. 349-356.

Sleath, J. F. A., 1970, Wave-induced pressures in beds of sand: Am. Soc. Civil Engineers Proc., Jour. Hydraulics Div., v. 96, no. HY2, p. 367-378.

MANUSCRIPT RECEIVED BY THE SOCIETY SEPTEMBER 12, 1973
REVISED MANUSCRIPT RECEIVED JANUARY 21, 1974

Influence of Grain Shape on the Selective Sorting of Sand by Waves

DAVID POCHÉ
AND
H. G. GOODELL

Department of Environmental Sciences
University of Virginia
Charlottesville, Virginia 22903

ABSTRACT

Nearshore-zone quartz sands were examined to determine the effect of grain shape upon their transport and sorting. A grain-shape index was used to determine the mean Corey shape factor (SF) of each sample in a traverse that extended seaward from a Florida beach. The grand mean shape factor shows that the previously assumed value of 0.70 is slightly too high and that a better estimate is 0.65.

Calculation of two sediment-movement-threshold formulations indicates that the critical value had not been reached for the data set studied. Field observations indicate otherwise. It is suggested that for grains of lower-than-average shape factor (SF < 0.65), the movement threshold is considerably lower than that predicted by either formulation. This discrepancy could be the result of the subjective nature of the critical threshold value or a failure of the swing cradle and wave tank to duplicate fully the natural condition.

INTRODUCTION

Selective transport and sorting of sand grains within the nearshore zone result from the interplay of wave-induced turbulence and grain size, shape, and density. Wave-induced turbulence generates complex velocity fluctuations which are responsible for grain transport. Grains with the lowest settling velocity are transported farthest and include those of low mass and sphericity. These grain parameters respond to progressive variations in the wave-induced forces across the nearshore zone and result in predictable distributional patterns. Most sand studies have considered only the effects of grain size; grain shape has been largely neglected because of the difficulties encountered in its quantification.

Grains in transport consist of suspension and traction populations of various sizes and shapes. A systematic study of the variation in these populations may be approached by (1) assuming a constant grain shape and allowing size to vary (Ippen and Eagleson in Ippen, 1966), or by (2) assuming a constant size and allowing grain shape to vary. The distribution of grain shape in the nearshore zone has not been previously determined, and little data exist on the mechanics of movement of nonspherical grains.

Systematic variation in size and size sorting across the beach and nearshore zone has been observed by several investigators. Inman (1955) pointed out that these textural lineations occur because the process responsible for them also parallels the beach. It would seem reasonable that shape properties of sediment also have a similar alignment as a result of the same processes.

This paper is concerned with (1) the determination of the occurrence of grain shapes within the nearshore zone and (2) the relation between nearshore processes and the variation and distribution in these populations. Two methods were employed to delineate the variation of grain shape based upon triaxial measurements: (1) the Corey shape factor and (2) the Zingg shape classification. Both are based on axial ratios.

In order to isolate the effect of shape from the effect of size, an area was selected where the offshore grain-size variation is minimal. Studies of the beach and offshore area of the Florida gulf coast between Panama City and Pensacola have documented that the sediments exhibit strikingly uniform size ranges (Goodell and Poché, 1968; Hopkins and others, 1969; Gorsline, 1964, 1966). Grain-density differences are also minimal because the mineralogy of the sediment of this region is predominantly quartz, with only 1 to 2 percent accessory minerals and shell fragments. Two offshore bars are commonly present: the outer one (150 to 180 m offshore) is continuous with and parallel to the coast; the inner bar (30 to 60 m offshore) is often sinuous and is frequently interrupted by rip channels. Although both bars act as "wave filters," the inner bar is commonly the plunge point. The sample set used for this study was taken from a 213-m pier located at Crystal Beach near Destin, Florida (Fig. 1).

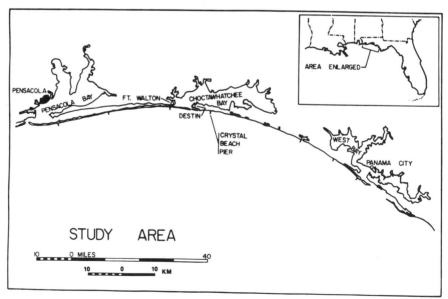

Figure 1. Map showing area of study.

THEORY

The probability of entrainment or capture of sand grains by wave-induced flow is proportional to the ratio of their settling velocity to the turbulent uplift. The forces acting on any grain in the presence of turbulence are those of gravity (weight, downward; bouyancy, upward), turbulent uplift, and drag.

The effect of shape on settling velocity and drag is usually approximated by either assuming that all of the grains in transport have identical shapes (spheres) or by assuming that the individual shape effects may be compensated by a coefficient of resistance to flow, called the coefficient of drag (C_d). The problem arises in the accurate determination of the lift and drag forces on the grains in the presence of complex turbulent flow. Graf (1971) reviewed the work of several investigators and concluded that the effects of turbulence upon the coefficient of drag are essentially unknown.

Difficulties in the measurement of bottom velocity, coefficient of drag, and lift on bottom materials under the influence of oscillatory flow have led either to the application of statistical mechanics or to a return to empirical observation to solve the problem of grain transport.

Einstein (1972) applied probability theory to the movement of gravel-sized particles in unidirectional flow. He examined the effect of shape on the transport rate and distinguished three classes of particle shapes based upon the work of Zingg (1935). These classes were spherical, flat, and the transitional shapes between these two end members. Zingg empirically observed that spherical clasts in traction transport moved about three times faster and three times farther than flat ones.

Mashima and others (1965, 1966) studied the transport of sand grains under the influence of oscillatory flow. They reasoned that whenever the drag forces exceed the resistance force of the bottom, grains begin to move. They observed that the relative frequency of sand grains in transport (P) could be calculated by

$$P = \frac{N}{S} \cdot \frac{D^2}{A}, \tag{1}$$

where N is the number of sections of areas A of the total number S in which sand grains of diameter D were suspended more than one grain diameter above the bed. They concluded that the transport of sand grains could best be described in terms of the average drag moment acting on the grains overbalancing the resistance moment. This relation may be written as

$$\frac{\overline{M}_s + \overline{M}_p}{M_r} > 1.0, \tag{2}$$

where $\overline{M}_s$ and $\overline{M}_p$ are the temporal mean moments of shear and pressure drag, respectively, acting over a quarter of a wavelength (assuming finite wave theory) and M_r is the moment of bottom resistance.

Since the ratio is an index of the threshold of the movement of all grains as a function of wave energy, its correlation with variations in grain-shape parameters should indicate selective transport of populations of grain shapes. In general, there is a shoreward increase in the ratio reflecting corresponding changes in topography and estimated bottom velocity. As the ratio becomes less than the value 1.0, grains with the lowest settling velocities (flat grains) are transported preferentially over more spherical grains of the same size and density because of their larger coefficients of drag. The ratio becomes less than one under low wave-height conditions.

Other aspects of nearshore sediment transport, such as the formation of ripples, have been successfully predicted by Harms (1969) and others by the use of dimensionless shear stress first introduced by Shields (1936).

Dimensionless shear stress (θ) is given for grains of uniform size (presumably uniform shape) by

$$\theta = \frac{\tau}{(\rho_s - \rho_f)gD}, \qquad (3)$$

where τ is the shear stress or drag per unit area of the bottom; $(\rho_s - \rho_f)$ is the effective density of the grains; g is the acceleration due to gravity, and D is the diameter of the grain. Bagnold (1966) used θ in conjunction with grain size to show the initiation of both traction and suspension transport of grains of uniform size in the presence of unidirectional flows. Bagnold observed that the suspension of grains of mixed size involves the spectra of turbulent uplift (eddy) velocity and settling velocity and that suspension would be initiated in the smaller sizes first. In a group of mixed sizes and shapes, this would presumably include suspension of nonspherical grains first, inasmuch as they have a larger coefficient of drag than their spherical counterparts of the same volume.

Komar and Miller (1973) observed that the application of equation (3) to oscillatory water-wave motion can lead to considerable error because an accelerating bottom current exerts a greater shear stress than does a constant flow, even though the velocities at the given instant are the same. Similarly, the stress in the decelerating case is somewhat less. They have shown that, unlike the results of unidirectional flow, the threshold of grain movement cannot be expressed as a single curve showing the relative stress variation with grain size (Shields' curve), but rather as a family of curves dependent upon wave period. Komar and Miller proposed that the threshold of grain movement under oscillatory wave motion is given by the relation

$$\theta_\tau = \frac{\rho_f U_m^2}{(\rho_s - \rho_f)gD} = 0.30 \left(\frac{d_0}{D}\right)^{1/2}, \qquad (4)$$

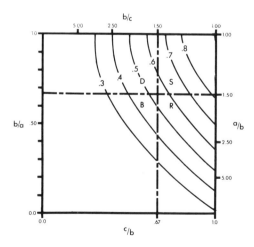

Figure 2. Shape classification diagram. Various axial ratios are shown at side of diagram. S, D, R, and B refer to spheres, disks, rods, and blades, respectively. Curved lines represent values of Corey shape factor.

where U_m is the maximum bottom velocity found from finite-amplitude wave theory and d_0 is the orbital diameter of the water motion near the bottom. This relation is in good agreement with the observations of several workers and is limited to grain sizes less than 0.50 mm (medium sand and finer).

The key phrase in the preceding discussion has been "the diameter of the grain." Graf (1971) pointed out that although geometrically regular shapes have been studied analytically and experimentally, irregular shapes, such as natural sand grains, have been investigated only experimentally. Wadell (1934) pointed out that irregular-shaped solids, like those found in nature, do not have diameters in the geometric sense and that any computation which deals with irregular-shaped particles must be related to a sphere. One such relation is the nominal diameter (D_n), which is the diameter of a sphere with the same density and settling velocity as that of the irregular particle (= sedimentation diameter; Interagency Committee on Water Resources, 1957).

The nominal diameter expresses the shape effect in terms of a sphere of varying size, and most equations of sediment-transport mechanics, including the Mashima transport ratio, use spherical grain shapes because of the simplicity of their geometry.

McNown and Malaika (1950, 1951) defined the nominal diameter of the ellipsoid as

$$D_n = 2 \sqrt[3]{abc}, \tag{5}$$

where a, b, and c were its semiaxes. They felt that the drag coefficient over a wide range of shapes could be estimated within 10 percent by using their empirical results for ellipsoids, provided that the principal-axes ratio of a grain is determined.

The nominal diameter of a mixture of heterogeneous shapes may be estimated from the mean settling velocity of the mixture if some average grain shape is assumed. Corey (1949) and McNown and Malaika (1950) devised an index of grain shape known as the shape factor (SF) which they related to the coefficient of drag and the nominal diameter.

$$SF = \frac{c}{\sqrt{ab}}, \tag{6}$$

where a, b, and c are the long, intermediate, and short axes of the grain, respectively. A grain with a shape factor of 1.0 would be a sphere; grains with shape factors of less than 1.0 would be nonspherical. Commonly, most naturally worn sand grains are assumed to have a Corey shape factor of 0.7. When values of shape factor are plotted on Zingg's classification (Zingg, 1935), they form a family of concentric curves around 1.0/1.0 (Fig. 2).

Various shape parameters, such as the Corey shape factor and nominal diameter, may be used to investigate the variation in the nearshore sediment. These measures in turn may be related to equation (4) and equation (2) in order to show relations between shape and the threshold of sediment movement.

PROCEDURES

Sampling and Other In Situ Measurements

Crystal Beach Pier was profiled with tape and lead weight with an estimated accuracy of 0.03-m error. A sampling set consisting of 20 samples was taken every

9.1 m, beginning with the swash zone and continuing to the end of the pier. Measurements of the index of refraction in the swash zone and at the end of the pier were taken with an A-O optical salinometer. The average index of refraction (based upon 20 measurements) and the in situ temperature measurements were used to determine the salinity, chlorinity, and density of the sea during sampling. The dynamic viscosity was estimated from the chlorinity and sea temperature by a method suggested by Miyake and Koizumi (1948). Other measurements and observations made during sampling were part of the U.S. Army Corps of Engineers LEO Data Bank (Berg, 1969) and included breaker type, wave height and period, direction of wave approach, and position of plunge point. Mean shape of sediment samples was determined prior to granularmetric analysis by splitting them down to about 2 g, from which 200 grains were selected for triaxial measurements. The axial lengths were measured with the aid of a Lietz microprojector and a low-power microscope mounted in the plane of the stage of the microprojector. The projected axes were measured to the nearest 0.05 mm; the axis was measured with the microscope to the nearest 0.02 mm. The Corey shape factor was calculated from the axes for each grain using equation (5). A total of 4,000 grains was measured.

Populations of shape were made by using the classification of Zingg (1935) by using the axial ratios b/a and c/b. The classification provides four shape forms: rods, disks, spheres, and blades (Fig. 2). The percentage of each Zingg shape form was calculated for every sample.

Granularmetric analysis was performed using 12.7-cm-diameter settling tubes, a transducer as sensor, and 2 g of sample per run. The recorder output consisted of a differential pressure versus time curve that was integrated to determine the mean time of fall of the sediment through 1 m, from which the mean settling velocity was calculated. The nominal diameter was estimated by using a computer table look-up routine assuming a mean shape factor of 0.70. This form of analysis assumes that the concentration of grains falling in the tubes is low enough so that no particle-wall or particle interference occurs during analysis. Study of the concentration problems has indicated that 2 g represents a low enough concentration (Brezenia, personal commun.; Gibbs, 1972).

Tabulated values (Skjelbreia, 1959) were used to approximate the root-mean-square value of the horizontal component of the bottom velocity. The velocity estimates were used in the calculation of the Mashima transport ratio. The transport ratio (Mashima and others, 1965, 1966) is given by

$$\frac{\overline{M}_s + \overline{M}_p}{M_r} = 1.567 \times 10^{-8} R^{2.448}$$

$$\cdot \left[\frac{\pi^2 H^2}{4T^2 \sinh^2(kh)} + \frac{4\pi^2 H^3}{3T^2 L \sinh^3(kh)} \right] \cdot \left[\frac{(\rho_s - \rho_f)}{\rho_s} gD_n \tan \theta \right]^{-1}, \quad (7)$$

where H is the wave height, T is the period, k is the wave number, and R is the sediment Reynolds number given by

$$R = \frac{D_n U}{\upsilon}, \quad (8)$$

where D_n is the nominal diameter estimate from granularmetric analysis, U is the root-mean-square estimate of the bottom velocity from finite wave theory, and υ is the kinematic viscosity.

RESULTS AND DISCUSSION

In Situ Measurements

The measured profile taken on the sampling date (November 29, 1969) is shown in Figure 3. It indicates the presence of one offshore bar and an extensive bar-trough area just seaward of the plunge point of the wave. (A second offshore bar is seaward of the last profile point and was neither profiled nor sampled.) The estimated wave height was 15 cm, and the period of the incoming waves was 1.5 seconds. The three-day average wave height prior to sampling was 30 cm. The index of refraction of the sea water was estimated at 1.3393, and the in situ temperature was 12° C. The density of the sea water was calculated to be 1.0243 g per cm^3; the salinity was 34.8 °/oo; the kinematic viscosity was 13.26 poise; and the chlorinity was estimated at 19.3 °/oo. The estimated longshore current was 10.5 cm per second to the east.

Bruno (1971) found that approximately 62 percent of the waves observed at Destin during that year were within the range of 0 to 1.5 m, and during the sampling month he estimated the mean wave height to be 60 cm, with an average wave period of 4.45 seconds.

Grain-Shape Distributions

The distribution of grain shapes within the Destin nearshore zone (Poché, 1973) is based upon the distribution of Corey shape factor (Table 1). The distribution is normal, with a mean of 0.653 and a standard deviation of 0.141. This measure of the central tendency of the distribution agrees with other shape-factor measurements taken in the nearshore zone by Harrison and Morales-Alamo (1964). A better estimate of the mean shape factor for sand grains appears to be 0.65 rather than the previously assumed value of 0.70. The mean shape factor of each Zingg population was found to be 0.8 for spheres, 0.7 for rods, 0.5 for disks, and 0.4 for blades.

The mean percent of the Zingg forms for all of the samples is spheres, 44.2 percent; disks, 23.5 percent; blades, 5.9 percent; and rods, 26.4 percent. This indicates that spheres commonly make up approximately one-half of the grain shapes at any one location. It should be noted that the classification is based upon axial ratios. Thus, a perfect cube and a perfect sphere of the same axial dimensions would both be classed as a sphere. Certainly neither the coefficient of drag nor the settling velocity would be the same for the two forms.

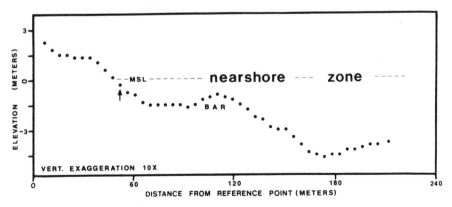

Figure 3. Profile of Crystal Beach Pier (November 29, 1969). Arrow indicates observed plunge point of incoming waves.

TABLE 1. DISTRIBUTION OF MEASURED SHAPE FACTORS

SF midpoint	Probability	Cumulative probability	SF midpoint	Probability	Cumulative probability
0.34	0.010	0.025	0.72	0.112	0.733
0.39	0.022	0.046	0.77	0.090	0.823
0.43	0.029	0.076	0.81	0.073	0.896
0.47	0.052	0.128	0.85	0.066	0.963
0.51	0.071	0.199	0.89	0.017	0.980
0.55	0.095	0.294	0.94	0.010	0.990
0.60	0.090	0.384	0.98	0.009	1.000
0.64	0.132	0.516			
0.68	0.106	0.622	$n = 4,000$	$\bar{x} = 0.653$	$\sigma = 0.141$

Grain-Shape Variations

Examination of the mean axial lengths (Fig. 4) indicates a grain-shape variation across the nearshore zone. There is a tendency for the inverse slope of the regression line of the mean long axis to be steeper than that of the other two axes. This indicates that the offshore samples are more equidimensional, whereas the shoreward samples are more rodlike, particularly in the swash zone.

The average shape factor across the nearshore zone does not vary significantly (Fig. 5). However, low shape-factor grains tend to collect in the trough, and higher shape-factor grains tend to be present in the swash zone and bar crest. Other data (Poché, 1973) indicate that the mean shape of the grains in the swash zone is in part determined by wave height. Larger waves plunge farther seaward and concentrate low shape-factor grains in the swash zone.

Nominal diameter variation is small, about 0.15 mm (Fig. 6). This means that, for practical purposes, the grains can be considered to be hydraulically equivalent in size but not in shape. In general, the well-known tendency of the swash-zone sediment to be coarser than the more seaward sediment is present.

Other shape-sorting trends are seen in the Zingg populations (Fig. 7). Blade shapes progressively increase from 2 to 6 percent across the zone, whereas disk (SF = 0.5) and blade (SF = 0.4) percentages are higher in the trough. No overall trends are present in the higher-than-average shape-factor populations (sphere, SF = 0.8; rods, SF = 0.7).

Local sorting by shape is indicated by systematic changes in shape factor from

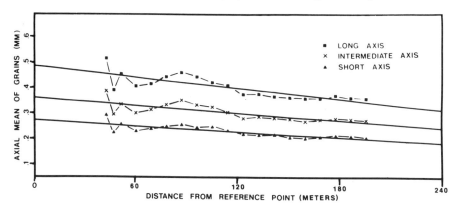

Figure 4. Change of mean axis of grains with distance from shore. Straight lines represent regression of data.

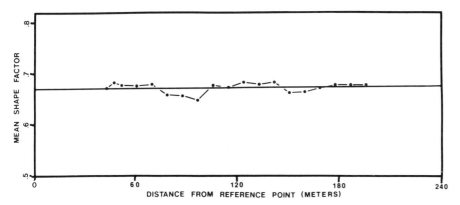

Figure 5. Change in mean shape factor (SF) with distance from shore. Straight line represents regression of data.

sample to sample. This implies relative increases and decreases in the proportions of high and low shape-factor grains making up the population of grains at any sampling location (Fig. 7). The proportion of spheres remains relatively stable across the zone, whereas grains of average and below-average shape (SF < 0.65) exchange proportions. The addition of disks and blades in the trough regions is apparently made by dilution of the rod population (SF = 0.7). Such dilution is the result of selective transport of low shape-factor grains due to their larger coefficient of drag. Mashima and others (1966) based their threshold of grain movement on the theoretical movement of a sphere (SF = 1.0). Since the sphere exhibits the least drag to a flow, equation (7) represents the limiting case for the general movement of all grains at the sediment interface. Selective transport by shape should best be seen where the transport ratio is less than 1.0. Under these conditions, the drag forces do not exceed the resistence forces for a spherical grain. However, lower-than-average-shaped grains (nonspherical) will be transported because of their higher drag coefficients.

Komar and Miller (1973) re-evaluated the works of Bagnold (1946) and Manohar (1955) to show that for grains less than 0.50 mm, the threshold of grain motion is reached while the flow in the boundary layer is still laminar. The constant of their formulation represents an average based upon the observations of Bagnold and Manohar. However, the determination of that threshold was, of necessity, subjective in the finer grain sizes.

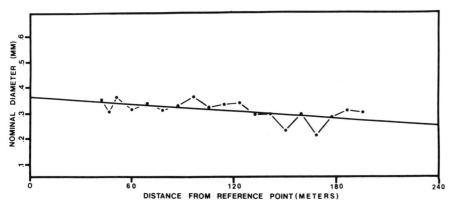

Figure 6. Change in nominal diameter (D_n) with distance from shore. Straight line represents regression of data.

The threshold of grain movement is not only a function of grain size and various wave parameters, but also of grain shape. The threshold of movement will be exceeded first for a particular wave height, period, and size for those grains that exhibit the largest drag or possess the lowest shape factor. The abundance of low shape-factor grains in the trough region of the profile (Figs. 3 and 5) suggests dilution by nonspherical grains within this region. The lower percentages of nonspherical grains near the bar crest and seaward side of the nearshore bar (Fig. 7) suggest selective removal and transport of grains into the trough region even at very low wave heights. This can be readily seen (Fig. 7) in the variation in the percentage as disk-shaped grains on either side of station 350. The Mashima transport ratio, equation (7), and the Komar-Miller threshold, equation (4), were calculated for the sediments of this region of the profile. The values (Table 2) of these parameters progressively increase in a shoreward direction. For waves of period 1.5 seconds and grains of nominal diameter 0.30 mm, the Komar-Miller threshold is $\theta_t \geq 4.5$. In a like manner, the Mashima transport ratio should be greater than 1.0. Neither of these values is reached (Table 2). This suggests either that their formulations probably best describe the movement of grains of average shape (SF = 0.65) or that their observations on the threshold for grains of such small size are indeed subjective. For grains of lower-than-average shape factor (SF = 0.65), the threshold of movement is considerably lower than predicted by either equation (4) or (7).

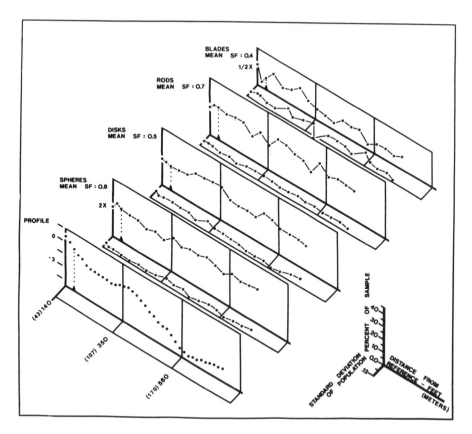

Figure 7. Near-shore variation in Zingg forms, Crystal Beach Pier, Florida, November 29, 1969.

TABLE 2. VARIATION IN VARIOUS PARAMETERS NEAR THE CREST OF THE INNER BAR

Station*	D_n	SF	R	MTR§ × 10^{-2}	θ_t
203	0.031	0.676	33.6	7.02	2.8
233	0.034	0.679	12.3	0.07	0.3
263	0.031	0.658	11.3	0.06	0.3
293	0.033	0.656	12.0	0.07	0.3
323	0.036	0.648	13.2	0.08	0.3
353†	0.032	0.677	14.4	0.58	0.5
383	0.033	0.673	14.9	0.62	0.4
413	0.034	0.683	6.8	0.02	0.1
443	0.029	0.679	1.5	..	0.03

*Station numbers refer to distance seaward from reference point in feet.
†Bar crest occurs between stations 353 and 383.
§MTR = Mashima transport ratio.

CONCLUSIONS

Selective sorting and transport of nonspherical grains take place as evidenced by addition and subtraction of low shape-factor populations to an unvarying population of spherical grains across the nearshore zone. It is suggested that grains are being removed and transported from the crest and seaward side of the nearshore bar portion of the profile and are being deposited in the bar trough even under very low wave-height conditions. This is particularly true of disks (SF = 0.5) and rods (SF = 0.7).

Calculation of two sediment-threshold formulations indicates that the critical value had not been reached. Since the formulations are in part based upon the theory of movement of spheres, it is suggested that for grains of lower-than-average shape factor (SF < 0.65), the threshold of movement is considerably lower than that predicted by either formulation. It is possible that the swing cradle and wave tank do not fully duplicate the natural condition.

ACKNOWLEDGMENTS

Field financial support was provided by the Mobile District Office of the U.S. Army Corps of Engineers. Richard Bruno aided with field work; Bruce Taylor and Ann Jones helped in the laboratory. Laboratory financial support was provided by the Research Policy Council, the University of Virginia.

REFERENCES CITED

Bagnold, R. A., 1946, Motion of waves in shallow water; interaction between waves and sand bottoms: Royal Soc. [London] Proc., v. 187, ser. A, p. 1-15.

———1966, Approach to the sediment transport from general physics: U.S. Geol. Survey Prof. Paper 422-I, 37 p.

Berg, D. W., 1969, Systematic collection of beach data: Proc. 11th Conf. on Coastal Engineering, London, p. 273-297.

Bruno, R. O., 1971, Longshore current system Panama City to Pensacola, Florida [M.S. thesis]: Tallahassee, Florida State Univ., 169 p.

Corey, A. T., 1949, Influence of shape of the fall velocity of sand grains [M.S. thesis]: Fort Collins, Colorado State Univ., 102 p.

Einstein, H. A., 1972, Bed load transport as a probability problem, in Shen, H. W., ed., Sedimentation: Fort Collins, Colorado State Univ., App. C, 105 p.

Gibbs, R. J., 1972, Accuracy of particle-size analyses utilizing settling tubes: Jour. Sed. Petrology, v. 42, p. 141-145.

Goodell, H. G., and Poché, D., 1968, Computer simulated topography and its relationship with sedimentary characteristics [abs.]: Am. Geophys. Union Trans., v. 49, p. 222.

Gorsline, D. S., 1964, Beach studies in West Florida, U.S.A., in Van Straaten, L. M., ed., Developments in sedimentology—Vol. 1, Deltaic and shallow marine deposits: Internat. Sedimentological Cong., 6th, Amsterdam and Antwerp, 1963, Proc., p. 144-147.

——1966, Dynamic characteristics of West Florida gulf coast beaches: Marine Geology, v. 4, p. 187-206.

Graf, W. H., 1971, Hydraulics of sediment transport: New York, McGraw-Hill Book Co., 513 p.

Harms, J. C., 1969, Hydraulic significance of some sand ripples: Geol. Soc. America Bull., v. 80, p. 363-396.

Harrison, W., and Morales-Alamo, R., 1964, Dynamic properties of immersed sand Virginia Beach, Virginia: U.S. Army Corps Engineers Coastal Eng. Research Center Tech. Memo. 9, 52 p.

Hopkins, E. M., Goodell, H. G., Chesser, S. A., May, J. P., and Poché, D., 1969, Relict nature of the sediments and submarine topography off Alligator Harbor, Florida: Gulf Coast Assoc. Geol. Socs. Trans., v. 19, p. 445-463.

Inman, D. L., 1955, Areal and seasonal variations in beach and nearshore sediments at La Jolla, California: U.S. Army Corps Engineers Coastal Eng. Research Center Tech. Memo. 39, 134 p.

Interagency Committee on Water Resources, 1957, Measurements and analysis of sediment loads in streams: Some fundamentals of particle size analysis: Washington, D.C., U.S. Govt. Printing Office, 55 p.

Ippen, A. T., 1966, Estuary and coastline hydrodynamics: N.Y., McGraw-Hill Book Co., 744 p.

Komar, P. D., and Miller, M. C., 1973, The threshold of sediment movement under oscillatory water waves: Jour. Sed. Petrology, v. 43, p. 1101-1110.

Manohar, M., 1955, Mechanics of bottom sediment movement due to wave action: U.S. Army Corps Engineers Coastal Eng. Research Center Tech. Memo. 75, 121 p.

Mashima, Y., Ikeuti, M., and Shigemura, T., 1965, Motion of sand grains by wave action: Proc. 15th Japan Natl. Cong. Appl. Mechanics, Tokyo, p. 164-168.

——1966, On the incipient sway of sand grains by wave action: Proc. 16th Japan Natl. Cong. Appl. Mechanics, Tokyo, p. 274-277.

McNown, J. S., and Malaika, J., 1950, Effects of particle shape on settling velocity at low Reynolds numbers: Am. Geophys. Union Trans., v. 31, p. 74-82.

——1951, Particles in slow motion: La Houille Blanche, v. 6, p. 701-722.

Miyake, Y., and Koizumi, M., 1948, Measurements of the viscosity coefficient of seawater: Jour. Marine Research, v. 7, p. 62-68.

Poché, D., 1973, Selective sorting of sediment by waves: The influence of grain shape [Ph.D. dissert.]: Charlottesville, Univ. Virginia, 100 p.

Shields, A., 1936, Anwendung der Ähnlichkeitsmechanik und der Turbulenzforschung auf die Geschiebebewegung: Mitteil., preuss. Versuchsanstalt für Wasserbau und Schiffbau, Berlin, no. 26, p. 3-26.

Skjelbreia, L., 1959, Gravity waves Stokes' third order approximation tables of functions: Council on Wave Research, Engineering Corp., California Research Corp., Berkeley, 337 p.

Wadell, H. A., 1934, Coefficient of resistance as a function of Reynolds number for solids of various shapes: Franklin Inst. Jour., v. 217, p. 458-490.

Zingg, T., 1935, Beitrag zur Schotteranalyse: Schweizer. Mineralog. u. Petrog. Mitt., v. 15, p. 39-140.

MANUSCRIPT RECEIVED BY THE SOCIETY MAY 23, 1973
REVISED MANUSCRIPT RECEIVED NOVEMBER 12, 1973

GEOPHYSICS AND HYDRODYNAMICS

Generation of Standing Edge Waves Through Nonlinear Subharmonic Resonance

G. E. Birchfield

Departments of Engineering Sciences and Geological Sciences
Northwestern University
Evanston, Illinois 60201

AND

Cyril J. Galvin, Jr.

Coastal Processes Branch
Coastal Engineering Research Center
Washington, D.C. 20016

ABSTRACT

Experimental and theoretical investigations of the generation of edge waves along a beach in the presence of gravity waves with crests parallel to the beach are presented. Subharmonic, resonant edge-wave generation is dependent on the beach slope and the period and amplitude of the primary wave. A simple, theoretical inviscid model which invokes the shallow-water approximation is studied. A standing wave with crests parallel to the coast and an edge-wave perturbation give rise to a Hill equation. Approximate determination of the stability properties indicates features similar to those observed in the experiments.

INTRODUCTION

Gravity waves trapped on a sloping beach are characterized by offshore amplitude decay from a maximum at the shore; they vary periodically and can propagate in either direction along the beach. These "edge" waves were first studied in a special case by Stokes (see Lamb, 1945), and a comprehensive theoretical and

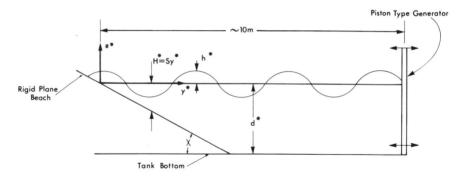

Figure 1. Schematic cross section of wave tank with rigid beach.

experimental study was made by Ursell (1952). In recent years there have been a number of investigations of edge waves of various horizontal scales in the natural environment. Bowen (1969, 1971), Bowen and Inman (1969), and Tait (1970) referred to the role of edge waves in rip currents, with respect to crescentic bars and to the generation of edge waves in the laboratory. Huntley and Bowen (1973) reported direct observations of standing edge waves on the coast of Devon.

Except for Ursell's (1952) work, the experiments described below (carried out in 1964 and 1965) are the earliest known efforts to study edge waves directly in the laboratory. The intent of these studies was to explore possible nonlinear generation of edge waves occurring at one-half the frequency of progressive waves approaching normal to a beach.

The possible similarity between edge waves in the presence of waves with crests parallel to a sloping beach and the appearance in the laboratory of deep-water, subharmonic "cross waves" at a plunger-type wave generator that is creating waves with crests parallel to the wave generator has been pointed out by several investigators of cross waves (for example, McGoldrick, 1968; Garrett, 1970). This mechanism is "parametric resonance" (Bogoliubov and Mitropolsky, 1961). Garrett's study was of cross waves at a wave generator opposite a rigid wall and is a discrete mode analysis; cross waves in a long or infinite channel were studied theoretically by Mahony (1972) and experimentally by Barnard and Pritchard (1972). Mahony's analysis involved a narrow but continuous spectrum; he was unable to find resonant interactions in the presence of a purely progressive wave. The experiments described in this paper were conducted in a long wave tank with a wave generator at the end opposite the beach. Although the primary wave is mainly progressive, it probably is not a purely discrete wave; some reflection occurs at the rigid plane beach.

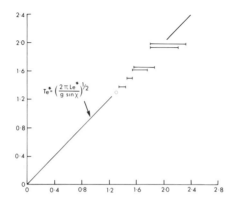

Figure 2. A comparison of observed and predicted edge-wave periods. Abscissa is the observed and ordinate is the predicted. Observed resonant period is exactly twice the period of the wave generation. L_e^* is twice the tank width.

In an effort to explore theoretically with minimal complexity, an inviscid model is discussed below in which the primary wave is a standing shallow-water wave on a beach of constant slope. Since both field and experimental evidence supports the presence of standing edge waves (even on a long beach), the perturbation wave in the model proposed here is considered a standing edge wave. An asymptotic analysis is made to determine possible subharmonic resonant instabilities in a standing perturbation edge wave.

EXPERIMENTS

Several sets of experiments on the generation of edge waves were carried out in the facilities of the Coastal Engineering Research Center from January 1964 to August 1965. The ones discussed here were run in a 22-m wave tank with a rigid plane beach installed as shown schematically in Figure 1. Waves were generated at one end of the tank by a simple piston-type generator with variable period and stroke. Periodic waves were produced that traveled to the beach with their crests parallel to the shoreline of the beach, located approximately 10 m from the generator. The depth in the uniform part of the tank ranged between 18 and 30 cm for different runs, with wavelengths between 1 and 2 m for forcing periods near 1 second. Hence, waves entering the variable depth zone, which was approximately 1 wavelength long, are not shallow-water waves nor are they strictly deep-water waves. The width of the tank (45.7 cm) determines the wavelength, $L_e^* = 91.4$ cm, of a standing Stokes edge wave and, with the beach angle χ, determines the frequency of the fundamental edge wave (Lamb, 1945; Ursell, 1952):

$$\sigma_e^{*2} = 2\pi g \sin \chi / L_e^*, \tag{1}$$

where g is the acceleration of gravity.[1] For the beach slopes used in the experiments, the period of the edge wave T_e^* was of the order of 2 seconds.

A number of experiments were run with several different slopes, in which the generator was driven at or near the period T_e^*. Nothing that could be attributed to resonant edge waves was observed. When the generator was driven at a period near $T_e^*/2$, however, resonant edge waves could be observed at the shoreline. Resonance is most easily described in terms of its effect on the runup limit, which is defined as the trace of the uppermost excursion of the water on the beach in one period of the primary wave. Primary waves from the generator initially had an approximately horizontal runup limit. Appearance of resonant waves occurred when the runup limit began to vary so that it reached a maximum against one wall of the tank while having a minimum at the other wall. After an initially imperceptible difference of runup limit across the tank, it gradually increased to an easily observed steady-state value. The maximum and minimum limits alternated from one wall to the other at each period of the primary wave; that is, the resonant wave displayed a periodicity twice that of the primary wave and thus represents a subharmonic resonance.

The range of edge-wave periods for which resonance was observed was measured for beach slopes ranging from 8.5° to 21° and is shown in Figure 2; the range generally increases as the slope decreases. Appearance of the resonant edge was dependent on the amplitude of the primary wave at the shoreline. If the amplitude

---

[1]In this paper, symbols bear stars (*) to signify dimensionalized variables; stars are omitted from symbols for nondimensionalized variables.

was too small or too large, for a given forcing frequency, the resonant wave could disappear. Resonance on slopes near 8.5° was examined in a series of experiments in terms of the generator period and primary wave height at the shoreline. A diagram showing the points of resonant or nonresonant conditions is shown in Figure 3. The ordinate is the nondimensional amplitude at the shoreline

$$\epsilon = h_0^* \sigma^{*2}/gs^2, \qquad (2)$$

where h_0^* is the amplitude of the primary wave at the shore, σ^* is the frequency of the primary wave (the generator frequency), and s is the beach slope. The abscissa is the wavelength L^* of an edge wave on a beach slope s, with a frequency $\sigma^*/2$, nondimensionalized by L_e^*, the resonant edge wavelength:

$$L = L^*/L_e^* = \frac{(8\pi g/\sigma^{*2}) \sin \chi}{(2\pi g/\sigma_e^{*2}) \sin \chi} = 4T^{*2}/T_e^{*2}, \qquad (3)$$

where $T^* = 2\pi/\sigma^*$ is the period of the primary wave. Thus, if the primary wave period is exactly one-half T_e^*, then $L^* = L_e^*$, $L = 1$. Although Figure 3 shows considerable intermingling of points with and without resonance, there appears to be a reasonably well-defined region where resonance does occur, surrounded by a region in which it is absent. The largest range of ϵ for which resonance occurs appears to be close to $L = 1$, with the range decreasing considerably more rapidly for $L < 1$ than for $L > 1$; the minimum values of ϵ for which resonance is observed also lie close to $L = 1$.

The theoretical model proposed below, although it is not a strict analogue of the experiments, hopefully contains enough of the essential elements to provide a possible mechanism for the observed subharmonic resonance.

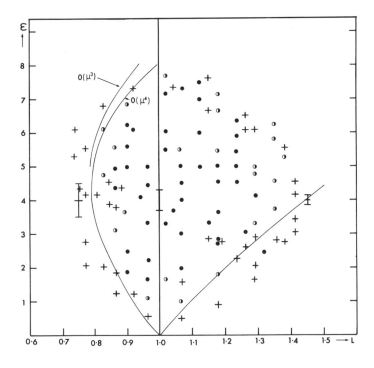

Figure 3. Resonant data from experiments with approximate neutral curve for the method of averaging greater than $W = 2\pi$. Solid circles represent self-generated resonant edge waves; half circles represent resonant edge waves appearing after periodically perturbing the water at the shoreline in the longshore direction; crosses denote observations where resonance did not appear even after perturbation; vertical bars denote limits of errors due to reading round off of the amplitude.

THEORETICAL MODEL

The experimental results above indicate that the amplitude at the shoreline and the period of the primary wave are the most important dynamic variables that govern the resonant edge-wave growth. In the model proposed here, it is assumed that only processes in the immediate vicinity of the coast are relevant; for example, the mechanism for generating the primary wave, assumed to be off the beach, is not specified. It is also assumed that the wavelengths of the primary wave at the shoreline and the resonant wave are long enough to justify the shallow-water approximation. The motion is also assumed to be irrotational and in the absence of friction. The dimensional equations of motion and continuity are

$$\frac{\partial \phi^*}{\partial t^*} + gh^* + \tfrac{1}{2}\nabla^*\phi^* \cdot \nabla^*\phi^* = 0,$$

$$\frac{\partial h^*}{\partial t^*} + s\frac{\partial \phi^*}{\partial y^*} + (h^* + sy^*)\nabla^{*2}\phi^* \cdot \nabla^* h^* = 0,$$

(4)

where $\mathbf{v}^*(x^*, y^*, t^*) = \nabla^*\phi^*$ is the horizontal velocity, and $h^*(x^*, y^*, t^*)$ is the free surface elevation above the mean level; x^*, y^*, z^* are right-handed cartesian coordinates with y^* normal to the beach on constant slope s (Fig. 1), and ∇^* is the horizontal del operator. These are nondimensionalized by horizontal-length scale gs/σ^{*2}, where σ^* is the frequency of the primary wave. We have

$$h = h^*/h_0^*,$$

$$\phi = \phi^*\sigma^*/gh_0^*,$$

$$\nabla = \frac{gs}{\sigma^{*2}}\nabla^*.$$

Using equation (2), equations (4) then become

$$\frac{\partial \phi}{\partial t} + h + \tfrac{1}{2}\epsilon \nabla \phi \cdot \nabla \phi = 0,$$

$$\frac{\partial h}{\partial t} + \frac{\partial \phi}{\partial y} + (\epsilon h + y)\nabla^2 \phi + \epsilon \nabla \phi \cdot \nabla h = 0.$$

(5)

If the nondimensional amplitude ϵ is set to zero, linearized equations are obtained. The solution is assumed to comprise two parts:

$$\phi = \phi_1 + \hat{\phi},$$

$$h = h_1 + \hat{h},$$

(6)

where (ϕ_1, h_1) represents a standing primary wave and $(\hat{\phi}, \hat{h})$ represents a standing perturbation edge wave, specifically

$$\phi_1(y, t) = -J_0(2y^{1/2}) \sin t \qquad \hat{\phi}(x, y, t) = a(t)e^{-ly} \cos lx,$$
$$h_1(y, t) = J_0(2y^{1/2}) \cos t \qquad h(x, y, t) = b(t)e^{-ly} \cos lx, \qquad (7)$$

where the primary-wave solution satisfies equations (5) on setting $\epsilon = 0$; $a(t)$ and $b(t)$ are the edge-wave perturbation amplitudes that are to be determined; $J_0(z)$ is the zero-order Bessel function.

Substitution of equations (6) into equations (5) yields, after using equations (7),

$$\hat{\phi}_t + \hat{h} + \epsilon \phi_{1y} \hat{\phi}_y = -\tfrac{1}{2} \epsilon (\hat{\phi}_{1y}^2 + \hat{\phi}_x^2 + \hat{\phi}_y^2)$$
$$\hat{h}_t + \hat{\phi}_y + \epsilon(\phi_{1yy}\hat{h} + \phi_{1y}\hat{h}_y + h_{1y}\hat{\phi}_y) = -\epsilon(\phi_{1y}h_{1y} + \phi_{1yy}h_1 + \hat{\phi}_x \hat{h}_x + \hat{\phi}_y \hat{h}_y), \qquad (8)$$

where letter subscripts denote differentiation. Because (ϕ_1, h_1) are known, equations (8) may be regarded as equations for $(\hat{\phi}, \hat{h})$. Since we seek a subharmonic solution, neither of the "forcing" terms on the right can contribute to such a solution to lowest order: the primary-wave terms cannot contribute because of the assumed form of time dependence in equations (7); the perturbation terms appear as a product and hence are of higher order. The size of the products of primary-wave terms relative to a linear primary-wave term can be described by an estimate of the maximum slope of the primary wave at the beach. This will be h_0^*/λ^*, where

$$\lambda^* = 1.45 gs/\sigma^{*2}$$

is the distance from $y = 0$ to the first node of the primary wave in equations (7), so that

$$h_0^*/\lambda^* = s\epsilon/1.45,$$

using equation (2). Since $s \ll 1$, "linearization" is reasonably valid for $\epsilon \sim O(1)$.

One asks then, can the homogeneous equations (8), regarded as equations for $a(t)$, $b(t)$ with coefficients depending on the primary wave, provide a mechanism for an exponentially growing, resonant edge wave? On using equations (7), equations (8) become

$$\theta \dot{a} + \theta b + \epsilon \theta_y \phi_{1y} a = 0,$$
$$\theta \dot{b} + \theta_y (1 + \epsilon h_{1y})a + \epsilon(\theta \phi_{1yy} + \theta_y \phi_{1y})b = 0, \qquad (9)$$

where $\theta = e^{-ly} \cos lx$. Because the coefficients in equations (9) depend on x, y in addition to time, the answer to the above question depends on successfully choosing means to satisfy the equations in some approximate sense. It is not clear what approach is the most appropriate, so several methods have been explored. In each case, the procedure is to reduce equations (9) to an ordinary differential equation for $a(t)$ with time-periodic coefficients and then to derive an asymptotic solution from which the approximate regions of stability and growth rates can be ascertained.

The simplest approach begins by eliminating x dependence in equations (9) by substituting for θ and canceling common factors, leaving only y, t-dependent coefficients. If one assumes interaction between the primary mode (1) is important only at the waterline, (2) expands the primary wave in a Taylor series about $y = 0$, and (3) retains only low-order terms, an equation for $a(t)$ is found:

$$\ddot{a} - \epsilon(2l + 1/2)\sin t \dot{a} + [l + 1/2 \epsilon^2 l(1/2 + l)$$

$$- 2\epsilon l \cos t - 1/2 \epsilon^2 l(1/2 + l)\cos 2t]a = 0.$$

By letting

$$Z = e^{1/2 A(t)} a(t), \quad A = \epsilon(1/2 + 2l)\cos t, \quad \text{and} \quad 2\xi = t,$$

this can be put in standard form

$$d^2 Z/d\xi^2 + (\alpha + \beta \cos 2\xi + \gamma \cos 4\xi)Z = 0, \tag{10}$$

where

$$\alpha = L - \mu^2, \quad \beta = 2\sqrt{2}(1 - L)\mu, \quad \gamma = \mu^2,$$

$$\mu = \epsilon/2\sqrt{2}, \quad \text{and} \quad L = 4l. \tag{11}$$

Equation 10 is a special case of Hill's equation (for example, see Magnus and Winkler, 1966). If $O(\mu^2)$ terms are neglected as was done by Garrett (1970), Mathieu's equation results. Approximate solutions valid for small μ can be found by the asymptotic method of "two-timing" (for example, see Cole, 1968, Chap. 3). If we let

$$L = 1 + \mu L_1 + \mu^2 L_2 + \mu^3 L_3 + \ldots$$

$$Z = Z_0 + \mu Z_1 + \mu^2 Z_2 + \ldots \tag{12}$$

$$Z_j = Z_j(\xi, \zeta), \quad \zeta \equiv \mu^3 \xi, \quad \mu \ll 1,$$

then $L_1 = 0$ and Z_0 grows exponentially, with growth rate $O(\mu^3)$, if

$$1 + \mu^2/8 - \mu^3/32 < L(\mu) < 1 + \mu^2/8 + \mu^3/32, \quad \mu \ll 1.$$

One observes that not only is there a slow growth rate, but resonance occurs only for a narrow region of width $O(\mu^3)$ and only for $L > 1$, at least to this approximation. Formally, these results come from the fact that, in equations (11), $\beta(\mu, L) \to 0, L \to 1$. This approach to equations (9) does not appear promising for the explanation of experimental results (see Fig. 2, dashed curves).

If θ is canceled from equations (9), an alternative hypothesis is to assume that an arithmetic average of the primary-wave coefficients in y over an appropriate distance W normal to the coast will incorporate the most important interaction between the two wave components. In this case equations (9) become

$$\dot{a} + b - \epsilon l \bar{\phi}_{1y} a = 0,$$

$$\dot{b} - (1 + \epsilon \bar{h}_{1y})la + \epsilon(\bar{\phi}_{1yy} - l\bar{\phi}_{1y})b = 0, \tag{13}$$

where

$$\overline{(\ )} \equiv \frac{1}{W}\int_0^W (\ )dy.$$

These are analogous to Garrett's (1970) equations (8) and (9).

Eliminating b and putting into standard form, we again obtain equation (10), but now, using equation (7),

$$\alpha = L - \mu^2,$$
$$\beta = 2\sqrt{2}(1 - kL)\mu, \qquad (14)$$
$$\gamma = \mu^2,$$

where

$$Z = \exp[(2l + 1/2k)\cos t]a(t), \qquad 2\xi = t,$$
$$\mu = m(W)\epsilon/\sqrt{2},$$
$$m(W) \equiv [1 - J_1(2W^{1/2})/W^{1/2}]/W, \qquad (15)$$
$$k(W) \equiv 1/2[1 - J_0(2W^{1/2})]/[1 - J_1(2W^{1/2})/W^{1/2}],$$

where $J_1(z)$ is the order-1 Bessel function.

For a fixed $W > 0$, if L, Z are expanded in μ as in equations (12), but now putting $\zeta = \mu\xi$, the two-timing solution of equations (10) and (14) indicates that Z grows exponentially at a rate $\sim\mu$, provided that

$$L_L(\mu, W) < L(\mu) < L_u(\mu, W), \qquad \mu \ll 1,$$

where L_L, L_u are curves on the L,μ plane that separates unstable solutions from periodic solutions. Once the growth rate and the possibility of unstable solutions have been established from the two-timing solution, the form of the curves L_L, L_u for equation (10) can be calculated to a higher approximation for $\mu \ll 1$ most easily by using the perturbation approach described in McLachlan (1947, Chap. 2). This was done for $W = \pi$, 2π, 4π, and 6π as shown in Figure 4. These values of W correspond to dimensional distances of 1/8, 1/4, 1/2, and 3/4 of the wavelength of an edge wave of frequency $\sigma/2$ (that is, $L = 1$); they correspond to 1/4, 1/2, 1, and 3/2 times the width of the tank. The nondimensional distances to the first and second nodes of the primary wave are found from equations (7) to be 1.4 and 7.6, respectively; the first wavelength distance is 12.2.

Figure 4 shows that the region of instability becomes small both for small and large W, at least for small μ. For large W the zone of stability decreases because for fixed ϵ in equations (15) $\mu \to 0$ as $W \to \infty$. If $W \to 0$ in equations (15), then $k \to 1$ and $\mu \to \epsilon/2\sqrt{2}$ so that equations (14) reduce to equations (11), and the first approach to equations (9) is recovered. Of the values of W selected, $W = 2\pi$, which corresponds to an average greater than approximately one-half the primary wavelength, shows the largest region of instability and has been superimposed on the experimental data in Figure 3. While the experimental data are fitted rather well by the choice of averaging distance W that gives the largest area of resonant instability (near $W = 2\pi$), it is not obvious why this should be; hence, a rationale for the choice of W is not obvious.

A second approach to equations (9) avoids this ambiguity. It is assumed that equations (9) are valid in an integral sense over a volume extending more than one-half of the tank width and from $y = 0$ to $y \to \infty$. Directly integrating equations (9) with reference to x, y and letting

$$\widetilde{(\ )} \equiv \int_0^{\pi/2l} dx \int_0^{\infty} (\ ) dy,$$

we have

$$\dot{a} + b - \epsilon l^3 \widetilde{\theta \phi_{1y}} a = 0, \tag{16}$$

$$b - [l + \epsilon l^3 \widetilde{\theta h_{1y}}] a + \epsilon l^2 [\widetilde{\theta \phi_{1yy}} - \widetilde{\phi_{1y} \theta l}] b = 0. \tag{17}$$

Following a procedure similar to that above, these may also be reduced to equation (10) with

$$\alpha = L - \mu^2 L^2 (1 - \psi)^2, \quad \beta = 2\sqrt{2} L (1 - 3\psi)\mu, \quad \gamma = \mu^2 L^2 (1 - \psi)^2,$$

$$Z = \exp[1/8\epsilon(1 + \psi) \cos t] a(t), \quad \psi = (1 - e)^{-4/L} L/4, \tag{18}$$

$$\mu = \epsilon/4\sqrt{2}, \quad \text{and} \quad 2\xi \equiv t.$$

L and Z are again expanded in powers of μ and the two-timing solution is constructed. The growth rate is $O(\mu)$; the approximate stability curves valid for $\mu \ll 1$ superposed on the experimental data are shown in Figure 5.

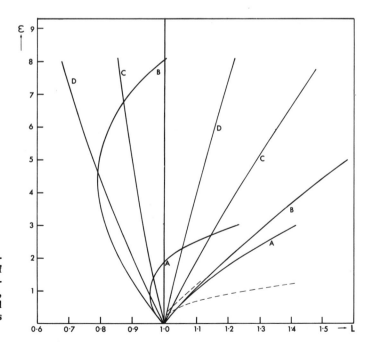

Figure 4. Approximate neutral curves for the method of averaging greater than W normal to the beach. (A) $W = \pi$, (B) $W = 2\pi$, (C) $W = 4\pi$, and (D) $W = 6\pi$. Dotted curve is for limit $W \to 0$.

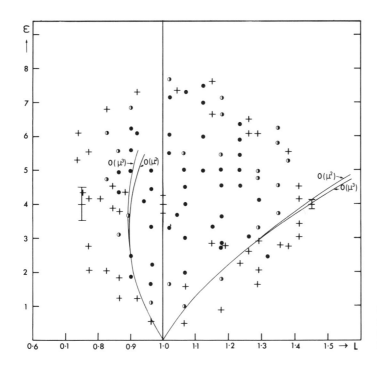

Figure 5. Same experimental results as in Figure 3, with approximate neutral curve for the method of average interaction.

DISCUSSION

Both methods of averaging introduced above result in equations that produce subharmonic, resonant edge waves for ranges of frequency and amplitude of the primary wave roughly corresponding to those observed in the experiments. This is demonstrated at least for smaller values of ϵ on which both the theoretical model and the method of approximate solution depend. Of the two methods, the second is the most readily justified a priori. In this formulation the coefficients in equations (18) for the Hill equation, as found from equations (16) and (17), depend explicitly on the interaction between the primary and perturbation waves through the integrals with reference to y. In the first method, however, no such interaction appears in the coefficients of equations (14), because averaging is done on the primary wave only (in equations (13)).

The prominent asymmetry of the unstable region about $L = 1$ observed in the data in Figures 3 and 5 appears in the second formulation as well as in the first if $W \leq 2\pi$. The disappearance in the experiments of resonant edge waves for $L < 1$ as ϵ increases seems to be portrayed in the recurving of the neutral curves in Figures 3 and 5, although the approximate solutions are obviously poor in this region. The disappearance of resonance in the upper right side of the experimental data, however, does not appear in the analytical model. This may be due to the nature of the asymptotic solution for $\mu \ll 1$, but it seems more likely that another mechanism, wave breaking, is directly involved.

The absence of resonance in the experiments for $\epsilon \leq 1$ appears attributable to the dissipation due to viscosity, in the manner demonstrated by McGoldrick (1968) for deep-water cross waves.

An energy equation can be formed for the total energy of the perturbation in an edge wave in the manner of Garrett (1970). Considering the last approach, if equation (16) is multiplied by $\epsilon^2 \pi^2 a/4l$ and then added to equation (17) multiplied by $\epsilon^2 \pi^2 b/4l^2$, after reduction,

$$\frac{d}{dt}(K + P) = \frac{\epsilon^3 \pi^2}{4} [\widetilde{l\theta h_{1y}} ab + l^2 \widetilde{\theta \phi_{1y}} a^2 + (\widetilde{l\theta \phi t_{1y}} - \widetilde{\theta \phi_{1yy}})b^2],$$

$$K + P = \frac{\epsilon^2 \pi^2}{8l^2}(la^2 + b^2),$$

where $K + P$ is the total energy of the edge-wave perturbation. A term of third order in the perturbation has been neglected. Therefore, there is work done by stresses consisting of an interaction between the primary wave and perturbation operating on strains proportional to the square of the perturbation.

In order to verify the consistency of this model, it is possible to use the two-timing solution and to calculate the rate of generation of energy in the growing wave. This has been done and it verifies that the right side of the energy equation is nonvanishing to low order in ϵ.

SUMMARY

Preliminary experiments in a long, shallow channel with a rigid beach at one end indicate that, for a wide range of amplitudes and periods of primary waves with crests parallel to the beach, subharmonic resonance can occur, producing a standing, fundamental edge wave with a wavelength twice the width of the channel. The primary wave is principally a progressive wave but has some reflection at the rigid beach. A theoretical model has been investigated which postulates that a standing primary wave can generate through parametric resonance, a subharmonic standing edge wave. As a result of approximately satisfying the equations of motion over a volume of water, a Hill-type differential equation for a perturbation edge-wave amplitude is generated, with coefficients either a function of properties of the primary wave or of interaction between the primary and perturbation waves, depending on how the "average" equations are derived. The stability properties were investigated asymptotically for small amplitudes. For the interaction approach, several features noted in the experiments seem to be present in the approximate solutions. An energy equation was derived illustrating the mechanism for generation of perturbation energy.

ACKNOWLEDGMENTS

The theoretical part of this research was carried out primarily at the Institute of Oceanographic Sciences, Wormley, Godalming, Surrey, where Birchfield was on sabbatical leave. We thank A. J. Bowen for a stimulating discussion on the subject. Part of the research was supported by National Science Foundation, Atmospheric Sciences Section, Grant GA-25992 to G. E. Birchfield.

REFERENCES CITED

Barnard, B.J.S., and Pritchard, W. G., 1972, Cross-waves, Pt. 2. Experiments: Jour. Fluid Mechanics, v. 55, p. 245-255.

Bogoliubov, N. N., and Mitropolsky, Y. A., 1961, Asymptotic methods in the theory of nonlinear oscillations: New York, Gordan & Breach, Science Pubs., Inc., 537 p.

Bowen, A. J., 1969, Rip currents. 1, Theoretical observations: Jour. Geophys. Research, v. 74, p. 5467-5478.

——1971, Edge waves and crescentic bars: Jour. Geophys. Research, v. 76, p. 8662-8671.

Bowen, A. J., and Inman, D. L., 1969, Rip currents. 2, Laboratory and field observations: Jour. Geophys. Research, v. 74, p. 5479-5490.

Cole, J. D., 1968, Perturbation methods in applied mathematics: Waltham, Mass., Blaisdell Pubs., 260 p.

Garrett, C.J.R., 1970, On cross waves: Jour. Fluid Mechanics, v. 41, p. 837-849.

Huntley, D. A., and Bowen, A. J., 1973, Field observations of edge waves: Nature, v. 243, p. 160-162.

Lamb, H., 1945, Hydrodynamics (6th ed.): New York, Dover Pubs., Inc., 738 p.

Magnus, W., and Winkler, S., 1966, Hill's equation: New York, Interscience Pubs., Inc., 127 p.

Mahony, J. J., 1972, Cross-waves, Pt. I, Theory: Jour. Fluid Mechanics, v. 55, p. 229-244.

McGoldrick, L. F., 1968, Faraday waves: The cross wave resonant instability: Chicago Univ. Dept. Geophys. Sci., Tech. Rept. no. 2, ONR Contract N-00014-67-A-0285-0002, 59 p.

McLachlan, N. W., 1947, Theory and application of Mathieu functions: New York, Oxford Univ. Press, Inc., 401 p.

Tait, R. J., 1970, Edge wave modes and rip current spacing [Ph.D. dissert.]: La Jolla, Univ. California, San Diego.

Ursell, F., 1952, Edge waves on a sloping beach: Royal Soc. [London] Proc., ser. A, v. 214, p. 79-97.

MANUSCRIPT RECEIVED BY THE SOCIETY SEPTEMBER 14, 1973
REVISED MANUSCRIPT RECEIVED APRIL 11, 1974

… Flexure as Shown by
… n of Glacial Lake
… orelines in
… British Columbia

R. J. FULTON

… ain Sciences Division
… ical Survey of Canada
… Energy, Mines and Resources
… a, Canada K1A OE4

R. I. WALCOTT

Gravity Division, Earth Physics Branch
Geological Survey of Canada
Department of Energy, Mines and Resources
Ottawa, Canada K1A OE4

ABSTRACT

Glacial lakes occupied many valleys in southern British Columbia during retreat of Fraser Glaciation ice. The strandline deformation that has taken place in response to removal of the ice load provides a clue to the flexural properties of the lithosphere. Surfaces fitted to four strandlines of former glacial lakes are tilted from 2.5 to 1.6 m per km, with the best documented section of strandline having a tilt of 1.8 m per km for an average slope of 0.0018. The Algonquin strandline, one of the most strongly developed proglacial strandlines of the Great Lakes area, has an average slope of 0.0010. The wavelength of flexure for the Algonquin strandline is 2π 180 km, and the indicated lithospheric thickness is about 110 km. If the Cordilleran ice load were the same as the Laurentide load, which caused deformation of the Algonquin strandline, a wavelength of flexure of 2π 100 km and a lithospheric thickness of about 50 km would be required to explain the greater Cordilleran deformation. But the Cordilleran ice load was probably only half that of the Laurentide area; so the lithosphere of the southern Canadian Cordillera apparently has a flexural parameter of about 50 km and a thickness of about 20 km.

INTRODUCTION

The deformed strandlines of the proglacial lakes formed about 12,000 years ago in the Great Lakes and Lake Winnipeg regions have been interpreted in terms of flexure of the lithosphere by the load of the Laurentide Ice Sheet (Walcott, 1970a, 1970b). This interpretation implies that the shape of the deformed strandline is determined by the flexural rigidity of the lithosphere, which in turn is largely governed by the thickness of the lithosphere. The deformation of several of the best developed strandlines indicates a lithospheric thickness of about 110 km, and this can be regarded as a value characteristic of stable continental platforms. In the Basin and Range province, however, the thickness of the lithosphere measured from Lake Bonneville flexure is less than 30 km (Walcott, 1970b), an anomalously low value but perhaps related to local high heat flow, recent volcanicity, and a shallow seismic-low-velocity mantle layer, all of which may be due to recent orogenesis. The lake strandline information in this paper makes it possible to compare the lithospheric properties of the Canadian Cordilleran region with those of the stable continental platform and those of the Basin and Range province.

The area discussed is part of the Interior Plateau of south-central British Columbia, lying 100 km north of the 49th parallel and 200 km northeast of Vancouver (see Fig. 1). It consists of hilly uplands with maximum elevations of about 1,700 m separated by broad valleys, with floor elevations ranging from 700 to 1,000 m.

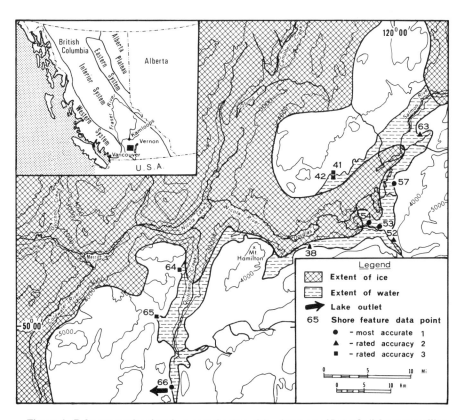

Figure 1. Paleogeography of study area at the time of development of Lake Quilchena strandline features.

The uplands consist of rock mostly covered by a thin and discontinuous mantle of glacial drift, whereas the valleys contain large areas of thick drift cover.

All parts of the area were covered by ice during the last (Fraser) glaciation, and according to the synthesis of information presented by W. H. Mathews on the first glacial map of Canada (Geological Association of Canada, 1958), the ice surface in the area discussed in this paper probably reached an elevation of about 2,400 m. During retreat, all major valleys were occupied by glacial lakes. Fraser ice appears to have advanced into the area about 19,000 C^{14} years B.P. (Fulton, 1971, Fig. 1). Parts of the area were ice-free 10,500 years ago, and all glacial lakes were drained by 9,000 years B.P.; so the period of retreat during which large glacial lakes were present could not have been longer than 1,500 years, and the life of any individual lake stage was probably shorter than 500 years.

GLACIAL LAKES

Three lakes and four identified water surfaces are included in this study. The precise chronology of lake history is not known, but Fulton (1969) discussed the relative chronology.

Lake Quilchena

Lake Quilchena (Mathews, 1944) is the highest lake included in this study. It developed on the periphery of ice that occupied the Nicola and adjacent valleys (Fig. 1), and many of the delta terraces built into it could be, in part, of ice-contact origin. The shore features of this lake are generally poorly developed. Drainage from this lake was to the south, via Quilchena Creek valley, into Otter Creek and the Similkameen River system. Traces of the lake are restricted to southern and eastern parts of the area.

Lake Hamilton

Lake Hamilton (Mathews, 1944) came into being when ice retreated from Salmon River valley permitting drainage to the east into Okanagan Valley. As the ice retreated, the lake expanded westward down the Nicola Valley and northward into Campbell Creek valley (Fig. 2). There is evidence for the former presence of two water planes about 7.5 m apart in many areas. In the data reduction and the following discussion, we refer to the upper level as level 1 and the lower level as level 2.

Lake Merritt

The lowest and youngest lake is Lake Merritt (Mathews, 1944). It was contained by ice in the lower Nicola Valley and discharged north over a rock sill into Campbell Creek and thence into the South Thompson River (Fig. 3).

OBSERVATIONS

Shore Features

Two main types of features were used to designate levels of former water planes. These are wave-cut benches and delta terraces.

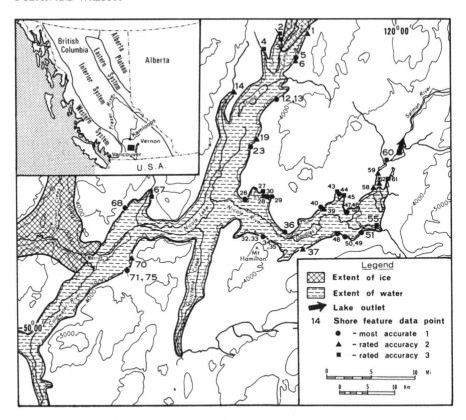

Figure 2. Paleogeography of study area at the time of development of Lake Hamilton strandline features.

Wave-cut benches are best developed on till-covered slopes that have been exposed to long fetches over open water. Well-developed benches are as much as 15 m wide and slope so that the bench front is 4.5 to 7.5 m lower than the break in slope at the rear. Well-developed benches are uncommon because the till cover on the valley walls is generally thin and because fetch was generally limited in the narrow valleys. The typical "cut" shore feature appears as a minor scallop in the hillside which is visible mainly because of the lighter colored vegetation that grows on the better drained, "washed" materials. At close range these shore features are virtually indistinguishable from the other minor changes in valley-wall slope, and they are often only recognized as traces of former shorelines when seen from a distance. The inflection of slope at the rear of the bench was taken as the part of the feature most closely approximating the level of the former water plane.

Delta terraces were constructed by the larger streams during the existence of the lakes. The deltas now have either been incised and exist as gently sloping terraces set inside tributary valleys, or they occur as gently sloping, fan-shaped deposits built into the main valleys at the mouths of side valleys. Determining the position of the former water plane on the delta landforms was a problem. Exposures that might have permitted locating the topset-foreset bed junction were generally absent; so it was necessary instead to measure the elevation of the terrace surface at the delta front. This part of the feature could have developed as much as 6 m below the water plane. There is no way of determining whether the break

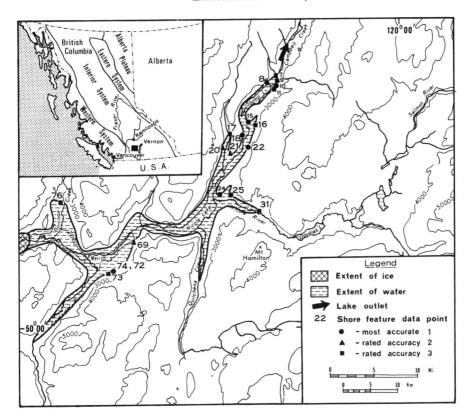

Figure 3. Paleogeography of study area at the time of development of Lake Merritt strandline features.

in slope at the front of the present-day feature is the original depositional delta front or the result of erosion. If the break in slope at the delta front is the original depositional break, the elevation obtained would be as much as 6 m below that of the former water plane. If the break is due to postdepositional erosion, the elevation obtained could lie from 6 m below the former water plane to 6 m or possibly more above it.

Measurement Technique and Accuracy

All elevations were measured by means of a single altimeter traverse tied to a bench mark. Most features were within 8 km of the nearest bench mark, and altimetry work was limited to early morning and late afternoon of days when weather conditions were stable. The instrument used was a Wallace and Tiernan surveying altimeter. Temperature corrections were made, but no adjustment was made for relative humidity, as these corrections generally would have been less than 0.2 percent. No measures of repeatability or other measures of accuracy were made, but past experience would indicate a standard error of measurement of 3 m for this technique used under favorable conditions.

Factors other than measurement techniques that affect accuracy are difficult to quantify. The problem of choosing the part of a feature that closely marks the level of the former water plane has already been discussed. Another complication is the annual fluctuation of water level that probably occurred. Uncontrolled lakes in this part of British Columbia currently undergo annual fluctuations of from 3 to 6.5 m. Water-level fluctuations during glacier recession may have been even greater. Delta terraces were probably constructed at times of high water (time of maximum inflow); whereas shore benches may have been cut during low-water stages (time of maximum storminess).

In an attempt to minimize and isolate the effects of these problems, the data were stratified. A three-point scale was used to assign a judged reliability to each observation. Accuracy 1 (most accurate) was assigned to shore-bench features where both identification of water plane and accuracy of measurement were judged to be reliable. Accuracy 2 was assigned to shore-bench data where there was a reason to question water-plane identification or measurement accuracy. Measurements obtained from delta terraces were assigned accuracy 3.

INTERPRETATION

Observational data were divided into four groups: (1) Quilchena observations, (2) Hamilton upper observations, (3) Hamilton lower observations, and (4) Merritt observations. Lake Hamilton surfaces are based on localities where shore development of the two water planes is visible. The data in each group were further subdivided as observations of rated accuracy 1, rated accuracy 1 plus 2, and rated accuracy 1 plus 2 plus 3.

To each of these 12 groups, a series of surfaces was fitted. The first-order surface was given by

$$E_i = A_0 + A_1 X_i + A_2 Y_i + \epsilon_i,$$

where E_i, X_i, and Y_i are the elevation and easterly and northerly positions (relative to an arbitrary datum and scale) of the ith observation, and A_0, A_1, and A_2 are parameters determined by minimizing the square of the residual elevations ϵ_i. The standard deviations of all parameters are obtained from the inverse matrix obtained in the least-squares procedure. The bearing of the upslope direction of the first-order surface is given by $\tan^{-1}(A_1/A_2)$.

Higher order surfaces were also fitted where warranted by the number of data in the group, but these surfaces were constrained to vary in one direction only along the bearing defined by the first-order surface. In no case was the improvement of fit sufficient to warrant anything more than the first-order surface to generalize the data.

Results

Lake Quilchena. The first-order surface fitted to data of rated accuracy 1 (four observations) has a bearing of 319° ± 24° and slope of 3.8 ± 1.5 m per km (quoted error here and hereafter is ±2σ). The largest single deviation from this plane is 3.3 m. Inclusion of observations of rated accuracies 2 and 3 (seven observations) results in a dramatically poorer fit, with the bearing becoming 164°, slope 0.6 m per km, and residual elevations ranging over ±15 m. If, therefore, Lake Quilchena shore features are tilted in the 320° direction indicated by the accuracy 1 data, some explanation is required for the wide scatter of the less accurate observations. It may be that some features are misidentified or that the data do not represent a single lake level.

The shore features of the lower lakes, Hamilton and Merritt, show a consistent upslope direction of tilting of 350° ± 10°, and the assumption is made that Lake Quilchena features are also tilted in that direction (see Fig. 4). The fit of Lake Quilchena data by a single plane is very poor, but most of the data cluster around two separate surfaces of about the same slope, 2.1 m per km, but separated vertically by about 75 m. The higher surface is defined by data in the vicinity of Quilchena Creek in the south (Fig. 1, observation 66 of rated accuracy 1 and observations 64 and 65 of rated accuracy 3). The lower surface is defined by data on the east in the vicinity of Douglas Lake (Fig. 1, observations 38, 53, 54, 57, and 63). This could suggest that there were two independent lakes separated by ice that lay in the vicinity of the Mount Hamilton region (Fig. 1). The lake in the south would have been controlled by the Otter Creek sill and that in the east by a lower outlet not yet located.

Three observations, 41, 42, and 52, do not fit with this explanation but lie above the lower plane by 16 m, 18 m, and 31 m, respectively. Thus, neither the postulate of a single lake level nor of multiple lake levels completely explains the observations, although the second postulate is preferred.

Lake Hamilton. For the upper level of Lake Hamilton, the first-order surface gives a bearing of 354° ± 5° and slope of 1.8 ± 0.3 m per km for five observations of rated accuracy 1, or 354° ± 4° and slope of 1.9 ± 0.2 m per km for all observations. The standard deviation of all data is 3.3 m.

For the lower level, the first-order surface gives a bearing of 347° ± 6° and slope of 1.6 ± 0.6 m per km and a standard deviation of 2.4 m for 11 observations of rated accuracy 1. Including all data in the analysis results in a poorer fit due to a very large deviation in one observation (6) which falls 13.5 m below the plane. This may be due to misidentification, or possibly it is a local feature related to the final drainage of Lake Hamilton.

The two levels appear to converge to the southeast (Fig. 4), but this cannot be regarded as significant with present data, as there is a considerable overlap in the confidence limits of the slopes of the two levels.

Lake Merritt. The first-order surface gives a bearing of 341° ± 14° with a slope of 1.8 ± 0.5 m per km for eight observations of rated accuracy 1 and a standard deviation of 3.3 m. Inclusion of lower grade data increases the standard errors considerably owing mainly to one observation (22), with a rated accuracy of 2, which lies 18 m below the surface.

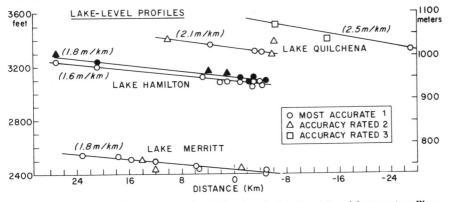

Figure 4. Location of shore features in relation to calculated position of former strandlines. The upslope edge of the plane of the diagram has a bearing of 350°. Accuracy-rated-3 data points have been plotted for Lake Quilchena only. Solid symbols are used for shore features related to the upper level of Lake Hamilton. Only those data clearly identifiable with upper or lower levels of Lake Hamilton are plotted in this diagram.

In summary, the observations of shore-feature levels can be adequately explained by five tilted planes with an upslope bearing of 350°. There is some question of the validity of the interpretation of the data defining the upper two planes, but the other three agree so closely that there can be little doubt that the interpretation is correct. One of the largest sources of uncertainty in work of this type is in the identification of a single synchronous strandline from scattered, often poorly preserved evidence that can be correlated in many different ways to give many different possible strandlines. However, in this study, the distinctive double Lake Hamilton strandline is the key to correlation. The broad areal distribution of observational data for the Hamilton strandlines and the close agreement of direction and amount of tilt of this pair with lower and higher strandlines provide strong additional support for the suggested interpretation. The amount and direction of upslope tilt of both levels of Lake Hamilton agree closely and show that over a distance of 40 km the average slope of the strandline is 1.8 m per km at a bearing of 350°.

DISCUSSION

The strandlines of the proglacial lakes were originally horizontal, and their tilting could, in principle, have occurred at any time since their formation and, indeed, could be continuing today. However, two lines of evidence suggest that the tilting occurred during the ice retreat and was largely, if not entirely, completed shortly after disappearance of the ice from the interior of British Columbia.

1. At the time of formation of Lake Hamilton strandlines, ice must have blocked Campbell Creek in the north and lower Nicola Valley in the west, with the outlet of the lake being controlled by a sill at the divide between Nicola River and Salmon River drainage (Fig. 2). The Kamloops, Fraser, Shuswap Lake, and Okanagan Valleys at this time were probably full of ice (Fulton, 1969, Fig. 3), and a 350° direction of downward tilt of the land is consistent with this pattern of ice loading.

2. Abandoned lake levels about 6.1 m higher than present on Kamloops and Shuswap Lakes probably represent lake levels controlled by flow into Okanagan Valley through a channel near Enderby (Fulton, 1969). A date from this channel indicates it was abandoned before 8,900 years B.P. The strandlines of these lakes show no evidence of tilting; therefore, most of the tilting of the older strandlines probably occurred before 8,900 years B.P.

Both the timing and direction of tilt suggest that the land was depressed by the ice load during the formation of the proglacial lakes and that during and following disappearance of the ice, the land rebounded, tilting the lake strandlines.

Deformed proglacial lake strandlines are found elsewhere in Canada but have been particularly well described in the Lake Winnipeg and Great Lakes regions. It has been suggested that the deformation was caused by the flexure of the lithosphere beneath and adjacent to the load of the Laurentide Ice Sheet. Using the wavelength of this flexure, Walcott (1970a, 1970b) calculated the flexural rigidity of the lithosphere. The flexural rigidity of an elastic sheet is determined to a great extent by its thickness, and the thickness calculated for the lithosphere in the vicinity of the Great Lakes and Lake Winnipeg was 110 ± 5 km.

If we had complete profiles of the strandlines from the ice edge to beyond the point where the deformed strandline flattens out, it would be possible to calculate the flexural rigidity and hence the thickness of the lithosphere in the interior of British Columbia, assuming that the deformation is an equilibrium lithospheric flexure. Unfortunately, we have only information on the tilt of the strandline within

50 km of the ice-load edge, and we do not know where the flexure flattens out. Unlike the flexure that is independent of the load, the amount of tilt is directly proportional to the load and is also dependent on the flexural rigidity. If the load can be assumed to be an ice sheet, the upper surface of which can be approximated by an exponential curve, the displacement W at distance X from the ice edge is (Walcott, 1970a)

$$W = A \exp(-X/\alpha) \left(2 \cos \frac{X}{\alpha} - \sin \frac{X}{\alpha}\right), \tag{1}$$

where the flexural parameter (α) is related to the flexural rigidity (D) as $\alpha^4 = 1.3 \times D \times 10^3$ and to the wavelength of flexure by $2\pi\alpha$, and A is a magnitude term (essentially proportional to the thickness of ice).

The slope of the deformed surface S at distance X where $X < \alpha/2$ is approximated to an accuracy of 3 percent by

$$S = dw/dx = -3.1 A \exp(-x/\alpha)/\alpha, \tag{2}$$

and the average slope $\bar{S}$ over the distance $0 < x < \alpha/2$ is

$$\bar{S} = -2.4 \ A/\alpha. \tag{3}$$

In Figure 5, equation (3) is plotted for values of $\bar{S}$ of 0.0010 and 0.0018. The latter is the average slope of the Lake Hamilton surfaces over a distance of approximately 40 km from the ice edge. It is compared with the slope of the Algonquin strandline—one of the most strongly developed proglacial lake strandlines of the Great Lakes—which over the first 100 km from the ice edge has a slope of 0.0010 (Walcott, 1972).

For the Algonquin strandline, there is a good estimate of the wavelength of flexure because observations of displacement are available from the ice edge to beyond the point where the strandline is deformed. This wavelength gives a value for α of 180 km (Walcott, 1970a). Using Figure 5, we find the appropriate value for Laurentide ice load magnitude is 0.075. If this same load magnitude was present in the interior of British Columbia, we would require a flexure parameter of about 100 km to explain the slope of 0.0018. However, because of the smaller size of the Cordilleran Ice Sheet and the relief of the underlying surface, it is probable that the ice thickness, and hence the value of α, was less in British Columbia

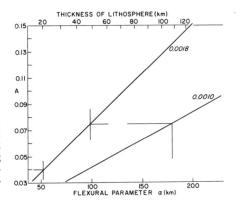

Figure 5. Flexural parameter (α) versus load magnitude (A) with the 0.0010 line representing values that would be associated with a shoreline deformed at that slope (Algonquin shoreline) and the 0.0018 line representing values of these parameters for shorelines deformed to 0.0018 slope (southern British Columbia shorelines).

than in areas covered by the Laurentide Ice Sheet. Using the information given in the introduction, we obtain estimated ice thicknesses of 2 km over the valleys and 0.8 km over the uplands. This suggests an average ice load of about 1.5 km, or somewhat less than half that of the Laurentide Ice Sheet. The magnitude factor A therefore may be less than half that of the Algonquin strandline, that is, <0.4, indicating a maximum value for the flexural parameter of 50 km.

Because of the uncertainties of the size of the load and the shape of its profile, this estimate of α is only approximate. However, it is most unlikely that either the ice was thick enough or its profile was steep enough to make A, the magnitude factor, as large as the Algonquin value. Thus, the steeper tilts on the proglacial strandline of the British Columbia interior indicate that the flexural parameter and, hence, flexural rigidity and lithospheric thickness of this area are much smaller. The maximum probable thickness of the lithosphere appropriate for this tilting of Cordilleran lake strandlines (obtained by using the Laurentide load factor) is less than 50 km; the thickness estimate obtained by using the probable maximum Cordilleran load factor of 0.4 is 20 km (Fig. 5).

According to Wickens (1971), surface wave studies indicate that there is a shallow, low-velocity channel under this part of the Cordillera. Seismic profiling shows a low Pn velocity of 7.8 km per sec over path lengths of several hundred kilometers (M. J. Berry, personal commun.); so the top of the low-velocity zone probably occurs at or near the Mohorovičić discontinuity at a depth of about 35 km. If this zone corresponds to the asthenosphere, then the seismically determined lithospheric thickness is about the same as that estimated from flexural studies.

Also, the electrical conductivity structure of the Cordillera, as determined by magnetotelluric and geomagnetic depth sounding, suggests that the lower crust and possibly upper earth are hydrated and of higher than normal temperature (Caner, 1970). The depth of melting of "wet" granite, according to this model of physical conditions deduced for the conductivity profile, is around 25 km. If this level corresponds to the base of the lithosphere, then the thickness of the lithosphere is near the lower value deduced from flexural studies.

Most of these various sources of information agree on the thickness of the lithosphere. Flexure, surface wave seismic profiling, and electromagnetic methods show that the interior of British Columbia and the Basin and Range region, as contrasted with the stable platforms, have a thinner lithosphere, shallower mantle low-velocity channel, and apparently higher temperatures at shallower depths. In terms of heat flow, the Cordillera has a higher value than the stable platform but less than that of the Basin and Range province (A. Jessop, personal commun.). Both the Basin and Range province and the Cordillera have experienced geologically recent volcanicity and orogeny, but the Basin and Range is seismically more active than the Canadian Cordillera today. All these points seem to be consistent with an upward bulge of the isogeotherms under the Canadian Cordillera, with a consequent decrease in lithospheric thickness and increase in conductivity.

In summary, the steepness of tilts of lake strandlines in the interior of British Columbia compared to those of strandlines formed at the same time on the stable continental platform, where the causative load was probably of greater intensity, suggests a much lower flexural rigidity in the Cordillera. We estimate the thickness of the lithosphere to be between 20 and 50 km and are inclined to accept the lower estimate rather than the higher one.

REFERENCES CITED

Caner, B., 1970, Electrical conductivity structure in western Canada and petrological interpretation: Jour. Geomagnetism and Geoelectricity, v. 22, p. 113-129.
Fulton, R. J., 1969, Glacial lake history, southern Interior Plateau, British Columbia: Canada Geol. Survey Paper 69-37, 14 p.
——1971, Radiocarbon geochronology of southern British Columbia: Canada Geol. Survey Paper 71-37, 28 p.
Geological Association of Canada, 1958, Glacial map of Canada: Geol. Assoc. Canada, Dept. Earth Sci., Univ. Waterloo, Waterloo, Ontario.
Mathews, W. H., 1944, Glacial lakes and ice retreat in south central British Columbia: Royal Soc. Canada Trans., v. 38, sec. IV, p. 39-58.
Walcott, R. I., 1970a, Isostatic response to loading of the crust in Canada: Canadian Jour. Earth Sci., v. 7, p. 716-727.
——1970b, Flexural rigidity, thickness and viscosity of the lithosphere: Jour. Geophys. Research, v. 75, p. 3941-3954.
——1972, Late Quaternary vertical movements in eastern North America: Quantitative evidence of glacio-isostatic rebound: Rev. Geophysics and Space Physics, v. 10, p. 849-884.
Wickens, A. J., 1971, Variations in lithospheric thickness in Canada: Canadian Jour. Earth Sci., v. 8, p. 1154-1162.

MANUSCRIPT RECEIVED BY THE SOCIETY JULY 2, 1973
REVISED MANUSCRIPT RECEIVED OCTOBER 29, 1973

Printed in the U.S.A.

Theory of Velocity of Earthquake Dislocations

JOHANNES WEERTMAN

Departments of Materials Science and Geological Sciences and
Materials Research Center
Northwestern University, Evanston, Illinois 60201

ABSTRACT

In a previous publication, I calculated the velocity at which earthquake dislocations should propagate if the friction law of a fault is dominated by a dependence on the slippage velocity. It was found that earthquake dislocations should travel at either the Rayleigh velocity (edge dislocations), or the shear-wave velocity (screw dislocations), or at a supersonic velocity. In this paper an investigation is made of the permitted earthquake dislocation velocities for faults whose friction law is dominated by a dependence on slippage displacement rather than slippage velocity. It is found that dislocations can move at arbitrary subsonic velocities and at least one transonic velocity.

INTRODUCTION

In this paper I calculate the velocity at which earthquake dislocations move on a fault when the dynamic friction-stress law of the fault differs from the type considered previously. In a previous calculation (Weertman, 1969a), the velocity V of uniformly moving dislocations was found for the dynamic friction law illustrated schematically in Figure 1. The friction stress in Figure 1 decreases with increasing velocity of slippage across the fault. For this friction law, it was shown that a group of smeared-out screw dislocations can move at the shear-wave velocity c or supersonically at an arbitrary velocity greater than c. It was further shown that a group of smeared-out edge dislocations can move subsonically at the Rayleigh velocity $c_r (c_r \approx 0.9\,c)$ or at an arbitrary supersonic velocity. For edge dislocations a supersonic velocity is any velocity greater than c_λ, where c_λ is the longitudinal sound velocity ($c_\lambda \approx 2\,c$).

In a number of earthquakes, the rupture velocity appears to have a velocity that is close to that of the shear-wave or Rayleigh-wave velocity (Press and others, 1961; Kanamori, 1970a, 1970b; Wu and Kanamori, quoted in Wu and others, 1972;

Bollinger, 1970; Bolt, 1972). However, both in these and in other earthquakes, rupture velocities smaller than the Rayleigh velocity are reported (Eaton, 1967; Fukao, 1972; Ben-Menahem and others, 1972; Abe, 1972a, 1972b; Canitez, 1972; Bollinger, 1968; Mikumo, 1973). One supersonic earthquake rupture velocity has been reported (Fukao, 1970). Recent laboratory experiments (Wu and others, 1972; Johnson and others, 1973) have shown that the stick-slip propagation velocity can have a value in the range from subshear to transonic velocities. (A transonic velocity for edge dislocations lies in the range $c \leq V \leq c_\lambda$.)

The theory of my original paper (Weertman, 1969a) is adequate to account for some of these observed earthquake propagation velocities but not for all of them. (It is, of course, possible that some of the discrepancies might be caused by the nonuniform motion of an earthquake dislocation. The dislocation first must accelerate and later must decelerate. The theory is developed for uniform motion only.) The results of the previous paper depend upon the assumed dynamic friction-stress law. In this paper I wish to show that with a different type of dynamic friction-stress law, different velocities are predicted for the earthquake dislocation.

Refer to Weertman (1969b), Weertman and Weertman (1974), and Savage (1974) for further information on the basic theory used in this paper.

THEORY

Consider smeared-out, straight, parallel dislocations moving with a uniform velocity in the x direction on a fault. Let the coordinate system move with the dislocations. The dislocation density $B(x)$ is defined by either

$$D(x) = \int_x^\infty B(x)\,dx \tag{1a}$$

or

$$B(x) = -dD(x)/dx, \tag{1b}$$

where $D(x)$ is the displacement across the fault at x. (Some authors in the earthquake literature regard the displacement $D(x)$ as the dislocation.) The displacement velocity (slippage velocity) $\dot{D} \equiv dD/dt$ across the fault for uniformly moving dislocations is given by the equation

$$\dot{D}(x) = VB(x). \tag{2}$$

The shear stress $\sigma(x)$ that acts across the fault plane and that is produced by the dislocations on the fault plane is given by the following equation, if the dislocations move at a subsonic velocity ($V \leq c$):

$$\sigma(x) = (\mu C/\pi) \int_{-\infty}^\infty B(x')\,(x - x')^{-1}\,dx', \tag{3}$$

where μ is the shear modulus and $C = \beta/2$ for screw dislocations and $C = (2c^2/\beta V^2)(\gamma\beta - \alpha^4)$ for edge dislocations. Here $\beta^2 = 1 - V^2/c^2$, $\gamma^2 = 1 - V^2/c_\lambda^2$, and $\alpha^2 = 1 - V^2/2c^2$.

If the dislocations move supersonically ($V \geq c$ for screw dislocations and $V \geq c_\lambda$ for edge dislocations), the stress is given by

$$\sigma(x) = -\mu C^* B(x), \tag{4}$$

where $C^* = \beta^*/2$ for screw dislocations and $C^* = (2c^2/V^2)(\gamma^* + \alpha^4/\beta^*)$ for edge dislocations. Here $\beta^{*2} = -\beta^2$ and $\gamma^{*2} = -\gamma^2$.

For transonic edge dislocations, the stress is equal to

$$\sigma(x) = -\mu C_t^* B(x) + (\mu C_t/\pi) \int_{-\infty}^{\infty} B(x')(x-x')^{-1} dx', \tag{5}$$

where $C_t^* = 2c^2\alpha^4/\beta^* V^2$ and $C_t = 2\gamma c^2/V^2$.

FRICTION STRESS

Suppose the friction stress σ_f of a fault is given by the general law

$$\sigma_f = \sigma_f(D, \dot{D}). \tag{6}$$

The friction stress thus is a function of the displacement and the displacement velocity across the fault.

Wherever there is slippage on the fault ($\dot{D} \neq 0$), the sum of the dislocation stress $\sigma(x)$ and the applied (tectonic) stress σ_a must equal the frictional stress. Thus,

$$\sigma(x) + \sigma_a = \sigma_f(D, \dot{D}). \tag{7a}$$

Wherever no slippage occurs on the fault, the stress across the fault must be less than the friction stress. Thus,

$$\sigma(x) + \sigma_a < \sigma_f(D, \dot{D}). \tag{7b}$$

The permitted dislocation velocities are found in principle by inserting equations (1) into equation (6), and equation (6) and one of equations (3), (4), or (5) into equations (7a and 7b).

For the dynamic friction law illustrated schematically in Figure 1, σ_f is a function only of $\dot{D}$. Using the above procedure with this law makes it possible to find the permitted velocities mentioned in the Introduction without knowing the exact shape of the curve of σ_f versus $\dot{D}$. (Solutions were found for the case in which the applied stress σ_a is smaller than the value of the friction stress for $\dot{D} = 0$ and larger than the friction stress for $\dot{D} = \infty$.)

Suppose the friction stress depends predominantly upon the displacement D rather than the displacement velocity $\dot{D}$. Figure 2 shows a simple functional dependence of σ_f on D. It is assumed that if $\dot{D} = 0$, the friction stress $\sigma_f = \sigma_1$, regardless of the value of D. On the other hand, if $\dot{D} \neq 0$, $\sigma_f = \sigma_1$ for a displacement in the range $0 \leq D \leq D_c$, where D_c is a critical displacement. For $D > D_c$, the friction stress $\sigma_f = \sigma_0$, where $\sigma_0 < \sigma_1$. We assume that once a dislocation propagates down the fault, the fault forgets its past history and D effectively is zero again.

DISLOCATION VELOCITIES

Suppose that the applied stress σ_a has a value in the range $\sigma_0 < \sigma_a < \sigma_1$. Suppose also that somehow a smeared-out packet of dislocations has been set into motion on the fault plane in the positive x direction. These dislocations exist in the region $-L \leq x \leq L$, where L is a constant to be determined (see Fig. 3a). The stress $\sigma(x)$ produced by these dislocations must equal $(\sigma_1 - \sigma_a)$ in the leading part of the distribution. In the trailing part of the distribution, the stress $\sigma(x)$ must equal $(\sigma_0 - \sigma_a)$. Figure 3b shows such a stress distribution. At $x = L'$, a constant to be determined, the stress changes from $(\sigma_1 - \sigma_a)$ to $(\sigma_0 - \sigma_a)$.

The value of the constant L' in Figure 3b is determined by the requirement that at $x = L'$ the displacement D must equal the critical displacement. Thus, L' is given by the equation

$$D_c = \int_{L'}^{L} B(x)\,dx. \tag{8}$$

It is impossible to satisfy the friction law of Figure 2 and the equations for a supersonic dislocation. Supersonic solutions of the problem do not exist.

A subsonic solution for $B(x)$ that produces the stress function $\sigma(x)$ shown in Figure 3b can be found by making a suitable linear combination of the dislocation density functions given by equations (6) and (9) in the paper of Weertman (1971a). [Note that equation (9) of Weertman, 1971a, should read $B(y) = (2\alpha\sigma_1/\pi\mu) \log|\{(L^2 - y^2)^{1/2} + (L^2 - L'^2)^{1/2}\}/\{(L^2 - y^2)^{1/2} - (L^2 - L'^2)^{1/2}\}|.$] The solution is

$$B(x) = 0 \tag{9a}$$

for $|x| \geq L$, and

$$B(x) = \{(\sigma_1 - \sigma_0)/2\pi\mu C)\}\{(|L'|/L')\,F(x,L,L') + G(x,L,L')\} \tag{9b}$$

for $-L \leq x \leq L$. The functions F and G are given by

$$F = \log|\{x(L^2 - L'^2)^{1/2} + |L'|(L^2 - x^2)^{1/2}\}/\{x(L^2 - L'^2)^{1/2}$$
$$-|L'|(L^2 - x^2)^{1/2}\}|, \tag{10}$$

and

$$G = \log|\{(L^2 - x^2)^{1/2} + (L^2 - L'^2)^{1/2}\}/\{(L^2 - x^2)^{1/2}$$
$$-(L^2 - L'^2)^{1/2}\}|. \tag{11}$$

The value of L' must satisfy the equation

$$L'/L = \sin\{\pi(\sigma_1 + \sigma_0 - 2\sigma_a)/2(\sigma_1 - \sigma_0)\}, \tag{12}$$

as well as equation (8). The ratio L'/L decreases in value from 1 to -1 as σ_a increases in value from σ_0 to σ_1.

The total displacement (or total length of the Burgers vector) D_t, defined by the equation

$$D_t = \int_{-L}^{L} B(x)\,dx, \tag{13}$$

that is produced when a group of dislocations moves across the fault plane is given by

$$2\pi\mu CD_t/L(\sigma_1 - \sigma_0) = 2\pi(1 - L'^2/L^2)^{1/2}. \tag{14}$$

If equation (9) is set into equation (8), the following equation is found:

$$2\pi\mu CD_c/L(\sigma_1 - \sigma_0) = R(L'/L), \tag{15}$$

where the function $R(z)$ is given by

$$R(z) = (1 - z^2)^{1/2}(\pi - 2\sin^{-1}z) + (|z| - z)\log z^2. \tag{16}$$

Combining equations (14) and (16) gives

$$D_t/D_c = 2\pi(1 - L'^2/L^2)^{1/2}/R(L'/L). \tag{17}$$

Equation (17) predicts that the total displacement D_t is a function only of the applied stress σ_a. It is not a function of the dislocation velocity. That is, once the value of σ_a is specified, equation (12) determines the ratio L'/L. The term D_c is a constant. Thus, D_t does not depend on V.

Figure 4 shows a plot of the ratio D_t/D_c versus the ratio σ_a/σ_0 for the case in which $\sigma_1 = 5\sigma_0$. It can be seen that D_t generally decreases in magnitude as σ_a increases. Thus, the largest earthquakes should occur when σ_a has a value very close to σ_0.

Let us examine next the conditions on the dislocation velocity V. Consider again equation (15). This equation gives the relation between the ratios C/L and L'/L. The ratio L'/L is fixed, once σ_a is specified in equation (12). Thus, equation (15) gives the relation between C/L and σ_a. The term C is, of course, velocity dependent. In Figure 4 the term $2\pi\mu CD_c/L(\sigma_1 - \sigma_0) = R(L'/L)$ is plotted against σ_a/σ_0 for the case in which $\sigma_1 = 5\sigma_0$.

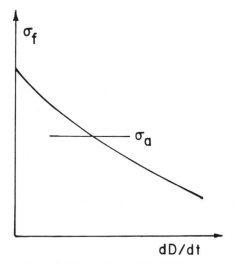

Figure 1. Schematic plot of the friction stress σ_f versus displacement (slippage) velocity dD/dt. An applied stress σ_a that can produce dislocation propagation also is shown.

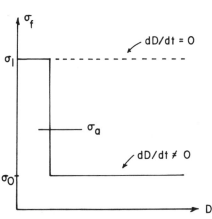

Figure 2. Schematic plot of friction stress σ_f versus displacement D for $dD/dt \neq 0$ and for $dD/dt = 0$. An applied stress σ_a that can produce dislocation propagation also is shown.

The term C has a finite, positive value at $V = 0$. This term decreases in magnitude and approaches zero in value as V approaches the Rayleigh velocity c_r in the case of edge dislocations, or the shear-wave velocity c in the case of screw dislocations. Since L can have an arbitrary value, according to our equations, subsonic solutions are possible at all velocities $0 \le V \le c$ for screw dislocations and $0 \le V \le c_r$ for edge dislocations. (Velocities in the range $c_r < V \le c$ are excluded for edge dislocations because C has a negative value at these velocities.) However, suppose L is considered to have a finite value. The function $R(L'/L)$ approaches zero in value when σ_a is approximately equal to either σ_0 or σ_1. Therefore, at these stress levels, the dislocation velocity has to be approximately equal to the Rayleigh velocity for edge dislocations or the shear-wave velocity for screw dislocations.

The equations of this section are greatly simplified for the limiting case in which $\sigma_1 \to \infty$ and $D_c \to 0$ in such a manner that the product $D_c \sigma_1$ is a constant. (This approximation is identical to one implicit in the Griffith fracture theory of a brittle solid. In that theory it is assumed that a solid can withstand an infinite stress but only for an infinitesimal displacement.) Set $D_c \sigma_1 = A$, where A is a constant. Equation (9) becomes

$$B(x) = \{(\sigma_a - \sigma_0)/\mu C\} \{(L + x)/(L - x)\}^{1/2} \tag{18}$$

for $-L \le x \le L$ and $B(x) = 0$ for $|x| > L$. Equation (14) reduces to

$$D_t = \pi L(\sigma_a - \sigma_0)/\mu C = A/(\sigma_a - \sigma_0) \tag{19}$$

and equation (15) to

$$L/C = A\mu/\pi(\sigma_a - \sigma_0)^2. \tag{20}$$

The ratio L'/L is, of course, equal to 1.

TRANSONIC SOLUTION FOR EDGE DISLOCATIONS

At the velocity $V = 2^{1/2} c$, the constant α^2 that appears in equation (5) is equal to zero. The supersonic type term thus drops out of this equation, and the equation is subsonic in character. (The velocity $2^{1/2} c$ is Eshelby's singular velocity. It is discussed in Weertman and Weertman, 1974.) A solution for dislocations propagating at transonic speeds can be found simply by substituting the term C_t that appears in equation (5) for the term C in the equations of the preceding section. The term L is fixed in value for this solution because C_t now is fixed. I believe that no solution exists at any other transonic velocity with the friction law of Figure 2. However, I have not been able to prove that no other solution exists.

DISCUSSION

In our earlier paper, which was based on a dynamic friction law in which the slippage velocity is the dominant factor, it was found that dislocations on faults should move either at the Rayleigh-wave velocity (edge dislocations), or at the shear-wave velocity (screw dislocations), or at a supersonic velocity. In this paper we have seen that for a dynamic friction law in which slip displacement is the dominant factor, the dislocations can move at any subsonic velocity (except for the range $c_r < V \leq c$ prohibited to edge dislocations) and, in the case of edge dislocations, at least one transonic velocity. Thus, the displacement-dominated friction law is capable qualitatively of accounting for the observed earthquakes that propagate with rupture velocities that are appreciably smaller than the Rayleigh velocity. It can also qualitatively account for the extremely slow propagation velocities of creep events on faults. The stress-displacement law for a fault creep event (Nason and Weertman, 1973, Fig. 7) is similar to the friction law of Figure 2. Dislocation velocities that are close to the Rayleigh and the shear-wave velocities are also possible in this theory. In fact, the larger the total Burgers vector of the earthquake dislocation, the more likely it is that the dislocation will move at a fast velocity. Paradoxically, provided that the dynamic friction stress is exceeded, the smaller the applied stress is on a fault, the larger the earthquake dislocation becomes and the more likely it is that the dislocation will move at a sound velocity. The higher the applied stress the smaller is the Burgers vector of a fault dislocation and, if the stress is not too great, the more likely it is that dislocations move slowly. This paradoxical result is a consequence of the fact that the dislocations are piled up against a frictional "barrier" of a constant strength. The larger the applied stress, the smaller the stress concentration produced by the piled-up dislocations has to be in order to overcome the barrier. Thus, the smaller the total Burgers vector of the piled-up dislocations, the larger the applied stress. We believe that because of this odd result, a moving group of dislocations may exhibit unstable behavior. The dislocations may split into separate groups or move in a nonuniform manner.

The friction laws considered in this and in our previous paper are undoubtedly too simple. The actual friction law, if there is one, is likely to depend upon both

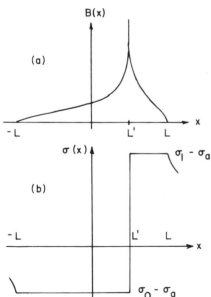

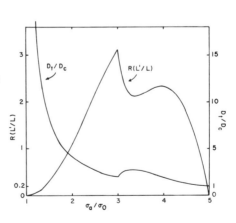

Figure 3. (a) Plot of dislocation density $B(x)$ calculated from equation (9), and (b) stress $\sigma(x)$ produced by the dislocation density $B(x)$ versus x.

Figure 4. Plot of D_t/D_c versus σ_a/σ_0 and plot of $2\pi\mu CD_c/L(\sigma_1 - \sigma_0) = R(L'/L)$ versus σ_a/σ_0. Both plots for $\sigma_1 = 5\sigma_0$. The discontinuity in slope occurs at $L'/L = 0$.)

the displacement D and the displacement velocity $\dot{D}$. The integral equations that must be solved for a friction law containing both these terms are so difficult that it appears impossible to deduce by analytic techniques the permitted dislocation velocities.

We have not discussed mixed dislocations in this paper. (A mixed dislocation is a combined edge and screw dislocation.) The equations of our previous paper do not permit a solution for a uniformly moving mixed dislocation. The screw and edge components of the dislocation move at different velocities, and the mixed dislocation would separate into a pure-edge and a pure-screw dislocation. There is no difficulty in applying the analysis of this paper to the problem of mixed dislocations. Thus, mixed dislocations also can propagate at arbitrary subsonic velocities.

ACKNOWLEDGMENT

This research was supported by the National Science Foundation and the Advanced Research Projects Agency of the Department of Defense through the Northwestern University Materials Research Center.

REFERENCES CITED

Abe, K., 1972a, Focal process of the South Sandwich Islands earthquake of May 26, 1964: Physics Earth and Planetary Interiors, v. 5, p. 110-122.

——1972b, Mechanisms and tectonic implications of the 1966 and 1970 Peru earthquakes: Physics Earth and Planetary Interiors, v. 5, p. 367-379.

Ben-Menahem, A., Rosenman, M., and Israel, M., 1972, Source mechanism of the Alaskan earthquake of 1964 from amplitude of free oscillations and surface waves: Physics Earth and Planetary Interiors, v. 5, p. 1-29.

Bollinger, G. A., 1968, Determination of earthquake fault parameters from long-period P waves: Jour. Geophys. Research, v. 73, p. 785-807.

——1970, Fault length and fracture velocity for the Kyushu, Japan, earthquake of October 3, 1963: Jour. Geophys. Research, v. 75, p. 955-964.

Bolt, B. A., 1972, San Fernando rupture mechanism and the Pacoima strong-motion record: Seismol. Soc. America Bull., v. 62, p. 1053-1061.

Canitez, N., 1972, Source mechanism and rupture propagation in the Mudurnu Valley, Turkey, earthquake of July 22, 1967: Pure and Appl. Geophysics, v. 93, p. 116-124.

Eaton, J. P., 1967, Instrumental seismic studies in the Parkfield-Cholame, California, earthquake of June-August, 1966: U.S. Geol. Survey Prof. Paper 579, p. 57-65.

Fukao, Y., 1970, Focal process of a deep focus earthquake as deduced from long period P and S waves: Earthquake Research Inst. Bull., v. 48, p. 707-727.

——1972, Source process of a large deep-focus earthquake and its tectonic implications—The western Brazil earthquake of 1963: Physics Earth and Planetary Interiors, v. 5, p. 61-76.

Johnson, T., Wu, F. T., and Scholz, C. H., 1973, Source parameters for stick-slip and for earthquakes: Science, v. 179, p. 278-280.

Kanamori, H., 1970a, Synthesis of long-period surface waves and its application to earthquake source studies—Kurile Islands earthquake of October 13, 1963: Jour. Geophys. Research, v. 75, p. 5029-5040.

——1970b, The Alaska earthquake of 1964—Radiation of long-period surface waves and source mechanism: Jour. Geophys. Research, v. 75, p. 504-505.

Mikumo, T., 1973, Faulting processes of the San Fernando earthquake of February 9, 1971, inferred from static and dynamic near-field displacements: Seismol. Soc. America Bull., v. 63, p. 249-269.

Nason, R., and Weertman, J., 1973, A dislocation theory analysis of fault creep events: Jour. Geophys. Research, v. 78, p. 7745-7751.

Press, F., Ben-Menahem, A., and Toksöz, M. N., 1961, Experimental determination of earthquake fault length and rupture velocity: Jour. Geophys. Research, v. 66, p. 3471-3485.

Savage, J. C., 1974, Dislocations in seismology, in Nabarro, F.R.N., ed., Dislocation theory: A treatise: New York, Marcel Dekker, Inc., chap. 10 (in press).

Weertman, J., 1969a, Dislocation motion on an interface with friction that is dependent on sliding velocity: Jour. Geophys. Research, v. 74, p. 6617-6622.

——1969b, Dislocations in uniform motion on slip or climb planes having periodic force laws, in Mura, T., ed., Mathematical theory of dislocations: Am. Soc. Mech. Engineers, p. 178-202.

——1971a, Theory of water-filled crevasses in glaciers applied to vertical magma transport beneath oceanic ridges: Jour. Geophys. Research, v. 76, p. 1171-1183.

Weertman, J., and Weertman, J. R., 1974, Moving dislocations, in Nabarro, F.R.N., ed., Dislocation theory: A treatise: New York, Marcel Dekker, Inc., chap. 9 (in press).

Wu, F. T., Thomson, K. C., and Kuenzler, H., 1972, Stick-slip propagation velocity and seismic source mechanism: Seismol. Soc. America Bull., v. 62, p. 1621-1628.

MANUSCRIPT RECEIVED BY THE SOCIETY MAY 23, 1973
REVISED MANUSCRIPT RECEIVED OCTOBER 1, 1973

Printed in the U.S.A.

GEOCHEMISTRY

ate of a diagenetic
Diagenetic Reactions as Stochastic Processes: Application to the Bermudian Eolianites

G. Michel Lafon

Department of Earth and Planetary Sciences
The Johns Hopkins University
Baltimore, Maryland 21218

AND

H. L. Vacher

Department of Geology,
State University of New York at Binghamton,
Binghamton, New York 13901
and
Consultant Geologist/Hydrologist
Public Works Department
Hamilton, Bermuda

ABSTRACT

Diagenetic reactions in rock bodies are best described as stochastic processes. When the diagenetic environment is fairly constant and when destruction of an unstable mineral is the limiting step, we propose that the rate of a diagenetic reaction can often be expressed by the equation $\delta C(t) = \Lambda(t)\, C(t)\, dt$, where $C(t)$ denotes the concentration of reactant minerals and $\Lambda(t)$ is a random function characteristic of the diagenetic environment. In the simplest case, the initial concentration is a constant and Λ is a random variable independent of time. Then, the logarithm of the concentration of an unstable mineral decreases linearly in the mean with time, and its standard deviation is an increasing linear function of time. This model describes well the time dependence of carbonate alterations in the vadose zone of the Bermudian eolianites. We further suggest that, owing to the stochastic nature of fluid flow in natural rock bodies, our model may apply to many diagenetic and metamorphic reactions.

INTRODUCTION

The geologic record contains numerous instances of mineral assemblages persisting metastably in environments where other assemblages would be thermodynamically more stable. It is also well known that, "given enough time," more stable minerals or mineral assemblages gradually supersede less stable ones. The time scales involved, however, are quite variable, ranging from less than a few thousand to hundreds of millions of years. Generally relating the degree of advancement of a given diagenetic process to time is still a major unsolved problem. Ultimately, the answer must come from a good understanding of the detailed individual kinetic processes at work during diagenesis. Although progress is being made (Nielsen, 1964; Berner, 1971), we are only beginning to appreciate the great complexity of chemical reactions and kinetic processes in natural environments. There is little hope at present that models based on kinetic data obtained from laboratory experiments can explain or predict the time dependence of diagenesis in rock bodies.

Another approach to this problem is to recognize that chemical reactions in rocks are best described as stochastic processes. This is easily demonstrated if we consider a familiar parameter, the concentration of a given mineral species in a rock unit, which is a priori a four-dimensional function $C(x, y, z, t)$ of position and of time. A precise and exhaustive knowledge of the values of C can never be practically obtained because of material limitations. Rather, one usually analyzes a limited number of samples considered representative of the rock unit, and one adopts some kind of average value as *the* concentration at time t. Implicit in this approach is the belief that the values of $C(x, y, z, t)$ are clustered in such a way that they can be adequately represented by a few numerical parameters, for example, the first and second moments of a probability distribution. More formally, the procedure outlined above is equivalent to replacing the four-dimensional function C by a random function $C(t)$, that is, a family of random variables depending on the parameter time. To determine the actual values of the concentration, one trades off the theoretically greater precision of the complete description afforded by $C(x, y, z, t)$ against the less complete but more practical description given by $C(t)$.

Similarly, the other physicochemical properties affecting natural chemical reactions, such as surface area of reactant grains, rate of flushing of an aquifer, chemical affinity, and various diffusion coefficients, are all four-dimensional functions of position and time that can be practically described only as random functions because no accurate census of their numerical values is feasible. Thus, the kinetics of a chemical reaction consuming reactant A of concentration $C_A(t)$ will in general be described by one or more random differential equations such as

$$\delta C_A(t) = h[dt, t, C_{A(t)}, U_1(t), ..., U_k(t)], \qquad (1)$$

where $U_1(t), ..., U_k(t)$ are random functions characterizing the physicochemical environment in which the reaction takes place, and h is some functional dependence.

Laboratory experiments probably cannot reproduce all of the considerable variability of natural diagenetic processes and cannot even approach the time scales of most natural reactions. We suggest that it may be more rewarding to study actual field occurrences of diagenetic transformations in cases where the diagenetic environment is well known, can be taken as reasonably constant through time, and the duration of the diagenetic process can be estimated. Such investigations are the analogs of laboratory experiments, albeit on much larger time and size scales. If we restrict our attention to relatively large systems evolving over long intervals of time, we may hope that much of the unknown variability will average out as a random noise component with mean close to zero and leave in evidence a few significant relations between the major parameters of the process. In other words, the very constraints forced upon us by the nature of the geologic record (limited sampling in time and space, incomplete record of events, necessary averaging for each recorded event) may help us distinguish between the important factors of diagenesis and those that can be lumped together as contributing to random noise.

Our approach can be used in two different, though not mutually exclusive, ways. First, we can obtain a better description of the time-dependent evolution of a diagenetic process than has been available before. We introduce a priori one or several unknown random functions to replace the unknown function h in equation (1), and we express the rate of the reaction in terms of these random functions. Their values are determined from the known changes with time of $C_A(t)$. Thus, we obtain a quantitative description of diagenesis based on random functions that are characteristic of the diagenetic process. Second, if we can identify the random functions $U_1(t)$, ..., $U_k(t)$, and the form of h needed to express the change with time of $C_A(t)$, then we have learned the identity of the physicochemical parameters (corresponding to the random functions) that significantly affect the diagenetic process, and we have an explicit model relating them to the variable of greatest interest—mineral concentration.

To illustrate these concepts, we develop below a simple stochastic model of diagenesis, which describes well the vadose alteration of calcareous eolianites in Bermuda. We suggest that similar treatment of comparable data may give us a better understanding of carbonate alterations and perhaps also of more general processes in diagenesis and metamorphism.

STOCHASTIC MODEL

Consider a mineral assemblage A that is thermodynamically unstable with respect to another assemblage B in the environment of interest. Assume further that there exist mechanisms which allow the reaction A → B to proceed. Let the volume concentration of A be a nonnegative random function $C_A(t)$, with initial distribution $C_A(0)$. We can describe the advancement of the reaction by means of the successive distributions of, say, $C_A(t)$ as t increases. The reaction is completed after a time T if we have, for all $t > T$, $\mathbf{P}\{C_A(t) > 0\} = 0$, where $\mathbf{P}\{E\}$ denotes the probability of event E.

To use equation (1) to determine explicitly the change of $C_A(t)$ with time, we need to know the form of h, and we must identify the parameters $U_1(t)$, ..., $U_k(t)$ in equation (1). A priori likely candidates include the surface areas of A and B that are available for reaction, the chemical affinity drive of the reaction, activation energies for the formation of excited states or for nucleation of B, possible diffusion gradients away from reacting boundaries, and frequency of exposure of a reactant grain to interstitial fluids. In a natural situation, the probability distributions of these quantities are generally unknown, and it is highly unlikely that equation (1) may be written explicitly. However, some general considerations lead to considerable simplification of the problem.

We suggest that, among the several random functions of interest, the mineral concentrations play a particular and privileged role because they can often be related to the rate of reaction in a simple manner and they are by far the most easily measured. The rate of diagenetic reaction can be written as $-\delta C_A(t)/dt$ or as $\delta C_B(t)/dt$. For the time and size scales we consider here, intermediate storage of material in a form that is neither A nor B (for example, in some interstitial fluid) is negligible and we have $-\delta C_A(t)/dt = \delta C_B(t)/dt$. Thus, the overall rate of reaction is controlled either by the rate of destruction of A or by the rate of production of B. In the former case, we rewrite equation (1) as

$$\delta C_A(t)/dt = h'[U_1(t), ..., U_k(t)]\, h''[C_A(t)]; \qquad (2)$$

in the latter, a similar equation can be written to relate $\delta C_B(t)$, $C_B(t)$, and the other random functions.

Because the problem of determining h' and the random functions other than the concentrations still appears extremely complex, we restrict ourselves to a descriptive approach. We replace $h'[U_1(t), ..., U_k(t)]$ by a single random function $\Lambda(t)$, which is now characteristic of the diagenetic environment as a whole. This is equivalent to assuming that, during the diagenetic process, the unknown environmental factors remain "fairly constant," and therefore that they can be lumped together and summarized quantitatively by $\Lambda(t)$. Furthermore, we shall only treat in the remainder of this paper the case where destruction of A is the limiting step. A similar, but not identical, treatment could be applied to the case where production of B is limiting.

It remains to find the form of the function h''. We suggest that there is often proportionality between $\delta C_A(t)$ and $C_A(t)$ so that $h''[C_A(t)] \equiv C_A(t)$. We illustrate this proposition by considering a rock volume uniformly permeated by a homogeneous fluid phase through which the reaction takes place, so that the velocity of the reaction is proportional to the surface area of the reactant. If the reactant minerals are present as grains with well-defined shapes and size frequencies, the reactant surface area can also be taken, in a first approximation, proportional to $C_A(t)$. A similar conclusion can result from a different situation in which two independent conditions must be fulfilled before the alteration reaction can actually proceed: presence of A, and reaching a suitable set of environmental conditions (for example, exceeding an activation energy barrier or providing an aqueous phase without which the reaction would be exceedingly slow). If the probability of fulfilling the second condition is independent of the presence or absence of A, it is clear that on the average, the velocity of the reaction will be proportional to $C_A(t)$, which is a measure of the probability of finding some A within a unit volume.

Having simplified the problem to this extent, we write the stochastic differential equation that describes the diagenetic process as

$$\delta C_A(t)/dt = \Lambda(t) C_A(t). \qquad (3)$$

Let us now assume that the random functions $\Lambda(t)$ and $C_A(t)$ are sufficiently well behaved that the operations of stochastic derivation and integration have a meaning (Lévy, 1948; Blanc-Lapierre and Fortet, 1953). We can rewrite equation (3) as

$$[1/C_A(t)]\,[dC_A(t)/dt] = \Lambda(t). \qquad (4)$$

Posing $Y(t) = \ln C_A(t)$, we have

$$dY(t)/dt = \Lambda(t), \qquad (5)$$

and symbolically,

$$Y(t) = \int_0^t \Lambda(t)\, dt.$$

In practical applications, the values of $Y(t)$ are known, and we wish to determine those of $\Lambda(t)$. Let $g(x, t)$ and $f(x, t)$ be the probability density functions of Y and Λ, respectively. By definition, we have

$$\mathbf{P}\{\Lambda(t) < \omega_1\} = \int_{-\infty}^{\omega_1} f(t, x)\, dx$$

and

$$\mathbf{P}\{Y(t) < \omega_2\} = \int_{-\infty}^{\omega_2} g(t, x)\, dx, \tag{6}$$

with the conditions that the initial distribution $Y(0)$ is known and that equation (5) is satisfied. Equating terms, we obtain the following:
(a) if $Y(0)$ is concentrated at one point Y_0, then

$$g(t, x) = \int_0^t f(t, x + Y_0)\, dt, \tag{7}$$

or (b) if $Y(0)$ has a probability density function $g(0, \xi)$,

$$g(t, x) = \int_{-\infty}^{+\infty} \left[\int_0^t f(t, x + \xi)\, dt \right] g(0, \xi)\, d\xi \tag{8}$$

where we have used equation (7) and assumed that $\Lambda(t)$ and $C_A(0)$ are stochastically independent.

To determine f knowing $g(t, x)$ and $g(0, \xi)$ is generally difficult (if not impossible) when g is an arbitrary function and the data are not closely spaced in time. Because most geologic information is often haphazardly distributed in time, it is worthwhile to investigate some special cases in which equation (8) takes a particularly simple form. We shall consider below the special case: $f(t, x) \equiv f'(x)$ and $g(0, \xi) = Y_0$, a constant. We then have

$$g(t, x) = t f'(x + Y_0), \tag{9}$$

or, redefining $Y(t)$ as $\ln[C_A(t)/C_A(0)]$,

$$g(t, x) = t f'(x). \tag{10}$$

If the mean and standard deviation of $Y(t)$ are μ_Y and σ_Y, the corresponding parameters for Λ are $(1/t)\mu_Y$ and $(1/t)\sigma_Y$, respectively. Conversely, if we assume that f' is known, we can predict the distribution of Y as a function of time.

The predicted distribution of concentrations of A in a rock body is somewhat analogous to that obtained from a first-order kinetic model. In $C_A(t)$ decreases in the mean linearly with increasing time, and its standard deviation is a linear function of time. One can also define a half-life parameter of the diagenetic process, that is, the time at which the mean logarithmic concentration of A is equal to the initial concentration minus ln 2.

These results, however, do not mean that the actual chemical reactions involved in the diagenetic process are first order. These reactions are certainly complex and involve incongruent solution, dissolution and reprecipitation, heterogeneous nucleation, crystal growth, and so forth. We do not expect them to be reducible to a simple exponential decay; on the contrary, we have deliberately neglected consideration of these small-scale phenomena to derive relations between statistical variables representative of large rock volumes. The well-known laws of chemical kinetics rest on an analysis of microscopic, molecular-scale phenomena. Our results, on the other hand, follow from the mathematical form of equation (3) and the subsequent simplifying assumptions discussed above. We caution the reader that our approach is not based on conventional chemical kinetics, in spite of superficial similarities, and that it cannot intuitively be compared with the latter.

More specifically, the simple exponential decay in the mean implied by equation (10) is the result of the following assumptions: 1. Equation (3) is valid. 2. The unknown random function $\Lambda(t)$ has a probability density function $f'(x)$ independent of time. 3. The initial concentration of A is a constant. All three assumptions are critical, and if the second or third were omitted, the mathematical result would not be similar to a first-order kinetic law. Note also that, because we use the random function $C_A(t)$, we average the spatial variability of the actual chemical processes at any given instant t. Assumption 2 is equivalent to assuming that this spatial average is independent of time and thus precludes application of our model to situations in which environmental factors are greatly variable with time. We could not treat a case where the diagenetic environment changes substantially with time (say, from subaerial to marine conditions). Determining how much variation of the diagenetic process with respect to time and space can be accommodated by our model is best done a posteriori, by comparison between predictions and data.

There are not many precise observations relating the change in composition of a rock body to the duration of a diagenetic process that satisfies the conditions stated above. To test the quantitative aspects of our theory, we have used recently collected data on the alteration of carbonate eolianites in Bermuda.

ALTERATION OF BERMUDIAN CALCARENITES BY VADOSE SEEPAGE

General Features

Bermuda is a much-studied group of limestone islands, located about 1,000 km east of the eastern coast of the United States, that represents only a small fraction of a 650 km^2 platform presently submerged to a depth of about 20 m. The exposed limestones are Pleistocene calcarenites, over 90 percent of which are of eolian origin. These cemented calcareous sands, deposited during successive Pleistocene submergences of the Bermudian platform (see Land and others, 1967, and Vacher, 1973, for a detailed account), are but an altered version of the sands currently being deposited at, or just off, the present-day shoreline. The altered limestones range from loosely cemented grains of aragonite and magnesium calcite to dense rocks consisting entirely of secondary calcite.

In a classic paper, Land and others (1967) used the parallel changes of several diagenetic features, including mineralogy, to define a succession of five diagenetic grades spanning the spectrum of Bermudian lithologies. These authors recognized a general progression in diagenetic grade as a function of geologic age, but they also pointed out that part of the considerable diagenetic variability could be associated with the action of different diagenetic environments (for example, vadose versus phreatic; see also Land, 1970) and microenvironments (for example, areas of vadose flow versus vadose seepage). More recently, extensive petrographic and mineralogic investigations of Bermudian rocks (Ristvet, 1972; Ristvet and Vacher, in prep.) have shown that, if attention is restricted to a single diagenetic environment, the change in mineralogy as a function of depositional age is indeed systematic, there being little compositional overlap between successive stratigraphic units (Vacher, 1973, Fig. 5).

Diagenetic Reactions

Nearly all of the calcarenite material originally consisted of high-magnesium calcite and aragonite, which are both thermodynamically unstable with respect to low-magnesium calcite in their natural environment. Although diagenetic alterations have not been reproduced experimentally at low temperatures, Thorstenson and others (1972) have shown that CO_2-rich water percolating through piles of material similar to the original carbonate sand preferentially dissolves the less stable minerals and precipitates low-magnesium calcite cements that simulate natural textures. The systematic and pronounced mineralogic change in the eolianites as a function of their age is further proof that their present composition is largely the result of the following symbolic reactions: high-magnesium calcite $\rightarrow$ low-magnesium calcite + Mg^{++}, and aragonite $\rightarrow$ low-magnesium calcite.

Diagenetic Environment

The largest portion, by far, of the presently exposed Bermudian calcarenites is being and has been altered exclusively in the vadose zone (Vacher, 1973), a volume where environmental conditions are rather uniform and that can be recognized or accurately located in the rocks.

Today, the water table, which defines the lower limit of the vadose environment, has a maximum elevation at the widest sections of the islands of about 0.5 m above mean sea level. Because the elevation of the water table in islands varies directly with recharge rate and island width and inversely with hydraulic conductivity (Bear, 1970), it is reasonable to conclude that during times when Pleistocene sea level was above present sea level and Bermuda was of comparable width or smaller than present, the water table was no more than a few meters above the sea level at that time. Thus the Bermudian sea level curve inferred from stratigraphic relations of marine limestones and eolianites (Land and others, 1967; Vacher, 1971, 1973) can be used to infer, in turn, the location of eolianite volumes that have been subjected only to vadose diagenesis.

The intermediate vadose environment (Domenico, 1972) extends from below the surficial layer penetrated by roots to the capillary fringe of the water table. Water passing through this hydrologic environment has in general previously resided in the overlying soil, and it is en route to the water table. Downward movement of water is along preferred channels (vadose flow) and by a more general percolation (vadose seepage; Thrailkill, 1968). In Bermuda, the overwhelming bulk of the intermediate vadose environment is one of vadose seepage. Preferred channels are easily identified and are, in general, anomalously well cemented, resistant,

TABLE 1. SAMPLED UNITS

Stratigraphic unit	Provenance	
	South series (largely reef-derived)	North series (largely lagoon-derived)
Paget Formation		
Upper Paget	S_I (Southlands eolianite)	
Lower Paget	S_{II} (Spice Hill eolianite)	N_{II} (Barker's Hill eolianite)
Sub Paget	S_{III}	N_{III}

and chalky; they commonly extend downward from large, carrot-shaped, soil-filled structures—called palmetto stumps, solution pipes, or soil pipes in the Bermudian literature.

Vadose seepage through the well-sorted Bermudian sand is probably very similar to that shown in laboratory experiments by Smith (1967). Water is held against gravity by capillary forces and occurs largely as intergranular pendular bodies, the percentage volume of which is the field capacity parameter. Field capacity is exceeded in a succession of downward-moving wetting fronts, each corresponding to past events of infiltration from the overlying soil (see also Remson and others, 1960).

Current hydrologic investigations enable quantification of certain features of the Bermudian vadose environment. Application of the Penman technique for computing potential evapotranspiration (Penman, 1963) and the Thornthwaite technique for constructing a water-balance inventory (Thornthwaite and Mather, 1955) suggests that 15 to 20 cm per year of water infiltrates the vadose environment from the overlying soil. The annual infiltration, amounting to 10 to 15 percent of the annual rainfall, occurs seasonally, principally from October to May. Laboratory and field investigations by Plummer (1970) indicated that the water infiltrating from the soils is at equilibrium with a partial pressure of CO_2 of about 0.01 atm. Depending on duration of residence in the soil, the infiltrating water is alternately saturated and undersaturated with low-magnesium calcite. Most of the diagenetic reactions probably involve the aqueous phase and consist of successive dissolution and reprecipitation (Purdy, 1965).

Environmental conditions in the vadose zone appear to be uniform enough so that equation (3) may provide a good description of the natural diagenetic processes. In the following, we restrict our discussion to the case of vadose alteration only.

TABLE 2. MINERALOGIC COMPOSITION OF SOUTH-SHORE BEACHES

Sample no.*	Wt %		
	Aragonite	High-Mg calcite	Calcite
2	33	57	10
3	31	59	10
4	31	57	12
Average	31.67	57.67	10.67

*In Ristvet (1972).

Diagenetic Data

The mineralogic data used here are selected from the quantitative x-ray diffraction results of Ristvet (1972) and are listed in the Appendix Table 1. Ristvet's data are reported in weight percent, and his reproducibility is ±5 percent. Although our theory applies strictly to concentrations in volume percent, we used the data without further manipulations because, in view of all the other possible errors involved in the calculations, we consider the difference negligible here.

All samples are from a map unit of Vacher (1973, unpub. data) and were selected to ensure that their diagenetic environment has been solely one of vadose seepage. Selection criteria were as follows:

1. The samples are from large exposures, such as quarries and roadcuts, and, on the basis of field observations during sampling, they are considered representative of the sampled face; specifically, samples are not from areas of anomalous cementation related to concentrated flow.

2. They are from elevations above subsequent sea-level high stands, according to the sea-level history worked out by Vacher (1971, 1973). Thus the samples were never subjected to phreatic submergence.

3. They are from at least 2 m below modern or fossil soils, so that diagenetic effects of present and former upper vadose zones are avoided. In addition, all selected samples are from the foreset wedge of the eolianites and well below the rollover area (see Vacher, 1973, Figs. 10 and 12). Thus, all samples occupied similar positions within the lithified dune ridges.

The sampled units can be classified according to both provenance and stratigraphic age (Table 1). This dual classification allows isolation of two important diagenetic parameters: initial composition and duration of diagenesis.

The provenance classification follows from the recognition of two distinct Bermudian source-sink sedimentary systems that led to the deposition of shoreline calcareous dunes (Vacher, 1973). The South Shore eolianites (S-series) are the sink of a reef-to-dune system and are composed of reef-suite (Upchurch, 1970) constituent particles, that is, fragments of *Homotrema*, gastropods, coralline algae, and corals. The North Shore eolianites (N-series) were formed by a lagoon-to-dune system and contain mostly lagoon-suite (Upchurch, 1970) constituent particles, that is, fragments of *Halimeda* and infaunal molluscs. The South Shore reef-dune system is presently active; the North Shore lagoon-dune system is not. Owing to the different mix of constituent particles, the lagoon-derived calcarenite was originally more aragonitic than the reef-derived sand, a difference that can still be observed in time-correlative units (Vacher, 1973). There is remarkably little variation among the mineralogic compositions of present-day beaches and no evidence of secular change in the initial compositions of the N- and S-series.

On the basis of these observations, we have set in our model the initial concentrations of aragonite and high-magnesium calcite at constant values determined from field and laboratory observations. Three samples of Ristvet (1972) are from South Shore beaches that grade into modern dunes. Their mineralogic compositions are nearly identical (Table 2), and we have selected the averages as an estimate of the initial composition of the S-series. There are no modern lagoon-derived beaches that supply modern dunes analogous to the N-series. We chose as an estimate of the initial composition of the N-series a petrographic constituent analysis of the Barker's Hill eolianite (Table 3) by Ristvet (1972, written commun.).

Our estimates of the duration of diagenesis follow directly from the stratigraphic classification. The investigated calcarenites were all deposited in their diagenetic environment (on the lee side of rapidly accreting dunes) at the time of formation of the eolianites and remained there above the water table (in the vadose zone) until sampling. Thus the duration of vadose diagenesis for a given sample is numerically equal to the depositional age of the sampled eolianite.

The stratigraphic classification shown in Table 1 is based on that of Vacher (1973). The Upper and Lower Paget members of the Paget Formation are the two stratigraphically highest soil-bracketed limestone units in the Bermuda column. They represent the last two Pleistocene submergences of the Bermuda platform. Southlands, Spice Hill, and Barker's Hill eolianites are informal map units (Vacher, 1973, Fig. 4). Each is a single, contiguous, lithified dune ridge composed of laterally interfingering foreset lobes that represent the individual shoreline dunes. Thus, each of the eolianites is essentially an isochronous unit. Samples denoted S_{III} and N_{III} are from eolianite lying directly below a Lower Paget unit. These samples are mostly (if not entirely) from the uppermost soil-bracketed limestone unit in the large (as yet not formally subdivided) section between the Walsingham Formation of Land and others (1967) and the Paget Formation of Vacher (1973).

The Lower Paget eolianites are the only units that can be dated with fair accuracy. The Lower Paget member includes a marine facies that contains corals dated by the uranium-series method at about 125,000 years B.P. (Land and others, 1967). Firmly dated deposits of this age are recognized in many coastal Pleistocene sections (Broecker and Van Donk, 1970). We therefore estimate the duration of diagenesis for S_{II} and N_{II} as 125,000 ± 10,000 years.

The Southlands and other Upper Paget eolianites correspond to a late Pleistocene platform submergence distinctly younger than the one dated at 125,000 years B.P. The Upper Paget contains a marine facies located slightly above present sea level. Thus, the Bermudian Upper Paget is like the latest Pleistocene deposits of many other coastal areas: There is evidence of a sea-level high stand at about the present datum after the widely recognized high stand of 125,000 years ago. Mörner (1971) has reviewed the published ocean-level information for the time interval following the last full interglacial in the light of glaciologic and climatologic data, and concluded that the post-Eemian high stand is correlative with the Brörup Interstadial. On the basis of Mörner's synthesis, we estimate the duration of diagenesis for S_I as 65,000 ± 15,000 years.

We can only guess at the ages of units S_{III} and N_{III}. The stratigraphic section between the Walsingham and Paget Formations is very complex and contains more than one separate marine tongue above present sea level (Vacher, 1971, and unpub. data). Richards and others (1969) have reported a uranium-series date on corals from one marine deposit in this interval to be greater than 300,000 years B.P. The volumes of S_{III} and N_{III} are comparable to those of the Lower Paget eolianites and presumably represent, like them, nearshore sediment production during a full interglacial. The age of the interglaciation that preceded the Eemian is unknown

TABLE 3. CONSTITUENT-PARTICLE ANALYSIS OF BARKER'S HILL EOLIANITE*

High-Mg calcite (%)		Aragonite (%)		Reworked grains (%)	Unidentified (%)
Homotrema	8.6	Corals	6.2		
Other forams	8.0	Molluscs	21.3		
Coralline algae	20.1	*Halimeda*	18.9		
Total	36.7		46.4	3.8	13.1

Note: Composition adopted: aragonite, 50.25%; high-Mg calcite, 39.75%; calcite, 10%.
*Average of six samples (Ristvet, 1972, written commun.).

(Flint, 1971). An estimate of 375,000 ± 75,000 years for the duration of diagenesis undergone by S_{III} and N_{III} would appear to cover most of the possibilities.

Calculations and Results

To test the validity of equations (3) through (5), we transformed the variables listed in Appendix Table 1 into the new variables $\log[C_{Ar}(t)/C_{Ar}(0)]$ and $\log[C_{Mg\text{-}c}(t)/C_{Mg\text{-}c}(0)]$, where Ar and Mg-c stand for aragonite and high-magnesium calcite, and log denotes the common logarithm. The means and standard deviations of these new variables are presented in Table 4. We have omitted parameters for high-magnesium calcite in S_{III} and N_{III}, owing to the large number of zero values present in the original data.

To carry the analysis further, it is useful to obtain estimates of the probability density functions of the variables. The shapes of the data histograms strongly suggest that the transformed concentrations may be closely approximated by normal variates. The number of samples for each unit, however, is quite small, so that standard comparison techniques such as the χ^2 goodness-of-fit technique (Krumbein and Graybill, 1965) cannot be used in this case. We compared the logarithmic variables with normal variates by means of the Kolmogorov-Smirnov test for goodness of fit (Massey, 1951; Lilliefors, 1967). Agreement was always excellent, indicating a significance level for Type 1 error of at least 20 percent in most instances and at least 5 percent in the least favorable case, as shown in Figure 1 (for comparison, the commonly used cutoff is 5 percent). To the extent that our data permit the determination of statistical distributions, we can accept the hypothesis that the concentrations of aragonite and high-magnesium calcite are lognormally distributed.

Equation (10) predicts that the diagenetic processes affecting a given reaction can be reduced to a single random variable whose probability density function is simply related to that of the logarithmic concentration. If this hypothesis is correct, the logarithmic concentrations of aragonite and high-magnesium calcite, when divided by time, should each yield one single random variable distributed normally because of the lognormal distribution of the concentrations. The data for aragonite span the longest interval of time, providing the most stringent test of our model, and were therefore examined in detail. For each eolianite unit, we calculated $X = (1/t)\log[C_{Ar}(t)/C_{Ar}(0)]$, which should belong to a single normal population. The values of t used were our best estimates, and we did not consider the time uncertainty in the formal statistical analysis because we lacked information about its probability distribution. We tested the hypothesis of single constant mean with an F-test (Dixon and Massey, 1969); the results are summarized in Table 5. We think that the hypothesis can be accepted if we take

TABLE 4. MEANS AND STANDARD DEVIATIONS OF THE LOGARITHMIC CONCENTRATIONS, $\log[C_{Ar}(t)/C_{Ar}(0)]$ AND $\log[C_{Mg\text{-}c}(t)/C_{Mg\text{-}c}(0)]$

Unit	Number of samples	Aragonite Mean	Standard deviation	High-Mg calcite Mean	Standard deviation
S_I	7	−0.0286	0.0865	−0.301	0.2256
S_{II}	15	−0.139	0.1064	−0.596	0.2473
S_{III}	10	−0.715	0.4357		
N_{II}	10	−0.1947	0.1008	−0.717	0.2140
N_{III}	6	−0.4354	0.2150		

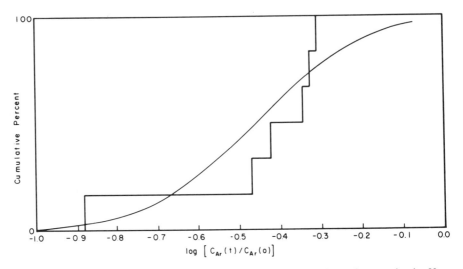

Figure 1. Kolmogorov-Smirnov test for logarithmic concentrations of aragonite in N_{III}. Theoretical curve is N(0.435, 0.0462). Maximum deviation observed here is d = 0.31, which, for a sample of six observations, corresponds to a probability of Type 1 error of at least 5 percent (Lilliefors, 1967) or at least 20 percent (Massey, 1951) if one rejects the hypothesis that the two distributions are identical.

into account (a) our neglecting the time uncertainties, and (b) the fact that S_I contributes one-half of the mean-square variability within groups, whereas it is one of the smallest samples. To test argument (b), we evaluated $\bar{X}_{S_I} - (1/4)(\bar{X}_{S_{II}} + \bar{X}_{S_{III}} + \bar{X}_{N_{II}} + \bar{X}_{N_{III}})$ with the F-statistic (Dixon and Massey, 1969, p. 166) and found it equal to -1.055 ± 1.29 at the 95 percent significance level. We can therefore accept the hypothesis that S_I is not significantly different from the other eolianite units.

The best estimates of the mean and variance of X, which are also those of Λ, are probably those based on S_{II} and N_{II}, for which the time uncertainty is least. Pooling the data for these samples, we obtain

$$\bar{X} = \bar{\Lambda}_{Ar} = -1.3 \times 10^{-6} \text{ year}^{-1}$$

and

$$s_X^2 = s_{\Lambda_{Ar}}^2 = 0.74 \times 10^{-12} \text{ year}^{-2}.$$

Summarizing our analysis of the aragonite data, we propose that they are consistent with equations (3) and (10). The means of the logarithmic concentrations fit well a linear function of time (Fig. 2) and the random parameter Λ_{Ar} is found to be

TABLE 5. ANALYSIS OF VARIANCE FOR X IN BERMUDIAN EOLIANITES

	Sum of squares	Degrees of freedom	Mean square
Between groups	11.363×10^{-12}	4	2.8532×10^{-12}
Within groups	40.386×10^{-12}	43	0.9392×10^{-12}
Total	51.749×10^{-12}	47	

Note: Observed F-ratio: 3.03. Theoretical values are $F_{0.95}(4,43) = 2.60$; $F_{0.975}(4,43) = 3.11$; $F_{0.99}(4,43) = 3.80$.

independent of time and location. The corresponding half-life of aragonite in the rocks is about 230,000 years.

A similar analysis can be conducted for the high-magnesium calcite data, although the lack of points for times greater than 125,000 years introduces additional uncertainties. Figure 3 is a plot of the mean logarithmic concentrations as a function of time. Within the accuracy of the data, a straight line again represents the relations very well. We obtain

and
$$\Lambda_{Mg-c} = -5.24 \times 10^{-6} \text{ year}^{-1}$$
$$s^2_{\Lambda_{Mg-c}} = 3.7 \times 10^{-12} \text{ year}^{-2},$$

corresponding to a half-life of about 60,000 years. Using these parameters, we calculate that the logarithmic mean and standard deviation of the high-magnesium calcite concentration in S_{III} and N_{III} should be, respectively, about -1.9 and 0.75 (that is, a concentration of 0.6 percent). This is in good agreement with the reported concentrations.

DISCUSSION

We have shown that the simple stochastic model developed earlier is consistent with one set of carefully selected data. For the Bermudian eolianites, the diagenetic conversion of aragonite and high-magnesium calcite to calcite in the vadose zone can be described quantitatively by only two random variables, Λ_{Ar} and Λ_{Mg-c}, respectively, which are both independent of time. The lack of significant differences between the S-series and the N-series demonstrates that Λ_{Ar} and Λ_{Mg-c} are also independent of the initial compositions of the rocks. These random variables reflect solely the general environment of diagenesis as it relates to the given natural reactions. The diagenetic environment, here the vadose seepage zone in Bermuda, is defined by all the relevant physicochemical variables other than concentrations, including in particular the grain-size distributions in the rocks, their permeability, and the hydrologic regime. We strongly emphasize that the statistical distributions of Λ_{Ar} and Λ_{Mg-c} as derived above apply strictly only to the case we have documented.

The Bermudian example represents the only set of data we know that are sufficiently precise to provide a test of our model. In view of the great simplicity of our results, we cannot claim that they have general validity. As we stressed earlier, the exponential decay in the mean of the mineral concentrations follows

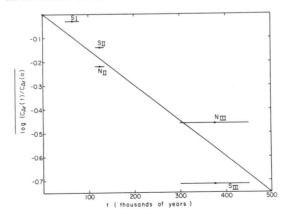

Figure 2. Plot of means of logarithmic aragonite concentrations versus time, for all eolianites studied. Horizontal bars represent respective time uncertainties.

from the time independence of the diagenetic random variables and from the choice of constant initial concentrations. Before the Bermudian model can be extended to other diagenetic situations, we need more well-documented examples of a similar nature. We hope that further work and accumulated experience will in the future more clearly define the range of applicability of our model.

Many diagenetic reactions can proceed at an appreciable rate only through the medium of an aqueous interstitial fluid (Berner, 1971). This appears to be true in particular for the vadose alteration of carbonates (Purdy, 1965; Fyfe and Bischoff, 1965). Although the reactions altering aragonite and high-magnesium calcite to low-magnesium calcite have been shown in the laboratory to be complex and not of the first order, we have seen that in a large-scale natural situation, the concentrations of aragonite and high-magnesium calcite follow a law of exponential decay in the mean. This suggests to us that a major control of the rate of diagenetic alteration in natural environments is the availability of a suitable aqueous phase where reactions can take place. As was emphasized by Scheidegger (1964), flow in natural porous media is a stochastic phenomenon. If presence of water is the limiting step, this may contribute to the distribution function of Λ much more than the actual reaction mechanisms studied so far.

Although the general equation (1) appears formidable and is perhaps of little practical help, the simplified equation (3),

$$\delta C(t) = \Lambda(t) C(t) dt$$

is much more amenable to study and may often be of widespread applicability. It allows the replacement of cumbersome arrays of compositional-mineralogical variables by a few random functions or variables that can be compared simply by use of standard statistical techniques. The stochastic approach can thus be used advantageously to analyze and summarize data pertaining to diagenetic processes. More specifically, our model, if found by later work to hold in general for these processes, will greatly simplify quantitative comparison of different field situations. By using half-lives, for example, we can readily identify the characteristic time scale of a diagenetic reaction with respect to other natural time scales. Furthermore, description of a diagenetic environment by means of Λ can lead to isolation of the physicochemical parameters that significantly affect the process. If the distributions of the parameters thought to be relevant are known, we can explain the variability of Λ in terms of these parameters (for example, by using a regression analysis) and determine which ones are not needed to describe the process. With respect to carbonate eolianite diagenesis, it would be particularly

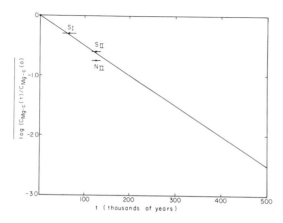

Figure 3. Plot of means of logarithmic high-magnesium calcite concentrations for S_I, S_{II}, and N_{II} versus time. The horizontal bars represent respective time uncertainties.

interesting to compare the Bermuda rocks with similar deposits that are being altered under different climatic conditions and to investigate the relative contributions of rainfall, CO_2 production, permeability, and hydraulic regime to the rates of the diagenetic reactions.

Finally, our approach is not restricted to specific initial conditions or diagenetic environments. It can in principle be applied to well-known diagenetic processes such as the submarine conversion of high-magnesium calcite into aragonite, the dolomitization of calcareous rocks, and others. In particular, it can also be applied to those metamorphic reactions for which the differential kinetic equation takes the form of equation (3). The kinetic data reported by Greenwood (1963) for the breakdown of talc may be consistent with this viewpoint because one would not a priori expect the concentration of talc to follow a first-order law when it is one of five phases involved in three simultaneous heterogeneous reactions.

SUMMARY

Because natural diagenetic processes are very complex, they are best described in terms of stochastic variables and equations. We propose that, in cases where the diagenetic environment is relatively constant and where destruction of the reactant is the limiting step, the rate of a diagenetic reaction can be described by the equation

$$\delta C(t) = \Lambda(t) C(t) dt,$$

where $\Lambda(t)$ is a random function that is characteristic of the particular diagenetic environment.

In the simplest theoretical case considered, Λ is a random variable (independent of t), and the initial concentration is a constant. Then, the logarithm of reactant concentration decreases linearly in the mean with time, and its standard deviation increases linearly with time.

This simple model describes well the alteration kinetics of high-magnesium calcite and aragonite in the Pleistocene eolianites of Bermuda, which have remained in the zone of vadose seepage. The half-lives of the alteration reactions are about 230,000 years for aragonite and about 60,000 years for high-magnesium calcite.

Our model appears to apply best in cases where stochastic phenomena, rather than deterministic parameters, are the major control of reaction rates. Because fluid flow in natural porous media is a stochastic process, and because many diagenetic and metamorphic reactions cannot proceed in the absence of an aqueous phase, our model should be widely applicable to large-scale chemical reactions of geologic interest.

ACKNOWLEDGMENTS

We are particularly grateful to Byron Ristvet, Northwestern University, for the use of his mineralogic and petrographic data. David Pyne's help with the statistics and Alan Karr's advice on random equations were invaluable. Hans P. Eugster, Thomas A. Jones, and Blair F. Jones critically read early drafts of the manuscript, and their comments led to much clarification and improvement. However, responsibility for the opinions and conclusions expressed in this paper is our own.

William C. Krumbein showed us that many geological processes are best described in probabilistic terms. It gives us great pleasure to contribute this paper to a volume in his honor.

APPENDIX TABLE 1. COMPOSITIONS OF SAMPLES USED

Sample no.*	Wt % aragonite	$\log[C_{Ar}(t)/C_{Ar}(0)]$	Wt % Mg-calcite	$\log[C_{Mg\text{-}c}(t)/C_{Mg\text{-}c}(0)]$
		S_I		
8	21	−0.1784	43	−0.1275
9	32	0.00455	50	−0.06195
12	32	0.00455	23	−0.3992
14	39	0.09045	11	−0.7195
15	28	−0.0534	23	−0.3992
25	26	−0.0856	40	−0.15885
27	33	0.0179	33	−0.2424
		S_{II}		
1	24	−0.1204	22	−0.4185
4	25	−0.1027	26	−0.34595
5	24	−0.1204	13	−0.6470
7	28	−0.05345	8	−0.8578
13	24	−0.1204	20	−0.4600
23	32	0.00455	17	−0.5305
24	29	−0.0382	40	−0.1589
26	21	−0.1784	10	−0.7609
28	22	−0.1582	12	−0.68175
30	24	−0.1204	9	−0.8067
33	23	−0.1389	6	−0.9828
35	12	−0.4214	11	−0.7195
36	18	−0.2453	24	−0.3807
37	30	−0.0235	23	−0.3992
40	18	−0.2453	5	−1.06195
		S_{III}		
2	16	−0.2965	0	
3	2	−1.1996	0	
6	4	−0.89855	0	
29	3	−1.0235	2	−1.4599
31	13	−0.38665	2	−1.4599
32	12	−0.4214	3	−1.2838
34	12	−0.4214	1	−1.7609
38	1†	−1.5	0	
39	20	−0.1996	1	−1.7609
83	5	−0.8016	0	
		N_{II}		
48	36	−0.1448	17	−0.3689
49	37	−0.1329	7	−0.7542
50	30	−0.2240	6	−0.8212
51	32	−0.1960	5	−0.9004
53	23	−0.3394	8	−0.69625
55	31	−0.2098	4	−0.9973
61	26	−0.2862	6	−0.8212
65	25	−0.3032	18	−0.3441
91	40	−0.0991	9	−0.6451
94	49	−0.0109	6	−0.8212
		N_{III}		
52	25	−0.3032	0	
54	24	−0.3209	0	
56	20	−0.4001	1	−1.5993
60	7	−0.8560	0	
90	18	−0.4460	1	−1.5993
95	26	−0.2861	0	

*Sample numbers are those of Ristvet (1972), who also gave description of sample locations.
†Original 0 percent aragonite in this sample was replaced by 1 percent to permit statistical calculations. In view of experimental precision, difference is not significant.

REFERENCES CITED

Bear, J., 1970, Two liquid flows in porous media, in Advances in hydroscience, Vol. 6: New York, Academic Press, Inc., p. 141-252.

Berner, R. A., 1971, Principles of chemical sedimentology: New York, McGraw-Hill Book Co., 240 p.

Blanc-Lapierre, A., and Fortet, R., 1953, Théorie des fonctions aléatoires: Paris, Masson et Cie, 693 p.

Broecker, W. S., and Van Donk, J., 1970, Insolation changes, ice volumes, and the O^{18} record in deep-sea cores: Rev. Geophysics and Space Physics, v. 8, p. 169-198.

Dixon, W. J., and Massey, F. J., 1969, Introduction to statistical analysis (3d ed.): New York, McGraw-Hill Book Co., 638 p.

Domenico, P. A., 1972, Concepts and models in groundwater hydrology: New York, McGraw-Hill Book Co., 405 p.

Flint, R. F., 1971, Glacial and Quaternary geology: New York, John Wiley & Sons, Inc., 892 p.

Fyfe, W. S., and Bischoff, J. L., 1965, The calcite-aragonite problem, in Pray, L. C., and Murray, R. C., eds., Dolomitization and limestone diagenesis: Soc. Econ. Paleontologists and Mineralogists Spec. Pub. 13, p. 3-13.

Greenwood, H. J., 1963, The synthesis and stability of anthophyllite: Jour. Petrology, v. 4, p. 317-351.

Krumbein, W. C., and Graybill, F. A., 1965, An introduction to statistical models in geology: New York, McGraw-Hill Book Co., 475 p.

Land, L. S., 1970, Phreatic vs. vadose meteoric diagenesis of limestones: Evidence from a fossil water table: Sedimentology, v. 14, p. 175-185.

Land, L. S., Mackenzie, F. T., and Gould, S. J., 1967, The Pleistocene history of Bermuda: Geol. Soc. America Bull., v. 78, p. 993-1006.

Lévy, P., 1948, Processus stochastiques et mouvement Brownien: Paris, Gauthier-Villars, 365 p.

Lilliefors, H. W., 1967, On the Kolmogorov-Smirnov test for normality with mean and variance unknown: Am. Statistical Assoc. Jour., v. 62, p. 399-402.

Massey, F. J., Jr., 1951, The Kolmogorov-Smirnov test for goodness-of-fit: Am. Statistical Assoc. Jour., v. 46, p. 68-78.

Mörner, N. A., 1971, The position of the ocean level during the interstadial about 30,000 years B.P.—A discussion from a climatic-glaciologic point of view: Canadian Jour. Earth Sci., v. 8, p. 132-143.

Nielsen, A. E., 1964, Kinetics of precipitation: New York, Macmillan Co., 151 p.

Penman, H. L., 1963, Vegetation and hydrology: Farnham Royal, England, Commonwealth Agr. Bur., Tech. Comm. 53, 124 p.

Plummer, L. N., 1970, Soils and the diagenesis of Bermuda carbonate sands: St. George's West, Bermuda, Bermuda Biol. Research Sta. Spec. Pub. 7, p. 56-98.

Purdy, E. G., 1965, Diagenesis of recent carbonate sediments, in Pray, L. C., and Murray, R. C., eds., Dolomitization and limestone diagenesis: Soc. Econ. Paleontologists and Mineralogists Spec. Pub. 13, p. 169.

Remson, I., Randolph, J. R., and Barksdale, H. C., 1960, The zone of aeration and ground-water recharge in sandy sediments at Seabrook, New Jersey: Soil Sci., v. 89, p. 145-156.

Richards, H. G., Abbott, R. T., and Skymer, T., 1969, The marine Pleistocene molluscs of Bermuda: Notulae Naturae, no. 425.

Ristvet, B. L., 1972, The progressive diagenetic history of Bermuda: St. George's West, Bermuda, Bermuda Biol. Research Sta. Spec. Pub. 7, p. 118-157.

Scheidegger, A. E., 1964, Statistical hydrodynamics in porous media, in Advances in hydroscience, Vol. 1: New York, Academic Press, Inc., p. 161-181.

Smith, W. O., 1967, Infiltration in sands and its relation to groundwater recharge: Water Resources Research, v. 3, p. 539-555.

Thornthwaite, C. W., and Mather, J. R., 1955, The water balance: Public Climatology 8(1), p. 1-86.

Thorstenson, D. C., Mackenzie, F. T., and Ristvet, B. L., 1972, Experimental vadose and phreatic cementation of skeletal carbonate sand: Jour. Sed. Petrology, v. 42, p. 162–167.

Thrailkill, J., 1968, Chemical and hydrologic factors in the excavation of limestone caves: Geol. Soc. America, Bull., v. 79, p. 19–46.

Upchurch, S. B., 1970, Sedimentation on the Bermuda platform: Detroit, U.S. Lake Survey, Research Rept. 2-12, 172 p.

Vacher, H. L., 1971, Late Pleistocene sea-level history: Bermuda evidence [Ph.D. dissert.]: Evanston, Ill., Northwestern Univ., 153 p.

——1973, Coastal dunes of younger Bermuda, *in* Coates, D. R., ed., Coastal geomorphology: Binghamton, N.Y., State Univ. of New York at Binghamton, Pubs. in Geomorphology, p. 355–391.

MANUSCRIPT RECEIVED BY THE SOCIETY SEPTEMBER 12, 1973
REVISED MANUSCRIPT RECEIVED JANUARY 25, 1974

Printed in the U.S.A.

Modeling of Geochemical Cycles: Phosphorus as an Example

A. LERMAN
AND
F. T. MACKENZIE

Department of Geological Sciences
Northwestern University
Evanston, Illinois 60201

R. M. GARRELS

Department of Oceanography
University of Hawaii
Honolulu, Hawaii 96822
and
Department of Geological Sciences
Northwestern University
Evanston, Illinois 60201

ABSTRACT

Analysis of various transient situations in a geochemical cycle helps to focus attention on those physical and chemical mechanisms that are more important to the understanding of the cycle. The geochemical cycle of an element can be modeled as a system of n reservoirs (each of mass M_i) and interreservoir fluxes (F_{ij}, in units of mass/time; $i, j = 1, ..., n$). Transient states of the geochemical cycle have been treated by a simple model based on first-order fluxes

$$\frac{dM_i}{dt} = \sum_j k_{ji} M_j - M_i \sum_j k_{ij},$$

where k_{ij} are rate constants (time^{-1}). The phosphorus cycle was used as an example, and the following two transient states were considered: (1) Cessation of photosynthetic productivity on Earth (doomsday scenario)—in this scenario, phosphorus from the dead biota would be redistributed among other reservoirs within 200 years. The most significant result of such a redistribution is an increase in the dissolved phosphorus content of surface ocean layer, making the phosphorus concentration throughout the ocean more uniform than at present. (2) Man's use of phosphorus from mining at an exponentially increasing rate (industrial scenario)—an accelerated rate of mining and fertilizer use that adds phosphorus to land and streams may, during the next 60 years, result in significant increases in the oceanic biomass (40 percent) and in the dissolved phosphorus content of surface ocean water (30 percent). According to the model, restoration of the earlier conditions (that is, those of the present day) will take 150 years.

INTRODUCTION

The geochemical cycle of an element is a model of its dynamic behavior in the shells of the Earth. In the system comprising the lithosphere, hydrosphere, biosphere, and atmosphere, transformations of the materials within each sphere and transport between spheres underscore the cyclic nature of material flow. When the system is in a steady state, the content of individual spheres or reservoirs and the fluxes between reservoirs are constant in time. At steady state, the reservoir content and the fluxes of materials between reservoirs are sufficient to account for the material balance in the system. However, knowledge of the steady-state fluxes and reservoir content tells us nothing about the history of the system prior to attainment of a steady state nor about the future evolution of the system in the case of departure from the existing steady state.

Such events as, for example, changes in the magnitude of fluxes between reservoirs and changes in size and content of reservoirs might have occurred throughout the Earth's history, and such changes might have been responsible for significant departures from the existing steady-state conditions. When a system is in a nonsteady or transient state, the geologically important questions are, How long will it take the system to attain a new steady state? and What will be the new reservoir contents? Conversely, one can assume some initial conditions for the system in the past and simulate its evolution to its present steady state.

In this paper we consider applications of the steady-state and transient-state models to the geochemical cycle of phosphorus. We chose phosphorus as an example owing to its occurrence and involvement in inorganic and biological processes in such major shells of the Earth as the hydrosphere, lithosphere, and biosphere.

In setting up a model of the geochemical cycle for any element (Fig.1), the foremost questions that should be asked are, What are the reservoirs and what are the fluxes responsible for distribution of the element among the reservoirs? The decisions as to how to subdivide a natural system into several reservoirs are based on the scale of the processes under consideration. Thus, if interest lies in the longer term processes, it may be irrelevant to include in the model any processes of very short duration. Similarly, the size and nature of a reservoir are defined to fit the problem: such reservoirs as "land biota" and "land," shown in Figure 2 and Table 1, are fairly big units, useful only insofar as the model aims at those aspects of the geochemical behavior of phosphorus that are representative of each reservoir as a whole. Clearly, the reservoirs of land biota and land can be subdivided into a greater number of reservoirs reflecting the different states of phosphorus in the components of the soils and in the variety of biological taxa. Our approach, however, was to treat the smallest number of reservoirs that was compatible with the broader features of the geochemical behavior of phosphorus in the hydrosphere, lithosphere, and biosphere. Hence, we chose the reservoirs shown in Figure 2 and the use of the quantity "total P" that we consider as being transported from one reservoir to another.

Discussion of the phosphorus cycle will be preceded by a brief summary of mathematical fundamentals that apply to an analysis of steady-state and transient-state cyclic processes.

SOME FUNDAMENTALS

Steady-State or Stationary Cycle

Figure 1 is a diagrammatic representation of a system of four reservoirs with two-way fluxes of material between any two reservoirs. At a steady state, the amount of material in each reservoir (M_i, in units of mass) and the fluxes between reservoirs (F_{ij}, in units of mass/time) are constant. In a closed system, such as the one shown in Figure 1, no material enters the system from the outside and none leaves the system. For a closed system of n reservoirs, the maximum possible number of interreservoir fluxes is $n(n-1)$. To maintain a material balance in the entire system, the amount and rate of input to each reservoir must be equal to the amount and rate of removal from it. For conservative chemical species (that is, those that neither decay nor leave the reservoirs by other processes), the input fluxes to the ith reservoir (F_{ji}, in the notation of Fig. 1) are balanced by the output fluxes (F_{ij}):

$$\sum_j F_{ji} = \sum_j F_{ij} \quad (i \neq j). \tag{1}$$

It follows from (1) that if one of the interreservoir fluxes changes, at least one of the remaining fluxes must also change, or the system would be out of balance.

In the geochemical cycles, the major agents responsible for transport of materials from one geochemical reservoir to another are streams, ground-water flow, ice, and wind. These transport agents of global nature are independent of the amount (or concentration) of the species in the geochemical reservoirs; in particular, the agents are independent of those species that do not affect the major physical characteristics of the reservoirs and fluxes. Thus, the simplest of all approaches is to think of the magnitudes of the fluxes as proportional to the amount (or concentration) of the species in the reservoirs:

$$F_{ij} = k_{ij} M_i, \tag{2}$$

where k_{ij} is a rate constant whose dimensions are time^{-1}. Rapid turnover of material in the reservoir corresponds to large k_{ij}.

Using (2), the material balance given in (1) can be written as

$$\sum_j k_{ji} M_j = M_i \sum_j k_{ij} \quad (i \neq j). \tag{3}$$

For a geochemical cycle in a steady state, when either the reservoir masses (M_i, Fig. 1) or the transport-rate constants (k_{ij}) are known, one of the two sets of parameters can be obtained by solution of a system of simultaneous linear

TABLE 1. PHOSPHORUS CONTENT OF GEOCHEMICAL RESERVOIRS AND INTERRESERVOIR FLUXES SHOWN IN FIGURE 2

Reservoir	Used in this paper	Other estimates	References and remarks
	P content (metric tons)		
1. Sediments	4×10^{15}		Van Wazer (1961, p. 1285)
		7.8×10^{14}	Poldervaart (1955)
		8.2×10^{14}	Ronov and Korzina (1960)
		8.37×10^{14}	Stumm (1972)
2. Land	2×10^{11}		Computed from land area (total land less ice, 133×10^6 km^2), assumed soil thickness (60 cm), density (2.5 g/cm^3), and mean P content of crustal material (0.1 wt %; Taylor, 1964): $133 \times 10^6 \times 6 \times 10^{-4} \times 2.5 \times 10^9 \times 1 \times 10^{-3} = 2.0 \times 10^{11}$ tons P. P content of soils reportedly ranges from 0.02 to 0.83 wt % (Fuller, 1972)
3. Land biota	3×10^9		Computed from N amount in land biota (12×10^9 tons N; Delwiche, 1970) and mean P/N atomic ratio in land plants (1.8/16; Deevey, 1970). $12 \times 10^9 \times 1.8 \times 31/(16 \times 14) = 3.01 \times 10^9$ tons P
		1.41×10^9	Computed from C amount in land biota (450×10^9 tons C; Bolin, 1970) and mean P/C atomic ratio in land plants (1.8/1,480; Deevey, 1970). Computation as in preceding estimate
		3.43×10^8	Computed from CO_2 amount in living matter on land (4×10^{11} tons CO_2; Lemon, 1968) and mean P/C atomic ratio in terrestrial organic matter (1.8/1,480; Deevey, 1970). Computation as above
		1.95×10^9	Stumm (1972)
4. Oceanic biota	1.38×10^8		Computed from N amount in oceanic biota (1×10^9 tons N; Vaccaro, 1965) and mean P/N atomic ratio in oceanic biota (1/16; Redfield and others, 1963; compare Ryther and Dunstan, 1971). $1 \times 10^9 \times 1 \times 31/(16 \times 14) = 1.38 \times 10^8$ tons P
		1.22×10^8	Computed from C amount in oceanic biota (5×10^9 tons C; Bolin, 1970) and mean P/C atomic ratio in oceanic biota (1/106; Redfield and others, 1963). Computation as in preceding estimate
		1.99×10^8	Computed from CO_2 amount in living oceanic biota (3×10^{10} tons CO_2; Lemon, 1968) and P/C atomic ratio of 1/106 (Redfield and others, 1963). Computation as above
		1.24×10^8	Stumm (1972)
5. Surface ocean	2.71×10^9		Computed from assumed mean concentration of total dissolved P (25 mg/m^3; compare Sverdrup and others, 1942, p. 241) in 300-m-thick water layer and ocean surface area (3.61×10^8 km^2). $25 \times 0.3 \times 3.61 \times 10^8 = 2.71 \times 10^9$ tons P
6. Deep ocean	8.71×10^{10}		Computed from assumed mean concentration of total dissolved P (80 mg/m^3; compare Sverdrup and others, 1942, p. 241) in 3,000-m-thick layer and ocean surface area. Computation as in preceding estimate.
		8.1×10^{10}	From Stumm's (1972, Fig. 2) presentation of Broecker's (1971) data.
7. Mineable resource	1×10^{10}		Assumed from within range of estimates given below
		3.1×10^{10}	Stumm (1972)
		1.8×10^9	Cited in Ronov and Korzina (1960)
		4.67×10^{10}	Van Wazer (1961, p. 966)

TABLE 1. (Continued)

Flux	Used in this paper	Other estimates	References and remarks
	(P metric tons/yr)		
F_{12}	2×10^7		Computed from combined rates of mechanical and chemical denudation of continents (approx. 2×10^{10} tons/yr; Garrels and Mackenzie, 1971, p. 120) and mean P content of crustal material (0.1 wt %; Taylor, 1964). F_{12} includes dissolved and detrital forms of P. Fraction of F_{12} in solution is 1.7×10^6 tons/yr, as explained in entry F_{25}
F_{23}	6.3×10^7		Computed from Bolin's (1970) figure for total amount of C fixed annually by land plants (20 to 30 billion tons) and mean P/C atomic ratio in land biota (1.8/1,480; Deevey, 1970). $2 \times 10^{10} \times 1.8 \times 31/(1,480 \times 12) = 6.3 \times 10^7$ tons/yr
		2.29×10^8	Stumm (1972)
		6.26×10^7	Computed from rate of photosynthesis on land (7.3×10^{10} tons CO_2/yr; Lemon, 1968) and mean P/C ratio (Deevey, 1970) in land biota. Computed as above
F_{25}	1.7×10^6		Garrels and others (1973, p. 80); excludes agricultural and other man-produced contributions to streams and coastal ocean
		2×10^6	Stumm and Morgan (1970, p. 550)
		1.86×10^6	Stumm (1972); includes agricultural and other man-produced contributions
		4.4×10^5	Cited in Ronov and Korzina (1960)
F_{32}	6.35×10^7		Estimate balanced by flux F_{23}
F_{54}	1.04×10^9		Computed from rate of N fixation by oceanic biota (7.5×10^9 tons N/yr; higher value cited by Vaccaro, 1965, 1×10^{10} tons N/yr) and mean P/N atomic ratio in oceanic biota (1/16; Redfield and others, 1963; compare Ryther and Dunstan, 1971). $7.5 \times 10^9 \times 1 \times 31/(16 \times 14) = 1.04 \times 10^9$ tons P/yr
		9.75×10^8	Computed from rate of C fixation by oceanic biota (4×10^{10} tons C/yr; Bolin, 1970) and mean P/C atomic ratio in oceanic biota (1/106; Redfield and others, 1963). Computed as above
		9.61×10^8	Stumm (1972)
		3.06×10^9	Computed from the CO_2 fixation rate by oceanic biota (4.6×10^{11} tons CO_2/yr; Lemon, 1968) and mean P/C atomic ratio, as above
F_{45}	9.98×10^8		Computed on assumption that 96% of living oceanic biota is being recycled within upper 300-m-thick water layer
F_{46}	4.2×10^7		Differences between the fluxes F_{54} and F_{45}
F_{56}	1.8×10^7		Computed from mean concentration of total dissolved P in surface ocean (25 mg/m^3; see entry "Surface ocean" in this table), water-exchange rate between surface and deep ocean layers (2 m/yr; Broecker, 1971), and surface area of ocean (3.61×10^8 km^2). $25 \times 2 \times 10^{-3} \times 3.61 \times 10^8 = 1.8 \times 10^7$ tons P/yr
F_{65}	5.8×10^7		Computed as for the flux F_{56}, using 80 mg/m^3 for P concentration in deep ocean
F_{61}	1.7×10^6		Flux value assumed as balanced by stream flux to ocean (F_{25})
F_{72}	1.2×10^7		Stumm (1972)
		1.5×10^7	Meadows and others (1972, p. 26; as PO_4^{3-}?)

equations. For a cycle involving n reservoirs ($n = 4$ in Fig. 1), the n simultaneous equations to be solved are of the type

$$\sum_j k_{ji} M_j - M_i \sum_j k_{ij} = 0 \qquad (i \neq j;\ i, j = 1, \ldots, n), \qquad (4)$$

where the minuend represents the total input flux of materials to the ith reservoir and the subtrahend is the total flux out of the reservoir. Because the amount of material in a closed cycle is constant ($\sum_i M_i$ = constant), the number of simultaneous equations given in (4) can be reduced by one.

Equations of the type given in (4) have been extensively used in computations of steady-state masses and rate constants in systems of simultaneous chemical reactions, and in studies of mass balance in "box models" as used in oceanography, geochemistry, and meteorology, and other fields (for example, Denbigh, 1958; Frost and Pearson, 1961; Craig, 1957; Broecker, 1963; Lal and Peters, 1967; Eriksson, 1971).

Transient or Nonsteady-State Cycle

In a transient state, the rate of input to a reservoir is not balanced by the rate of removal, and consequently, the amount of material in the reservoir changes with time:

$$\frac{dM_i}{dt} = \sum_j k_{ji} M_j - M_i \sum_j k_{ij} \qquad (i \neq j;\ i, j = 1, \ldots, n). \qquad (5)$$

When the rate of change in the reservoir content is zero ($dM_i/dt = 0$), a steady state has been attained, and (5) becomes identical with (3).

For a system of n reservoirs, the amount in each reservoir (M_i) as a function of time can be obtained by solving n simultaneous differential equations of the type given in (5). Even with constant coefficients (k_{ij}), solution by hand (Boas, 1966) of more than two or three simultaneous differential equations is laborious and not convenient. Much more powerful methods are available that utilize analog and digital computers and that are aimed at solution of equations more general than (5). For example, in a more general case, the rate constants of the reservoirs (k_{ij}) may be some functions of time; there may be additional time-dependent inputs to reservoirs from outside the system; there may be chemical reactions within the reservoirs, in the course of which a chemical species may be added or removed at a rate proportional to some power of its concentration M_i. A variety of computing techniques for solution of simultaneous differential equations of varying degrees of complexity has been developed for engineering, environmental, and ecological problems (Carnahan and others, 1969; Patten, 1971; Pytkowicz, 1971; Machta, 1972; Morgan and Weinberg, 1972; Lerman, 1972; Lerman and Childs, 1973).

With reference to our treatment of the transient states of the phosphorus cycle, we solved the system of simultaneous equations of the type given by (5) using a program routine written for a CDC-6400 digital computer (Stein, 1972). The computational routine BSSODE (Stein, 1972), as well as other similar routines available in the libraries of computation centers, is based on stepwise integration of a system of equations whose solutions are being sought. In brief, given the initial values of the reservoir contents (M_i at time $t = 0$), the solutions of the equations are returned by the computer as the values of M_i at successive values of time t. The solution method applies to a one-boundary value problem: only some initial values of the independent and dependent variables (for example, t and M_i) must be stipulated; then the solution can be carried out for practically any length of time t or, in general, up to large values of the independent variable. This feature of the computational program BSSODE allows the user to test its performance by computing the values of the quantity sought (M_i in this case) up to large values of time t when a steady state essentially has been attained. An agreement between such a solution of equation (5) and a direct solution of equation (4) is usually very good (better than 1 percent). It is also pertinent that the numerical solution program performs satisfactorily even when there are differences of several orders of magnitude between the reservoir values as well as between the residence times of phosphorus in individual reservoirs (Fig. 2 and Table 2).

PHOSPHORUS CYCLE

A diagram of the geochemical cycle of phosphorus is shown in Figure 2. The phosphorus content of the geochemical reservoirs and the fluxes shown in Figure 2 have been taken or computed from literature sources, as explained in Table 1. The major reservoirs considered (Fig. 2) are sediments, land (soil), land biota, oceanic biota, surface ocean, deep ocean, and mineable phosphorus deposits. Excluding the mineable phosphorus reservoir and its fluxes, the phosphorus cycle as shown in Figure 2 is in a steady state. The balances between inputs and outputs within individual reservoirs determine whether the entire cycle is in a steady or transient state. Some of the estimates of reservoir content and interreservoir fluxes (listed from various sources in Table 1) differ by as much as a factor of three or four. Among the values we used, several had to be assumed in order to balance the cycle: this applies to fluxes F_{61}, F_{21}, F_{32}, and F_{46}, as explained more fully below and in Table 1. These four fluxes, however, were taken at the values balancing the respective reservoir content when the values of complementary input or output fluxes were obtained by other means.

The weathering of sediments is responsible for transport of phosphorus in solution and in solid materials. The flux F_{25} (Fig. 2) represents the amount of phosphorus transported by streams from land to the surface ocean (water depth less than 300 m); flux F_{25} is only a fraction of the total phosphorus flux (F_{12}) that includes both the detrital and dissolved matter than leaves the sediment reservoir. In terms of Figure 2, 20×10^6 tons/year of phosphorus in solid materials is transported from the land reservoir; of this, 18.3×10^6 tons/year is redeposited as sediment. The exact path of return is not important as long as significant amounts of phosphorus are not lost from the solid materials en route; the return path may be visualized as the phosphorus-bearing suspended load that is carried to the nearshore ocean bottom and becomes part of the sediment reservoir.

The only flux from land to surface ocean shown in Figure 2 is the flux due to streams. There may be additional fluxes from certain types of oceanic sediments, because higher concentrations of phosphorus develop in the interstitial water in the course of decomposition of organic matter. We had no sufficient data to estimate such fluxes.

Other inputs of phosphous to the surface ocean are by water exchange with the deep ocean (flux F_{65}) and by recovery of phosphorus taken up by the oceanic organisms (flux F_{45}).

In terms of the data given in Figure 2, the residence time of phosphorus in the land biota is significantly longer (46 years) than in the oceanic biota (0.14 years).

The flux F_{61} that removes phosphorus from the deep ocean to sediments was assumed equal to the flux F_{25} that inputs phosphorus from land to the ocean (Fig. 2). If the removal mechanism is primarily precipitation of calcium phosphate minerals in oceanic sediments, then the removal flux F_{61} may be very different from the simple proportionality relationship given in equation (2). The present evidence on the possible precipitation of calcium phosphate in the ocean is inconclusive (as summarized by Gulbrandsen and Roberson, 1973), and the changes in the phosphorus concentration in the deep ocean that arise in some of the processes we discuss are only of the order of a few percent. Therefore, our assumption of the simple first-order relationship between the flux out of the ocean and the phosphorus content of the deep ocean (F_{61}) should not distort the picture of the cycle even if there were calcium phosphate precipitation.

TRANSIENT STATES

We now consider two kinds of perturbations in the presumably steady-state cycle of phosphorus: (1) complete cessation of photosynthetic productivity on Earth and (2) man-produced contribution of phosphorus to land and streams from the mining reservoir. Either event or process perturbs the existing steady state and induces transient changes, the outcomes of which may be informative as guidelines for similar limiting situations.

Doomsday Perturbation

As one limiting case of perturbation of the steady-state cycle, we considered a complete cessation, achieved instantaneously, of photosynthetic productivity on land and in the ocean. In terms of the cycle diagram shown in Figure 2, the doomsday scenario corresponds to no uptake of phosphorus by the reservoirs of land and oceanic biota (the fluxes F_{23} and F_{54} and the rate constants k_{23} and k_{54} become zeros). The standing crop of the biota that has become sterile or dead continues to decay so that the decay rate constants (k_{32}, k_{45}, and k_{46} in Table 2) remain unchanged.

The outcome of the doomsday scenario is shown in Figure 3. The oceanic biota has decayed within less than 1 year. The amount of land biota has decreased to 1/100 of its original value in 200 years. It is appropriate to point out that the decay curves for the oceanic and land biota should be regarded as of orientational value, for the reason that they are based on constant decay rates, k_{32}, k_{45}, and k_{46}. Cessation of photosynthesis may result in development of new bacterial decay mechanisms or in a greater role of inorganic oxidation, either of which may take place at rates very different from those of the present. Thus, the curves for the oceanic and land biota may tend to some new steady-state levels, or they may begin to rise again in response to a hypothetical development of nonphotosynthetic biota. All this, however, is difficult to include in a rational manner in the model.

During the 200 years plotted in Figure 3, the phosphorus content of the surface ocean increases by a factor of about 2.5. The changes in the phosphorus content of the land and deep ocean are minuscule by comparison with that of the surface ocean. There is virtually no change in the sediment reservoir. The result of the doomsday perturbation after 200 years is that the total P concentrations in the surface ocean and deep ocean come closer to one another: in the surface ocean, the total P concentration increases from 25 mg/m^3 to 66 mg/m^3; in the deep ocean, there is a 5 percent decrease from 80 mg/m^3 to 76 mg/m^3. At a new steady state, with no terrestrial and oceanic biota left, the amounts of phosphorus in the land, deep ocean, and sediment reservoirs return to the values shown in Figure 2 ($M_1 = 4 \times 10^{15}$, $M_2 = 2 \times 10^{11}$, and $M_6 = 8.7 \times 10^{10}$ tons). The amount in the surface ocean, however, increases to 8.97×10^{10} tons, which corresponds to the concentration value of 83 mg/m^3. Thus, at a new steady state, the phosphorus concentration in the ocean is nearly uniform. This feature emphasizes the role played by the terrestrial and oceanic biota in the global phosphorus cycle. The amount of oxygen consumed in the oxidation of the land and oceanic biota is a negligibly small fraction of the amount available in the atmosphere.

Industrial Perturbation

Man's contribution to the phosphorus cycle is his removal of phosphorus from the mineable reservoir (7 in Fig. 2) and his unloading of phosphorus in the form of fertilizers and detergents in the land reservoir. Before presenting the geochemical consequences of phosphorus addition to land and rivers, we will discuss a few points pertaining to the paths of man-produced phosphorus in the industrial scenario.

Estimates of the phosphorus amount in mineable deposits (Table 1) vary from 1.8×10^9 tons in "known phosphorite deposits" (Ronov and Korzina, 1960) to an amount that is "for all practical purposes unlimited" (Emigh, 1973). We used the value of 1×10^{10} tons for the mineable phosphorus reservoir and the value of $F_{72} = 12 \times 10^6$ tons per year for the present rate of mining (flux F_{72} in Fig. 2).

The world consumption of fertilizers doubles every 10 years (Meadows and others, 1972, p. 26). Accordingly, we also considered an alternative mining rate of phosphorus that doubles every 10 years, starting with the present value. In mathematical terms, this increasing flux from mining becomes $F_{72} = 12 \times 10^6 \times e^{0.07t}$, where t is time in years.

Even if all the phosphorus mined were used in fertilizers, there would still be an uncertainty in what fraction of the fertilizer is taken up by the terrestrial biota. The agricultural experience and experimental evidence cited by Russell (1961, p. 57–63, 505–511) suggest that the fraction of fertilizer phosphorus taken up by plants depends to a considerable extent on the nature of the vegetation, soil conditions, and duration of fertilization. Depending on these factors, uptake of phosphorus by plants varies from a few percent to almost 100 percent of the amount added to soils.

As limiting cases, we considered two extremes of phosphorus utilization by plants.

In the first case, all the phosphorus from mining is assumed to be taken up by the land biota. In terms of the cycle diagram in Figure 2, the flux F_{73} replaces F_{72}. In the second case, all the phosphorus from mining is considered to be added to the land (flux F_{72}), where it mixes with phosphorus present in the land reservoir, and is subsequently taken up by the land biota through the flux F_{23}, as shown in Figure 2.

Considering the simpler situation first, addition of 12×10^6 tons/year to the existing flux that supplies the land biota reservoir will increase the mass of biota by approximately 20 percent: $(63.5 + 12) \times 10^6/0.021 = 3.6 \times 10^9$ tons of phos-

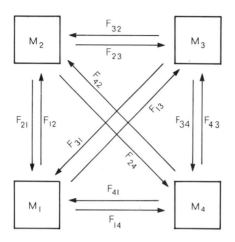

Figure 1. Scheme of a closed geochemical cycle. Reservoir content (M_i, in units of mass) and fluxes between reservoirs (F_{ij}, in units of mass/time).

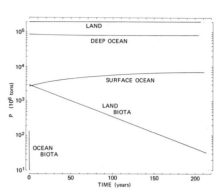

Figure 3. Doomsday scenario. Changes in phosphorus content of reservoirs after all photosynthetic productivity ceased on Earth at time 0 (virtually no change in reservoirs not shown).

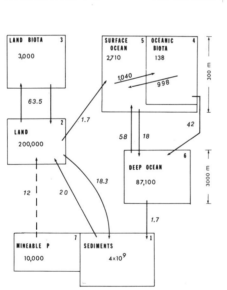

Figure 2. The phosphorus cycle (sources of data and remarks given in Table 1). Reservoir number given in upper right corner of each reservoir. Roman figures (such as 3,000, in reservoir 3) are phosphorus content, in millions of metric tons. Italic figures adjacent to arrows (such as *1,040*) are interreservoir fluxes, in millions of metric tons/year. Assumed depths of reservoirs 5 and 6 are shown.

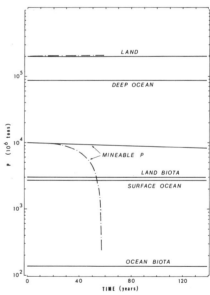

Figure 4. Changes in phosphorus content of reservoirs under conditions of (1) constant rate of phosphorus mining (solid lines), $F_{72} = 12 \times 10^6$ tons/year, and (2) exponentially increasing rate of phosphorus mining (dashed lines), $F_{72} = 12 \times 10^6 \times e^{0.07t}$ tons/year. Under condition (2), mineable reservoir would be depleted in approximately 60 years from the present (virtually no change in reservoirs not shown).

TABLE 2. RATE CONSTANTS OF INTERRESERVOIR FLUXES

$k_{12} = 4.28 \times 10^{-10}$	$k_{32} = 2.12 \times 10^{-2}$	$k_{54} = 3.84 \times 10^{-1}$	$k_{65} = 6.70 \times 10^{-4}$
$k_{23} = 3.16 \times 10^{-4}$	$k_{45} = 7.23$	$k_{56} = 6.70 \times 10^{-3}$	$k_{72} = 1.20 \times 10^{-3}$
$k_{25} = 8.51 \times 10^{-6}$	$k_{46} = 3.04 \times 10^{-1}$	$k_{61} = 1.96 \times 10^{-5}$	

Note: k_{ij} (in yr^{-1}) computed from data in Table 1, $k_{ij} = F_{ij}/M_i$. k_{12} computed as $k_{12} = (F_{12} - F_{21})/M_1$, see text.

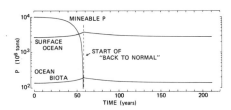

Figure 5. Changes in phosphorus content of reservoirs under conditions of an exponentially increasing rate of phosphorus mining and removal of phosphorus by stream flux ($F_{72} = 12 \times 10^6 \times e^{0.07t}$, and $F_{25} = 1.7 \times 10^6 \times e^{0.07t}$). Accelerated rate of mining does not essentially affect the phosphorus content of other reservoirs that are not shown in the figure. At 58 years from the present, the mineable phosphorus reservoir is depleted, and the conditions begin to return to their previous (initial) state.

phorus. This 20 percent increase in the mass of land biota and the flux F_{32} will have little effect on the phosphorus content of the reservoirs located downcycle from the land biota, because the amount of phosphorus in the land reservoir (2 × 10^{11} tons) is large by comparison with the increase in the terrestrial biota.

Similarly, there is virtually no effect on the phosphorus content of the bigger reservoirs if the constant flux from mining ($F_{72} = 12 \times 10^6$ tons/year) inputs phosphorus to the land reservoir (Fig. 4, solid lines) instead of directly to the land biota. Under these conditions, the mass of mineable phosphorus decreases by about 12 percent over a 100-year period, but hardly any changes are noticeable in the other reservoirs.

If the phosphorus flux from mining grows at an exponential rate, doubling in value every 10 years, then the mineable reservoir of 1 × 10^{10} tons would be depleted in approximately 60 years (dashed line, Fig. 4). Despite the rapidly growing amount of phosphorus contributed to the land reservoir under these conditions, the phosphorus content of the land increases by only 4 percent, and there are no noticeable changes in other reservoirs.

The lack of significant increases in the phosphorus content of the reservoirs supplied from land is due to the large amount of phosphorus in the land reservoir by comparison with the total amount added from the mineable reservoir. Also, the long residence time of phosphorus in the land reservoir (2 × 10^{11}/(1.71 × 10^6) ≃ 1 × 10^5 year) is responsible for the small magnitude and slow rate of changes in the oceanic reservoirs that are supplied from land.

A more severe perturbation of the phosphorus cycle in the presence of an exponentially growing flux from mining may be achieved if the stream flux (F_{25}) of phosphorus grows in proportion to the mining contribution to land. In this case ($F_{25} = 1.71 \times 10^6 \times e^{0.07t}$, where t is time in years), as shown in Figure 5, major changes take place in the surface ocean and oceanic biota during the 60-year period of accelerated phosphorus mining. The amount of phosphorus in the surface ocean increases by 38 percent, and the amount of oceanic biota increases by 30 percent. For reservoirs not shown in Figure 5, the increase in the phosphorus content is 3 percent for the land and a fraction of 1 percent for the land biota and the deep ocean.

The 30 percent increase in the amount of oceanic biota shown in Figure 5 may be considered a valid estimate only if biological productivity in the ocean is not limited by the abundance of some other element (for example, nitrogen). If nitrogen is a limiting nutrient in the ocean, then it is conceivable that under the conditions of greater supply of phosphorus from land, the biological composition of oceanic plankton may change toward greater abundance of organisms capable of more efficient nitrogen fixation. Then, the mean rate of phosphorus uptake by the biota (flux F_{54} in Fig. 2) may become significantly different from the value we use. In the absence of information on such possible changes, the effects on the phosphorus cycle cannot be reasonably estimated.

If the biological productivity in the ocean is limited by some nutrient other than phosphorus, then there might be no increase in the ocean biomass despite a greater input of phosphorus to the surface ocean. The effect of this scenario on the phosphorus concentration in the ocean is small: The 30 percent increase in the oceanic biomass amounts to $138 \times 10^6 \times 0.30 = 41 \times 10^6$ tons phosphorus; if the biomass does not increase, this amount would be distributed among the much larger amounts of phosphorus in the surface ocean and deep ocean.

It is instructive to ask, What would be the course of the phosphorus cycle after the mineable reservoir has been depleted? Some answers to this are shown in Fig. 5, to the right of the dashed line that delineates the 58-year mark of depletion of the mineable reservoir. The course of events leading "back to normal" takes, according to our computation, approximately 150 years. The mineable reservoir remains depleted, and the amount of the oceanic biomass and the phosphorus content of the surface ocean decrease to the initial (present-day) conditions, owing to the removal of the phosphous to the deep ocean.

Further Discussion

Our model of the phosphorus cycle in transient states, induced by various perturbations discussed in the preceding sections, gives the magnitudes of some possible changes and their time scales in the individual geochemical reservoirs.

We conclude our discussion by pointing out that knowledge of the nature of the interreservoir fluxes, and their dependence on the physical and chemical conditions of the cycle are critically important to the correctness of the results of transient-state models. In the preceding sections, it has been pointed out how the contents of the geochemical reservoirs (Fig. 2) may be affected by mineral-forming reactions in soils and oceanic sediments, as well as by changes in the oceanic productivity. Occurrence of such processes in nature dictates a particular need to consider interactions among the geochemical cycles of more than one element, all of which may be responsible for the picture. Application of steady-state and transient-state models to several interacting geochemical cycles is a matter of extension of established methods. A careful review of the results obtainable for transient-state models, no matter how tentative, points out those parts of the geochemical cycles that need additional information on a priority basis, as required for an analysis of their past and future evolution.

ACKNOWLEDGMENTS

We thank David J. Toth of Northwestern University for his help in data processing and preparation of materials for this paper, and Donald L. Graf and L. T. Kurtz of University of Illinois for critical reading of the paper. The research was supported by the Oceanography Section, National Science Foundation, NSF Grant GA-30769.

REFERENCES CITED

Boas, M. L., 1966, Mathematical methods in the physical sciences: New York, John Wiley & Sons, Inc., 778 p.
Bolin, B., 1970, The carbon cycle: Sci. American, v. 223, no. 3, p. 125-132.
Broecker, W., 1963, Radioisotopes and large-scale oceanic mixing, in Hill, M. N., ed., The Sea, Vol. 2: New York, Interscience Pubs., Inc., p. 88-108.
──── 1971, A kinetic model for the chemical composition of sea water: Quaternary Research, v. 1, p. 188-207.
Carnahan, B., Luther, H. A., and Wilkes, J. O., 1969, Applied numerical methods: New York, John Wiley & Sons, Inc., 604 p.
Craig, H., 1957, The natural distribution of radiocarbon and the exchange time of carbon dioxide between atmosphere and sea: Tellus, v. 9, p. 1-17.
Deevey, E. S., Jr., 1970, Mineral cycles: Sci. American, v. 223, no. 3, p. 149-158.
Delwiche, C. C., 1970, The nitrogen cycle: Sci. American, v. 223, no. 3, p. 137-146.
Denbigh, K. G., 1958, The thermodynamics of the steady state: New York, John Wiley & Sons, Inc., 103 p.
Emigh, G. D., 1973, Economic phosphate deposits, in Griffith, E. J., Beeton, A., Spencer, J. M., and Mitchell, D. T., eds., Environmental phosphorus handbook: New York, John Wiley & Sons, Inc., p. 97-116.
Eriksson, E., 1971, Compartment models and reservoir theory: Ann. Rev. Ecology and Systematics, v. 2, p. 67-84.
Frost, A. A., and Pearson, R. G., 1961, Kinetics and mechanism (2d ed.): New York, John Wiley & Sons, Inc., 405 p.
Fuller, W. H., 1972, Phosphorus element and geochemistry, in Fairbridge, R. W., ed., The encyclopedia of geochemistry and environmental sciences: New York, Van Nostrand Reinhold Co., p. 942-946.
Garrels, R. M., and Mackenzie, F. T., 1971, Evolution of sedimentary rocks: New York, W. W. Norton & Co., Inc., 397 p.
Garrels, R. M., Mackenzie, F. T., and Hunt, C. A., 1973, Man's contributions to natural chemical cycles: Evanston, Ill., Northwestern Univ., Dept. of Geol. Sci., Authors' copyright, mimeogr., 104 p.
Gulbrandsen, R. A., and Roberson, C. E., 1973, Inorganic phosphorus in seawater, in Griffith, E. J., Beeton, A., Spencer, J. M., and Mitchell, D. T., eds., Environmental phosphorus handbook: New York, John Wiley & Sons, Inc., p. 117-140.
Lal, D., and Peters, B., 1967, Cosmic ray produced radioactivity on the Earth, in Flügge, S., ed., Handbuch der Physik: Berlin, Springer-Verlag, v. 46, no. 2, p. 551-612.
Lemon, E. R., 1968, Carbon fixation and solar energy budget, in Altman, P. L., and Dittmer, D. S., eds., Metabolism (biological handbooks): Bethesda, Md., Federation of Am. Socs. for Experimental Biology, p. 487.
Lerman, A., 1972, Strontium 90 in the Great Lakes: Concentration-time model: Jour. Geophys. Research, v. 77, p. 3256-3264.
Lerman, A., and Childs, C. W., 1973, Metal-organic complexes in natural waters: Control of distribution by thermodynamic, kinetic and physical factors, in Singer, P. C., ed., Trace metals and metal-organic interactions in natural waters: Ann Arbor, Mich., Ann Arbor Science Pubs., p. 201-235.
Machta, L., 1972, The role of the oceans and biosphere in the carbon dioxide cycle, in Dyrssen, D., and Jagner, D., eds., The changing chemistry of the oceans: New York, John Wiley & Sons, Inc., p. 121-145.
Meadows, D. H., Meadows, D. L., Randers, J., and Behrens, W. W., III, 1972, The limits to growth: New York, Universe Books, 205 p.
Morgan, R. E., and Weinberg, R., 1972, Computer simulation of world systems: Biogeochemical cycles: Internat. Jour. Environ. Studies, v. 3, p. 103-118.

Patten, B. C., 1971, Systems analysis and simulation in ecology: New York, Academic Press, v. 1, 607 p.

Poldervaart, A., 1955, Chemistry of the Earth's crust: Geol. Soc. America Spec. Paper 62, p. 119.

Pytkowicz, R. M., 1971, The chemical stability of the oceans: Oregon State Univ. Dept. Oceanography Tech. Rept. 214, Ref. 71-20, 24 p.

Redfield, A. C., Ketchum, B. H., and Richards, F. A., 1963, The influence of organisms on the composition of seawater, in Hill, M. N., ed., The Sea, Vol. 2: New York, Interscience Pubs., Inc., p. 26-77.

Ronov, A. B., and Korzina, G. A., 1960, Phosphorus in sedimentary rocks: Geochemistry, no. 8, p. 805-829.

Russell, E. W., 1961, Soil conditions and plant growth (9th ed.): New York, John Wiley & Sons, Inc., 688 p.

Ryther, J. H., and Dunstan, W. M., 1971, Nitrogen, phosphorus, and eutrophication in the coastal marine environment: Science, v. 171, p. 1008-1013.

Stein, J., 1972, Computer program BSSODE: Evanston, Ill., Northwestern Univ., Vogelback Computing Center, Program Library No. NUCC227, mimeogr., 6 p.

Stumm, W., 1972, The acceleration of the hydrogeochemical cycling of phosphorus, in Dyrssen, D., and Jagner, D., eds., The changing chemistry of the oceans: New York, John Wiley & Sons, Inc., p. 329-346.

Stumm, W., and Morgan, J. J., 1970, Aquatic chemistry: New York, Interscience Pubs., Inc., 583 p.

Sverdrup, H. U., Johnson, M. W., and Fleming, R. H., 1942, The oceans: Englewood Cliffs, N.J., Prentice-Hall, 1087 p.

Taylor, S. R., 1964, Abundance of chemical elements in the continental crust: A new table: Geochim. et Cosmochim. Acta, v. 28, p. 1273-1285.

Vaccaro, R. F., 1965, Inorganic nitrogen in sea water, in Riley, J. P., and Skirrow, G., eds., Chemical oceanography: New York, Academic Press, v. 1, p. 365-408.

Van Wazer, F., ed., 1961, Phosphorus and its compounds: New York, Interscience Pubs., Inc., v. 2, 1091 p.

MANUSCRIPT RECEIVED BY THE SOCIETY OCTOBER 22, 1973
REVISED MANUSCRIPT RECEIVED FEBRUARY 25, 1974

Geological Society of America
Memoir 142
© 1975

Mixing of Sea Water with Calcium Carbonate Ground Water

L. N. PLUMMER*

*Department of Geological Sciences
State University of New York at Buffalo
Buffalo, New York 14207*

ABSTRACT

The mixing of sea water and (or) saline subsurface water with fresh calcium carbonate ground water can be of major importance in the chemical diagenesis of carbonate rocks and sediments. A computational scheme for calculation of the theoretical distribution of dissolved species in mixtures of two solutions was developed and used to evaluate the effect of mixing saline water with fresh calcium carbonate ground water. The calculations include data on (1) mixtures of surface sea water (pH 8.15) with solutions in equilibrium with calcite at CO_2 partial pressures from 10^{-4} to 1.0 atm at 5°, 15°, 25°, and 35° C; (2) mixtures of surface sea water with solutions twofold saturated with calcite; and (3) mixing of actual saline water and fresh carbonate ground water from central Florida and the Yucatan Peninsula.

The results show that the amount of undersaturation in mixtures is a function of P_{CO_2}, temperature, ionic strength, degree of calcite saturation, and pH of end-member solutions prior to mixing. All mixtures of surface sea water and saturated calcium carbonate ground water investigated containing 0 to 10 percent sea water by volume were undersaturated with calcite. The range of undersaturation increased greatly with increasing P_{CO_2} (above 10^{-3} atm) in calcite-saturated water and with decreasing temperatures. No undersaturation occured in mixtures of surface sea water and supersaturated calcium carbonate ground water (twofold saturated) having CO_2 partial pressures less than 10^{-2} atm at 25° C. More than 10 percent surface sea water mixed with supersaturated calcium carbonate ground water (twofold saturated) at CO_2 partial pressures greater than 10^{-2} atm was required to cause undersaturation. Arbitrarily lowering the pH of sea water increased the range of undersaturation in mixtures, but this range decreased in mixtures of carbonate ground water and sea water at pH values lower than 8.15 when the calcite

---

*Present address: U.S. Geological Survey, National Center, Mail Stop 432, Reston, Virginia 22092.

supersaturation is maintained constant and equal to that of surface sea water at pH 8.15.

The results of these calculations define geochemical environments that favor (1) development of increased porosity and permeability in limestone aquifers and (2) fresh-water dolomitization of calcium carbonate rocks.

INTRODUCTION

In recent years there has been considerable interest in the mixing of natural waters, particularly as applied to the diagenesis of carbonate sediments and rocks. The mixing of two waters saturated with calcite at differing CO_2 partial pressures can produce solutions that are undersaturated with calcite, and thus gain an increased capacity to dissolve more calcite (Bogli, 1963, 1964; Arntson, 1964; Ernst, 1964; Howard, 1964; Thrailkill, 1968; Runnells, 1969; Badiozamani, 1972, 1974). Because the rate of change of mineral solubility is a function of change in ionic strength, Runnells (1969) showed that it is possible to mix undersaturated waters resulting in supersaturated solutions, and to mix supersaturated waters resulting in undersaturation. This mixing effect can be seen, in part, in changes in individual ion-activity coefficients as a function of mixing. Runnells (1969, Fig. 2, p. 1194), for example, showed from the experimental data of Frear and Johnston (1929) and Shternina and Frolova (1945) that the solubility of calcite in NaCl solutions at 0.97 atm CO_2 and 25°C is expressible as a downward-concave parabolic function with increasing NaCl content. The mixing of two solutions using this calcite-solubility curve is a linear function of the total concentrations of calcium and carbonate in solution; however, the activity coefficients of Ca^{++} and CO_3^{--} ions are smaller in the resulting mixture than those predicted from a linear relation of activity coefficients between the two solutions. Although the ion-activity product of calcite (IAP_c) is equal to the equilibrium constant of calcite (K_c) along the calcite-solubility curve in NaCl solutions, the non-linear function of γ_i with changing ionic strength causes IAP_c to be less than K_c in mixtures, resulting in undersaturation. By definition, $IAP_c = a_{Ca^{++}} a_{CO_3^{--}}$, and $a_i = m_i \gamma_i$, where a_i and is the thermodynamic activity of the ith ion in solution, and m_i and γ_i are the molality and activity coefficients of the ith ion in solution. This mixing effect is important to the process of diagenesis of carbonate sediments and rocks where there is mixing of calcium bicarbonate waters of differing ionic strengths; the resulting mixtures may have an increased capacity to dissolve calcium carbonate.

Mathews (1971) suggested that the mixing of sea water with fresh-water aquifers in carbonate environments may produce brackish waters undersaturated with calcite. Badiozamani (1972, 1974) showed that the mixing of fresh carbonate water with sea water favors a dolomitization mechanism, because mixtures can dissolve calcite and aragonite but are supersaturated with dolomite and, thus, have a thermodynamic potential for replacement.

In addition to changes in ion-activity coefficients related directly to mixing sea water with calcium bicarbonate solutions, important changes in carbonate equilibria result from (1) differing CO_2 partial pressures of initial solutions and (2) changes in pH owing to the formation of complex ions in mixtures. Whereas these latter two factors are not important to the mixing of calcium bicarbonate solutions with NaCl solutions at 0.97 atm CO_2, they are of major importance in the mixing of sea water with calcium bicarbonate ground water. When these factors, as well as the effects of temperature, pH of saline solution, and saturation state of end-member solution, are taken into account, the mixing effect becomes important when low-temperature, saturated, fresh carbonate ground water at high CO_2 partial pressures is mixed with saline subsurface water.

Mixing of supersaturated ground water from carbonate rocks at CO_2 partial pressures lower than $10^{-1.8}$, such as the mixing of some central Florida ground waters with sea water, does not cause calcite undersaturation. Thus, some carbonate ground-water environments have a potential for limestone-cavern excavation (and fresh-water dolomitization) related to mixing with sea water, whereas other environments do not.

The computational scheme presented in this paper allows the prediction of geochemical environments in which calcite undersaturation results from mixing natural waters. A theoretical calculation scheme to compute the pH and chemical equilibria in mixtures of two solutions was developed and used to evaluate (1) the mixing of ground waters in equilibrium with calcite at differing partial pressures of CO_2 with sea water at 5°, 15°, 25°, and 35°C; (2) the mixing of sea water with solutions twofold saturated with calcite; (3) the dependence of the mixing effect on ionic strength, pH, and saturation index of end-member waters; and (4) the mixing of actual saline and fresh ground water from central Florida and the Yucatan Peninsula.

CALCULATION PROCEDURE

Chemical Model

Calculation of carbonate equilibria that result from mixing two aqueous solutions requires a chemical model for computation of activities of inorganic ions and ion pairs in solution. The chemical model used was modified from the PL/1 version of WATEQ (Truesdell and Jones, 1974) and programmed in FORTRAN IV for a CDC 6400 computer. Twenty-seven inorganic ions and complex species of calcium, magnesium, sodium, potassium, chlorine, sulfur, and carbon were considered in the equilibria calculations (Table 1). Reference thermodynamic data at 25°C were computed to other temperatures by means of the van't Hoff equation and, in some cases, from analytical expressions derived by Truesdell and Jones (1974) from experimental data in the literature (Table 1).

The distribution of aqueous species was solved by a successive-approximation routine involving the simultaneous solution of a set of mass-action and mass-balance equations (Garrels and Thompson, 1962; Garrels and Christ, 1965) that represent the aqueous model. Individual ion-activity coefficients of charged species were computed from the extended Debye-Hückel theory used by Truesdell and Jones (1973), and activity coefficients of neutral species were calculated from the relation $\gamma_i^0 = 10^{0.1I}$ where I is the ionic strength of the solution.

pH of Mixtures

The pH of mixtures is critical to calculation of the mixing effect on carbonate equilibria and cannot be estimated accurately by simple means, especially when mixing solutions such as sea water with carbonate ground water. The calculation procedure used is based on the principle that mixtures of electrically balanced solutions are also electrically balanced, so that the pH of mixtures can be computed through charge-balance relations. This method of calculation is theoretically correct and imposes no constraint on the results other than the assumptions implicit in the aqueous model.

The two waters to be mixed were each forced to convergence on mass balance in the aqueous model to approximately 10^{-11} (actually $10^{-3} \times 10^{-pH}$, which allowed pH of mixtures to be computed to at least four significant figures). For example, mass-balance convergence on carbon is defined by

$$[\text{Total CO}_2]_{\text{actual}} - [m_{H_2CO_3} + m_{HCO_3^-} + m_{CO_3^{2-}}$$
$$+ m_{CaCO_3^0} + m_{MgCO_3^0} + m_{MgHCO_3^+}$$
$$+ m_{NaHCO_3^0} + m_{NaCO_3^-} + m_{Na_2CO_3}]_{\text{computed}} \leq 10^{-pH} 10^{-3}.$$

Similar mass-balance relations were written for chloride and sulfate species. After

TABLE 1. THERMOCHEMICAL DATA

Reaction	ΔH_r^{0*}	Log $K_{(25)}$[†]
$CaOH^+ = Ca^{++} + OH^-$	-1.19	-1.40
$CaSO_4^0 = Ca^{++} + SO_4^{--}$	-1.65	-2.309
$CaCO_3^0 = Ca^{++} + CO_3^{--}$	-3.13	-3.20
$MgOH^+ = Mg^{++} + OH^-$	-2.14	-2.60
$MgSO_4^0 = Mg^{++} + SO_4^{--}$	-4.92	-2.238
$MgHCO_3^+ = Mg^{++} + HCO_3^-$	-10.37	-0.928
$MgCO_3^0 = Mg^{++} + CO_3^{--}$	-0.058	-3.398
$NaSO_4^- = Na^+ + SO_4^{--}$	-2.229	-0.226
$Na_2SO_4^0 = 2Na^+ + SO_4^{--}$	2.642	-1.512
$NaHCO_3^0 = Na^+ + HCO_3^-$	..[§]	0.250
$NaCO_3^- = Na^+ + CO_3^{--}$	-8.911	-1.268
$Na_2CO_3^0 = 2Na^+ + CO_3^{--}$	..[§]	-0.672
$NaCl^0 = Na^+ + Cl^-$	..[§]	1.602
$KCl^0 = K^+ + Cl^-$	..[§]	1.585
$H_2O = H^+ + OH^-$	13.345	-13.998
$CaMg(CO_3)_{2(\text{dolomite})} = Ca^{++} + Mg^{++} + 2CO_3^{--}$	-8.29	-17.00

Reaction	Analytical expression
$CaCO_{3(\text{calcite})} = Ca^{++} + CO_3^{--}$	Log $K(T)$[#] = $13.870 - 0.04035T - 3059/T$
$H_2CO_3^0 = H^+ + HCO_3^-$	Log $K(T)$[**] = $14.8435 - 0.032786T - 3404.71/T$
$HCO_3^- = H^+ + CO_3^{--}$	Log $K(T)$[††] = $6.498 - 0.02379T - 2902.39/T$
$KSO_4^- = K^+ + SO_4^{--}$	Log $K(T) = -3.106 + 673.6/T$
$HSO_4^- = H^+ + SO_4^{--}$	Log $K(T) = 5.3505 - 0.0183412T - 557.2461/T$
$CO_2 + H_2O = H_2CO_3^0$	Log $a_{H_2CO_3^0} = \log P_{CO_2} - 14.0184 + 0.015264T + 2385.73/T - I(0.84344 - 0.004471T + 0.00000666T^2)$[§§]

Note: The thermodynamic data, except where indicated, have been taken from the recent compilation by Truesdell and Jones, 1974, in which references to the original data sources are given. The $CaHCO_3^+$ complex has been ignored for reasons discussed by Plummer and Mackenzie (1974).

*Standard enthalpy of reaction (kcal/mole).
[†] Log of equilibrium constant, K, for the reaction at 25° C.
[§] No value of ΔH_r^0 is known. Log $K_{(25)}$ has been used at all temperatures considered.
[#] Jacobson and Langmuir (1974).
[**] Harned and Davis (1943).
[††] Harned and Scholes (1941).
[§§] Analytical expression for log $a_{H_2CO_3^0}$ as a function of log P_{CO_2}, temperature, $T°K$, and I, the ionic strength of the solution (Truesdell and Jones, 1974). This relation has been used in computing calcite solubility as a function of temperature and CO_2 partial pressure.

each convergence on mass balance for the initial solutions, charge balance was checked and adjusted by addition or subtraction of small amounts of chloride ion. Both mass and charge balance in the chemical model were driven to convergence near 10^{-11} in successive approximations for the initial solutions. These two electrically balanced solutions were then mixed in the desired volumetric proportions, and the mixture was forced to convergence on mass balance to 10^{-11} using the pH of the dominant starting solution as an initial approximation of the new pH of the mixture. Because changes in charge balance in computed mixtures after convergence on mass balance are due only to changes in pH from mixing effects, the pH of the mixture was then adjusted in successive approximations until both the chemical model and charge balance converged to approximately 10^{-11}. This procedure accurately estimates the pH of mixtures of known solutions and allows accurate calculation of mineral saturation in mixtures. All ions tabulated in Table 1 were considered in the calculations.

Figure 1 illustrates the necessity of determining the effect of ion-pair formation on the pH of mixtures. Figure 1 compares the pH of mixtures of surface sea water (pH 8.15) and theoretical solutions in equilibrium with calcite at $10^{-2.5}$ and 1.0 atm CO_2 as computed by three procedures: (1) mixing the molalities of free hydrogen ion of end-member solutions in volumetric proportions, "volumetric m_{H^+}," ignoring changes in pH caused by ion-pair formation and assuming $\gamma_i = 1$; (2) mixing the molalities of free hydrogen ion of end-member solutions in volumetric proportions, computing the activity of H^+ from the resulting molality of H^+, and determining an activity coefficient of H^+ from the ionic strength of the mixture, "volumetric pH," similar to the method of Badiozamani (1972, 1974); and (3) considering the effect of ionic strength and ion pairing, utilizing the charge-balance procedure described above. Mixing curves that assume $\gamma_i = 1$ (volumetric m_{H^+}) are shown for comparison with curves in which γ_i was considered (volumetric pH) in order to isolate the importance of activity coefficients on pH (Fig. 1). Both at $10^{-2.5}$ and 1 atm, the differences between volumetric m_{H^+} and volumetric pH increase as percent of sea water in the mixture increases owing to decreasing values of γ_i.

The pH of mixtures computed by the charge-balance method described above differs markedly from the approximations of volumetric m_{H^+} and volumetric pH.

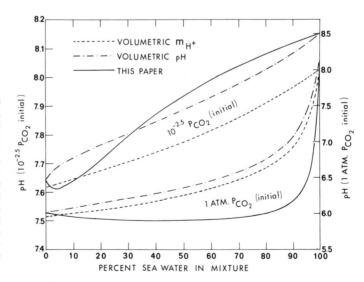

Figure 1. The pH of mixtures of sea water and solutions in equilibrium with calcite at $10^{-2.5}$ and 1 atm CO_2 at 25° C and 1 atm total pressure. "Volumetric m_{H^+}" ignores activity-coefficient and ion-pairing effects on pH, and "volumetric pH" neglects ion-pairing effects on pH. The solid curve defines the pH as computed in this paper and introduces no approximations other than the assumption of the existence of the aqueous model.

The differences are not simple to explain and represent the combined effects of multiple shifts in chemical equilibria on mixing. Some of these effects can be summarized as follows: (1) The CO_3^{--}/HCO_3^- ratio in sea water is tenfold that of a solution saturated with calcite at $10^{-2.5}$ atm CO_2, so that on mixing, a greater proportion of CO_3^{--} is added to the mixture that reacts with H^+ to form additional HCO_3^-, causing pH to increase. (2) The increased ionic strength of mixtures from the addition of sea water causes γ_{H^+} to decrease, again increasing pH in mixtures. (3) The formation of MCO_3^0 complexes, particularly $MgCO_3^0$, decreases the amount of free CO_3^{--} in solution, which causes HCO_3^- to dissociate, forming additional CO_3^{--}; the release of H^+ decreases pH. (4) The formation of $MHCO_3^+$ complexes consumes free HCO_3^-. If H_2CO_3 dominates CO_3^{--}, the dissociation of H_2CO_3 to HCO_3^- and H^+ decreases pH, but if CO_3^{--} is more abundant than H_2CO_3,

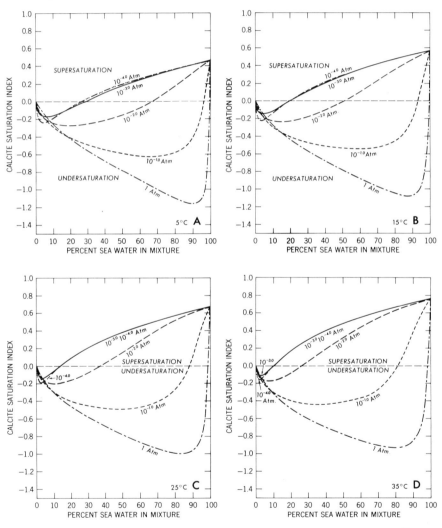

Figure 2. Saturation index of calcite in mixtures of sea water and solutions in equilibrium with calcite at differing CO_2 partial pressures and temperatures. Numbers shown on curves are log P_{CO_2} of solutions in equilibrium with calcite prior to mixing. A, 5° C; B, 15° C; C, 25° C; D, 35° C.

the association of CO_3^{--} and H^+ causes an increase in pH. An important reaction in the mixing of sea water with solutions saturated with calcite at 1 atm CO_2 is the formation of $MgHCO_3^+$. Because $m_{H_2CO_3} \gg m_{CO_3^{--}}$ in solutions saturated with calcite at 1 atm CO_2, H^+ is released by the formation of $MgHCO_3^+$ in mixtures, which to a major extent accounts for the low pH computed (Fig. 1). (5) The formation of MOH^+ species on the addition of sea water releases H^+, causing a decrease in pH. (6) The addition of SO_4^{--} from sea water to lower pH solutions increases the formation of HSO_4^-, which consumes H^+ and causes an increase in pH. The formation of HSO_4^- alone accounts for 60 percent of the observed differences between the volumetric pH curve and the curve designated "this paper" in a mixture containing a calcite-saturated solution at $10^{-2.5}$ atm CO_2 and 70 percent sea water (Fig. 1). (7) There are many other effects—such as the formation of MSO_4^0, which reduces the relative amount of free SO_4^{--} and M^{++} in mixtures; reduction of the amount of SO_4^{--} and M^{++} decreases the effectiveness of the formation of HSO_4^-, MOH^+, $MHCO_3^+$, and MCO_3^0 to change pH.

The charge-balance calculation procedure used in this paper takes into account all possible effects on pH that can occur in the defined aqueous model. The accuracy of the results calculated in this paper is dependent upon the accuracy of the equilibrium constants used, the accuracy of the calculated activity coefficients, and the assumption that the ionic species considered are the only important species in the aqueous phase. A test of the charge-balance procedure for calculation of pH of mixtures was made by predicting pH values of K_2CO_3-HCl-$MgCl_2$ solutions identical to the pH observed by Reardon and Langmuir (1974) when their individual values of the equilibrium constant for the dissociation of $MgCO_3^0$ were used in the aqueous model.

All results presented in this paper apply to mixing in closed systems; that is, the mixing calculations do not reflect change in mass of the constituent or constituents in mixtures owing to exchange or interaction with adjacent systems, such as the exchange of CO_2 between a surface water body and the atmosphere. The calculations presented here have direct application to many examples of subsurface mixing that, for all practical purposes, can be considered closed systems. The results also describe the thermodynamic potential for dissolution or precipitation reactions to occur in closed, heterogeneous systems at the time of mixing, but they do not actually reflect the effect of such reactions on the composition of mixtures.

RESULTS

P_{CO_2} and Temperature Dependence

Figure 2 (A through D) shows the effect of mixing solutions in equilibrium with calcite at 5°, 15°, 25°, and 35° C at CO_2 partial pressures from $10^{-4.0}$ to 1 atm (Table 2) with surface sea water at pH 8.15 (Table 3). These results are plotted in terms of percent of sea water by volume in the mixture and the resulting saturation index of calcite (SI_c). The SI_c, as defined by Langmuir (1971), is the log of the ratio of the ion-activity product of calcite (IAP_c) in mixtures to the equilibrium constant (ion product) of calcite (K_c); that is, $SI_c = \log IAP_c/K_c$. Values of SI_c greater than zero indicate supersaturation, and values less than zero correspond to undersaturation.

The calculations show (Fig. 2) that the addition of 10 percent or more sea water by volume to all solutions saturated with calcite at $10^{-4.0}$ to 1 atm P_{CO_2} causes undersaturation. The degree of resulting undersaturation and the range of sea-water content of undersaturated mixtures increase with decreasing temperature and (or) increasing CO_2 partial pressure in saturated calcite solutions. The effect of temperature on the degree of undersaturation is seen more in the range of sea-water content in undersaturated solutions than in the actual magnitude of the undersaturation. For example, a solution saturated with calcite at $10^{-2.0}$ atm CO_2 and 5° C containing 67 percent sea water is undersaturated, whereas at 35° C and $10^{-2.0}$ atm CO_2, undersaturation occurs with only 27 percent sea water in the mixture. For the same example at 5° and 35° C respectively, however, the SI_c differs by only 0.1 SI_c units. Mixing curves at and above 25° C are essentially identical for saturated calcite solutions at $10^{-3.0}$ and $10^{-4.0}$ atm CO_2 that contain more than 10 percent sea water.

The range of sea-water content in undersaturated mixtures of sea water and solutions saturated with calcite as a function of temperature and P_{CO_2} of saturated solutions prior to mixing is summarized in Figure 3. All mixtures above an

TABLE 2. THEORETICAL COMPOSITION OF SOLUTIONS IN EQUILIBRIUM WITH CALCITE AT TEMPERATURES AND CO_2 PARTIAL PRESSURES OF INTEREST

log P_{CO_2}	pH	Calcium*	Carbon†	log P_{CO_2}	pH	Calcium*	Carbon†
5°C pK_c = 8.351§				15°C pK_c = 8.373§			
0.0	6.05277	12.86006	88.22792	0.0	6.03102	10.86071	66.64054
−0.5	6.37384	8.33271	36.56137	−0.5	6.35284	7.05371	28.38540
−1.0	6.69667	5.43266	17.17961	−1.0	6.67633	4.60986	13.74566
−1.5	7.02104	3.56380	9.12488	−1.5	7.00128	3.03123	7.49132
−2.0	7.34674	2.35184	5.33067	−2.0	7.32745	2.00504	4.45620
−2.5	7.67352	1.56112	3.31399	−2.5	7.65459	1.33410	2.80183
−3.0	8.00108	1.04254	2.13750	−3.0	7.98237	0.89331	1.81958
−3.5	8.32899	0.70113	1.40839	−3.5	8.31028	0.60285	1.20420
−4.0	8.65653	0.47606	0.94037	−4.0	8.63740	0.41150	0.80665
25°C pK_c = 8.420§				35°C pK_c = 8.491§			
0.0	6.01464	9.10175	51.98090	0.0	6.00315	7.57664	41.61494
−0.5	6.33720	5.92618	22.57639	−0.5	6.32643	4.94614	18.28555
−1.0	6.66134	3.88268	11.16100	−1.0	6.65119	3.24886	9.15279
−1.5	6.98683	2.55926	6.18840	−1.5	6.97722	2.14668	5.12831
−2.0	7.31346	1.69686	3.72566	−2.0	7.30429	1.42665	3.11069
−2.5	7.64094	1.13177	2.36049	−2.5	7.63210	0.95386	1.98074
−3.0	7.96892	0.75992	1.54040	−3.0	7.96021	0.64233	1.29688
−3.5	8.29675	0.51477	1.02275	−3.5	8.28782	0.43700	0.86291
−4.0	8.62326	0.35358	0.68667	−4.0	8.61327	0.30260	0.57981

Note: Computed from the thermochemical data of Table 1 and specified partial pressure of CO_2. As with the mixing calculations, the $CaHCO_3^+$ complex has been ignored (Table 1). Although the numbers may not be accurate beyond two significant figures due to uncertainties in the thermochemical data for calcite (Jacobson and Langmuir, 1974; Plummer and Mackenzie, 1974), they are reported to six significant figures because this precision was used to initiate the theoretical mixing calculations. Six-figure precision minimizes charge imbalances so that only small amounts of chloride ion ($\simeq 10^{-5}$ meq/l) were required to adjust the charge balance of initial solutions to within 10^{-11}.

*Total calcium in solution in mmoles/kg H_2O equal to the sum (in mmoles/kg H_2O) of Ca^{++}, $CaCO_3^0$, and $CaOH^+$.

†Total carbon in solution in mmoles/kg H_2O equal to the sum (in mmoles/kg H_2O) of $H_2CO_3^0$, HCO_3^-, CO_3^{--}, and $CaCO_3^0$.

§The term pK_c is the negative log of the equilibrium constant for the reaction $CaCO_{3\,(calcite)}$ = Ca^{++} + CO_3^{--}, $pK_c = -\log K_c$.

isotemperature line (Fig. 3) are supersaturated and those below are undersaturated with calcite for that given temperature and P_{CO_2} of saturated calcite solution. Figure 3 shows that the maximum percent of sea water causing undersaturation is essentially independent of P_{CO_2} below $10^{-3.0}$ atm, and that nearly all volumes of sea water in saturated solutions at greater than $10^{-0.5}$ atm cause undersaturation. The maximum temperature dependence on the range of undersaturation occurs over the region of maximum P_{CO_2} dependence, which is $10^{-1.0}$ to $10^{-3.0}$ atm CO_2 (Fig. 3).

pH Dependence

The pH of surface sea water (pH 8.15 used in the calculation of Fig. 2) may be higher than many subsurface saline waters that mix with fresh carbonate ground water in natural environments. For example, Sutcliffe and Joyner (1968) report laboratory pH values of saline waters from southern Florida near 7.6. Even though measured in the laboratory, these values are probably near field values owing to the strong buffering of the carbonate equilibria in the solutions. Therefore, the effect of lower pH saline water on mixing should be considered.

Figure 4 shows the result of arbitrarily lowering the pH of surface sea water and then mixing that solution with a calcite-saturated solution at $10^{-2.5}$ atm CO_2 and 25° C. The amount of undersaturation increased, as was expected, with decreasing pH of sea water. All such mixtures with sea water at a pH slightly lower than 7.5 were undersaturated. These calculations are not particularly satisfactory, however, because there is little evidence of a mechanism for arbitrary change in pH of saline solutions without accompanying mineral dissolution or precipitation—that is, in natural environments, changes in pH are often in response to mineral dissolution and precipitation. Figure 4 maximizes the possible changes in saturation indices in response to pH changes, because the possible effects of dissolution and precipitation were not accounted for.

In order to examine more realistically the pH dependence of sea water on the mixing effect, a modified sea water at pH 7.5 was calculated by adding 4.7 mmoles of $CaCO_3$ per kg of water (Table 4), so that the saturation index of calcite was near that of surface sea water (pH 8.15). This modified sea water was mixed with solutions in equilibrium with calcite at CO_2 partial pressures from $10^{-3.0}$ to 1 atm at 25° C (Fig. 5). Comparison of Figure 5 with Figure 2C shows that

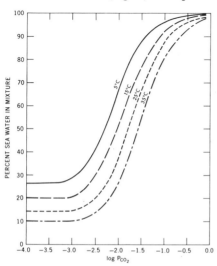

TABLE 3. COMPOSITION OF SEA WATER

Constituent	ppm	mmoles/kg H_2O
Calcium	413.	10.68
Magnesium	1,294.	55.16
Sodium	10,760.	485.0
Potassium	387.	120.6
Chloride	19,353.	565.7
Sulfate	2,712.	29.26
Bicarbonate	142.	2.412

Note: From Culkin, 1965. The pH of surface sea water was assumed to be 8.15.

Figure 3. Maximum percent of sea-water content (by volume) in undersaturated mixtures of sea water and solutions in equilibrium with calcite as a function of temperature and P_{CO_2} of saturated solution prior to mixing.

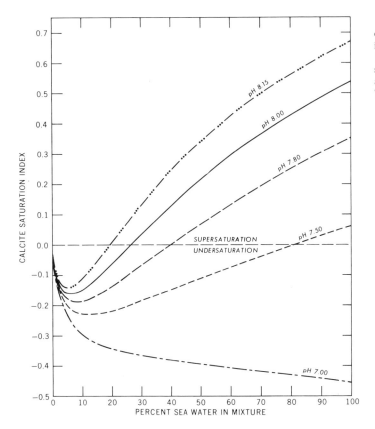

Figure 4. Saturation index of calcite in mixtures of a solution in equilibrium with calcite at $10^{-2.5}$ atm CO_2 and 25° C and sea water in which pH was arbitrarily set at 7.00, 7.50, 7.80, 8.00, and 8.15.

both the degree of undersaturation and the range in percent of sea water causing undersaturation is decreased in mixtures containing modified sea water (pH 7.5). The decrease is due principally to the higher CO_2 partial pressure of the modified sea water (Table 4).

Saturation Effect

The above calculations apply to the mixing of saline water with calcium bicarbonate solutions that are in equilibrium with calcite. While many ground waters from carbonate rocks appear to be near calcite saturation, there are examples of both undersaturation and supersaturation in carbonate aquifers (Back, 1963; Barnes and Back, 1964; Back and Hanshaw, 1970; Thrailkill, 1970, 1972; Langmuir, 1971; Plummer and others, in prep.).

The mixing of undersaturated carbonate ground water with saline water will increase undersaturation, but the mixing of supersaturated carbonate ground water with saline water can cause undersaturation only in certain mixtures. Ground water from carbonate rock in which minerals more soluble than calcite (such as gypsum and magnesian calcites) are dissolved (Back and Hanshaw, 1970) are often as much as twofold saturated with calcite (SI_c = 0.3). This supersaturation may correspond to the nucleation-free energy required for the precipitation of calcite (Plummer and Mackenzie, 1974) in low-temperature natural environments. Figure 6 shows the effect of mixing surface sea water (pH 8.15) with calcium bicarbonate solutions twofold saturated with calcite at CO_2 partial pressures from $10^{-3.0}$ to

1 atm and 25° C. These calculations show that CO_2 partial pressures greater than $10^{-1.8}$ atm in carbonate ground water twofold saturated with calcite at 25° C are required to cause undersaturation in mixtures with sea water. Although there is a tendency toward undersaturation in all mixtures, the calculations indicate that no undersaturation would occur by mixing twofold saturated carbonate ground water at CO_2 partial pressures less than $10^{-1.8}$ atm with sea water. It is shown below that even higher CO_2 partial pressures in supersaturated ground water are required to cause undersaturation in mixtures with saline water at ionic strengths less than sea water.

Figure 7 summarizes results from Figures 3, 5, and 6 at 25° C showing the maximum percent of sea water causing undersaturation in mixtures with calcium bicarbonate ground water. Curve 1 shows the maximum percent of sea water causing undersaturation in mixtures of water in equilibrium with calcite at differing CO_2 partial pressures and surface sea water at pH 8.15. Curve 2 shows the result of mixing similar waters with modified sea water at pH 7.5 in which the SI_c is near that of surface sea water at pH 8.15. Curve 3 shows the maximum percent of sea water causing undersaturation in mixtures of water twofold saturated with calcite at differing CO_2 partial pressures with surface sea water (pH 8.15). Curve 3 indicates that all mixtures containing less than 10 percent sea water remain supersaturated, and that greater than $10^{-1.8}$ atm CO_2 in carbonate ground water twofold saturated with calcite is required to cause undersaturation in mixtures containing more than 10 percent sea water. Comparison of curves 1 and 2 (Fig. 7) shows that lower pH of saline waters decreases the range of undersaturation in mixtures.

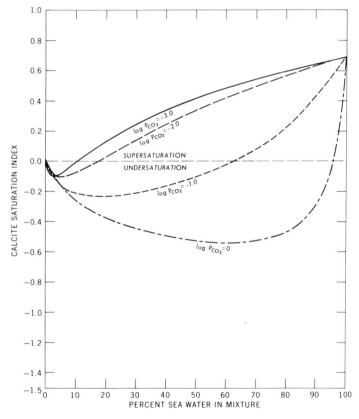

Figure 5. Saturation index of calcite in mixtures of solutions saturated with calcite at differing CO_2 partial pressures and 25° C and modified sea water at pH 7.50 in which SI_c was near that in surface sea water at pH 8.15 (Table 4).

Mixing Natural Waters

Figure 8 (A through D) shows the effect of mixing actual carbonate ground water from central Florida and the Yucatan Peninsula (Back and Hanshaw, 1970) with three saline subsurface waters from southern Florida (Sutcliffe and Joyner, 1968), surface sea water (pH 8.15), and modified sea water (pH 7.5). The carbonate ground waters selected from Back and Hanshaw (1970) are 1.3- to 1.5-fold saturated with calcite and have CO_2 partial pressures between $10^{-2.9}$ and $10^{-1.3}$ atm (Table 4). The saline ground waters from Sutcliffe and Joyner (1968) are from test well 8 sampled at depths of 25.6 to 34.1, 103.9 to 119.5, and 103.9 to 125.9 m. These waters are 1.9- to 2.7-fold supersaturated with calcite, have similar CO_2 partial pressures, near $10^{-2.6}$ atm, and have variable ionic strengths (Table 4). Similar properties of surface sea water and modified sea water are shown in Table 4 for comparison.

The differences between the mixing curves shown in Figure 8 can be explained, for the most part, in terms of (1) P_{CO_2} of calcium carbonate ground water, (2) the saturation state of saline water, and (3) the ionic strength of saline water. Even though less than twofold saturated with calcite, no undersaturation in any mixture of saline water with Polk City or Arcadia waters occurs (Figs. 8A, 8B), because both carbonate ground waters have relatively low CO_2 partial pressures. If these carbonate ground waters were in equilibrium with calcite, rather than supersaturated, considerable undersaturation could be expected in mixtures with saline waters. The mixing of surface sea water with carbonate ground water from Puerto Juarez (Fig. 8C) causes undersaturation due to the high CO_2 partial pressure of the fresh water ($10^{-1.6}$ atm), the high ionic strength of sea water, and the low P_{CO_2} of sea water. No undersaturation occurs in mixtures with water less saline

TABLE 4. SELECTED CHARACTERISTICS OF NATURAL WATERS OF INTEREST

Sample	$T°C$*	pH	mmoles/kg H_2O Ca†	mmoles/kg H_2O CO_2§	Log P_{CO_2}	I#	SI_c**	Reference
Libre Union	28.0	6.95	3.00	8.24	−1.34	0.015	0.12	1
Puerto Juarez	24.9	7.13	2.77	7.01	−1.59	0.022	0.13	1
Arcadia	26.3	7.44	2.65	3.63	−2.14	0.016	0.14	1
Polk City	23.8	8.00	0.85	2.14	−2.90	0.003	0.18	1
25.6–34.1	..	7.7	12.51	2.53	−2.67	0.444	0.43	2
103.9–119.5	..	7.7	11.21	3.04	−2.58	0.357	0.50	2
103.9–125.9	..	7.6	7.48	2.77	−2.50	0.242	0.27	2
Sea water	..	8.15	10.68	2.41	−3.21	0.666	0.67	3
Modified sea water	..	7.5	15.38	7.11	−2.05	0.678	0.69	4

References are as follows: 1, Back and Hanshaw (1970); 2, Sampling depth in meters Test Well 8 of Sutcliffe and Joyner (1968); 3, Culkin (1965); 4, Same as reference 3 except that the pH has been set at 7.5 and 4.7 mmoles of $CaCO_3$ have been added to bring SI_c near that of surface sea water at pH 8.15.

*Temperature in °C. In mixing saline water with the carbonate ground waters of Back and Hanshaw (1970), the temperature of the saline water was assumed to be that of the fresh water. The calculated P_{CO_2}, ionic strength, and SI_c of saline waters have been computed at 25°C.

†Total concentration of calcium in solution.

§Total concentration of carbon in solution.

#Ionic strength. $I = 1/2 \sum_i m_i z_i^2$.

**Saturation index of calcite, see text.

than sea water, even though these saline waters are less supersaturated with calcite than surface sea water, nor does undersaturation occur in mixtures with modified sea water that has a higher P_{CO_2} than surface sea water.

Undersaturation occurs to some extent in all mixtures involving Libre Union waters (Fig. 8D) owing to its high CO_2 partial pressure ($10^{-1.34}$ atm). Surface sea water causes the greatest degree of undersaturation because of its high ionic strength and low CO_2 partial pressure. Modified sea water causes the least amount of undersaturation owing to its higher CO_2 partial pressure and higher calcite supersaturation. While the actual degree of undersaturation in mixtures with saline water from southern Florida is never great, the range of undersaturation extends from approximately 10 percent to as much as 80 percent saline water in the mixture owing to the lower degree of calcite saturation in these solutions.

DISCUSSION

The calculations presented above indicate that the mixing of high-ionic strength, low-P_{CO_2} saline water with near-saturation, low-temperature, high-P_{CO_2} carbonate ground water favors the formation of increased porosity and permeability in carbonate aquifers, as well as the possibility of deep phreatic and coastal caves along margins of limestone aquifers that come into contact with saline and (or) sea water. An example of this effect, as suggested by Vernon (1969), may be the "boulder zone" of Florida, well known to drillers as a cavernous zone of high permeability generally present below 366 m throughout peninsular Florida. In general, this boulder zone lies within the mixed zone between overlying fresh carbonate ground water (average $P_{CO_2} = 10^{-2.5}$ atm, computed from the data of Back and Hanshaw, 1970) and underlying sea water or hypersaline waters (Vernon, 1969). The results of the mixing calculations indicate that either the excavation of the boulder zone in Florida is not active, or that mixing conditions more extreme than those shown in Figure 8 exist or have existed in the Floridian aquifer—such as higher CO_2 partial pressures in carbonate ground water at depth and (or) mixing with saline waters at ionic strengths greater than that of sea water.

For chemical conditions that do cause calcite undersaturation in mixtures, there is a potential for dolomitization of limestone by replacement, as discussed by Hanshaw and others (1971), Badiozamani (1972, 1974), Folk and Land (1972), and Land (1973a, 1973b). Many mixtures of sea water and carbonate ground water are supersaturated with dolomite and undersaturated with calcite, so that there is a thermodynamic potential for dolomitization by replacement. Figure 9 shows the saturation index of dolomite (SI_d) in mixtures of sea water and solutions in equilibrium with calcite at differing CO_2 partial pressures and 25° C. Dolomite supersaturation increases with lower CO_2 partial pressures in carbonate ground water. Sea water mixing with saturated calcium bicarbonate solutions at $10^{-1.0}$ atm causes dolomite undersaturation in all mixtures except those with high sea-water content (not shown in Fig. 9). The dolomite saturation curve for sea water in mixture with a solution in equilibrium with calcite at $10^{-4.0}$ atm CO_2 and 25° C is almost identical to the curve at $10^{-3.0}$ atm CO_2 and is not shown in Figure 9.

Hanshaw and others (1971) have proposed that dolomitization occurs in the brackish water mixing zone in central Florida where magnesium to calcium molal ratios are greater than unity. They point out from a survey of the literature that many modern dolomites are associated in natural waters with Mg/Ca ratios greater than 3.0. Figure 10 shows Mg/Ca ratios in mixtures of sea water and solutions

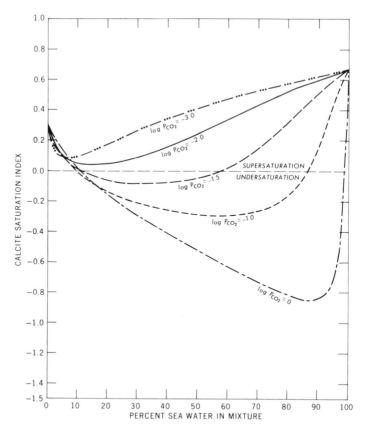

Figure 6. Saturation index of calcite in mixtures of solutions twofold saturated with calcite at differing CO_2 partial pressures and 25° C with surface sea water (pH 8.15).

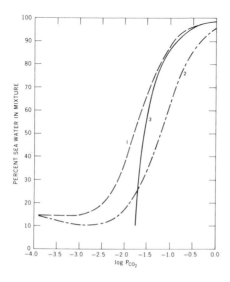

Figure 7. Maximum percent of sea water causing undersaturation in mixtures with calcium bicarbonate solutions at 25° C; 1, mixtures of surface sea water (pH 8.15) and solutions saturated with calcite; 2, mixtures of modified sea water (pH 7.50, $SI_c = 0.69$) and solutions saturated with calcite; 3, mixtures of surface sea water (pH 8.15) and solutions twofold saturated with calcite.

in equilibrium with calcite from $10^{-1.0}$ to $10^{-4.0}$ atm CO_2 at 25° C. If dolomitization requires Mg/Ca ratios greater than 1.0, then sea-water mixtures of more than 4 percent are required for dolomitization to occur in carbonate ground water at CO_2 partial pressures lower than $10^{-2.0}$ atm. If Mg/Ca ratios must be greater than 3.0 for fresh-water dolomitization to occur, however, ground water at less than $10^{-2.0}$ atm CO_2 (in equilibrium with calcite) would require a mixture of more than 18 percent sea water.

Comparison of Figures 2C, 3, 9, and 10 points out constraints on mixing conditions favorable for dolomitization of limestone by replacement. For example, if a

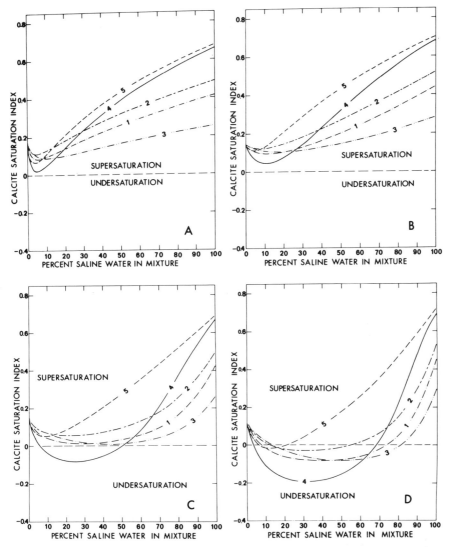

Figure 8. Saturation index of calcite in mixtures of actual carbonate ground water from central Florida and the Yucatan Peninsula with saline water from southern Florida, modified sea water, and sea water. The saline waters are identified as follows: 1, 25.6 to 34.1 m; 2, 103.9 to 119.5 m; 3, 103.9 to 125.9 m; 4, sea water (pH 8.15); 5, modified sea water (pH 7.50). See Table 4 for description of waters. A, Polk City water; B, Arcadia water; C, Puerto Juarez water; D, Libre Union water.

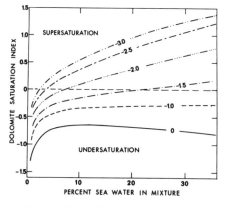

Figure 9. Saturation index of dolomite in mixtures of surface sea water (pH 8.15) and solutions in equilibrium with calcite at differing log CO_2 partial pressures and 25° C.

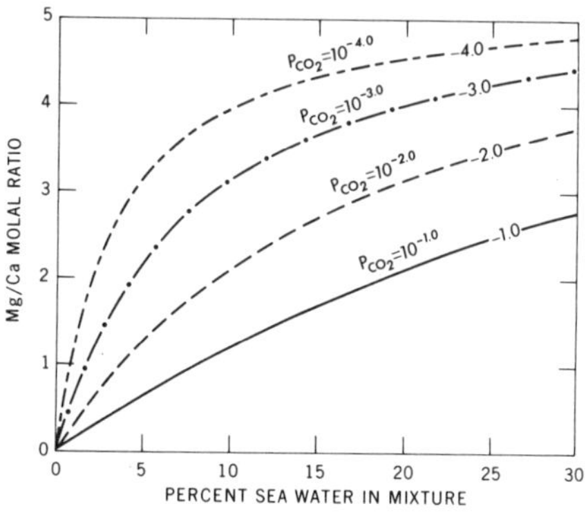

Figure 10. Magnesium/calcium molal ratio in mixtures of sea water and solutions in equilibrium with calcite at $10^{-1.0}$ to $10^{-4.0}$ atm CO_2 and 25° C.

carbonate ground water in equilibrium with calcite at $10^{-2.0}$ atm CO_2 mixes with sea water, a mixture containing as much as 37 percent sea water can dissolve calcite, but more than 7 percent sea water in the mixture is required for dolomite supersaturation; therefore, if the Mg/Ca ratio must be greater than 3.0, a mixing zone capable of replacing limestone with dolomite can contain from 18 to 37 percent sea water. If carbonate ground water saturated with calcite at $10^{-3.0}$ atm CO_2 mixes with sea water, and Mg/Ca ratios greater than 3.0 are required for the formation of dolomite, only mixtures containing from 9 to 15 percent sea water are capable of replacing limestone with dolomite. As calcite supersaturation in carbonate ground water increases, the potential for dolomitization by replacement decreases further (Fig. 6).

Although more calcite undersaturation occurs when solutions in equilibrium with calcite at high CO_2 partial pressures are mixed with sea water, lower levels of dolomite supersaturation occur in the mixture (Figs. 2C, 3, 9), which again may constrain the mechanism or mechanisms of nucleation and growth of dolomite. There may be some optimum geochemical conditions most favorable for ground-water dolomitization that are a function of calcite saturation, partial pressure of CO_2, and temperature of the ground water; percent of sea water in the mixture; and the pH, saturation state, and salinity of saline waters. But until the kinetics of nucleation and growth of dolomite are understood, these chemical conditions cannot be defined.

Both Vernon (1969) and Dieke (1973, oral commun.) reported the presence of abundant small euhedral crystals of dolomite on exposed faces of caves and debris in the boulder zone of central Florida. The fact that this boulder zone even exists, however, demonstrates the inefficiency of the replacement process for dolomitization of limestone by sea water mixing with carbonate ground water. The rate of calcite dissolution in the boulder zone of Florida must far exceed the rate of nucleation and growth of dolomite.

Sea water mixing with carbonate ground water may only create the porosity and permeability in carbonate rocks and sediments required for the transport of large volumes of solutions that are capable of forming dolomite. It would be a unique chemical environment in which the rate of dolomite growth equaled the rate of limestone dissolution.

SUMMARY AND CONCLUSIONS

The mixing of sea water and (or) saline subsurface water with fresh calcium bicarbonate ground water can result in calcite undersaturation in mixtures, even though both solutions may be saturated or supersaturated with calcite prior to mixing. The amount of undersaturation in mixtures increases with the following: (1) low-ionic-strength water mixing with high-ionic-strength solutions, (2) high-P_{CO_2}, fresh carbonate water mixing with low-P_{CO_2} (high-pH) saline water, (3) lower temperatures, and (4) near calcite saturation in both solutions prior to mixing. The mixing calculations define certain geochemical environments that favor formation of increased porosity and permeability in carbonate aquifers and dolomitization of limestones by replacement.

ACKNOWLEDGMENTS

I am grateful to B. F. Jones and T.M.L. Wigley for valuable discussions during this research. The manuscript has been improved by the criticisms of B. F. Jones, F. T. Mackenzie, B. B. Hanshaw, D. Langmuir, and E. J. Reardon. This work was supported by funds from the Water Resources Division of the U.S. Geological Survey.

REFERENCES CITED

Arntson, R. H., 1964, Effect of temperature and confining pressure on the solubility of calcite at constant CO_2 concentrations: Geol. Soc. America, Abs. for 1963, Spec. Paper 76, p. 6-7.

Back, W., 1963, Preliminary results of a study of calcium carbonate saturation of ground water in central Florida: Internat. Assoc. Sci. Hydrology Bull. v. 8, p. 43-51.

Back, W., and Hanshaw, B. B., 1970, Comparison of chemical hydrogeology of the carbonate peninsulas of Florida and Yucatan: Jour. Hydrology, v. 10, p. 330-368.

Badiozamani, K., 1972, The Dorag dolomitization model—Application to the Middle Ordovician of Wisconsin: Geol. Soc. America, Abs. with Programs (Ann. Mtg.), v. 4, no. 7, p. 440.

———1974, The Dorag dolomitization model—Application to the Middle Ordovician of Wisconsin: Jour. Sed. Petrology, v. 43, p. 965-984.

Barnes, I., and Back, W., 1964, Dolomite solubility in ground water: U.S. Geol. Survey Prof. Paper 475D, p. D179-D180.

Bogli, A., 1963, Beitrag zur Entstehung von Karsthöhlen: Vienna, Die Höhle, v. 14, p. 63-68.

———1964, Mischungskorrosion; ein Beitrag zum Verkarstungs problem: Erdkunde, v. 18, p. 83-92.

Culkin, F., 1965, The major constituents of sea water, in Riley, J. P., and Skirrow, G., eds., Chemical oceanography: New York, Academic Press, Inc., v. 1, p. 121-161.

Ernst, L., 1964, On the question of mischungkorrosion (solution by mixing): Vienna, Die Höhle, v. 15, p. 71-75; transl., 1965, Nittany Grotto News: University Park, Pennsylvania State Univ., v. 13, p. 133-136.

Folk, R. L., and Land, L. S., 1972, Mg/Ca vs salinity: A frame of reference for crystallization of calcite, aragonite, and dolomite: Geol. Soc. America, Abs. with Programs (Ann. Mtg.), v. 4, no. 7, p. 508.

Frear, G. L., and Johnston, J., 1929, The solubility of calcium carbonate (calcite) in certain aqueous solutions at 25° C: Am. Chem. Soc. Jour., v. 51, p. 2082-2093.

Garrels, R. M., and Christ, C. L., 1965, Solutions, minerals and equilibria: New York, Harper & Row, Pubs., 450 p.

Garrels, R. M., and Thompson, M. E., 1962, A chemical model for sea water at 25° C and one atmosphere total pressure: Am. Jour. Sci., v. 260, p. 57-66.

Hanshaw, B. B., Back, W., and Deike, R. G., 1971, A geochemical hypothesis for dolomitization by ground water: Econ. Geology, v. 66, p. 710-724.

Harned, H. S., and Davis, R., Jr., 1943, The ionization constant of carbonic acid in water and the solubility of carbon dioxide in water and aqueous salt solutions from 0 to 50°: Am. Chem. Soc. Jour., v. 65, p. 2030-2037.

Harned, H. S., and Scholes, S. R., Jr., 1941, The ionization constant of HCO_3^- from 0 to 50°: Am. Chem. Soc. Jour., v. 63, p. 1706-1709.

Howard, A. D., 1964, Processes of limestone cave development: Internat. Jour. Speleology, v. 1, p. 47-60.

Jacobson, R. L., and Langmuir, D., 1974, Dissociation constants of calcite and $CaHCO_3^+$ from 0 to 50° C: Geochim. et Cosmochim. Acta, v. 38, p. 301-318.

Land, L. S., 1973a, Holocene meteoric dolomitization of Pleistocene limestones, North Jamaica: Sedimentology, v. 20, p. 411-424.

——1973b, Contemporaneous dolomitization of middle Pleistocene reefs by meteoric water, North Jamaica: Bull. Marine Sci., v. 23, p. 64-92.

Langmuir, D., 1971, The geochemistry of carbonate ground waters in central Pennsylvania: Geochim. et Cosmochim. Acta, v. 35, p. 1023-1046.

Mathews, R. K., 1971, Diagenetic environments of possible importance to the explanation of cementation fabric in subaerially exposed carbonate sediments, *in* Bricker, O. P., ed., Carbonate cements: Baltimore, Johns Hopkins Univ. Press, p. 127-132.

Plummer, L. N., and Mackenzie, F. T., 1974, Predicting mineral solubility from rate data: Application to the dissolution of magnesian calcites: Am. Jour. Sci., v. 274, p. 61-83.

Reardon, E. J., and Langmuir, D., 1974, Thermodynamic properties of the ion pairs $MgCO_3^0$ and $CaCO_3^0$ from 10 to 50° C: Am. Jour. Sci., v. 274, p. 599-612.

Runnells, D. D., 1969, Diagenesis, chemical sediments, and the mixing of natural waters: Jour. Sed. Petrology, v. 39, p. 1188-1201.

Shternina, E. B., and Frolova, E. V., 1945, The system $CaCO_3$-$CaSO_4$-$NaCl$-CO_2-H_2O at 25°: Acad. Sci. USSR Comptes Rendus, v. 47, p. 33-35; *in* Linke, W. F., and Seidell, A., eds., 1958, Solubilities of inorganic and metal organic compounds: Princeton, N.J., D. van Nostrand, Inc., p. 544.

Sutcliffe, H., Jr., and Joyner, B. F., 1968, Test well exploration in the Myakka River Basin area, Florida: Florida Div. Geology Inf. Circ. 56, 61 p.

Thrailkill, J., 1968, Chemical and hydrologic factors in the excavation of limestone caves: Geol. Soc. America Bull., v. 79, p. 19-46.

——1970, Solution geochemistry of the water of limestone terrains: Univ. Kentucky Water Res. Inst., Research Rept. 19, 125 p.

——1972, Carbonate chemistry of aquifer and stream water in Kentucky: Jour. Hydrology, v. 16, p. 93-104.

Truesdell, A. H., and Jones, B. F., 1974, WATEQ, a computer program for calculating chemical equilibria in natural waters: Natl. Tech. Inf. Service, PB-220464 (1973), U.S. Geol. Survey Jour. Research, v. 2, p. 233-248.

Vernon, R. O., 1969, The geology and hydrology associated with a zone of high permeability (boulder zone) in Florida: Soc. Mining Engineers, Preprint 69-AG-12, 24 p.

MANUSCRIPT RECEIVED BY THE SOCIETY SEPTEMBER 10, 1973
REVISED MANUSCRIPT RECEIVED MARCH 18, 1974

PETROLOGY

# Three-Dimensional Polynomial Trend Analysis Applied to Igneous Petrogenesis

Geoffrey W. Mathews

Department of Earth and Space Sciences
Indiana University at Fort Wayne
Fort Wayne, Indiana 46805

J. Allan Cain

Department of Geology
University of Rhode Island
Kingston, Rhode Island 02881

Philip O. Banks

Department of Geology
Case Western Reserve University
Cleveland, Ohio 44106

ABSTRACT

Three-dimensional polynomial trend-surface analysis has been applied to whole-rock measurements of specific gravity, SiO_2, Fe_2O_3 + MgO, and Na_2O + K_2O from each of the four principal intrusive units of the Audubon-Albion stock, Colorado. Except in the earliest unit (orthoclase gabbro), the three-dimensional variability of these parameters indicates a simple pattern of crystallization dominated by gravitative settling with little or no convective motion. Petrography and petrochemistry of the orthoclase gabbro indicate that it originated mainly as a crystal cumulate, whereas trend-surface analysis suggests that it was rotated and rafted into its present position during a later period of intrusion. Hence, statistical treatment of subtle petrologic variations offers insight into otherwise obscure petrogenetic processes and, in favorable situations, produces information that can be used to interpret subsequent structural history.

INTRODUCTION

Oldham and Sutherland (1955) and Grant (1957) introduced the application of polynomial trend-surface analysis to gridded data, and Krumbein (1959) extended the technique to irregularly spaced observations. Peikert (1962) expanded Krumbein's method to permit inclusion of elevation as a third geographic coordinate. Krumbein (1966) compared polynomial and Fourier trend-surface models in map analysis, and Whitten (1970) described a method for calculating orthogonal polynomial trend surfaces for irregularly spaced data. Summaries of the principles of trend-surface analysis, illustrations of surfaces, and lists of major literature sources were given by Krumbein and Graybill (1965), Harbaugh and Merriam (1968), Koch and Link (1971), and Davis (1973).

The importance of areal variability in igneous petrology was emphasized by Whitten (1959) in his recognition of ghost stratigraphy in the Donegal granite, and by Dawson and Whitten (1962) in their analysis of the Lacorne massif, Ontario. Although Peikert (1963) subsequently demonstrated that interpretations based on

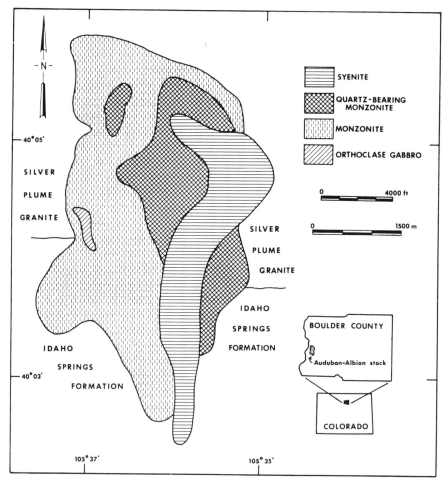

Figure 1. Geologic map of the Audubon-Albion stock, adapted from Wahlstrom (1940) and Mathews (1970).

variability expressed only in terms of two-dimensional map coordinates may be misleading, there are few illustrations of three-dimensional trend analysis (Koch and Link, 1971, p. 41). Examples from igneous petrology include Peikert's (1962) study of specific-gravity measurements, and Whitten and Boyer's (1964) investigation of the San Isabel granite.

In the present study, three-dimensional polynomial analysis was applied to chemical and specific-gravity data for the Audubon-Albion stock, Boulder County, Colorado (Wahlstrom, 1940; Mathews, 1970). The stock is a multiple intrusion emplaced in the Precambrian Idaho Springs Formation and Silver Plume Granite during Laramide orogenesis. It consists of five distinct intrusive units: from oldest to youngest, they are orthoclase gabbro, monzonite, quartz-bearing monzonite, syenite, and small granite dikes. Except for the granite intrusion at Jamestown, 13 mi (21 km) to the east-northeast, the stock is the northernmost Laramide intrusion exposed in the Colorado mineral belt (Fig. 1).

The Audubon-Albion stock was selected for this investigation for the following reasons: First, if the Laramide intrusions of the Colorado mineral belt are genetically related, as suggested by numerous authors, a detailed petrologic investigation of one might provide a petrogenetic working model applicable to others. Second, the Audubon-Albion stock has escaped extensive postintrusive mineralization, hence complicating effects of mineralogenesis are avoided. The importance of this was underscored by Lipman and others (1971) who indicated that their attempts to reconstruct the geometry of the latest Cretaceous-Paleocene subduction zone had been hampered by the extensive alteration of many of the analyzed Laramide igneous rocks. Finally, the stock is suitable for three-dimensional trend-surface analysis because it is located in an area of high relief (>800 m); its height to width to length ratio is 1:4:8.

Field, textural, major and trace element, mineralogical, and specific-gravity data were used in interpreting the origin and history of the Audubon-Albion stock. Also employed was a statistical modeling technique that Professor Krumbein played a major part in developing. The purpose of this paper is to indicate how the application of this technique offers insight into otherwise obscure petrogenetic processes and thus provides an additional tool in igneous petrologic investigations.

PREVIOUS GEOLOGICAL WORK

Although literature on the Colorado mineral belt is extensive, most explanations of its origin fall into one of two general categories: the mineral belt as a magmatic feature or as a tectonic feature.

Crawford (1924) believed that the volcanism, tectonism, and mineralization were consequences of the emplacement of a quartz monzonitic batholith at depth. Lovering and Goddard (1938) proposed that differential anatexis of a dioritic or gabbroic basement complex produced the magma(s) from which the Laramide intrusives were derived, and Doe (1967) found the lead isotope data consistent with this hypothesis. Lovering and Goddard (1938, 1950) suggested that the Laramide intrusives might represent two separate lines of magmatic evolution. Phair and Fisher (1962, p. 479) supported this interpretation and concluded that "consistent differences in composition, thermal state, and paragenesis of the feldspars confirm the subdivision of the rocks into two related lines of descent." Pearson and others (1962), Tweto (1968), and Wallace and others (1968) showed that the epizonal plutons of the mineral belt fall into two groups with respect to radiometric ages, the younger of which is Cenozoic and therefore not traditional "Laramide." Tweto and Case (1972) gave geophysical evidence that a batholith underlies the mineral belt and suggested that the range in composition of epizonal stocks is a result of differentiation of that batholith.

Tweto and Sims (1963) concluded that the mineral belt is an expression of Laramide orogenic forces whose pattern was controlled by a Precambrian zone of weakness. Lipman and others (1971, 1972) interpreted the history of igneous and tectonic activity in the western United States, including the Colorado mineral belt, within the framework of plate tectonics theory.

The Audubon-Albion stock was described in some detail by Wahlstrom (1940) who proposed that the composite intrusion evolved from a parental magma having the composition of orthoclase gabbro. Hart (1960) obtained a K-Ar age of 66.3 ± 1.1 m.y. from the monzonite phase of the stock, placing it in the older group of intrusives of traditional "Laramide" age.

COLLECTION AND TREATMENT OF DATA

Not all petrographic and petrochemical data accumulated during this investigation can be presented here. To illustrate the broad utility of three-dimensional polynomial trend analysis, whole-rock attributes were selected, as these have the advantages of being obtained from large, presumably more representative samples and of requiring relatively simple laboratory preparation.

The variates dealt with are specific gravity, SiO_2 content, total iron plus magnesium (Fe_2O_3 + MgO), and total alkalis (Na_2O + K_2O). This choice was governed mainly by classical petrologic theory; that is, the ferromagnesian constituents are taken to represent, in an approximate manner, the early-stage crystallization products, whereas SiO_2 and the alkalis complementarily represent the later stage crystallization products. Correlations between ferromagnesian content and crystallization stage depend on oxygen fugacity, for where the fugacity is low, Fe tends to remain in the liquid until the later stages of crystallization (for example, Wyllie, 1971, p. 56). In all units of the Audubon-Albion stock, however, the presence of hydrous phases and magnetite argues that oxygen fugacity was not exceptionally low in the magmas from which these rocks crystallized.

Specific gravity, an especially simple quantity to measure, was included to test its correspondence with chemical composition. Moreover, specific-gravity values are independent of problems associated with closed-table data and, as pointed out by Hyndman (1972, p. 129), variations in specific gravity may provide useful petrologic information.

Various aspects of the sampling scheme, analytical procedures, and statistical tests are described below, and were more fully explained by Mathews (1970).

Sampling

After determining outcrop availability within each rock unit, areas to be sampled were purposefully chosen so that the distribution of sampling sites would be fairly uniform throughout the stock. Where there were numerous exposures in an area, outcrops to be sampled were selected randomly. Wherever possible, samples were taken at the corners of a square approximately 1 m on a side. Five sampling sites were then selected randomly from each unit in order to test within- versus between-site variability. All four samples from these sites were analyzed, but for each remaining site a random-number table was used to determine which one of the four samples would be analyzed. A similar procedure determined which of the four analyses at each replicated test site would be included in the final three-dimensional trend analyses. Distribution of the sampling sites is shown in Figure 2.

Analytical Procedures

The analytical error involved in specific-gravity measurements and the decimal place to which variability in specific gravity in the Audubon-Albion stock should be expressed were determined in a pilot study. From each unit of the stock, five randomly selected samples (each approximately 200 g) were measured to four decimals five separate times. Values used in this pilot study were averages of separate measurements on two halves of each sample. The standard deviation of specific-gravity determinations for a single sample was confined to the third decimal place. Analysis of variance indicated that significant differences between samples exist at the second decimal level. Therefore, subsequent specific-gravity measurements were carried to two decimals, as given in Table 1.

Geochemical parameters used in this study were obtained by x-ray spectroscopy. Major-constituent analysis was accomplished by fusing a 1:1 mixture of powdered sample and transparent flux (lithium tetraborate) and making a boric acid backed wafer. X-ray emission intensities were compared to standard W-1 for samples of orthoclase gabbro and to standard G-1 for samples of quartz-bearing monzonite and syenite. Both standards, W-1 and G-1, were used in the analysis of monzonite samples, and the reported figures are an average of the results using each standard in turn. Matrix effects were at least partially minimized by the dilution of the

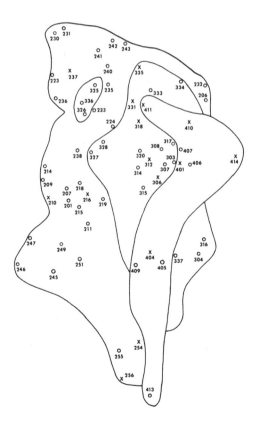

Figure 2. Distribution of sampling sites for major units of Audubon-Albion stock. Sampling sites for orthoclase gabbro are shown in Figure 3. Sites with replicate analyses for testing within- versus between-site variability are indicated by crosses; all others are open circles.

samples with lithium tetraborate. Further correction was made mathematically using the procedure outlined by Hower (1959) and Czamanske and others (1966).

Analysis of Variance

Analysis of variance techniques were used to test between-site variability against within-site variability. The five randomly selected replicated sampling sites from each unit were used for this purpose. Any measured variate whose between-site variability is not significantly greater than the within-site variability will not produce petrologically meaningful trend surfaces.

A single-factor analysis of variance was used for the measured variates from each of the major intrusive units of the Audubon-Albion stock. This assumes that all variability of a parameter may be partitioned into a between-site effect and a within-site effect, used here as the error term against which the between-site factor was tested. It is assumed that any within-sample variability is negligible. Indeed, homogenization of samples during preparation for x-ray spectrographic analysis and the averaging of two specific-gravity determinations per sample mitigated any within-sample variability. Table 1 includes data from the replicated sampling sites used in the analysis of variance; Table 2 shows that there is significant between-site variability for all measured variates at the 95 percent confidence level (or 0.05 level of significance).

Three-dimensional Polynomial Trend Analysis

Descriptions of regression analysis of data in three dimensions (three geographic coordinates) are given in several of the references cited earlier (for example, Koch and Link, 1971, p. 36). Davis (1973, p. 352) pointed out that the resulting trend surface is variously called a "hypersurface," "isopleth envelope," "UVW trend," or "four-dimensional trend surface," and that the surfaces have the "form of a solid (representing the space defined by the three coordinates) containing nested contour sheets or envelopes. These sheets are interpreted in the same manner as contour lines in a conventional map."

As described by Krumbein and Graybill (1965, p. 327), regression surfaces of different degrees can be fitted to data by the method of least squares so that the computed values of a dependent variable lie on the surface generated. This procedure partitions the total sum of squares associated with any given variate into two component parts: that associated with a regional trend and that due to deviations from the trend. The latter comprises that part of any systematic variability that might be described by higher order surfaces as well as random deviations from the computed surface.

Mandelbaum (1963) pointed out that the surface that best separates regional from local variability is the one for which the mean-square deviation is a minimum, which is not necessarily equivalent to the surface that accounts for the highest percentage of the total sum of squares (for example, Peikert, 1962).

The number of data points available for each unit in the Audubon-Albion stock limited the highest order surfaces to first-degree for the orthoclase gabbro and syenite, and to third-degree for the monzonite and quartz-bearing monzonite. These third-degree surfaces account for the highest percentage of the total corrected sum of squares and also are those for which the mean-square deviation is a minimum. Inasmuch as lower order surfaces for the monzonite and quartz-bearing monzonite do not meet Mandelbaum's criterion, they are not discussed further.

TABLE 1. SPECIFIC GRAVITY AND SELECTED CHEMICAL MEASUREMENTS, AUDUBON-ALBION STOCK*

Sample site	Specific gravity	SiO_2	$Fe_2O_3 + MgO$	$Na_2O + K_2O$
Orthoclase gabbro				
101	2.96	49.75	18.16	5.37
102-1	2.95	51.38	16.07	5.91
102-2	2.92	50.17	16.71	5.52
102-3	2.90	51.33	16.28	6.22
102-4	2.87	50.93	15.87	6.45
103-1	3.00	48.01	21.76	4.93
103-2	3.00	47.90	21.44	5.59
103-3	3.01	48.19	21.39	4.39
103-4	2.97	50.19	18.84	5.68
104-1	3.03	48.08	22.30	4.21
104-2	3.02	48.51	21.85	4.37
104-3	3.02	49.08	20.62	5.01
104-4	2.97	49.91	19.57	4.82
105-1	2.89	51.24	14.40	5.97
105-2	2.89	51.15	14.88	6.27
105-3	2.90	50.70	14.89	5.62
105-4	2.91	50.58	15.62	5.54
107-1	2.88	51.12	17.58	5.65
107-2	2.90	50.14	16.42	5.57
107-3	2.92	50.81	17.87	5.06
107-4	2.90	49.13	18.64	5.16
108	2.86	52.03	16.00	5.95
109	2.98	49.09	19.79	4.80
110	2.85	51.92	14.87	6.02
Monzonite				
201	2.71	55.57	7.81	9.16
207	2.79	53.15	10.80	8.67
209	2.73	56.28	8.16	9.23
210-1	2.85	51.84	12.76	7.35
210-2	2.83	52.66	11.59	7.75
210-3	2.76	53.50	9.89	8.38
210-4	2.80	53.77	10.66	8.25
211	2.79	47.79	16.60	6.72
214	2.76	53.47	9.50	8.67
215	2.78	53.62	10.17	8.55
216-1	2.72	58.93	6.30	10.76
216-2	2.71	58.85	5.97	10.96
216-3	2.68	60.68	4.91	10.98
216-4	2.71	58.49	6.40	10.44
218	2.75	55.72	8.37	9.12
219	2.79	55.10	9.53	8.57
223	2.74	58.39	7.60	10.60
224	2.66	64.15	3.54	11.01
230	2.79	55.56	10.74	7.94
231	2.76	57.20	9.26	8.57
232	2.75	57.89	7.51	8.81
233	2.65	63.44	3.94	9.93
235	2.66	62.69	4.19	9.74
236	2.64	62.81	3.93	9.81
237-1	3.13	45.75	21.79	4.52
237-2	2.94	50.36	15.12	6.02
237-3	3.02	48.05	17.65	4.51
237-4	2.99	48.19	17.82	4.62
238	2.83	54.85	10.91	7.65
240	2.67	64.18	4.02	10.17
241	2.84	55.81	11.94	5.72
242	2.72	60.77	6.92	8.52
243	2.79	52.77	11.20	8.67
245	3.18	42.48	23.90	2.17
246	3.23	42.08	26.45	2.84
247	2.83	50.87	13.15	6.56
249	2.75	57.43	7.51	9.09
251	2.69	59.54	5.46	10.70
254-1	2.77	55.60	8.41	8.37
254-2	2.77	55.35	8.70	8.03
254-3	2.79	55.66	8.47	8.32
254-4	2.76	55.35	8.62	8.26
255	2.71	57.05	7.13	10.23
Monzonite (continued)				
256-1	2.75	56.59	8.21	8.76
256-2	2.75	56.65	7.78	8.84
256-3	2.75	57.05	7.68	9.09
256-4	2.75	57.11	7.75	8.84
Quartz-bearing monzonite				
303	2.58	64.27	3.03	12.74
304	2.70	56.79	6.73	10.71
306-1	2.57	64.26	1.59	13.00
306-2	2.57	68.40	1.09	12.04
306-3	2.58	64.60	1.59	14.15
306-4	2.57	65.75	1.42	13.06
307	2.55	63.93	1.51	13.85
308	2.66	59.68	5.41	11.17
312-1	2.63	60.54	4.89	11.91
312-2	2.62	60.58	3.79	11.91
312-3	2.70	59.05	6.47	10.92
312-4	2.63	59.88	4.34	11.06
314	2.61	62.58	2.72	11.75
315	2.58	62.90	2.37	11.91
316	2.63	61.65	5.80	10.25
317	2.56	64.99	1.73	12.51
318-1	2.63	62.19	2.81	11.69
318-2	2.62	62.38	2.68	12.08
318-3	2.61	62.32	3.48	12.04
318-4	2.62	61.82	2.51	10.99
320	2.62	63.35	3.12	10.41
325	2.66	61.28	4.76	9.53
326	2.65	61.67	4.10	11.15
327	2.64	62.37	4.07	11.12
328	2.67	61.30	5.50	11.35
331-1	2.62	62.67	2.81	10.97
331-2	2.63	62.64	3.02	11.83
331-3	2.63	63.25	2.65	12.26
331-4	2.63	62.27	3.02	11.52
333	2.67	61.16	4.37	11.05
334	2.64	63.90	4.22	9.75
335-1	2.66	63.54	4.68	10.80
335-2	2.66	63.42	4.53	10.03
335-3	2.65	63.51	4.86	10.28
335-4	2.66	62.36	4.69	9.64
336	2.67	60.56	6.32	9.91
337	2.68	62.20	5.88	9.47
Syenite				
401-1	2.55	64.14	1.21	13.13
401-2	2.54	64.28	1.22	14.11
401-3	2.58	68.59	1.44	12.87
401-4	2.56	64.68	1.26	13.35
404-1	2.59	63.77	1.43	10.86
404-2	2.58	64.05	1.31	12.97
404-3	2.58	64.27	1.30	12.09
404-4	2.59	63.79	1.38	12.45
405	2.58	68.26	1.36	11.04
406	2.58	64.02	3.55	13.81
407	2.55	65.32	1.35	13.18
409	2.59	61.94	1.81	12.66
410-1	2.55	70.24	1.92	10.96
410-2	2.58	69.77	2.22	11.70
410-3	2.61	63.29	3.47	10.77
410-4	2.59	63.04	3.45	11.90
411-1	2.64	62.81	3.69	10.78
411-2	2.65	61.84	4.00	11.85
411-3	2.62	63.37	3.21	10.78
411-4	2.64	62.61	3.84	11.57
413	2.69	57.90	6.48	10.86
414-1	2.56	63.23	1.69	13.91
414-2	2.58	63.53	1.92	14.36
414-3	2.59	63.61	2.37	14.02
414-4	2.59	62.96	1.74	13.08

*Chemical constituents expressed as weight percent.

TABLE 2. SUMMARY OF RESULTS FROM ANALYSIS-OF-VARIANCE TESTS FOR
BETWEEN- VERSUS WITHIN-SITE VARIABILITY

Source of variance	Sum of squares	Degrees of freedom	Mean square	F*	Source of variance	Sum of squares	Degrees of freedom	Mean square	F*
Orthoclase gabbro					Quartz-bearing monzonite				
		sp gr					sp gr		
Between	0.0480	4	0.0120		Between	0.0170	4	0.0043	
Within	0.0076	15	0.0005	24.00	Within	0.0045	15	0.0003	14.33
Total	0.0556	19			Total	0.0215	19		
		SiO_2					SiO_2		
Between	20.2863	4	5.0716		Between	68.9226	4	17.2307	
Within	8.9925	15	0.5995	8.46	Within	12.7981	15	0.8532	20.20
Total	29.2788	19			Total	81.7207	19		
		$Fe_2O_3 + MgO$					$Fe_2O_3 + MgO$		
Between	120.6024	4	30.1506		Between	33.1393	4	8.2848	
Within	13.7728	15	0.9182	32.84	Within	4.8672	15	0.3245	25.53
Total	134.3752	19			Total	38.0065	19		
		$Na_2O + K_2O$					$Na_2O + K_2O$		
Between	5.1778	4	1.2945		Between	16.6728	4	4.1682	
Within	2.6068	15	0.1738	7.45	Within	5.4444	15	0.3630	11.48
Total	7.7846	19			Total	22.1172	19		
Monzonite					Syenite				
		sp gr					sp gr		
Between	0.2405	4	0.0601		Between	0.0136	4	0.0034	
Within	0.0254	15	0.0017	35.35	Within	0.0036	15	0.0002	17.00
Total	0.2659	19			Total	0.0172	19		
		SiO_2					SiO_2		
Between	289.9606	4	72.4902		Between	40.7211	4	10.1803	
Within	16.1106	15	1.0740	67.50	Within	62.0973	15	4.1398	2.46
Total	306.0712	19			Total	102.8184	19		
		$Fe_2O_3 + MgO$					$Fe_2O_3 + MgO$		
Between	360.2373	4	90.0593		Between	16.6124	4	4.1531	
Within	28.9898	15	1.9327	46.60	Within	2.6594	15	0.1773	23.42
Total	389.2271	19			Total	19.2718	19		
		$Na_2O + K_2O$					$Na_2O + K_2O$		
Between	71.9406	4	17.9852		Between	22.3087	4	5.5772	
Within	2.6194	15	0.1746	103.01	Within	5.9740	15	0.3983	14.00
Total	74.5600	19			Total	28.2827	19		

Note:
Critical values of F with 4 and 15 degrees of freedom:

α	F
0.005	5.8029
0.010	4.8932
0.050	3.0556
0.100	2.3614

*$F = MS_{between}/MS_{within}$.

DISCUSSION OF RESULTS

Orthoclase Gabbro

This is the smallest and most poorly exposed unit in the Audubon-Albion stock. On the basis of 29 thin sections, the average orthoclase gabbro consists of labradorite (44 percent), diopsidic augite (24 percent), biotite (17 percent), orthoclase (11 percent), and opaques (4 percent). The first-degree hypersurfaces for specific gravity, Fe_2O_3 + MgO, Na_2O + K_2O, and SiO_2 are shown in Figure 3. Cross sections of these surfaces indicate a steep northeasterly dip with higher specific gravity and Fe_2O_3 + MgO values to the southwest, and higher Na_2O + K_2O and SiO_2 values to the northeast.

The texture of the gabbro suggests that augite and labradorite are, at least in part, cumulate phases and that the biotite and orthoclase are products of postcumulus crystallization of the residual intercumulus liquid. Variation diagrams for both major and trace elements (Mathews, 1970) show that data from the gabbro plot off the trends generated by data from the other units of the stock (Fig. 4), thus implying that the gabbro deviates significantly from the main line of liquid descent (Mathews and others, 1968). This singular character of the gabbro also is indicated by three-dimensional trend-surface analysis of the other units of the stock, for only the gabbro has hypersurface patterns that depart markedly from the topographic contours.

These dipping hypersurfaces must be interpreted with caution because of the small number of sampling sites and the low percentages of total corrected sums of squares accounted for by the first-degree surfaces. Nevertheless, a reasonable interpretation, based on classical fractionation sequences and gravitational separation, is that the more felsic northeast portion represents the top (or youngest part) of the gabbro, and the more mafic southwest part, the bottom (or oldest). If this is so, it follows that subsequent to solidification the gabbroic cumulate was rotated from horizontal, presumably during intrusion of the enveloping monzonite magma. This petrogenetic interpretation is consistent with the information that would be available without consideration of three-dimensional variability.

An appropriate question is whether these variations could not be detected in the field or by manual contouring of the raw data. The gabbro is not layered in the usual sense of that term, and the changes are too subtle to recognize in the field. Simple contouring of the data gives ambiguous map patterns for many variates, as would be expected because it offers no criterion for separating "noise" from meaningful, systematic changes. Moreover, variability with respect to elevation is not taken into account. It is, of course, to overcome precisely these problems that one uses trend analysis. As Chayes and Suzuki (1963, p. 309) stated, trend-surface maps present a "picture of a tendency whose expression is *obscured* in the rock."

Monzonite

The monzonite is mineralogically variable, but compositionally and texturally distinct from the orthoclase gabbro. Modal averages of 96 thin sections are andesine, 43 percent; orthoclase, 41 percent; augite, 8 percent; hornblende, 5 percent; and accessories, 3 percent. Textures within the unit are dominantly hypidiomorphic-granular, but locally range to porphyritic. Compositionally, the porphyritic phase may be either more mafic or more felsic than the nonporphyritic variety.

This unit has the longest contact zone with the country rocks. However, structures in the surrounding Idaho Springs Formation and Silver Plume Granite have not been disturbed by its emplacement. Xenoliths in the monzonite are only at the highest elevations or near the contacts; no intrusion breccias were found and fluxion structures, only moderately well developed, are restricted to its margins.

Third-degree hypersurfaces for specific gravity, Fe_2O_3 + MgO, Na_2O + K_2O, and SiO_2 are shown in Figure 5. These surfaces account for high percentages of the total corrected sum of squares. In contrast to those of the orthoclase gabbro, the isopleths conform closely to the topographic contours, indicating that the surfaces are essentially horizontal.

Compositional heterogeneity of the monzonite could be due to assimilation, strong differentiation, and heterogeneities in the original magma. The two principal textural types (hypidiomorphic-granular and porphyritic, with the porphyritic phases, in general, containing more hornblende than augite) suggest heterogeneities also in the conditions under which crystallization occurred. Higher partial pressure of water, for example, would favor the observed development of hornblende over diopsidic augite and growth of feldspar phenocrysts in the porphyritic phases. Alternatively, the two textural varieties may represent two different intrusive episodes. A third possibility is that the distribution of textural types is due to convective movement during crystallization, bringing the phenocrysts to their present position near the western side of the monzonite unit. A fourth interpretation is that the phenocrysts represent an earlier (lower?) part of the intrusion exposed by tilting.

The hypersurfaces in Figure 5 indicate little systematic horizontal variability,

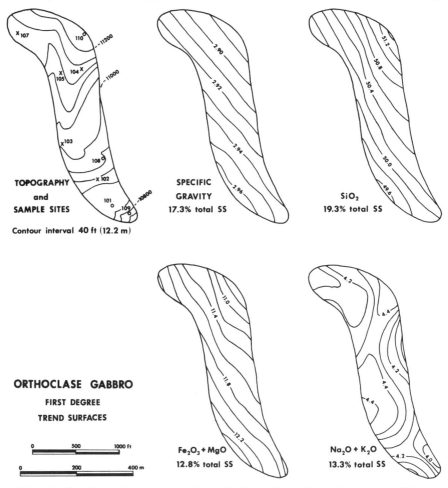

Figure 3. First-degree hypersurfaces for orthoclase gabbro. Raw data are in Table 1; distribution of 9 sample sites is shown in upper left illustration. Total SS refers to percentage of total corrected sum of squares accounted for by depicted surfaces.

but, instead, emphasize a vertical compositional variability in the monzonite. This stratification suggests that little or no movement occurred in the monzonite magma subsequent to intrusion. The surfaces show a somewhat anomalous increase in mafic content toward the top of the unit which is not readily detectable in hand specimen. This is consistent with the presence of a few xenoliths at high elevations and suggests that assimilation was an important factor for part of the unit. Dickson's (1958) concept of zone melting may be applicable here: as roof rocks are melted, somewhat higher temperature (more mafic) minerals crystallize toward the base of the magma chamber, and the latent heat of crystallization is transferred to the top of the chamber by movement of volatiles. Such volatile movement might account for the observed local compositional and textural inhomogeneities, and would permit assimilation of mafic phases in the roof at lower temperatures, giving the patterns shown by the hypersurfaces.

To estimate the crystallization temperature of the monzonite, normative albite, anorthite, and orthoclase for 24 samples were plotted on the projection of the quaternary system $NaAlSi_3O_8$-$KAlSi_3O_8$-$CaAl_2Si_2O_8$-H_2O (Yoder and others, 1957). With allowances for partial pressure of water and the presence of nonfeldspathic components, a temperature range of 850° to 700° C was indicated, making selective assimilation possible.

Three-dimensional trend analysis contributes significantly to an understanding of the petrogenesis of the monzonite. Isopleth patterns emphasize vertical rather than horizontal variability and hence argue against the dual instrusion, the tilting, or the convection models, but are consistent with assimilation or zone melting near the top of the unit. Petrogenetic inferences made without using hypersurfaces could not so readily discriminate among these models.

Quartz-bearing Monzonite

The mineralogical composition of the quartz-bearing monzonite is also highly variable. Modal averages of 45 thin sections are orthoclase, 47 percent; oligoclase, 44 percent; quartz, 3 percent; and augite, hornblende, and accessory minerals about 2 percent each. The quartz-bearing monzonite, however, does not display the wide textural variability that characterizes the monzonite; nearly everywhere, it has a medium-grained hypidiomorphic-granular texture.

The unit is exposed in three disconnected areas separated by other units of the stock (Fig. 1). Local intrusion breccias are present, but are cut by dikes of the quartz-bearing monzonite; there are no fluxion structures.

The third-degree hypersurfaces for specific gravity, Fe_2O_3 + MgO, Na_2O + K_2O, and SiO_2 shown in Figure 6 have similar patterns but different gradients. In constructing this figure, the three outcrop areas of quartz-bearing monzonite were treated as a single entity by extending fictitious contacts across the intervening areas and considering the latter as portions of the unit for which no data were available. All surfaces account for a high percentage of the total corrected sums of squares, and isopleth surfaces are essentially horizontal as indicated by the close similarity between isopleth and topographic patterns.

Hyndman (1972, p. 129), summarizing work on the Sierra Nevada batholith, California, stated that for many plutons "concentric zoning . . . is best determined in the laboratory by measurement of the specific gravities of systematically collected hand specimens." In the present study there is strong indication of the influence of gravitative stratification in the magma and the weak effect of horizontal crystallization (zonation) patterns: low-value regions on the specific gravity and

Fe_2O_3 + MgO surfaces correspond with topographic highs. Specific-gravity measurements, therefore, could be misleading if variations with elevation are not taken into account.

The assumed consanguineity of the three separated parts of the quartz-bearing monzonite is supported by the high percentage of the total corrected sums of squares accounted for by hypersurfaces generated by treating the three parts as one unit.

Syenite

The last major intrusive unit of the stock is characterized by hypidiomorphic-granular texture and is mineralogically the most simple: the average of 27 thin sections is perthite, 60 percent; oligoclase, 36 percent; and accessories, 4 percent. This small mass shows chilled margins against the quartz-bearing monzonite; contact breccias are exposed but fluxion structures are absent.

First-degree hypersurfaces for specific gravity, Fe_2O_3 + MgO, Na_2O + K_2O, and SiO_2 are shown in Figure 7. The close agreement between different trend-surface isopleth patterns and the topographic contours again shows that the first-degree surfaces are essentially horizontal. High values of specific gravity and Fe_2O_3 + MgO occur toward the lower elevations of the unit, and high values of Na_2O + K_2O and SiO_2 toward the top. This pattern is again consistent with the model of passive crystallization under the influence of gravity (Fig. 3).

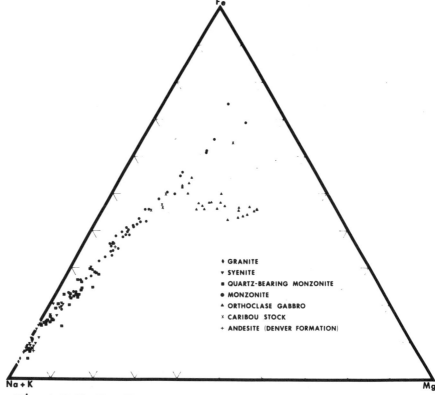

Figure 4. Fe-Mg-(Na + K) ternary diagram for rocks of Audubon-Albion stock. Also shown are some analyses from nearby, possibly related, units.

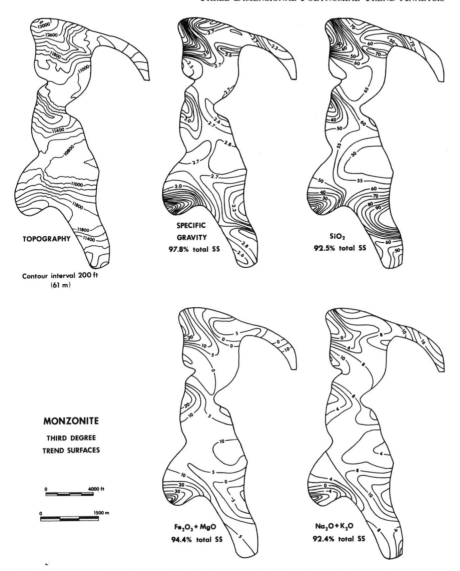

Figure 5. Third-degree hypersurfaces for the monzonite. Raw data are in Table 1; distribution of 32 sample sites is shown in Figure 2. Total SS refers to percentage of total corrected sum of squares accounted for by depicted surfaces.

CONCLUSIONS

The petrogenetic model deduced for the Audubon-Albion stock (Mathews, 1970) is that it, probably along with several other similar epizonal plutons of the Front Range, was derived from a large, fairly deep-seated reservoir of anatectically derived monzonitic magma in which the orthoclase gabbro formed by crystal accumulation. This cumulate subsequently was rafted to its present position in the stock by the intrusive movement of the monzonite magma. Continuing differentiation altered the liquid remaining at depth in a progressive pattern of magmatic evolution dominated by fractional crystallization. Intrusion at different stages of evolution, or perhaps from different levels in the magma reservoir, produced first the quartz-bearing monzonite and then the syenite. The few small granite dikes in the syenite represent the last stages of the differentiation process.

Two principal conclusions in this study can be attributed directly to trend-surface analysis. First, the analysis indicated the probable rotation and rafting of the orthoclase gabbro and hence revealed a unique aspect of the evolution of that unit. Second, it called attention to subtle vertical stratification of variates and emphasized the otherwise nonobvious major role that gravity and assimilation played in the crystallization of the magmatic bodies. Concomitant, and of importance in current considerations of global lithospheric behavior, is the implication from the dominantly horizontal hypersurfaces that the crustal unit of which the Audubon-Albion stock is a part has not been rotated significantly from (or has regained) a horizontal position during the past 65 m.y. In areas where stratigraphic or other evidence such as layered intrusives is absent, trend analysis could be a valuable structural tool.

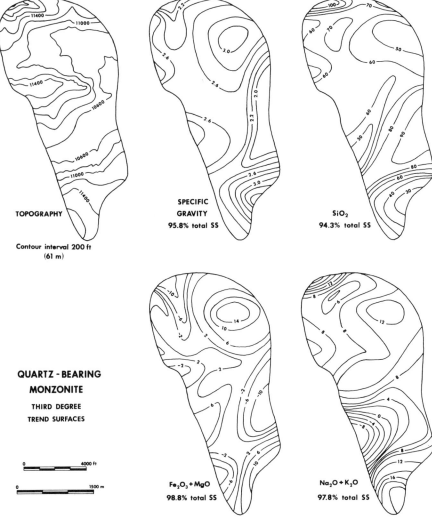

Figure 6. Third-degree hypersurfaces for the quartz-bearing monzonite (22 sample sites). Details same as Figure 5.

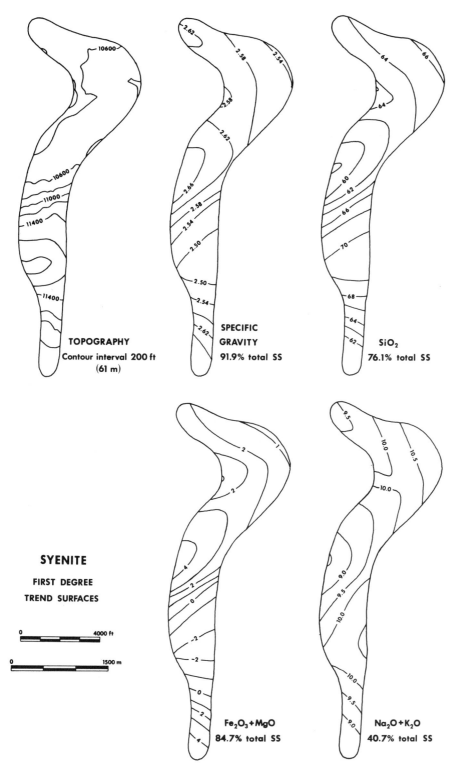

Figure 7. First-degree hypersurfaces for the syenite (10 sample sites). Details same as Figure 5.

Three-dimensional trend analysis contributes to igneous petrology by permitting a determination of spatial compositional variability, in addition to presumed temporal changes (represented by variation diagrams) that have long been used in petrologic investigations. Figure 4 shows that, except for the anomalous orthoclase gabbro, the trend of crystallization in the Audubon-Albion stock is much more similar to the later part of the sequence (illustrated by Hyndman, 1972, p. 80) for tholeiitic flood basalts, diabase sills, and mafic layered intrusions (postorogenic, no assimilation, low P_{O_2} environment) than to the sequence for the basalt, andesite, rhyolite association (orogenic, assimilation, high P_{O_2}). To some extent this may reflect assimilation of Fe-rich Precambrian country rocks, as described for the monzonite unit. Nevertheless, it is a somewhat surprising result, and perhaps the dominantly vertical variability (stratification) in the stock, as expressed by hypersurfaces, is related to this possible tholeiitic affinity.

The evolution of the Audubon-Albion magma has broader, although more tenuous, implications regarding global tectonics of the western United States during the Cretaceous-early Cenozoic. The tholeiitic-type Fe-Mg-(Na + K) diagram (Fig. 4) and current hypotheses of plate tectonics could be taken together to suggest that the parental monzonitic magma was derived through subduction of the Farallon plate along a Benioff zone under the North American plate (Atwater, 1970; Lipman and others, 1971). Scholz and others (1971) used this hypothesis to explain volcanic activity in the Great Basin, Nevada and Utah, some 40 m.y. ago. Extrapolation to the Colorado Front Range requires a long, shallow-dipping Benioff zone and must somehow account for the Colorado Plateau, perhaps by the imbricate subduction model proposed by Lipman and others (1971). However, the suggested partial melting of the upper surface of the downgoing oceanic slab, followed by diapiric rise of the derived material into the lithosphere and its entrapment by the reaction with strong sialic crust, is an attractive explanation of the origin of the parental magma and its seemingly mafic line of descent.

The chemical data also may be used, in the manner described by Hatherton and Dickinson (1969), to estimate depth to the center of the proposed subduction zone. Variation diagrams were made for K_2O against SiO_2 for 133 samples. The derived K_2O contents at 55 and 60 percent SiO_2, when plotted on Hatherton and Dickinson's extrapolated curves, indicate a depth to the center of the subduction zone of approximately 420 km. This is in reasonable agreement with depths calculated for nearby areas of Colorado by Lipman and others (1971, p. 823), and it supports their conclusion that "the few Laramide igneous suites for which reliable compositional trends can be drawn fit the trends defined by middle Cenozoic igneous suites. This suggests that the major shift in the geometry of subduction [from steep to gentle dip] occurred between about 80 and 70 million years ago."

ACKNOWLEDGMENTS

We are indebted to W. C. Krumbein for introducing us to trend-surface analysis and to E. H. T. Whitten for demonstrating applications of the technique in igneous petrology. John Hower gave invaluable assistance in the x-ray spectrographic analyses of the samples. During the spring of 1973, the Department of Geological Sciences, Harvard University, provided Cain with facilities and congenial surroundings, which aided preparation of this paper. The manuscript was much improved by suggestions from Leila S. Cain and O. Don Hermes. Financial support was provided by National Science Foundation Grant GP-3670.

REFERENCES CITED

Atwater, T., 1970, Implications of plate tectonics for the Cenozoic tectonic evolution of western North America: Geol. Soc. America Bull., v. 81, p. 3516-3536.
Chayes, F., and Suzuki, Y., 1963, Geologic contours and trend surfaces: A discussion: Jour. Petrology, v. 4, p. 307-312.
Crawford, R. D., 1924, A contribution of the igneous geology of central Colorado: Am. Petroleum Geologists Bull., v. 53, p. 714-715.
Czamanske, G. K., Hower, J., and Millard, R. C., 1966, Non-proportional, non-linear results from x-ray emission techniques involving moderate-dilution rock fusion: Geochim. et Cosmochim. Acta, v. 30, p. 745-756.
Davis, J. C., 1973, Statistics and data analysis in geology: New York, John Wiley & Sons, Inc., 550 p.
Dawson, K. R., and Whitten, E. H. T., 1962, The quantitative mineralogical composition and variation of the Lacorne, LaMotte, and Preissac granitic complex, Quebec, Canada: Jour. Petrology, v. 3, p. 1-37.
Dickinson, F. W., 1958, Zone melting as a mechanism of intrusion—A possible solution of the room and superheat problems (abs.): Am. Geophys. Union Trans., v. 39, no. 3, p. 513.
Doe, B. R., 1967, The bearing of lead isotopes on the source of granitic magmas: Jour. Petrology, v. 8, p. 51-58.
Grant, F., 1957, A problem in the analysis of geophysical data: Geophysics, v. 22, p. 309-344.
Harbaugh, J. W., and Merriam, D. F., 1968, Computer applications in stratigraphic analysis: New York, John Wiley & Sons, Inc., 282 p.
Hart, S. R., 1960, Extracts from the thesis investigation of S. R. Hart, in Variation in isotopic abundance of Sr, Ca, and Ar and related topics, in 8th annual progress report for 1960: Boston, Massachusetts Inst. Technology, NYO-3841, p. 87-197.
Hatherton, T., and Dickinson, W. R., 1969, The relationship between andesitic volcanism and seismicity in Indonesia, the Lesser Antilles, and other island arcs: Jour. Geophys. Research, v. 74, p. 5301-5310.
Hower, J., 1959, Matrix corrections in the x-ray spectrographic trace element analysis of rocks and minerals: Am. Mineralogist, v. 44, p. 19-32.
Hyndman, D. W., 1972, Petrology of igneous and metamorphic rocks: New York, McGraw-Hill Book Co., Inc., 533 p.
Koch, G. S., and Link, R. F., 1971, Statistical analysis of geological data, Vol. 2: New York, John Wiley & Sons, Inc., 438 p.
Krumbein, W. C., 1959, Trend surface analysis of contour type maps with irregular control-point spacing: Jour. Geophys. Research, v. 64, p. 823-834.
——1966, A comparison of polynomial and Fourier models in map analysis: U.S. Office Naval Research, Tech. Rept. no. 2, ONR Task no. 388-078, Contract NONR-1228(36), 45 p. (Clearinghouse AD635476).
Krumbein, W. C., and Graybill, F. A., 1965, An introduction to statistical models in geology: New York, McGraw-Hill Book Co., Inc., 475 p.
Lipman, P. W., Prostka, H. J., and Christiansen, R. L., 1971, Evolving subduction zones in the western United States, as interpreted from igneous rocks: Science, v. 174, p. 821-825.
——1972, Cenozoic volcanism and plate-tectonic evolution of the western United States. I. Early and middle Cenozoic: Royal Soc. London Philos. Trans., Ser. A, v. 271, p. 217-248.
Lovering, T. S., and Goddard, E. N., 1938, Laramide igneous sequence and differentiation in the Front Range, Colorado: Geol. Soc. America Bull., v. 49, p. 35-68.
——1950, Geology and ore deposits of the Front Range, Colorado: U.S. Geol. Survey Prof. Paper 223, 319 p.
Mandelbaum, H., 1963, Statistical and geological implications of trend mapping with nonorthogonal polynomials: Jour. Geophys. Research, v. 68, p. 505-519.

Mathews, G. W., 1970, The petrology and geochemical variability of the Audubon-Albion stock, Boulder County, Colorado [Ph.D. thesis]: Cleveland, Case Western Reserve Univ., 261 p.

Mathews, G. W., Banks, P. O., and Cain, J. A., 1968, Chemical variation and evidence of a crystal cumulate in a composite Laramide intrusion: Geol. Soc. America, Abs. for 1967, Spec. Paper 115, p. 140-141.

Oldham, C. H. G., and Sutherland, D. B., 1955, Orthogonal polynomials; their uses in estimating regional effects: Geophysics, v. 20, p. 295-306.

Pearson, R. C., Tweto, O., Stern, T. W., and Thomas, H. H., 1962, Age of Laramide porphyries near Leadville, Colorado, in Short papers in geology and hydrology: U.S. Geol. Survey Prof. Paper 450-C, p. C78-C80.

Peikert, E. W., 1962, Three-dimensional specific-gravity variation in the Glen Alpine stock, Sierra Nevada: Geol. Soc. America Bull., v. 73, p. 1437-1442.

——1963, IBM 709 program for least-squares analysis of three-dimensional geological and geophysical observations: U.S. Office Naval Research, Tech. Rept. no. 4, ONR Task no. 389-135, Contract NONR-1228(26), 72 p. (Clearinghouse AD420274).

Phair, G., and Fisher, F. G., 1962, Laramide comagmatic series in the Colorado Front Range: The feldspars, in Petrologic studies (Buddington volume): New York, Geol. Soc. America, p. 479-522.

Scholz, C. H., Barazangi, M., and Sbar, M. L., 1971, Late Cenozoic evolution of the Great Basin, western United States, as an ensialic interarc basin: Geol. Soc. America Bull., v. 82, p. 2979-2990.

Tweto, O., 1968, Leadville district, Colorado, in Ridge, J. D., ed., Ore deposits of the United States, 1933-1967 (Graton-Sales volume): New York, Am. Inst. Mining, Metall., & Petroleum Engineers, v. 1, p. 681-705.

Tweto, O., and Case, J. E., 1972, Gravity and magnetic features as related to geology in the Leadville 30-minute quadrangle, Colorado: U.S. Geol. Survey Prof. Paper 726-C, 31 p.

Tweto, O., and Sims, P. K., 1963, Precambrian ancestry of the Colorado mineral belt: Geol. Soc. America Bull., v. 74, p. 991-1014.

Wahlstrom, E. E., 1940, Audubon-Albion stock, Boulder County, Colorado: Geol. Soc. America Bull., v. 51, p. 1789-1820.

Wallace, S. R., Muncaster, N. K., Jonson, D. C., MacKenzie, W. B., Bookstrom, A. A., and Surface, V. E., 1968, Multiple intrusion and mineralization at Climax, Colorado, in Ridge, J. D., ed., Ore deposits of the United States, 1933-1967 (Graton-Sales volume): New York, Am. Inst. Mining, Metall., & Petroleum Engineers, v. 1, p. 605-640.

Whitten, E. H. T., 1959, Composition trends in granite; modal variations and ghost-stratigraphy in part of the Donegal granite, Eire: Jour. Geophys. Research, v. 64, p. 835-848.

——1970, Orthogonal polynomial trend surfaces for irregularly spaced data: Jour. Internat. Assoc. Math. Geol., v. 2, no. 2, p. 141-152.

Whitten, E. H. T., and Boyer, R. E., 1964, Process-response models based on heavy-mineral content of the San Isabel granite, Colorado: Geol. Soc. America Bull., v. 75, p. 841-862.

Wyllie, P. J., 1971, The dynamic earth: Textbook in geosciences: New York, John Wiley & Sons, Inc., 416 p.

Yoder, H. S., Stewart, D. B., and Smith, S. R., 1957, Ternary feldspars, in Annual report of the director of the geophysical laboratory: Carnegie Inst. Washington Year Book, no. 56, p. 206-214.

MANUSCRIPT RECEIVED BY THE SOCIETY SEPTEMBER 12, 1973
REVISED MANUSCRIPT RECEIVED JANUARY 24, 1974

Petrogenetic Significance of Grain-Transition Probabilities, Cornelia Pluton, Ajo, Arizona

WILLIAM B. WADSWORTH

Department of Geology
Whittier College
Whittier, California 90608

ABSTRACT

Grain-transition probabilities among four mineral phases are used to analyze variation in grain sequence among 60 samples distributed throughout the zoned Cornelia pluton of southwestern Arizona. Six basic patterns in grain sequence are recognized, and similarity among them defines three major groups. These three predominantly record (1) plagioclase replacement by K-feldspar; (2) primary, magmatic intergrowth of K-feldspar and quartz; and (3) an extreme poikilitic habit in which quartz encloses sodic plagioclase.

Variation in these patterns among units of the pluton is responsible for initiating a completely new petrologic understanding of the Cornelia stock. Detailed correspondence is found with modern models of pegmatite derivation, which involves inward crystallization of early, anhydrous phases; concentration of volatiles beneath the encroaching roof; evolution of potassium-rich aqueous fluids that impregnated chamber walls, rose to concentration in the stock's apex, and left quenched residue behind; and consequent enrichment in sodium and silicon in the rest magma leading to evolution of hydrothermal fluids.

The three textural features above, which gave rise to the grain-sequence groups, are genetically related to (1) impregnation of rocks along the roof of the conduit that was followed by evolving pegmatitic fluid, (2) orthomagmatic crystallization at depth in the rest magma, and (3) mixing of crystallization products from the potassium-depleted residual magma and a newly developed aqueous fluid.

INTRODUCTION

Textural studies of granitic rocks have been slow to develop, probably because of the complexity of the granitic fabric and the difficulty in defining and measuring meaningful textural attributes. One promising approach—recently introduced—is through analysis of grain-contact frequencies. These data can be obtained easily and precisely and, if structured appropriately, can be tested for correspondence with statistical models.

In their simplest form, analyses of grain-transition records can test for randomness of sequence along a linear traverse. If the transitions depart significantly from random sequence, then a range of tests exist to establish the type of "memory" exhibited by the data, specifically, whether the transition sequence could have derived from a Markov process.

Some studies of grain-contact frequencies in rocks other than granite have achieved useful results simply by establishing departure from randomness of grain sequence (Flinn, 1969; Kretz, 1969). Vistelius (1966a, 1972) introduced a stochastic model of crystallization in granites in which the process results in grain sequences having the properties of simple Markov chains. In other reports (Vistelius, 1966b, 1967, 1971; Vistelius and Faas, 1971), several plutons have been tested against this model. Vistelius' method is admirable in the rigor of mathematical development and in derivation of a conceptual petrogenetic model from crystallization theory. Collection of data to compare with that model, however, requires recognition of contacts among grains of the same mineral species and involves a number of other operational definitions, all of which lead to doubt that the data can be collected with precision.

This paper is believed to be the first attempt to structure grain-transition data according to the model of embedded Markov chains, which avoids the necessity of recognizing grain contacts among like species. This structure has long been recognized as useful in stratigraphic sequence studies, and some contrasts between the Markov chain and embedded Markov chain models are presented by Krumbein (1967) and Krumbein and Dacey (1969).

In the present study, grain-transition probabilities are derived from 60 samples areally distributed throughout the Cornelia pluton, Arizona. Many samples have nonrandom grain-transition probabilities, but the Markov property, if present, involves a high order of complexity that has not been adequately defined yet. Thus, the analysis proceeds by informal treatment of the transition probabilities and finds similarity among groups of samples that appear to have great petrogenetic significance. Areal distribution of texturally similar samples has led to a revised theory of petrogenesis for the Cornelia stock, in which it is recognized as the source area for separation of one or more aqueous fluids. These results, and inferences in several other recent works concerning episodes of fluid separation (Fournier, 1967; Nielsen, 1968; Moore, 1973), demonstrate substantial agreement between the properties of porphyry-copper plutons and a modern model of pegmatite genesis (Jahns and Burnham, 1969). The latter appears to provide a detailed and unifying framework for understanding these plutons, and the Cornelia stock is particularly important in testing concepts of this model because it displays some of the deepest levels of exposure known among the copper-bearing porphyries (Lowell and Guilbert, 1970).

FRAMEWORK OF THE STUDY

The Cornelia pluton, a Tertiary copper-bearing stock located at Ajo, Arizona, is the source of samples for this work. Detailed petrographic descriptions and general petrologic history were presented by Gilluly (1942, 1946) and Wadsworth (1968). Figure 1 and Table 1 serve to summarize essential petrographic characteristics and relations. Some changes in my 1968 map have been incorporated in Figure 1 on the basis of more detailed sampling and qualitative study in the interim. Sixty of the 66 original sample localities, spaced at 0.25-mi (0.4-km) intervals, are utilized below, but more than 300 samples from 123 sites have been used for current map construction.

The 1968 report defined and mapped five varieties of the plutonic rocks on the basis of modal or textural properties, and modal analyses were presented that include all samples used here. Discrete episodes of emplacement have been proposed for all units except the porphyritic quartz monzonite, and the continuous trend in modal composition throughout has been used to infer nearly complete representation of the differentiation spectrum in the samples studied. The quartz diorite unit has not been studied areally; only two samples near the granodiorite contact were employed in this report. A few sample localities had to be excluded because feldspar discrimination was prevented by degree of alteration or indeterminate staining results (in cases of extremely fine grain size).

Records obtained for analysis are of identity-change type (Chayes, 1956), in which numbers of contacts between different, major mineral species are recorded along linear traverses in thin section. Such records previously have been used in tests of this variate as a coarseness measure throughout the compositional range of the Cornelia samples (Wadsworth, 1969). In that report, the mafic minerals were treated as an aggregate element and added to the coarseness variate as a fourth "species" to be tallied. Through sequential multiple regression and analysis of variance, it was found that a 50-mm traverse length in thin section was sufficient to establish differences in coarseness among samples within each unit of the stock. In addition, the coarseness measure could be accepted as a Gaussian frequency distribution in all cases. These results are pertinent here because coarseness, as applied above, is the total number of transitions among the four states of plagioclase, K-feldspar, quartz, and mafic minerals along a traverse of unit length. Total transition tallies in a 50-mm traverse range from 90 to 345 (as an average of eight traverses of this length per sample) in these specimens. In the absence of other data, the smaller value may provide a minimum tally magnitude that should be obtained between pairs of states when interest centers on areal variation in state-transition probabilities. Table 2 shows that the mafic suite contribution most often falls short of this value. In contingency table tests for Markov character using the chi-square criterion, all that is usually required is about five tallies per cell. Previous designs for study of grain-transition sequences in granites appear to have utilized less than 1,000 total tallies (in some cases less than 150), even when areal variability was of interest. In those cases total tallies were distributed among 9 grain-transition categories, by comparison with the 12 categories that can exceed zero in the design used here.

Identity change records can be compiled most easily by assigning a number key on a typewriter to each state. Traverses 25 mm in length were made on standard-size thin sections, and 16 successive traverses were spaced at 2-mm intervals. The total traverse length of 400 mm is distributed over an area that is approximately that needed for a good modal analysis of a medium-grained granite (Chayes, 1956). Each traverse was made in the same direction (which permits tests for preferred direction of grain sequence), so that end corrections must be applied. The only correction was to begin each traverse by recording the transition between the first and second state encountered; that is, the last element of a traverse is not exited and the first of the new traverse is not entered. This procedure

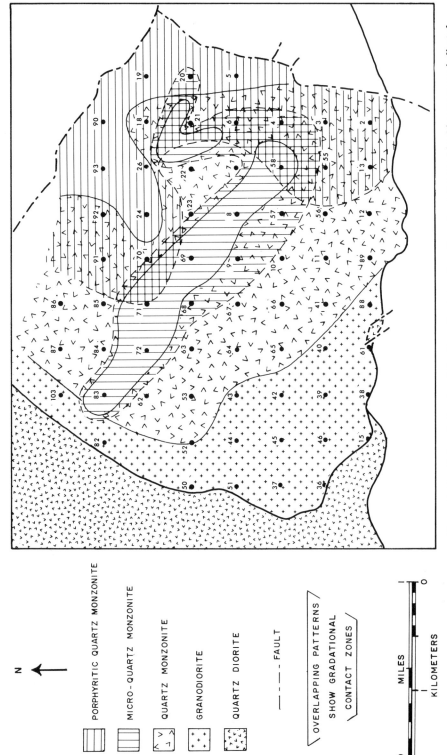

Figure 1. Detailed geologic map of the Cornelia pluton. Locations of samples used in this report are indicated.

TABLE 1. CHARACTERISTICS OF THE CORNELIA STOCK

Inferred relative age	Unit	Defining and important characteristics	Geologic relations
Youngest	Porphyritic quartz monzonite	Coarser grained than other quartz monzonites; hornblende-bearing and biotite-bearing; deuteric effects most prominent here. Large, round, and poikilitic quartz grains with myriads of plagioclase inclusions increase toward this unit. Seriate porphyritic in plagioclase and K-feldspar. Plagioclase aggregates large and very complex	Continuously gradational with other quartz monzonites. Sheared and recrystallized dioritic xenoliths of large size only occur here. Occurs at lowest level of magma chamber
	Porphyritic micro-quartz monzonite	Distinctly finer grained than other quartz monzonites; average modal composition closest to the ternary minimum. Biotite-bearing; quartz and K-feldspar replacement textures; hiatal porphyritic in plagioclase and (or) K-feldspar in some examples; others equigranular	Basically a complex of dikes. Broadly gradational with the quartz monzonite unit through veining and grain-size transition. Possible fault-zone control of emplacement. Was the deeper conduit of the orebody, now exposed by fault decapitation and tilting of the entire stock
	Quartz monzonite	Medium-grained; either hornblende or biotite dominant, and clinopyroxene relics occur. Compositionally intermediate between units above and below. Hypidiomorphic-granular grading to seriate porphyritic in plagioclase and K-feldspar	Contacts with units below are hiatal in modal composition, but gradational with granodiorite in some textural properties. Contact not observed with the latter; sharp and intrusive with quartz diorite. Small aplite dikes occur in this unit and units above
	Granodiorite	Medium-grained to coarse-grained; size increasing toward outer stock margin. Variable mafic suite includes substantial augite and hypersthene in some samples. Normal-order reaction relations. Color index relatively high here and in unit below. Hypidiomorphic-granular. Tendency to form plagioclase aggregates less than in units above. More local variability in modal composition than all other units	Sharp contact with quartz diorite to south and southwest; alteration and textural and modal convergence to northwest renders contact uncertain. Andesite and variable leucocratic dikes common, up to 6.1 m thick. Also occur in unit below
Oldest	Quartz diorite	Quartz content less than 6%; mafic suite as in unit above. Varies from very fine grained in south to coarse grained in northwest. K-metasomatism and biotite growth by recrystallization evident near granodiorite contact. Elongate, equisized plagioclase laths dominate fabric	Found in unmapped, thin, and discontinuous band all along southern stock margin. Possible ring-fracture control of emplacement

avoids introduction of spurious tallies but can cause minor discrepancies between row and column sums in the transition tally matrix. Spacing between traverses is slightly larger than twice the average diameter of single plagioclase grains in the coarsest rock unit (plagioclase commonly being the coarsest mineral), and it is somewhat greater than the average diameter of plagioclase aggregates in the unit where those are largest. Traverse spacing is important in Vistelius' design but is not thought to be so limiting in this case because the design causes tiny intergrowths to be treated equally with contacts between large crystals or aggregates. Area of analysis and tally magnitude are important in both cases because they determine, in different ways, adequacy of the sample data to represent the specimen.

The tally matrix of transitions was prepared by computer, as were most other matrices and calculations cited below. Programs for various Markov chain investigations have been published by Krumbein (1967, 1968a, 1968b).

Before discussion of data analysis, it is necessary to explain fully my proposal that, in grain-transition studies of granites, data collection according to the embedded chain structure is preferable to the simple Markov chain structure for reasons of precision, standardization, and ease of data acquisition. To obtain the Markov chain structure, it is necessary to record contacts between grains of a single state. Given the ubiquitous occurrence in granites of complex twinning, primary growth aggregates, and mechanical grain disruption during and after consolidation, it is my opinion that grain contacts within a single species cannot be recognized either simply or precisely. This was likewise the opinion of Chayes (1956, p. 61) when he wrote, ". . . in the granitic fabric it has so far proved impossible to specify, in any usable fashion, what is meant by a single grain." In the embedded chain structure, transitions from a given state into itself are not obtained. Errors in measurement are thus reduced to misidentification of mineral species, and with traditional staining procedures, such errors approach the irreducible minimum of a careful modal analysis.

Vistelius' (1972) approach to the general topic is laudable because it developed a petrogenetic model in which grain sequence is predicted from crystallization paths based on the work of Tuttle and Bowen (1958). Acceptance of such a model for comparison, however, further complicates the measurement process.

Big porphyritic crystals should be ignored during counting.

In the process of counting, it is necessary to count only grains of the main granite mass. Grains located in fissures, grains in granophyre textures, myrmekites, and cataclastic parts of the rock should be ignored. . . . One grain with twins should be counted as one grain. Thin pertitic intergrowths and similar antypertitic growths should be ignored, but big lenticular pertites or parts of feldspar grains composed by components of dissociation of solid solutions should be investigated specially. (Vistelius, 1972, p. 101-102)

Further restrictive definitions are involved, such as

Each individual of the mineral in the grain sequence along a straight line can be encountered only once. Thus such textures as rapakivi and hooked grains which are able to occupy two positions simultaneously in the same sequence as representatives of the same individual are ignored. We understand that some observations are contradicted but how important and what is the influence of these observations? (Vistelius, 1972, p. 93)

In addition to the doubt expressed here that such distinctions can, in fact, be made or that they can be made suitably precise, the attempt to do so must become sufficiently laborious that few would choose to pursue the method. Standardization among petrographers would certainly be very difficult.

Using an identity change type of record, which records only transitions between different mineral species, the measurement procedure is simple and acceptably rapid, and it is standardized among petrographers everywhere. No restrictions on textural variants were utilized for this study, but it should be noted that the Cornelia rocks have very minor myrmekite development, and perthite strings are so extremely narrow that they were not distinguished from K-feldspar.

The restrictive operational definitions employed by Vistelius (1972) were erected in an attempt to record only transitions that result from primary magmatic crystallization, so that sample data can be compared with a model of magmatic crystallization. Data collection methods employed in this study obtain all grain transitions, regardless of origin, and contain contributions from late magmatic and postmagmatic events. As Whitten (1963, p. 115) has stressed, however, an understanding of the geochemical evolution of granites may require study of the textural evidence of these later stages. The sampling design used produces transition frequencies that correspond directly to surface areas of contact among the states recognized, which has not been the case in previous studies of granites. Construction of conceptual models for this case will be difficult, since sufficient knowledge of the surface chemistry of solids in natural systems may not be available. Some feeling for the status of such research in rock systems may be gained from DeVore (1959), Spry (1969), and Ehrlich and others (1972).

TEST FOR MARKOV CHARACTER

The analytical procedure used by Gingerich (1969), parts of which were suggested also by Krumbein and Dacey (1969), is summarized below and uses sample number 12 (Fig. 1) as an example.

The matrix of observed, one-step transition frequencies (the tally matrix) is

		state entered (j = 1, 4)				Row sums (vector of state frequencies)
		P = 1	K = 2	Q = 3	M = 4	
state exited (i = 1, 4)	P = 1	0	221	108	38	367
	K = 2	226	0	155	117	498
	Q = 3	99	156	0	56	311
	M = 4	40	118	54	0	212
Column sums		365	495	317	211	1,388 Total tally

where the mineral states of the system are P = plagioclase, K = K-feldspar, Q = quartz, and M = mafic phases. Matrix cell (1, 3) recores 108 transitions from plagioclase into quartz. In cases in which traverse end-corrections leave no discrepancies and in which there is no preference for direction of transition between any two given states (no mineral layering, for example), the tally matrix would be symmetrical across the main diagonal on the average. Given expected sampling variation, the asymmetry observed above is little evidence for fabric preference of grain-sequence direction. Most samples show asymmetry similar to this; a few somewhat more (Table 2). The inference that mineral layering is absent is not unexpected in these rocks, but the possible application of this matrix in studies of cumulus textures is worth noting.

The transition-probability matrix is constructed from the tally matrix by expressing the count in each cell as a proportion of its row sum. This new matrix expresses the transition probabilities derived from the data and has the stochastic matrix property of row sums which equal one. The transition-probability matrix is

$$\begin{array}{c} & \begin{array}{cccc} P & K & Q & M \end{array} \\ \begin{array}{c} P \\ K \\ Q \\ M \end{array} & \left[\begin{array}{cccc} 0 & 0.602 & 0.294 & 0.104 \\ 0.454 & 0 & 0.311 & 0.235 \\ 0.318 & 0.502 & 0 & 0.180 \\ 0.189 & 0.557 & 0.255 & 0 \end{array} \right] \end{array}$$

Because each row suffers the constraint of closure to a constant sum, entries for reverse transitions may be very different although they were nearly identical as tally data. In using the Markov chain structure, which has possible nonzero probabilities along the main diagonal, Ivanov (1970) compared column sums among samples to avoid this constraint. In the embedded chain structure, such comparisons do not appear to permit simple interpretation.

The test for Markov character used here begins by comparing the transition-probability matrix with a matrix of probabilities derived under the hypothesis that the grain sequence results from a series of independent trials. In a first-order (or one-step) Markov chain, the probability of entering a given state depends on the state last entered; that is, it results from a process having a one-step memory. In independent trials, the probability of transition into a given state depends only on the relative frequency of occurrence of that state in the system. Thus, to calculate the independent-trials probabilities of a given transition, we compute the ratio between the frequency of occurrence of the state to be entered and the frequency of occurence of all states which could be entered. Row sums of the tally matrix provide a vector of occurrence frequencies for each state.

Thus, from the tally matrix,

$$\text{independent-trials probability (P/Q)} = \frac{\text{sum of row Q}}{\text{matrix sum} - \text{sum of row P}}$$

which yields, for the example, the independent-trials matrix:

$$\begin{array}{c} & \begin{array}{cccc} P & K & Q & M \end{array} \\ \begin{array}{c} P \\ K \\ Q \\ M \end{array} & \left[\begin{array}{cccc} 0 & 0.488 & 0.305 & 0.208 \\ 0.412 & 0 & 0.349 & 0.238 \\ 0.341 & 0.462 & 0 & 0.197 \\ 0.312 & 0.423 & 0.264 & 0 \end{array} \right] \end{array}$$

Statistical comparison of observed probabilities in the transition matrix with calculated probabilities under the independent-trials hypothesis can be accomplished using the chi-square criterion. Degrees of freedom for the test have been assigned as the number of nonzero entries in the matrix minus the number of constraints (Gingerich, 1969; Read, 1969). Each row has the constraint previously cited, so

TABLE 2. TALLY MATRICES

Sample no.	Grain-transition tallies											
	P/K	P/Q	P/M	K/P	K/Q	K/M	Q/P	Q/K	Q/M	M/P	M/K	M/Q
3	159	130	62	160	164	40	131	168	37	59	42	42
4	306	173	61	291	286	88	178	292	72	62	78	80
9	218	164	27	205	135	49	167	125	41	30	47	41
10	253	127	54	262	117	95	124	121	29	43	101	33
11	155	88	40	154	107	63	81	112	50	47	55	49
12	221	108	38	226	155	117	99	156	56	40	118	54
21	292	164	66	297	290	93	152	292	83	72	93	74
22	194	227	108	203	229	107	230	228	55	95	114	62
23	253	254	63	247	251	102	249	252	60	73	94	55
41	235	77	40	238	137	93	76	138	36	37	94	40
53	170	79	24	175	232	57	82	227	44	17	70	41
55	176	188	53	180	259	95	178	262	45	54	96	43
56	181	153	44	182	125	59	151	128	34	45	54	39
57	231	185	48	232	106	63	187	97	57	48	71	53
62	130	109	41	148	187	71	95	215	39	40	61	50
63	132	87	39	136	176	74	71	185	47	50	68	40
64	151	104	37	159	134	86	96	140	35	34	88	35
65	255	98	38	260	158	92	92	158	25	34	94	22
66	177	87	35	178	118	93	83	127	28	40	78	36
67	215	260	54	198	240	76	251	240	74	50	87	68
68	187	138	38	181	231	65	144	231	22	42	58	22
70	351	132	23	343	206	60	131	209	42	28	56	39
84	153	122	55	168	116	50	112	132	43	47	51	49
85	170	146	40	156	319	82	161	307	41	41	79	43
87	230	95	41	245	193	29	75	208	22	37	33	24
89	247	105	54	257	148	96	97	152	30	53	93	31
91	213	261	70	239	175	51	233	201	43	73	51	40
92	137	109	81	135	365	54	110	364	52	78	53	56
8	439	153	59	434	326	77	165	319	37	55	79	41
58	201	175	53	197	286	56	178	284	31	52	50	37
69	675	233	55	664	290	65	239	280	40	55	65	41
71	610	244	29	597	398	58	261	392	39	21	52	54
72	485	140	22	488	178	30	147	177	12	17	25	20
83	374	102	35	378	323	73	100	341	28	34	62	41
5	143	99	59	153	192	69	96	193	44	53	78	39
19	191	144	63	186	149	79	137	160	32	77	66	33
24	187	192	57	177	237	49	205	237	27	53	43	39
26	113	182	73	113	151	66	191	151	64	67	70	70
90	107	66	33	107	145	35	67	146	22	30	37	26
93	75	62	41	83	171	43	65	165	20	35	53	18
15	166	72	56	160	125	59	74	119	71	57	60	69
36	75	20	34	80	211	112	12	204	79	37	115	69
37	125	65	46	128	119	121	60	118	49	47	120	45
38	65	80	63	57	67	57	84	53	59	64	59	58
39	76	74	74	68	151	78	89	127	87	60	89	87
40	73	128	93	80	185	63	124	190	66	88	63	72
42	70	120	76	69	116	61	117	109	82	77	70	74
43	145	57	81	139	56	105	63	55	57	87	99	61
44	118	80	90	123	53	88	80	60	71	91	82	80
45	51	54	48	51	161	108	57	159	105	41	110	108
46	48	73	63	47	140	73	70	133	85	61	77	88
50	169	93	62	166	155	101	101	148	62	54	103	67
51	219	146	62	200	231	84	154	217	84	74	80	79
52	162	110	47	177	156	82	103	155	48	37	96	46
61	116	103	61	105	85	73	124	72	69	47	70	82
82	179	122	68	177	152	79	118	145	48	69	85	39
88	140	83	70	138	149	75	82	142	54	73	74	48
103	101	82	114	104	39	131	77	41	105	119	129	101
15a	374	298	431	379	11	34	305	12	63	419	40	70
49	49	41	68	56	82	57	38	76	65	58	70	59

that there are (12 − 4) 8 degrees of freedom for this case. The test criterion is the sum of contributions from each transition pair, each contribution being calculated as

$$\frac{(\text{observed frequency} - \text{expected frequency})^2}{\text{expected frequency}}$$

where the independent-trials frequencies are the expected frequencies, having been derived as the product of corresponding elements of the independent-trials matrix and the vector of observed-state frequencies (row sums of the tally matrix). The matrix of chi-square contributions is

	P	K	Q	M	Row sums
P	0	9.80	0.14	19.25	29.19
K	2.11	0	2.03	0.02	4.17
Q	0.47	1.06	0	0.45	1.98
M	10.33	8.95	0.07	0	19.35

Total chi square = 54.69

The null hypothesis (that observed data do not differ from the independent-trials model) is rejected in this case, because the total calculated chi square of 54.69 exceeds the theoretical value of 20.09 for eight degrees of freedom at the 0.01 significance level. The alternative hypothesis to be accepted provisionally (see below) is that the grain sequence does have the properties of embedded Markov chains.

Individual entries in the chi-square matrix can be useful as well because the total can be partitioned. By inspection, it seems clear that quartz contributes little to hypothesis rejection and that the important entries are for P/K, P/M, M/P, and M/K transitions. If one large value occurs, closure within each row results in at least one other large contribution in that row; thus, it is not possible to say whether the P/K or the P/M transition (or both) is actually most descriptive of the textural feature that composes the chain. Of two large values in a row, one will represent less than predicted frequencies, the other, more than predicted. On the average, closure causes two values in each row to derive from the former. This is exactly true over all data used here and, without further inquiry required for the present purpose, it is assumed to result from some form of the restraints of closure investigated by Chayes (1960), Krumbein (1962), and Chayes and Kruskal (1966), with probable relation here to the binomial distributions involved.

In any case, the rows on which one should concentrate can be selected by significance tests. Row sums of the observed chi-square matrix can be compared with the theoretical value for two degrees of freedom (three nonzero entries, minus one restriction), which has the value 9.21 at the 0.01 level. Thus, only the P and M rows significantly exceed the value expected under the independent-trials hypothesis. Doveton (1971) tests each entry for significance, individually, with one degree of freedom, through use of Yates's correction for continuity.

For the Cornelia stock, 50 of the 60 samples tested are assumed to have the Markov property on the basis of this test, subject to further tests to establish agreement with a specific order of Markov process. Those which agree with the independent-trials model are indicated in Figure 3, and comprise 2 of 28 quartz monzonite specimens, none of the porphyritic or micro-quartz monzonites, 1 of 2 quartz diorite samples, and 7 of 18 granodiorites.

As the frequency of transition sequences with possible Markov character decreases in the more mafic rock types, one might question what the results would have been if mafic mineral components had not been considered in the design. All matrices were recalculated for the three-state system, and only 31 specimens were then provisionally accepted as embedded Markov chains (those having probabilities in agreement with independent-trials expectations comprised 12 quartz monzonites, 1 micro-quartz monzonite, 5 porphyritic quartz monzonites, 1 quartz diorite, and 9 granodiorites). Considering the mafic phases has obviously had the important effect, particularly in those units for which color index is relatively low, of producing a memory in sequences that otherwise do not demonstrate it. This might not be expected, since many workers have shown that mafic phases in granites have little effect on crystallization sequence of the other major phases (Tuttle and Bowen, 1958; von Platen, 1965). It is important to be sure that inclusion of them in the design has not introduced some deleterious effect on the analysis. It is contended below that inclusion of this state in the design is of value because it permits much more discrimination among textural varieties than is otherwise possible. The effect, in fact, is to allow recognition of subdivisions within major textural groups (Fig. 3). The major categories, as shown in Figure 4 for the four-state system, are much the same when defined by either treatment. Whatever their influence, these minerals are abundant and integral parts of the textural framework of many samples studied, and it seems imprudent to ignore them.

All data sets were next tested for homogeneity (Vistelius, 1966a) because it is necessary to demonstrate that the grain sequence is stationary along its length if Markov properties are to be inferred. For this purpose, the data record was divided into two equal submatrices and tested for agreement with the total tally matrix by the chi-square test. Six samples rejected on this basis are indicated in Figure 3.

Further tests to establish Markov character involve determination of the specific order of the process—that is, determination of the number of steps prior to the ith event at which significant control on occurrence of the ith event is exercised. In a two-step process, for instance, the ith event is dependent upon event $(i - 2)$. The test for agreement with a first-order Markov chain is described clearly by Doveton (1971), and his stratigraphic data follow the embedded Markov chain structure. Selected samples from the study show that second- and third-order Markov processes commonly are involved. This complicates formal analysis considerably, and further investigation of higher order Markov properties in the data is continuing.

For present purposes, it was decided to proceed informally with the one-step transition probabilities and to look for similarities among samples that might permit their assignment to nominal classes. The procedure described below has been employed previously for data assumed or shown to have first-order Markov properties (Gingerich, 1969; Doveton, 1971), but its use here is not intended to imply an underlying Markov-1 scheme. It does imply, however, that the one-step transition probabilities observed can be used in comparing and discriminating among samples, even if dependence of higher order exists. This is a reasonable assumption because this matrix is descriptive of the contact frequencies measured directly from the samples. The method is somewhat crude, but that is perhaps acceptable in the first application of the embedded Markov chain structure to textural data. The result is regional comparison among samples through assignment to nominal classes.

RECOGNITION OF GRAIN-SEQUENCE TYPES

Gingerich (1969) introduced use of the difference matrix, calculated by subtracting the independent-trials matrix from the transition-probability matrix. For sample 12, this difference matrix is

$$\begin{array}{c c} & \begin{array}{cccc} P & K & Q & M \end{array} \\ \begin{array}{c} P \\ K \\ Q \\ M \end{array} & \left[\begin{array}{rrrr} 0 & 0.114 & -0.001 & -0.104 \\ 0.042 & 0 & -0.038 & -0.003 \\ -0.023 & 0.040 & 0 & -0.017 \\ -0.123 & 0.134 & -0.009 & 0 \end{array} \right] \end{array}$$

This matrix simplifies recognition of those transitions (with positive value) that have a higher probability of occurrence than if drawn from a random sequence. Of greater importance, departures from probabilities expected in a random sequence are put on an equal basis among all states through use of the difference matrix, regardless of differences in abundance of the mineral states employed (Read, 1969).

In Gingerich's application, positive entries were followed through the matrix to delineate what he termed the "fully-developed cycle" in the stratigraphic section studied. Read (1969) has termed such sequences the "preferred path" and Doveton (1971) calls them the "optimum transition scheme." By similar practice here, the fully developed grain sequence (the sequence of most likely repetitive occurrence among all four states) would be

$$M \rightarrow K \rightleftharpoons P$$

$$Q \rightarrow K \rightleftharpoons P$$

In the analytical method adopted here, it was decided that all four states of the system should appear in any sequences defined in this manner, and so the sequences were arranged to begin with those states to which no return could be made from any other state. Positive difference values less than 0.01 are ignored, but no other use is made of the relative magnitude of difference values. The sequences of sample 12 can be represented pictorially as in Figure 2B, where this sample is used to illustrate one type of sequence among the six that were defined by treating all 50 samples (those having nonrandom grain sequences) in this manner. Definition of the classes and sample assignment to them require judgment and choice by the investigator because most samples yield sequences that differ slightly from those chosen to typify the class.

The sequence cited in Figure 2 as representative of each class can be regarded as a kind of mode that the specific sequences of individual samples cluster around. For a few of the 50 samples, assignment to one or the other of two classes was difficult; for most, the assignment was fairly obvious once the basic classes had been recognized. For example, class B comprises three samples (numbers 41, 65, and 89) that have sequences identical to that of sample 12, and three variants exist:

Sample 8 $M \rightarrow K \rightleftharpoons P$ Sample 66 $M \rightleftharpoons K \rightleftharpoons P$ Sample 52 $M \rightarrow K \rightleftharpoons P$
           $Q \rightleftharpoons K \rightleftharpoons P$ $Q \rightarrow K \rightleftharpoons P$ $Q - K \rightleftharpoons P$
                                                                                                                                          $\uparrow\!\_\_\_ \leftarrow \_\_\!\lrcorner$

Variation within other classes can be greater than in class B. Dashed arrows (above) reflect the lesser of two positive difference values for transitions from a single state.

The six classes of Figure 2 are characterized by the following sequences:

A	B	C
$M \to Q \to K \rightleftharpoons P$	$M \to K \rightleftharpoons P$	$M \to K \rightleftharpoons P$
	$Q \to K \rightleftharpoons P$	$M \to Q \to P \rightleftharpoons K$

D	E	F
$M \to K \rightleftharpoons Q$	$M \rightleftharpoons P \to K \rightleftharpoons Q$	$M \to K \to Q \rightleftharpoons P$
$P \to K \rightleftharpoons Q$	$P \to M \to K \rightleftharpoons Q$	

All sequences but one begin in plagioclase or mafic minerals, which is just an expression of the fact that, in many systems, there is a low probability of entering these states. Mafic phases occur in a position other than first in only one of the sequences above. This accords with their observed occurrence in relatively large aggregates—which generally are not intergrown with other states—and confirms that crystallization of mafic phases in granites proceeds independently of other major components.

Each sequence representative of a class contains one pair of states for which positive difference values exist for transition in either direction. In each case, neither of these states has a second, positive difference value for transition into any other state. For simplicity of discussion, these portions of sequences will be termed intergrowths (the latter term refers to the probable existence of a repetitive occurrence of contacts between two states, regardless of how that phenomenon arose). Further similarity among samples is shown by the fact that the six classes can be combined into three groups of samples that display the same intergrowth type. Specifically, the three groups are based on the intergrowths K/P/K (classes A,B,C), K/Q/K (classes D,E), and Q/P/Q (class F).

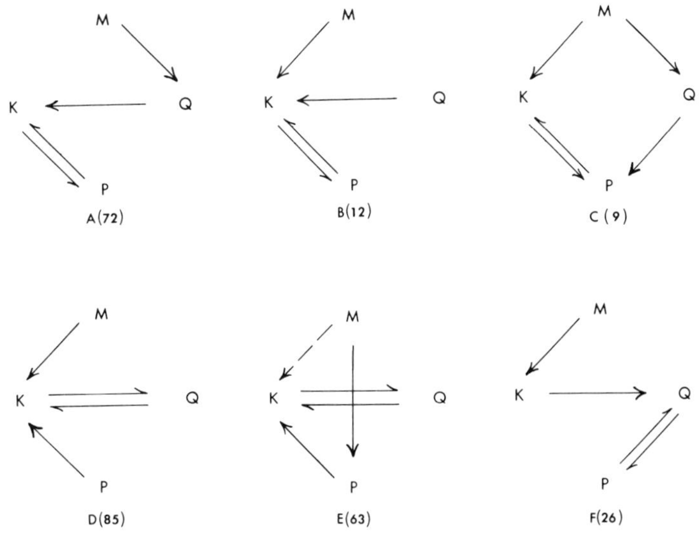

Figure 2. Schematic diagrams of grain sequences for classes A through F. Arrows denote transition directions that have positive values in the difference matrix. Dashed arrow is lesser of two positive values from same state. States M = mafic constituents, K = K-feldspar, P = plagioclase, and Q = quartz. Number of type sample for each class enclosed in parentheses.

WITHIN-PLUTON VARIATION IN GRAIN-SEQUENCE TYPE

Figure 3 presents the areal distribution of the grain-sequence classes A through F, and the three groups based on intergrowth type are shown in Figure 4. It is worth remarking that the textural classes are not distributed consistently over rock types. Class E, for instance, has representatives in each of the four lithic units mapped, and the two principal areas in which it is found are in granodiorite and porphyritic quartz monzonite. The latter pair of units share coarse grain size as the only other apparent similarity in modal composition or texture. This suggests that grain-sequence studies can provide nonredundant information. On the other hand, class A has eight of its nine-member samples distributed within the microquartz monzonite or within its gradational border zone. These borders are texturally defined, principally on the basis of grain size.

That these classes have possible geologic controls and significance is suggested by the relation of some of them to rock unit distribution, as above, and also by the clusters represented in Figure 3. There are three areas, each belonging to a different class, for which five proximate sample localities are identically classified. The inference is that similar grain sequences can exist over areas up

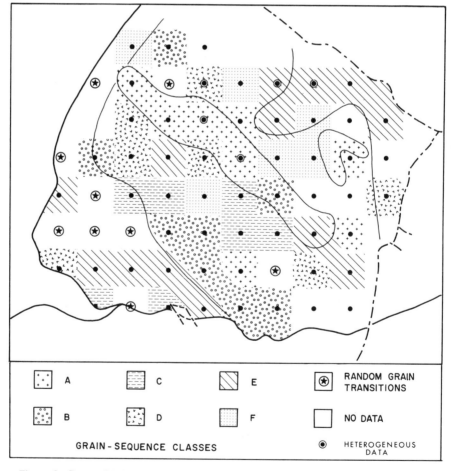

Figure 3. Geographic distribution of six grain-sequence classes in the Cornelia pluton, Arizona. Letters designating the classes refer to the grain sequences in Figure 2.

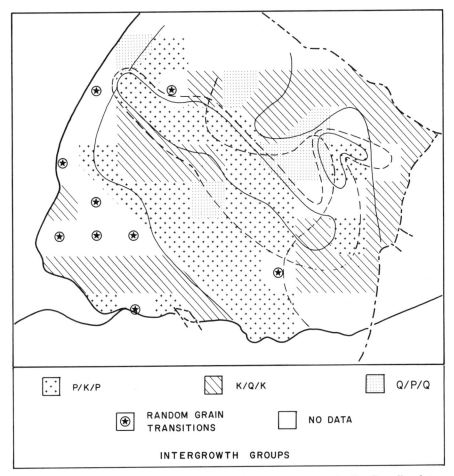

Figure 4. Geographic distribution of three grain-sequence groups in the Cornelia pluton, Arizona. Groups represent combinations of grain-sequence classes in Figure 3 on the basis of similar reverse transitions between final pair of states of sequence.

to approximately 0.5 mi² (1.3 km²). In other cases, a checkerboard pattern exists where areas of similar grain sequence may be no larger than 0.06 mi² (0.15 km²).

The three groups are more closely related to rock type (Fig. 4). K/P/K intergrowths typify the quartz monzonite and micro-quartz monzonite, and the porphyritic quartz monzonite predominantly displays the K/Q/K type. The largest area of Q/P/Q intergrowth is in an unusual and interesting location. It is situated within the quartz monzonite, in the re-entrant of the micro-quartz monzonite outcrop belt. The latter unit has been shown to be the probable conduit through which mineralizing solutions rose on their way to concentration in the Cornelia orebody (Wadsworth, 1968). Before block faulting occurred, the Q/P/Q intergrowth area was both roofed and partly floored by the conduit and was situated within its lateral closure.

PETROGRAPHIC INVESTIGATION OF GRAIN-SEQUENCE TYPES

Analysis of grain sequence will become useful as more than a didactic device only when it can provide, through numerical means, a clear picture of those parts of the textural framework that it describes. An attempt is made to begin this process by searching for textural evidence of the origin of grain-sequence classes previously defined.

In Figure 5, extensive K-feldspar replacement of plagioclase is inferred. Thin plagioclase seams between K-feldspar grains have poorly defined twinning or patchy extinction, but they most often appear to extinguish in zones elongate parallel to the seam length. The largest plagioclase grain appears to have been penetrated parallel to composition plane traces. Alternate interpretations of these relations are possible, but there is preponderant evidence of plagioclase replacement by K-feldspar throughout most samples of the stock. Other possible sources of the K/P/K intergrowth sequence representative of this class are present elsewhere in this sample and are displayed in Figures 6 and 7 for other specimens. The upper half of Figure 5 shows contacts between subhedral K-feldspar and anhedral quartz that are typical of much of the slide area and that may produce the Q/K transition of the class A sequence. No evidence of a source for excess M/Q transitions was recognized, and biotite surrounded and replaced by K-feldspar seems to be the most common relation. This is the only case, among all type sequences, in which no possible cause for a given transition could be inferred from textures observed.

Figure 6 represents another source of excess K/P/K intergrowths. Here, K-feldspar occurs as thin mantles around plagioclase, with quartz interstitial to these. Thus, beginning traverses in interstices would most often produce the sequence Q/K/P/K, which characterizes class B. Biotite and hornblende are mantled in a fashion similar to plagioclase and are somewhat replaced by K-feldspar. Being smaller than plagioclase, the dark mineral grains tend to occur in the interstices also. The sequence M/K/P/K may be accounted for by such assemblages. The K-feldspar mantles seem mostly to be discrete crystals elongate parallel to plagioclase and could be an effect of primary crystallization accounted for by epitaxy between the two feldspar phases. In many other samples, however, these mantles are anhedral and merge with the plagioclase gradationally, suggesting possible K-feldspar replacement of albite rims in a topotactic relation.

Figure 7 may account for the Q/P transition excess, which is the principal difference of class C from A and B. All the quartz in the area shown, except for that at the extreme margins of the field, is a single poikilitic crystal. Small plagioclase grains and fragments scattered throughout represent a minor development of the texture which dominates the grain sequences in class F.

Intergrowths of K-feldspar and quartz characterize classes D and E. In Figure 8, illustrative of class D, part of an interstitial area is shown to the upper right, ringed by large K-feldspar and plagioclase megacrysts. Interstices are predominantly K/Q intergrowths and, in these grain-sequence classes, either quartz or K-feldspar appears to display subhedral crystal form at their mutual contacts. Thus, this intergrowth appears to be more of a primary crystallization effect than are the K/P/K intergrowths in classes in which they dominated the sequence. Mantling of biotite, hornblende, and magnetite by K-feldspar is well displayed, and replacement of biotite along cleavage planes is extensive in some cases. These features presumably give rise to the excess M/K transitions. There may be no need to explain the cases of transitions that are not in excess of the frequency predicted for independent trials. In this sample, however, M/P transitions are not part of the sequence defined, and the dark mineral phases are significantly finer grained than the coarse feldspar bodies that dominate the fabric. Thus, mafic phases occur interstitially with quartz and fine-grained K-feldspar. Small plagioclase crystals are of relatively rare occurrence and have little opportunity to occur in contact with the dark minerals.

Figure 5. Sketch from thin section of sample 72 (grain-sequence class A). Field diameter = 1.4 mm. Stippled = K-feldspar, q = quartz, pl = plagioclase. Plagioclase replaced and penetrated along twin composition planes by K-feldspar. Subhedral K-feldspar and anhedral quartz relations in upper right.

Figure 6. Sketch from thin section of sample 12 (grain-sequence class B). Field diameter = 1.4 mm. Stippled (k) = K-feldspar, q = quartz, pl = plagioclase, b = biotite. Subhedral K-feldspar rims around plagioclase may be oriented by epitaxy. Quartz interstitial and separated from plagioclase by K-feldspar rims.

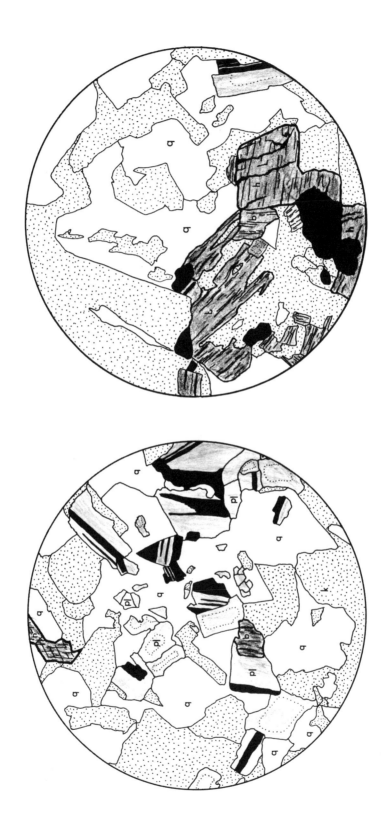

Figure 7. Sketch from thin section of sample 9 (grain-sequence class C). Field diameter = 1.4 mm. Stippled = K-feldspar, q = quartz, b = biotite, h = hornblende. Plagioclase poikilitically enclosed in single quartz grain in center of field.

Figure 8. Sketch from thin section of sample 85 (grain-sequence class D). Field diameter = 1.4 mm. Stippled = K-feldspar, q = quartz, b = biotite, h = hornblende. K-feldspar and quartz each may develop crystal form at mutual contacts. Mafic minerals enveloped by K-feldspar, which replaces biotite parallel to cleavage planes.

This could imply that there was nucleation of plagioclase before the mafic phases, or that crystallization was terminated on the K-feldspar–quartz cotectic, or both.

Class E differs from D principally in the excess incidence of M/P transitions. To show the probable source of these most easily, sample 39 is pictured in Figure 9 (rather than sample 63, which previously was cited as representative of the class). The sample is from the granodiorite unit and has much more abundant dark mineral components than sample 63. Figure 9A shows the typically small interstices of much larger plagioclase laths, in which K-feldspar and quartz are commonly intergrown with crudely "graphic" texture, possibly resulting from concomitant crystallization within a residual liquid fraction. Also, small hornblende grains and other dark minerals occur in these areas, and K-feldspar commonly envelops them. An attempt is made in Figure 9B to show a characteristically complex aggregate of dark minerals and fine-grained plagioclase. Plagioclase outside these aggregates is much coarser grained, suggesting that these aggregates formed early in growth. These relations are believed to account for the K/Q/K intergrowths and excess M/K and M/P transitions involved in the grain sequence of class E.

Figure 10 shows the poikilitic enclosure of plagioclase by quartz which gives the class F sequence its distinctive character. Because this is only a small part of one quartz pool, it is clear that the number of plagioclase inclusions can be very large. Their numbers can approach 100 within a single pool. There is no doubt in this case of the characteristic textural feature that defines the class. However, its geologic origin is uncertain. Immediately north and west of sample point 26, areas with large dioritic xenoliths are exposed locally. These occur in linear zones tens of meters long, with individual blocks up to several meters in diameter. Nowhere else in the stock have xenoliths been observed to exceed about 20 cm. One large xenolith from the area cited above has been sectioned and shows a boxwork of extremely long and slender actinolite needles, rigidly oriented in two directions that cross at a high angle. The rock is otherwise of altered, fine-grained diorite character. This structure is interpreted to represent shearing and reconstitution, perhaps in the final stages of consolidation of its magma host. On the basis of this observation and the similarity of plagioclase in size and shape in the quartz diorite unit and in quartz pools, it has been suggested that disaggregation of xenoliths

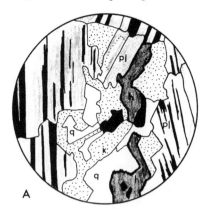 

Figure 9. Sketches from thin section of sample 39 (grain-sequence class E). Field diameter = 4 mm. Stippled = K-feldspar, q = quartz, pl = plagioclase, h = hornblende, b = biotite, c = chlorite, p = pyroxene. A. Interstitial area among large plagioclase laths. K-feldspar and quartz in crudely graphic relation. B. Mafic aggregate with small plagioclase included; larger plagioclase envelops aggregate.

276   W. B. WADSWORTH

is the source of this unusual texture (Wadsworth, 1968). But, at that time before the grain-sequence classes had been defined, this texture was used to help define the outer gradational limit of the porphyritic quartz monzonite unit (Fig. 1). Thus, the texture is not as areally limited as class F and can be recognized much more widely on a qualitative basis.

Samples within and near the porphyritic quartz monzonite outcrop area also show occasional development of chessboard twinning in plagioclase (Starkey, 1959). Opinion is not unanimous, but the favored view has been that this results from albitization of either a more calcic plagioclase or K-feldspar. Such alterations could aid in the disaggregation of primary plagioclase crystals. Lastly, these samples of class F contain features interpreted as large-scale replacement features of both feldspar species by quartz. Quartz grains and pools are large and characteristically round. Where several large quartz grains compose a pool, they often have thin K-feldspar seams discontinuously around their mutual borders, with thicker areas at irregular triple junctions. However, some optically countinuous quartz grains preserve identically curved K-feldspar seams entirely within them. Such pools

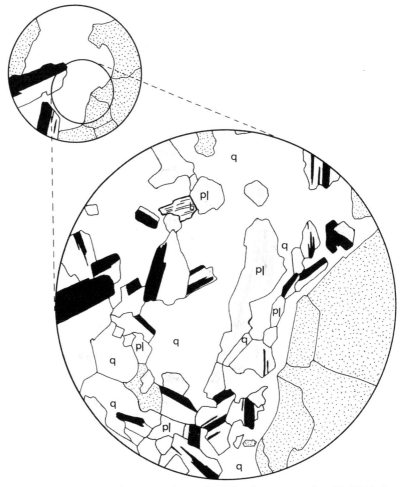

Figure 10. Sketch from thin section of sample 26 (grain-sequence class F). Field diameter = 1.4 mm. Stippled = K-feldspar, q = quartz, pl = plagioclase, b = biotite. Part of a large quartz crystal that poikilitically encloses swarms of small plagioclase grains.

contain few plagioclase inclusions. Replacement and corrosion of feldspar margins by tiny quartz blebs is well displayed in samples 71, 69, 6, and 20; the texture of these samples was used to correlate their rock unit with the mineralized area of the mine. This texture involves much larger grains and more extensive possible replacement, but it may well be a related feature.

CONCLUSIONS AND PETROLOGIC INFERENCES

Combining the petrographic features recognized in Figures 5 through 10 with the areal variation patterns of Figures 3 and 4 has shown that the Cornelia pluton provides an excellent and detailed match to the model of Jahns and Burnham (1969) for pegmatite derivation. This conclusion had not been reached, or even closely approached, before these results on grain sequences were analyzed by the present method.

Pluton geometry during crystallization conformed to a simple and highly significant pattern, but its recognition is hindered by the effects of postmagmatic block faulting. Gilluly (1942, 1946) first recognized the structural history used in the reconstruction described below. By analogy with the southern pluton boundary, which has a very steep dip to the north, all rock-unit contacts within the pluton are assumed to have near-vertical attitudes. Thus, the map view is also a suitable cross section for features such as the micro-quartz monzonite (the orebody conduit). By appropriate manipulation, Figure 1 can be reoriented to show the geologic features in their prefaulting attitudes, and Figure 11 has been prepared to aid this procedure. Restoration requires that the map (Fig. 11) be tilted toward the observer, through an angle of about 45°, using the Little Ajo Mountain fault trace at the bottom of the map as the axis of rotation. As viewed from the northeast, this is the original orientation of the map plane within the stock. The conduit can now be understood to have an upward-arching form, as do several other major rock-unit boundaries. These boundaries dip perpendicular to the plane of section and, considered as arches, plunge at 45° or so to the south or southwest.

The outermost unit boundaries within the map area are expressions of a sloping roof, and the porphyritic quartz monzonite occurs at the floor of the chamber (as far as that can be determined after fault truncation). In this original orientation, the conduit would have extended above and behind the observer for 1,000 to 3,000 m to the stock's inferred apex. It would later be truncated and dropped down to the left (still viewing from the northeast) on the Gibson fault, whose long trace marks the east edge of the section. That apex is exposed for study in the New Cornelia open pit mine 2 mi (3.2 km) to the east.

The extensive area of K/P/K intergrowth illustrated in Figure 4 encompasses most of the micro-quartz monzonite unit and extends entirely through the quartz monzonite to its border with the granodiorite and older envelope. Textures suggest K-feldspar replacement of plagioclase as the dominant cause, and the areal distribution pattern, by revealing the extent of this effect, suggests that the potassium source might have existed in the conduit. This is reasonable because the influence of gravity (buoyancy) could then explain the distribution of potassium throughout the quartz monzonite located higher in the chamber. The micro-quartz monzonite or conduit area seems to comprise an anastomosing set of leucocratic dikes emplaced in a quartz monzonite that displays a continuous grain-size gradient from medium-grained quartz monzonite into the fine-grained core of the conduit. In the core, one is seldom certain whether an outcrop is a distinct dike or not. The samples with most extreme corrosion textures are found here, as described above. Some rocks here are porphyritic (probably porphyroblastic in the case of K-feldspar) and some are not, but the main portion in all is a largely xenomorphic-granular and equivolume aggregate of plagioclase, quartz, and K-feldspar. In essence, these are microgranites or coarse-grained aplites, with modal compositions plotting nearer the ternary minimum than any other rock units of the stock. However, if outside additions of K-feldspar could be deleted, the latter attribute might be changed.

Thus, this unit can be well explained as the rapidly quenched residue of crystals and silicate fluid after separation of an aqueous fluid phase (Jahns and Tuttle, 1963; Jahns and Burnham, 1969). Such an aqueous phase is often potassium bearing, according to these writers, and can account, through upward migration, for the K/P/K intergrowth distribution and the corrosion textures locally developed in the micro-quartz monzonite unit itself.

Concentration of volatiles in this zone of the stock is easily accounted for by

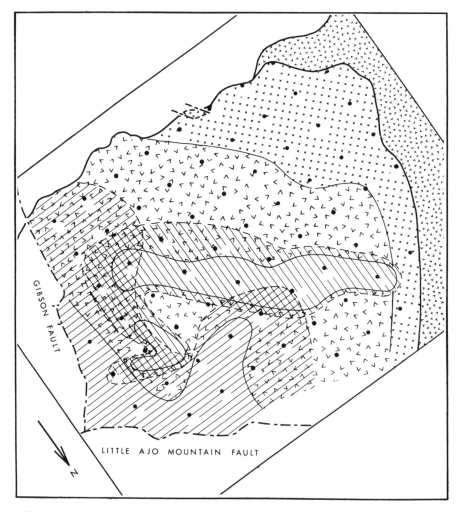

Figure 11. Geologic map of Cornelia pluton. Same as Figure 1, but reoriented to aid restoration of prefaulting attitudes. Rotation of this map plane toward observer through an angle of about 45°, using the Little Ajo Mountain fault trace as the axis of rotation, restores map section to its inferred position at time of crystallization. View then is from northeast.

crystallization inward from roof and walls of the stock—an inference strongly suggested by the shapes of contacts and relative age relations among older units. Outer units nearer the roof have a progressively more anhydrous and higher temperature mineral assemblage. Similar gradients exist within the quartz monzonite unit as its southern contact with envelope rocks is approached. No chamber floor is evident today, but volatiles streaming upward through magma and those being excluded during consolidation nearer the roof would have concentrated under the latter, even if not driven upward by crystallization from below.

Rapid quenching of the volatile-charged magma could have occurred either before or after attainment of saturation with H_2O, dependent only on rapid release of confining pressure on the magma system. Rapid evolution of volatiles would have left a fine-grained and volatile-depleted residue. From Gilluly's (1946) work in the mine area, the proposed sequence of postmagmatic events there consisted of (1) development of corrosion textures, (2) formation of north-northwest-trending fissures, (3) orthoclase-rich pegmatite emplacement parallel to these fissures and more widespread impregnations by quartz and orthoclase, and (4) pervasive shattering and introduction of metals.

At this deeper level, it seems probable that steps (1) and (3) are related in being different manifestations of the complicated processes that occurred on appearance of a second and mobile fluid phase. The pegmatites are inferred to have originated, at least for part of their volume, at this deeper level. Typical small dikes of aplite are present locally throughout all three quartz monzonite units, and they could represent these same fluids in cases where chilled within narrow fractures. Their composition would be of interest to establish affinity with either the aqueous or silicate phase, but no data are available. On migration to the apex, the aqueous fluid produced metasomatic effects throughout its conduit. We cannot know whether fracturing in the apex, prior to pegmatite emplacement, was caused by processes in the deeper magma, but surely that episode is related to formation of the quench residue.

The K/Q/K intergrowths appear to be related to primary magmatic crystallization and are concentrated in the porphyritic quartz monzonite beneath and along the eastern edge of the conduit. This unit, though coarse-grained and coarsely porphyritic, has none of the local segregations or similar features that would lead to its designation as pegmatitic granite. Rather, it may be another textural variant of the residual magma after aqueous phase separation. Such an origin would yield largely orthomagmatic textures, but (as shown by Jahns and Burnham, 1969) the residual melt often is relatively enriched in sodium and silica by the escape of an aqueous phase rich in potassium. Such melts can be quite active chemically. It is possible that this lower part of the chamber would have been somewhat depleted in potassium by upward diffusion of the latter through silicate fluid, even if saturation and subsequent quenching had not occurred above it. In any case, this unit contains several evidences of sodium and silica activity. Examples are greater albite rim development; large grains with chessboard twinning that suggests albitization of either K-feldspar or plagioclase; large, circular pools and individual crystals of quartz that appear to have replaced K-feldspar extensively; and the quartz of similar shape charged with small plagioclase inclusions.

The latter feature has perplexed me for some time, but a reasonable explanation may be found in acceptance of the models of Jahns and Burnham (1969). The largest area of Q/P/Q intergrowth in Figure 4 occurs within the crestal closure of the conduit and just beyond the upper limit of characteristic porphyritic quartz monzonite textures (its outer gradational limit in Fig. 1). Perhaps the rapid quenching of the micro-quartz monzonite formed a barrier effective in trapping fluids diffusing through the rest magma nearer the floor. Earlier, rapid diffusion of potassium from this lower magma might have caused complementary crystallization of more fine-grained sodic minerals (Jahns and Burnham, 1969, p. 858), and in the presence of a diffusing aqueous phase, these could have been incorporated in large crystals precipitated from that fluid. With the quenched unit acting as a barrier to further migration, such textures would be best developed in the trap, where the silicate and aqueous phases would come in closer contact. This texture is present throughout the gradational border zone of the porphyritic quartz monzonite (as recognized qualitatively). It is interesting to note that this zone fans out from the eastern edge of the stock; this is a route for diffusing fluids from depth that could have bypassed the proposed barrier presented by the quenched conduit. At some stage, sodium-bearing and silica-bearing fluids became sufficiently active to diffuse along a major joint set of the pluton, which would have had a northwest trend and vertical dip at that time, promoting a color-bleaching effect through albitization and silicification. More pervasive alteration has occurred over larger areas along east, north, and northwest borders of the stock, and various processes such as epidotization and chloritization were locally effective throughout the pluton.

This explanation of features in the porphyritic quartz monzonite and its border zone reverses the relative age relation with the micro-quartz monzonite previously inferred from little evidence except grain size and map pattern (Wadsworth, 1968). The new relation is supported by recognition of a probable xenolith of microgranite in the porphyritic quartz monzonite during recent study of thin sections. Previously, there had been no acceptable explanation for the differences among the various units of the Cornelia pluton except as a series of separate intrusions differentiated at depth. In the model of Jahns and Burnham (1969), each unit finds an explanation that agrees fully with previously observed structural and modal properties and with textural variation in grain sequence discerned in this study. Rather than resorting to controlling processes operating at deeper levels, it is probable that the depth of observation is sufficient here to observe the site of differentiation processes of the late magmatic and postmagmatic stages; some of the products remained, some diffused widely, and some migrated to concentrate in the apex. Within this framework, grain-sequence classes appear to discriminate among rest magma of orthomagmatic textures, rocks altered by potassium metasomatism due to impregnation by aqueous fluids, and rocks that represent hybrid effects of rest magma and aqueous agents rich in sodium and silica.

Lengthy attention has been given to comparison of the Cornelia stock with the model for pegmatite derivation simply because the fit is so good that inferences constructed from it must be at least generally true. Areal distribution of grain sequences provided the necessary information, and posed the required questions, to lead to recognition of this fundamental petrologic conclusion about the Cornelia stock.

ACKNOWLEDGMENTS

In 1971 and 1972, I was a Science Faculty Fellow of the National Science Foundation at Pomona College. This study was much advanced through that experience and I wish to express my gratitude to both institutions for their support. I particularly would like to acknowledge the efforts of all the staff of the Geology Department at Pomona College for making my tenure there both enjoyable and rewarding.

REFERENCES CITED

Chayes, F., 1956, Petrographic modal analysis—An elementary statistical appraisal: New York, John Wiley & Sons, Inc., 113 p.

——1960, On correlation between variables of constant sum: Jour. Geophys. Research, v. 65, p. 4185-4193.

Chayes, F., and Kruskal, W., 1966, An approximate statistical test for correlations between proportions: Jour. Geology, v. 74, p. 692-702.

DeVore, G. W., 1959, Role of minimum interfacial free energy in determining the macroscopic features of mineral assemblages: I. The model: Jour. Geology, v. 57, p. 211-226.

Doveton, J. H., 1971, An application of Markov chain analysis to the Ayrshire Coal Measures succession: Scottish Jour. Geology, v. 7, p. 11-27.

Ehrlich, R., Vogel, T. A., Weinberg, B., Kamilli, D. C., Byerly, G., and Richter, H., 1972, Textural variation in petrogenetic analyses: Geol. Soc. America Bull., v. 83, p. 665-676.

Flinn, D., 1969, Grain contacts in crystalline rocks: Lithos, v. 3, p. 361-370.

Fournier, R. O., 1967, The porphyry copper deposit exposed in the Liberty open-pit mine near Ely, Nevada. Part 1. Syngenetic formation: Econ. Geology, v. 62, p. 57-81.

Gilluly, J., 1942, The mineralization of the Ajo copper district, Arizona: Econ. Geology, v. 37, p. 247-309.

——1946, The Ajo mining district, Arizona: U.S. Geol. Survey Prof. Paper 209, 112 p.

Gingerich, P. D., 1969, Markov analysis of cyclic alluvial sediments: Jour. Sed. Petrology, v. 39, p. 330-332.

Ivanov, D. N., 1970, Analysis of sequences of mineral grains in granites of the Kyzyltas massif (central Kazakhstan) as a manifestation of the Markov process, in Romanova, M. A., and Sarmanov, O. V., eds., Topics in mathematical geology: New York, Consultants Bureau, 281 p.

Jahns, R. H., and Burnham, C. W., 1969, Experimental studies of pegmatite genesis: I. A model for the derivation and crystallization of granitic pegmatites: Econ. Geology, v. 64, p. 843-864.

Jahns, R. H., and Tuttle, O. F., 1963, Layered pegmatite-aplite intrusives: Mineralog. Soc. America Spec. Paper 1, p. 78-92.

Kretz, R., 1969, On the spatial distribution of crystals in rocks: Lithos, v. 2, p. 39-66.

Krumbein, W. C., 1962, Open and closed number systems in stratigraphic mapping: Am. Assoc. Petroleum Geologists Bull., v. 46, p. 2229-2245.

——1967, FORTRAN IV computer programs for Markov chain experiments in geology: Kansas Geol. Survey Computer Contr. 13, 38 p.

——1968a, Computer simulation of transgressive and regressive deposits with a discrete-state, continuous-time Markov model, in Merriam, D. F., ed., Computer applications in the earth sciences: Colloquium on simulation: Kansas Geol. Survey Computer Contr. 22, p. 11-18.

——1968b, FORTRAN IV computer program for simulation of transgression and regression with continuous-time Markov models: Kansas Geol. Survey Computer Contr. 26, 38 p.

Krumbein, W. C., and Dacey, M. F., 1969, Markov chains and embedded Markov chains in geology: Internat. Assoc. Mathematical Geology Jour., v. 1, p. 79-96.

Lowell, J. D., and Guilbert, J. M., 1970, Lateral and vertical alteration-mineralization zoning in porphyry ore deposits: Econ. Geology, v. 65, p. 373-408.

Moore, W. J., 1973, Igneous rocks in the Bingham mining district, Utah: U.S. Geol. Survey Prof. Paper 629-B, p. B1-B42.

Nielsen, R. L., 1968, Hypogene texture and mineral zoning in a copper-bearing granodiorite porphyry stock, Santa Rita, New Mexico: Econ. Geology, v. 63, p. 37-50.

Read, W. A., 1969, Analysis and simulation of Namurian sediments in central Scotland using a Markov-process model: Internat. Assoc. Mathematical Geology Jour., v. 1, p. 199-219.

Spry, A., 1969, Metamorphic textures: Oxford, Pergamon Press, 336 p.

Starkey, J., 1959, Chess-board albite from New Brunswick, Canada: Geol. Mag., v. 96, p. 141-145.

Tuttle, O. F., and Bowen, N. L., 1958, Origin of granite in the light of experimental studies in the system $NaAlSi_3O_8$-$KAlSi_3O_8$-SiO_2-H_2O: Geol. Soc. America Mem. 74, 153 p.

Vistelius, A. B., 1966a, Genesis of the Mt. Belaya granodiorite, Kamchatka (an experiment in stochastic modeling) (in Russian): Akad. Nauk SSSR Doklady, v. 167, p. 1115-1118.

——1966b, A stochastic model for the crystallization of alaskite and its corresponding transition probabilities (in Russian): Akad. Nauk SSSR Doklady, v. 170, p. 653-656.

——1967, Crystallization of alaskite from the Karakul' dzhur River (Central Tien Shan) (in Russian): Akad. Nauk SSSR Doklady, v. 172, p. 165-167.

——1971, Grain sequence of the main rock-forming minerals in the Carnmenellis Granite, Cornwall: Geocom Bull., v. 4, p. 145-149.

——1972, Ideal granite and its properties. I. The stochastic model: Internat. Assoc. Mathematical Geology Jour., v. 4, p. 89-102.

Vistelius, A. B., and Faas, A. V., 1971, The probability properties of sequences of grains of quartz, potassium feldspar, and plagioclase in magma granites (in Russian): Akad. Nauk SSSR Doklady, v. 198, p. 925-928.

von Platen, H., 1965, Experimental anatexis and genesis of migmatites, *in* Pitcher, W. S., and Flynn, G. W., eds., Controls of metamorphism: New York, John Wiley & Sons, Inc., p. 203-218.

Wadsworth, W. B., 1968, The Cornelia pluton, Ajo, Arizona: Econ. Geology, v. 63, p. 101-115.

——1969, Further data on Chayes' identity change number, a bulk coarseness estimator in granitic rocks: Geol. Soc. America, Abs. With Programs for 1969, Pt. 5 (Rocky Mountain Sec.), p. 85-86.

Whitten, E. H. T., 1963, Application of quantitative methods in the geochemical study of granitic massifs, *in* Shaw, D. M., ed., Studies in analytical geochemistry: Royal Soc. Canada Spec. Pub. 6, p. 76-123.

Manuscript Received by the Society September 14, 1973
Revised Manuscript Received January 24, 1974

Appropriate Units for Expressing Chemical Composition of Igneous Rocks

E. H. TIMOTHY WHITTEN

Department of Geological Sciences
Northwestern University
Evanston, Illinois 60201

ABSTRACT

Chemical analyses should be expressed as cation or oxide weights per unit volume (instead of the traditional weights percent) for almost all chemical, petrogenetic, and economic studies of rocks. For many purposes, the use of oxide or cation weights percent may yield misleading information about the interrelations between chemical variables of igneous rocks. Apparent trends among cogenetic igneous rocks differ, or are even reversed, when expressed as weights per unit volume and as weights percent. Specific illustrations are based on published data for the Palisades and Lugar Sills and for numerous other granitic and mafic plutonic igneous masses for which density data have been published.

INTRODUCTION

Current methods of describing rock compositions reflect a basic dichotomy in the use of measurement units. Modes are now commonly expressed as volumes per unit volume (that is, volumes percent), although weight-percent modes are still occasionally used; for the same samples, chemical analyses are, in most cases, recorded as oxide (or cation) weights per unit weight (that is, weights percent). The use of traditional units of measurement in petrology and petrography has led to numerous inconsistencies, and as a result, it would be useful to review many of the variables and diagrams that are extensively used by petrologists and petrographers. Although no specific reasons or explanations were given, Whitten (1972) suggested that, for most purposes, the oxide composition of igneous rocks should be expressed in grams per 100 cc, rather than as weight percentages. In this paper, special attention is given to the justification, need, and consequences of using cation weight or oxide weight per unit volume for rock analyses, instead of the standard and traditional oxide weight percentages.

MODES

Personal discussion with many geologists has shown that considerable uncertainty exists about the units that should be used by petrologists to express modes.

Cross and others (1903, p. 147) introduced the term "mode" for the actual mineral composition of a rock without expressly stating whether weight or volume percent should be used. It seems implicit, though not explicit, that weight percent was intended. They (p. 224) demonstrated how to calculate the mode from a chemical analysis; in addition, they (p. 226) tabulated observed volume percent data (based on micrometric counts) and used density values to calculate and tabulate equivalent weight percentages, without referring to either set of numbers as a mode. Although many experienced petrologists still claim to use weight percentages for modes, following the usage of Cross and others (1903), actual practice has varied considerably. In textbooks, Shand (1947) and Barth (1962), for example, cited proportions of minerals in rocks without mentioning the units. Shand (1947, p. 235) defined "color index" as the percentage by volume of minerals with density >2.8, commenting "(by volume, since it is easier to judge volume than weight)"; his book quoted many actual mineral percentages, but only rarely did he state whether a volume or a weight basis applied (for example, on p. 386, the weight-percent mode is cited for an andesite).

Over the past two decades, the increasing popularity of micrometric point counting has caused much more abundant modal data to be published, and they are almost always in volume percent. Recent texts reflect this trend; for example, Bayly (1968, p. 59) merely stated that modes are the percentages by volume of the actual minerals present, and Chayes (1956) dealt almost exclusively with volume percentages. In most recent papers, such as Exley (1959), Ragland and Butler (1972), Walker (1940), and White (1973), volume percent was used for modes, although for mafic igneous rocks (for which norms are more commonly used than for other rocks), weight-percent modes are still published. Occasionally, recent authors have used both units. Walker (1969, Table 5), for example, tabulated micrometric analyses for the Palisades Sill, New York, in both volume percent and in weight percent. Assessing the modes of accessory minerals presents special problems. For the San Isabel Granite in Colorado, Whitten and Boyer (1964) used weight percent for minerals separated from crushed rock. However, the widespread popularity of point counting for analyzing both thin sections and sawn hand specimens has led to the almost exclusive use of volume-percent modes in modern petrology. I contend that, for most purposes, this is the appropriate unit for modes.

CHEMICAL DATA

For the Stillwater Complex in Montana, the volumes of unit-weight samples of chromite-rich rock are significantly smaller than those of the least mafic rocks. To evaluate the economic prospects of mining and extracting Cr_2O_3, the costs involved depend on the weight of Cr_2O_3 per unit weight of rock; that is, on the Cr_2O_3 weight percentage in the ore. For petrographic and petrogenetic analyses, however, it would rarely be significant to compare the weight of FeO, for example, in samples of markedly dissimilar volume collected from successive layers of a differentiated mass. Nevertheless, this is done when sample analyses are expressed in weights percent. By contrast, use of grams of Fe or of FeO per 100 cc permits realistic assessment of the chemical variations and gradients within such a rock suite; this is because such units reflect the amount of each ion or oxide within a given volume of rock.

Chemical analyses of rocks have traditionally been expressed in weights percent or, occasionally, cations percent. When whole-rock density has been determined, the product of weight percent and sample specific gravity yields weight per unit volume. In a very small number of cases, authors have compared analyses on an equal-volume basis; most examples come from economic geology (for example, Boyle, 1961) or from studies in which possible equal-volume metasomatic changes were of interest. Wager (1929), Exley (1959), and Babcock (1973) provided examples of the problems that arise in the latter case. Throughout most of igneous petrology, however, the effect of rock density on petrological calculations and conclusions appears to have been ignored or overlooked. Tyrrell (1948, 1952) afforded a notable exception in his study of the differentiated Lugar Sill, Scotland. He weighted individual chemical analyses by sample density before calculating the bulk composition of the sill for comparison with the teschenitic chilled sill margin. For most petrological purposes, cation weight or oxide weight per 100 cc involves convenient units (Table 1). However, the density of chemically analyzed rock samples has been recorded in the literature very infrequently. It is therefore impossible to compute weights per unit volume for most published analyses. Several recent publications of the United States and Canadian Geological Surveys are significant exceptions in that they have included the specific gravities of many analyzed rocks.

Trace-element analyses are commonly expressed as parts by weight per million so that comparisons of rock compositions involve comparable problems to those with major-element analyses. In many purely geochemical studies, difficulties are avoided by using molarity (moles of solute per 1,000 ml of solution) instead of molality (moles per 1,000 g). For most petrological purposes, an appropriate unit for trace elements would be milligrams per 1,000 cc (representing the product of the standard parts per million and density of the analyzed rock).

For a variety of objectives, chemical analyses of rocks are plotted on two-

TABLE 1. REPRESENTATIVE CHEMICAL ANALYSES AND VARIANCES

Variable	Chemical analyses						Variances 59 Granitoid rocks§		
	Troctolite*			Granitoid†					
	oxide weight percent	oxide g/100 cc	cation g/100 cc	oxide weight percent	oxide g/100 cc	cation g/100 cc	oxide weight percent	oxide g/100 cc	cation g/100 cc
SiO_2 or Si	41.1	131.11	61.23	76.0	197.60	92.28	22.233	98.703	21.526
Al_2O_3 or Al	15.1	48.17	25.48	12.9	33.54	17.74	1.987	19.946	5.582
Fe_2O_3	1.2	3.83		0.42	1.09		0.474	3.735	
Fe			53.81			1.19			10.528
FeO	20.6	65.71		0.21	0.55		0.925	7.238	
MgO or Mg	11.1	35.41	21.25	0.21	0.55	0.33	0.618	4.849	1.743
CaO or Ca	6.6	21.05	15.03	0.55	1.43	1.02	1.617	12.812	6.534
Na_2O or Na	1.3	4.15	3.08	3.7	9.62	7.14	0.207	1.642	0.904
K_2O or K	0.18	0.574	0.477	4.3	11.18	9.28	0.863	5.568	3.838
H_2O^+	0.50	1.595	..	0.85	2.210	..	0.144	1.094	..
TiO_2 or Ti	2.0	6.38	3.83	0.08	0.208	0.125	0.060	0.468	0.168
P_2O_5 or P	0.31	0.989	0.432	0.01	0.026	0.011	0.010	0.078	0.015
MnO or Mn	0.24	0.766	0.186	0.03	0.078	0.023	0.001	0.008	..
CO_2	0.08	0.255	..	<0.05	<0.104	..	0.003	0.025	..
Specific gravity	3.19	3.19	3.19	2.60	2.60	2.60	0.003	0.003	0.003

*Sample 372 of Espenshade and Boudette, 1967, Table 6.
† Sample 3 of Lee and van Loenen, 1971, Table 5.
§ Set of 59 samples from Lee and van Loenen, 1971, Table 5, for which specific gravities were recorded.

and three-dimensional scatter diagrams (for example, Harker and ternary diagrams). Weight-percent and weight-per-unit-volume data can yield significantly different results. For instance, it is not easy to predict the differences between regression lines when these two dissimilar units are used; the slopes of regression lines commonly differ and, in certain cases, may even have opposite signs (Fig. 1). For example, if there is x weight percent of an oxide in the more dense rock (density ρ_{large}) and $(x + \Delta)$ weight percent in the less dense rock (density ρ_{small}), where $x < (x + \Delta)$, the conversion to grams per 100 cc gives

$$x \cdot \rho_{large} < (x + \Delta) \cdot \rho_{small};$$

however, the slope of the regression line is reversed if

$$x \cdot \rho_{large} > (x + \Delta) \cdot \rho_{small}; \qquad (1)$$

that is, if

$$x > \Delta \cdot \rho_{small} / (\rho_{large} - \rho_{small}).$$

For the density range on the abscissa of Figure 1, $x = 2.38\Delta$. Analogous relations hold if, instead of density, the abscissa represents any cation or oxide positively correlated with ρ. In general, major changes in regression-line slopes for igneous-rock data can occur when weight percent is used as ordinate instead of weight per unit volume; however, relatively few igneous-rock chemical data can occur for which the sign of the regression-line slopes will change.

Although Figure 1 refers to unrelated analyses, analogous results apply to data on a Harker diagram for rocks from a differentiated sequence (for example, Skaergaard Complex) or a theoretical differentiation series. A. L. Howland (1973, personal commun.) recently demonstrated that Fenner's (1926) classic variation diagram (Bowen, 1928, Fig. 26), illustrating the change of liquid composition during the crystallization of a liquid in the system diopside-anorthite-albite, is drastically modified if plotted in weight-per-unit-volume units instead of weight percentages; on this revised basis, the sample points actually occur in a different order.

To plot cation weights per unit volume, instead of oxide weights per unit volume, the oxide values parallel to the abscissa and ordinate are merely multiplied by the constant appropriate for each element. Because of the new units of length

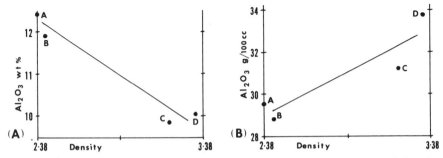

Figure 1. Comparison of Al_2O_3 density (ρ) plots for (A) weight percent and for (B) weight-per-unit-volume Al_2O_3 values. The samples plotted are as follows: A, perlite, sample 47, Thompson and White (1964); B, intrusive rhyolite glass, sample 15, Cornwall and Kleinhampl (1964); C, nepheline syenite, sample 128, Maxwell and others (1965); D, coronite, sample 219, Faessler (1962).

parallel to each axis, the regression lines for cation-weights-per-unit-volume data have different slopes (but not signs), although the correlation coefficients (r) are identical to those for the corresponding oxide-weights-per-unit-volume data.

When grams per 100-cc units are used instead of weights percent, regression lines and contours (for chemical components of igneous rocks) plotted on ternary diagrams show comparable changes to those that occur on Harker-type plots. In almost all cases, normative values and diagrams, and calculations based on norms and analogous values, should be based on analyses expressed in grams per 100 cc, rather than weight percentages.

SPECIFIC EXAMPLE BASED ON THE PALISADES SILL, NEW YORK

In the abstract, it is difficult to evaluate whether use of the several different units for chemical analyses discussed in the preceding section makes any significant difference to petrological and petrogenetic conclusions. As a specific example, the composition of the well-known Palisades Sill is considered on the basis of F. Walker's (1940) frequently cited data (compare Turner and Verhoogen, 1960, p. 212-216). Fortunately, Walker gave the densities of his analyzed rocks. The more recent, detailed data of K. R. Walker (1969) cannot be used because rock densities are not available.

Walker (1940, Table 3) published 12 new chemical analyses for the George Washington Bridge section of the sill; these have been used for computing compositional changes during crystallization and differentiation. For the following discussion, it is assumed that Walker's data are error free and that they include reasonable samples on the basis of which to analyze chemical variations within the sill. Following the example of Tyrrell (1948, Fig. 3), Walker (1940, Fig. 9), Turner and Verhoogen (1960, Fig. 33), and Walker (1969, Figs. 8 and 9), a plot of chemical components against height within the sill is used. Figure 2 shows that the smallest SiO_2 weight percentage (48.28 percent) occurs 18.29 m above the base of the sill, and that 51.46 percent SiO_2 occurs at 213.36 m. Correction to the more appropriate Si grams per 100 cc (or to SiO_2 grams per 100 cc) reveals that, in reality, the 213.36-m level is *less* silicic than the more olivine-rich, 18.29-m level. This reversed trend is not likely to be unique, but data to establish this are scarce. However, comparable (though much less complete) data were published by Tyrrell (1917, 1948) for the differentiated Lugar Sill in Scotland. Equally dramatic changes occur when these SiO_2 weight percentages are corrected to Si grams per 100 cc (Fig. 3). SiO_2 weight percentage is smaller at 17.86 m above the base of the sill than at the base, and is larger at 48.34 m than at 31.80 m above the base of the sill. A plot of Si grams per 100 cc (and of SiO_2 grams per 100 cc) shows that, in actual fact, both of these gradients are the reverse of those implied by SiO_2 weight-percentage data.

For new chemical analyses of the Englewood Cliffs and Union City sections through the Palisades Sill, K. R. Walker (1969, Fig. 8 and Table 9) plotted all oxide weight percentages in a manner similar to that used for SiO_2 weight percent in Figure 2. Planimetric estimates were then used by him to derive the mean oxide weight percentages for the whole section. The dissimilarities between SiO_2 weight percent and Si grams per 100-cc curves in Figure 2 suggest that average values based on planimetry of oxide-weight-percentage curves must introduce serious errors into calculated mean compositions. (In addition, the heights at which samples were plotted in Walker's [1969] Figure 8 reflect some drafting errors which must have introduced additional, independent errors in his mean estimates [Walker,

1969, Table 9, column 7].) In comparing the chilled margin of the Lugar Sill with the mean composition of that sill's interior, Tyrrell (1948, 1952) correctly weighted the chemical analyses by the rock densities.

The lack of range of silica values in differentiated sills has sometimes been cited as a reason for not using Harker diagrams to portray the chemical relations; however, such diagrams dramatically emphasize the differences that arise when traditional weight-percentage values are used instead of cation weights per unit volume (Fig. 4).

Walker (1969, p. 75-81) discussed the fractionation index

$$100 \times (FeO + Fe_2O_3) / (FeO + Fe_2O_3 + MgO);$$

this index has the same numerical value when weight-percent and weight-per-unit-volume data are used. In discussing differentiation trends, Walker wrote that the

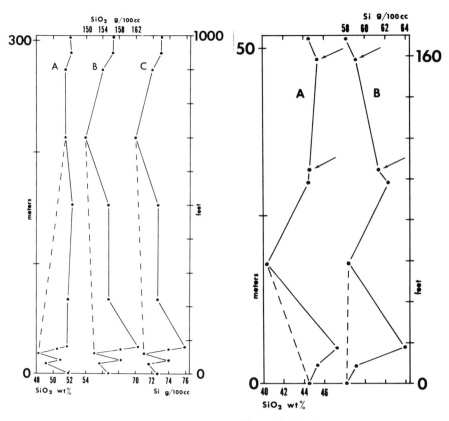

Figure 2. Chemical variation shown by 12 samples representing a complete vertical profile of the Palisades Sill (George Washington Bridge section), New Jersey; ordinate is height above lower sill contact. Data from Walker (1940). Solid lines connect successive analyses; broken lines inserted to emphasize dissimilar gradients between samples at 18.29 and 213.36 m above the base. A, SiO_2 weight percent; B, SiO_2 grams per 100 cc; and C, Si grams per 100 cc.

Figure 3. Chemical variation in a complete vertical profile of the Lugar Sill, Ayrshire, Scotland; ordinate is height above lower sill contact. Data from Tyrrell (1917, 1948). Solid lines connect analyses for successive sites in the sill; broken lines inserted to emphasize the dissimilar gradients between the chilled margin and the sample at 17.86 m above the sill base. The slopes between samples at 31.80 and 48.34 m (indicated by arrows) are opposite. A, SiO_2 weight percent; B, Si grams per 100 cc.

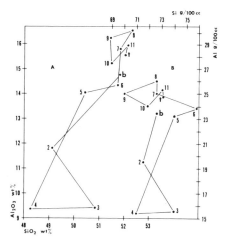

Figure 4. Harker-type diagrams for the Palisades Sill, New Jersey, based on data from Walker (1940); twelve samples through the George Washington Bridge section are labeled b (bottom), 2, 3, . . ., 10, 11, t (top). A, traditional Harker plot of Al_2O_3 weight percent against SiO_2 weight percent; B, Harker diagram replotted in terms of Al and Si grams per 100 cc.

"analyses (Table 8 and Fig. 8) also show that silica-enrichment was concurrent with alkali-enrichment." Density was not recorded by K. R. Walker, so his conclusion cannot be assessed accurately from his data, but F. Walker's analyses provide a reasonable basis for evaluation. For the latter data, both Na_2O and K_2O weight percentage are positively correlated with SiO_2 weight percentage (Fig. 5). Conversion to the more appropriate Na, K, and Si grams per 100 cc (or oxides in grams per 100 cc), however, reveals weak negative, rather than distinct positive, correlations. Special petrogenetic importance has been attached to the iron-enrichment of late fractions of differentiating basic magmas; (FeO + Fe_2O_3) weight-percentage data reflect such a trend for the Palisades Sill, but correction to Fe grams per 100 cc shows that rocks at the 213.36- and 274.32-m levels really have Fe contents close to the midpoint of the Fe range of the available analyses (Fig. 6).

These illustrations based on the Palisades Sill provide compelling evidence that significant differences in chemical interrelations appear when proper allowance is made for the density of the analyzed rocks—a conclusion recognized by Tyrrell. (If the Palisades and Lugar Sills were sampled differently, different variation patterns would almost certainly emerge. In this discussion, the sampling issue and [or] the representativeness of the analyzed samples are neither raised nor considered; until the samples employed have been shown to be an adequate sample, Figures 2 through 6 should be considered as showing the pattern of variation of the collected samples, rather than necessarily that of the sill represented.) The examples go a long way toward justifying the contention that significant errors of data appraisal and interpretation can be eliminated by utilizing grams per 100-cc units for the chemical composition of igneous rocks.

RELATIONS FOR SOME OTHER IGNEOUS-ROCK SUITES

Five data sets were assembled to investigate a wider range of igneous rocks (Table 2). For most pairs of oxides in these sets, the correlation coefficients (r) between oxides expressed as grams per 100 cc are only slightly different from those computed for weights percent. Correlations with SiO_2, Al_2O_3, or Na_2O show the greatest changes; there is a tendency for r to be larger in absolute value when weight percentages are used to correlate all other oxides with SiO_2, and for r to be closer to zero when weight percentages are used to correlate all other

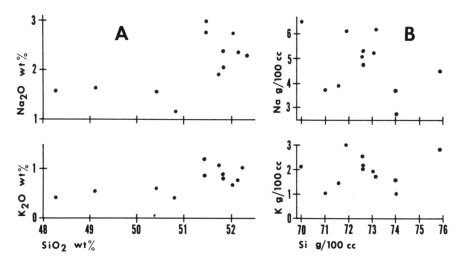

Figure 5. Harker-type diagrams for the Palisades Sill, New Jersey, based on data from Walker (1940). A, traditional Harker plots for K_2O and Na_2O weight percent against SiO_2 weight percent, showing weak positive correlations; B, Harker diagrams replotted in terms of K, Na, and Si grams per 100 cc, with consequent loss of the positive correlations.

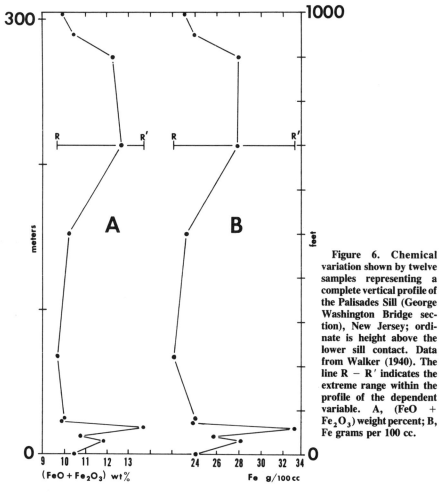

Figure 6. Chemical variation shown by twelve samples representing a complete vertical profile of the Palisades Sill (George Washington Bridge section), New Jersey; ordinate is height above the lower sill contact. Data from Walker (1940). The line $R - R'$ indicates the extreme range within the profile of the dependent variable. A, (FeO + Fe_2O_3) weight percent; B, Fe grams per 100 cc.

TABLE 2. CHEMICAL ANALYSES USED

Set name	Number of analyses*	Rock types	Source
Donegal	114 W	"Older granite" of Donegal, Eire; dioritic to granodioritic rocks collected roughly on 1/2-mi grid	Whitten, unpub. analyses
Malsburg	37 W	Malsburg Granite, southwest Germany	Whitten, 1972, Table 1
Snake	59 P	Hybrid granitoid rocks, south Snake Range, Nevada (115 analyses listed but only 59 with density)	Lee and van Loenen, 1971, Table 5
Nevada	39 W	Gabbroic, basaltic, peridotitic, and anorthositic rocks, Carson Sink region, Nevada	Speed, 1963, Table 1; 35 unpub. analyses
Nevada	6 W	Norite and troctolite, Moxie pluton, Maine (powder densities also given)	Espenshade and Boudette, 1967, Table 6
Australia	67	Granite-gabbro association rocks from Hartley Complex, New South Wales	Rhodes, 1969, Table 5.4

*W = whole-rock density; P = powder density.

oxides with Al_2O_3. It is difficult to generalize about the variable changes in r when Na_2O is involved (Table 3).

Table 3 suggests that the units used will result in markedly different significance levels for regression lines on variation diagrams (for example, ternary and Harker diagrams) when SiO_2, Al_2O_3, or Na_2O are involved. Figures 7 and 8 compare Harker-type plots for selected weight-percentage and grams-per-100-cc data from Table 2; these diagrams were chosen on the basis of clear visual dissimilarity and sizable difference in sum-of-squares reductions for the linear regression lines (when the two units of measurement are used). Plots for a majority of the other pairs of oxides show only trivial dissimilarities. Lee and van Loenen (1971, Fig. 3) plotted all their oxide weight percentages on Harker-type diagrams with CaO as abscissas; these diagrams barely change when replotted in grams-per-100-cc units, and pairs of r values (calculated for the two units of measurement) for CaO and each of the other oxides only differ in the second decimal place (on the basis of 59 of their 115 chemically analyzed rocks for which specific gravity was quoted).

The percentage of the sum of squares accounted for by a linear regression line for analyses expressed as oxide weight per 100 cc is

$$100\ r^2 = 100 \times \frac{\sum_{k=1}^{n} (\hat{w}_k \hat{\rho}_k)^2 - \left(\sum_{k=1}^{n} \hat{w}_k \hat{\rho}_k\right)^2 \cdot n^{-1}}{\sum_{k=1}^{n} (w_k \rho_k)^2 - \left(\sum_{k=1}^{n} w_k \rho_k\right)^2 \cdot n^{-1}} \qquad (2)$$

where w_k is weight percentage of an oxide in the kth of n rock samples, and the kth rock sample has density ρ_k, and $\hat{w}$ and $\hat{\rho}$ refer to computed as opposed to observed values (w and ρ). The corresponding value for weight-percentage data

TABLE 3. CHANGES IN LINEAR CORRELATION COEFFICIENTS (r)

Data set	Donegal $N = 114$	Malsburg $N = 37$	Snake $N = 59$	Nevada $N = 45$	Australia $N = 67$
SiO_2 with TiO_2	1			G	G
Al_2O_3	1			1	1 G
Fe_2O_3		1 G		1 G	
FeO		1		2	
MnO	n.a.	n.a.	G	2	
MgO					
CaO	1	1			
Na_2O			1 G	2	1
K_2O	1		1		
P_2O_5	1				G
H_2O^+	n.a.	n.a.	G	1 G	
CO_2	n.a.	n.a.	G	2 G	G
Density	2	2	1	2	
Al_2O_3 with SiO_2	G			1	G
TiO_2	G	G	G	G	G
Fe_2O_3		G	G	G	1 G
FeO	G	1 G	G	1	G
MnO	n.a.	n.a.	G	2	1 G
MgO	G	G	G	1	1 G
CaO	G	1 G	G	1 G	G
Na_2O	1 G	G	G	1	
K_2O			G		G
P_2O_5	G	G	G	G	G
H_2O^+	n.a.	n.a.	G	G	1 G
CO_2	n.a.	n.a.		G	G
Density	2 G	1 G	G	1	1 G
Na_2O with SiO_2			1 G	2	1
TiO_2	1 G	G	1 G		
Al_2O_3	1 G	1 G	G	1	
Fe_2O_3			1 G	G	
FeO	G	G	1 G		
MnO	n.a.	n.a.	G		
MgO	1 G	G	1 G	G	
CaO	1 G	G	1 G	G	
K_2O	S		1 G	G	1
P_2O_5	G	G	1 G	G	G
H_2O^+	n.a.	n.a.	G		
CO_2	n.a.	n.a.			S
Density	1 G	1 G	1 G		1

Note: G = r is larger with gram-per-100-cc data.
S = r is same for gram-per-100-cc and weight-percent data.
no letter = r is smaller with gram-per-100-cc data.
2 = r is changed by more than 0.2 when grams-per-100-cc data are used instead of weight percent.
1 = r is changed by more than 0.1.
no number = r is changed by less than 0.1.
N = number of samples in data set.
n.a. = no analysis.

is obtained by omitting the four ρ_k from equation (2). The changes in graphs like Figures 7 and 8, in linear correlation coefficients (r) and in percentage sum of squares associated with the regression lines, are not immediately predictable; the same is true of differences between results for cation weight percentage and cation weights per unit volume. In part, this is because, although each oxide or each cation can be expressed as a linear function of the sample's specific gravity, that is, $w_k = f(\rho_k)$, the relation tends to be weak for most sets of analyses.

HARKER DIAGRAMS AND RELATED GRAPHS

The objective in constructing a Harker diagram may simply be to record data in a pictorial manner. Commonly, when a Harker or ternary diagram is used to elucidate genetic interrelations between rocks, or to test a quantitative model, the model should prescribe variables to be used, operational definitions of the variables, and statistical constraints on the sampling plan; departures from the prescriptions may introduce significant biases and (or) errors. Without access to original analytical data (including rock densities), it is difficult to anticipate whether the gradient shown by a particular pair of oxides or cations (expressed as weights percent) in a data set will be significantly changed by expressing them as weights per unit volume.

The general case can be illustrated as follows. Let the weight of the kth sample be M_k g and its volume be V_k cc; then,

$$M_k = m_{ki} + m_{kj} + m_{kr}$$

where m_{ki} and m_{kj} are the weights (in grams) of two component oxides, and the total remaining oxides weigh m_{kr} g. The density of the rock sample, ρ_k, is

$$\rho_k = M_k/V_k = m_{ki}/V_k + m_{kj}/V_k + m_{kr}/V_k. \qquad (3)$$

Let the weight percent of the ith oxide be

$$w_{ki} \text{ percent} = 100 \cdot m_{ki}/M_k \text{ percent}$$

and the weight per unit volume of the ith oxide be

$$v_{ki} g = w_{ki} \rho_k = (100 \cdot m_{ki}/M_k) \cdot (M_k/V_k) \cdot 10^{-2} = m_{ki}/V_k. \qquad (4)$$

Now, $w_{ki} + w_{kj} + w_{kr} = 100$, so

$$dw_j/dw_i = -(1 + dw_r/dw_i),$$

and thus

$$w_j = -\int (1 + dw_r/dw_i) \, dw_i.$$

Suppose that the ith oxide is silica, the jth oxide is potash, and the variation of w_r with respect to w_i (that is, dw_r/dw_i) is approximately linear and $\simeq c_1$. In fact, for these particular oxides, c_1 will be <0 and, in most cases, have an almost constant value. Carrying out the integration yields

$$w_j = -(1 + c_1) w_i + C_1 \qquad (5)$$

where C_1 is the constant of integration.

Next consider the weights-per-unit-volume case. Using equations (3) and (4) gives the following:

$$\rho_k = v_{ki} + v_{kj} + v_{kr}$$

so that

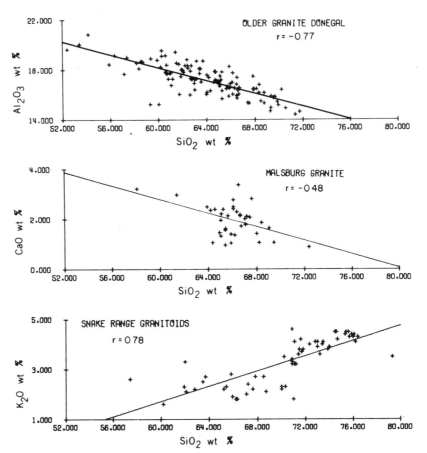

Figure 7. Harker diagrams (with least-squares linear regression lines) to illustrate the differences that occur when analyses are expressed in weight-percent and in weight-per-unit-volume

$$dv_j/dv_i = -(1 + dv_r/dv_i) + d\rho/dv_i,$$

and

$$v_j = -\int (1 + dv_r/dv_i - d\rho/dv_i)\, dv_i.$$

The behavior of dv_r/dv_i and $d\rho/dv_i$ is discussed below, but if, for simplicity, their values are assumed to be the constants c_2 and c_3, then

$$v_j = -(1 + c_2 - c_3)\, v_i + C_2 \tag{6}$$

where C_2 is another constant of integration. From equations (4) and (6),

$$w_{kj} = v_{kj}/\rho_k = -(1 + c_2 - c_3)\, v_{ki}/\rho_k + C_2/\rho_k. \tag{7}$$

Equations (5) and (7) can be graphed to illustrate the dissimilarities of the two units of measure (weight percent and weight per unit volume). Equation (5) represents a straight line with slope $-(1 + c_1)$. However, for equation (7), even if (1 +

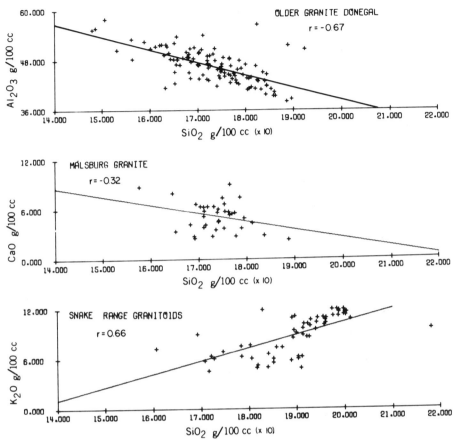

units; the analyses used are referenced in the first three entries of Table 2, r = linear correlation coefficient.

$c_2 - c_3$) is a constant, C_2/ρ_k is variable, except in the unusual case where the specific gravity of all samples is constant, that is, where ρ_k is a constant. Commonly, there is not a simple linear relation between SiO_2 percentage and rock density, but (as a first approximation) the simplifying assumption can be made that $\rho_k = aw_{ki} + b$, where a and b are constants. Then, substitution in equation (7) yields the quadratic expression:

$$w_{kj} = -(1 + c_2 - c_3) w_{ki} + C_2/(aw_{ki} + b), \qquad (8)$$

which is clearly dissimilar to equation (5). When one attempts to compare equations (5) and (8), $C_2/(aw_{ki} + b)$ could also be considered approximately constant, so that equation (8) becomes a straight line with slope $-(1 + c_2 - c_3)$. According to equations (5) and (8), weight percent and weight per unit volume then yield straight lines with slopes of $-(1 + c_1)$ and $-(1 + c_2 - c_3)$, which are not parallel unless

$$dw_r/dw_i = dv_r/dv_i - d\rho/dv_i = dw_r \rho/dw_i \rho - d\rho/dw_i \rho, \qquad (9)$$

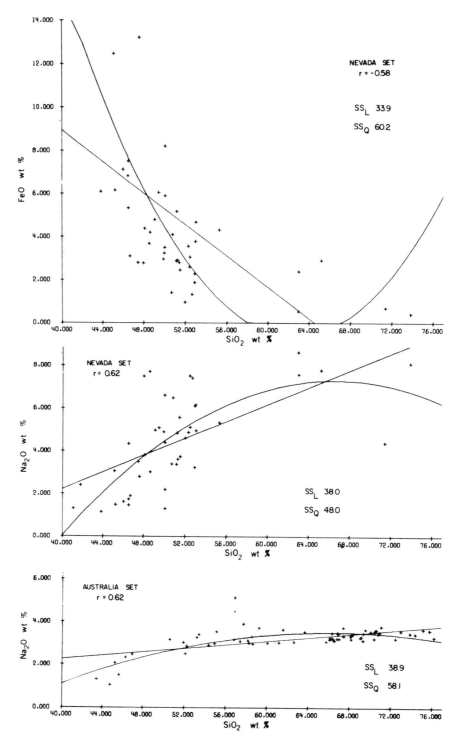

Figure 8. Harker diagrams to illustrate the differences that occur when analyses are expressed in weight-percent and weight-per-unit-volume units; the analyses plotted are referenced in the last two entries of Table 2. Both the least-squares linear and quadratic regression lines are

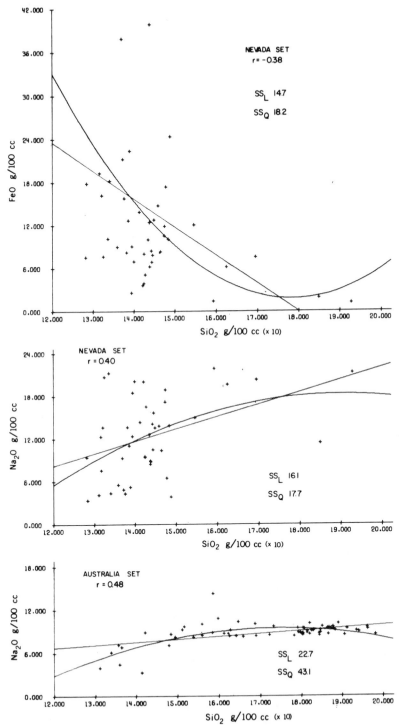

shown because the latter are associated with appreciable increments of the total sums of squares. The percentages of the corrected sums of squares accounted for by the linear (SS_L) and the quadratic (SS_Q) regression lines are shown; r = linear correlation coefficient.

using the substitutions from equation (4). The term dw_r/dw_i may be linear, but the other terms are markedly nonlinear. The term $d\rho/dw_i\rho$ is < 0, but is small by comparison with the other terms. Since w_i and ρ are negatively correlated and w_r and ρ are positively correlated, $dw_r\rho/dw_i\rho \gg dw_r/dw_i$, and the slope $-(1 + c_1)$ of the weight-percentage line is less than $-(1 + c_2 - c_3)$ of the weight-per-unit-volume line.

For this discussion, the ith and jth oxides were taken to be silica and potash, but analogous arguments hold for all other pairs of oxides or cations. The main point is that the graphs represented by equations (5) and (7) have different slopes and curvatures. Thus, regression lines based on weight-percentage and on weight-per-unit-volume data are, in general, dissimilar. It is difficult to anticipate the magnitude of possible differences without recourse to original data. This is unfortunate in practical situations. For example, Jakeš and White (1972, Fig. 1) published seven lines for SiO_2 and K_2O weight percentages on Harker diagrams for volcanic rocks of island arcs, and they concluded that the slope of such lines increases across an island arc toward the continent. These patterns may or may not be different if the more appropriate weight-per-unit-volume units were used. The analyses represented in Figure 9 are from six unrelated areas; they were plotted solely to illustrate the dissimilarity that can arise when the two dissimilar units of measurement are used.

Where justified by large increments of the sums of squares, degree-2 regression lines are included in Figure 8. Few authors describe how their lines were constructed on Harker diagrams. Most published diagrams have subjectively sketched curved lines, rather than lines fitted by least-squares or other objective methods. Notable exceptions were Fenner's (1926) diagrams for Katmai rocks, on which he appears to have drawn linear regression lines.

In addition to the desirability of using analytical values per unit volume instead of weight percentages for most problems, other important questions arise. For example:

1. The data are commonly strongly clustered with respect to one or both variables (Fig. 9) so that methods using the least-squares criterion yield a much better fit (and prediction) within the range of the clustered samples, and a much poorer fit in areas of less abundant information. Ideally, for a least-squares regression line on a Harker diagram, there would be equal-sample density over the whole range of both variables involved. For a particular igneous complex, such a distribution of analyses may be difficult to achieve. As an experiment, the array of 322 analyses (Table 2) was augmented by additional published analyses of igneous rocks until, as nearly as possible, every one-percent interval of SiO_2 weight-percent values was represented by 15 analyses. (Analyses with accompanying specific gravity values were abstracted from papers listed in Table 4, and 36 rocks [20 Donegal, 4 Malsburg, and 12 Snake] with 65 to 67 percent SiO_2 were deleted arbitrarily from the original set of 322 analyses.) The resultant set of 505 analyses was used for Figure 10. Whether this is the proper manner to weight the analyses depends on the reason or objective for making the plot, as is discussed in the following paragraph.

2. In constructing a regression line, standard procedures give equal weight to each plotted point and thus to each analyzed sample. The actual volume of rocks represented by an analyzed sample is commonly not determined, although this could materially affect the significance of the results. For example, for a granitoid pluton with minor gabbroic-noritic border zones, a Harker diagram based on 50 mafic rocks and 5 granitic samples would yield an unrealistic picture, because the analyses would not reflect the volumetric abundance at the sampled level of

exposure; insuring proportional representation would almost certainly lead to clustering of the data points for silica-rich rocks and, consequently, to a relatively poor regression-line fit and prediction for the silica-poor rock types. For most purposes, to portray an igneous lithic unit or volcanic association correctly, each analysis should represent an equal volume of rocks. Weighting analyses to obtain such representation is rarely, if ever, done or possible. However, for many objectives, significant biases may be introduced if such factors are overlooked. For these purposes, grams per 100 cc is again the appropriate unit, whereas weight percent is not.

MAPPED SPATIAL VARIABILITY

In mapping and preparing isoline maps for chemical components of an igneous massif, the target population of interest almost always involves all specimens of specified size that comprise the surface of the rock unit. For example, in drawing isolines for MgO content of the exposed surface of a pluton, one commonly intends to map the amount of MgO in samples of similar size (volume), rather than in samples of varying size. Grams per unit volume is an appropriate unit for this purpose, although the gross map patterns are not likely to be very dissimilar to those for weight percent. However, the question is not one of how big the differences are, but of which is correct for the stated objective under study. Figure 11 shows a trend-surface map for Al_2O_3 for 114 samples of a small part of the "Older Granite" of Donegal in Eire. Without raising the question of the propriety of

TABLE 4. SOURCE OF ADDITIONAL CHEMICAL ANALYSES USED IN FIGURE 11

Authors	Publication date	Number of samples used
Albers, J. P., and Robertson, J. F.	1961	1
Baragar, W. R. A.	1967	15
Boyle, R. W.	1965	1
Byers, F. M., Jr.	1959	20
Card, K. D.	1968	1
Coleman, R. G., and Lee, D. E.	1963	8
Cornwall, H. R., and Kleinhampl, F. J.	1964	6
Faessler, C.	1962	17
Hamilton, W.	1965	4
Hotz, P. E., and Willden, R.	1964	4
Huber, N. K., and Rinehart, C. D.	1967	2
Jenness, S. E.	1966	1
Jones, W. R., and others	1967	2
Larrabee, D. M., and others	1965	3
MacKevett, E. M., Jr.	1963	9
Maxwell, J. A., and others	1965	30
Nelson, A. E.	1966	1
Pye, E. G.	1968	6
Ratté, J. C., and Steven, T. A.	1967	1
Rinehart, C. D., and Ross, D. C.	1964	3
Sainsbury, C. L.	1969	6
Sims, P. K., and Gable, D. J.	1967	1
Smedes, H. W.	1966	3
Thompson, G. A., and White, D. E.	1964	26
White, D. E., and others	1964	5
Wilshire, H. G.	1967	43
Total		219

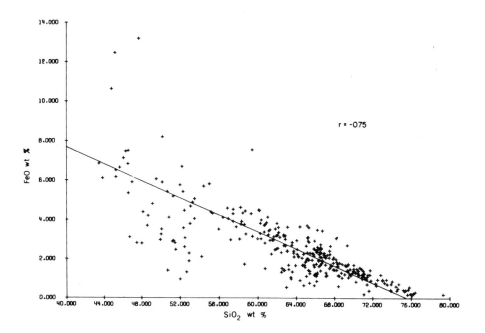

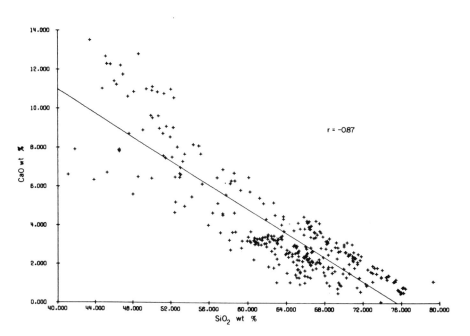

Figure 9. Harker diagrams (with linear regression lines) for all 322 analyses referenced in

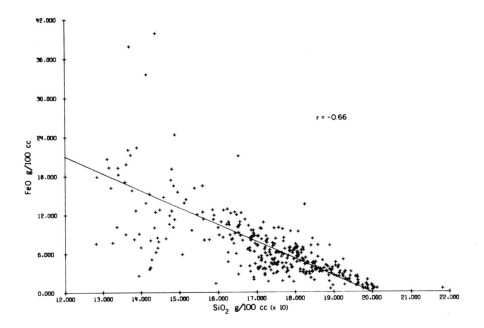

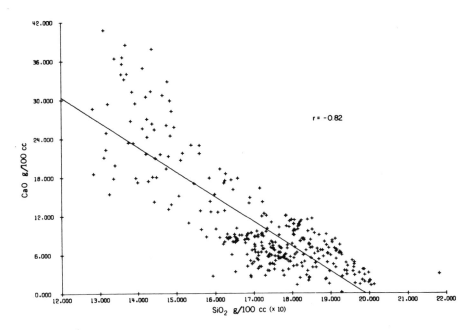

Table 2 to illustrate the differences that occur when weight-percent and weight-per-unit-volume units are used; r = linear correlation coefficient.

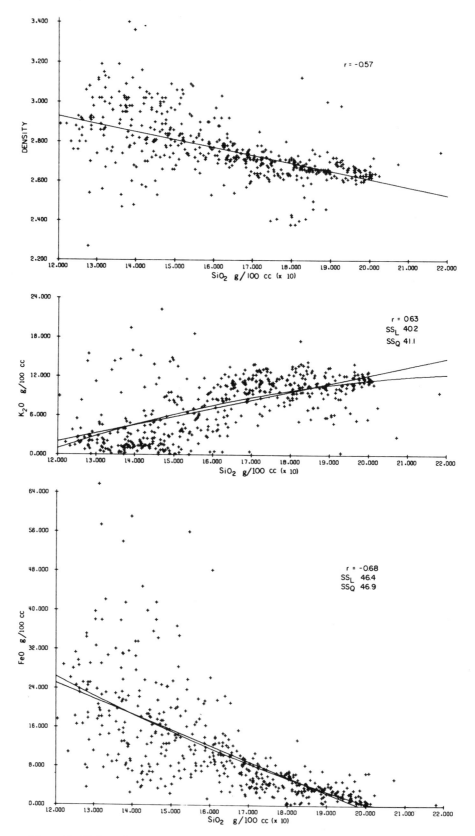

Figure 10. Harker diagrams (with linear regression lines) for the augmented set of 505 igneous-rock analyses; as nearly as possible, 15 rocks per one percent interval of SiO_2 weight percent are included. Quadratic regression lines are included on four of the plots, where justified

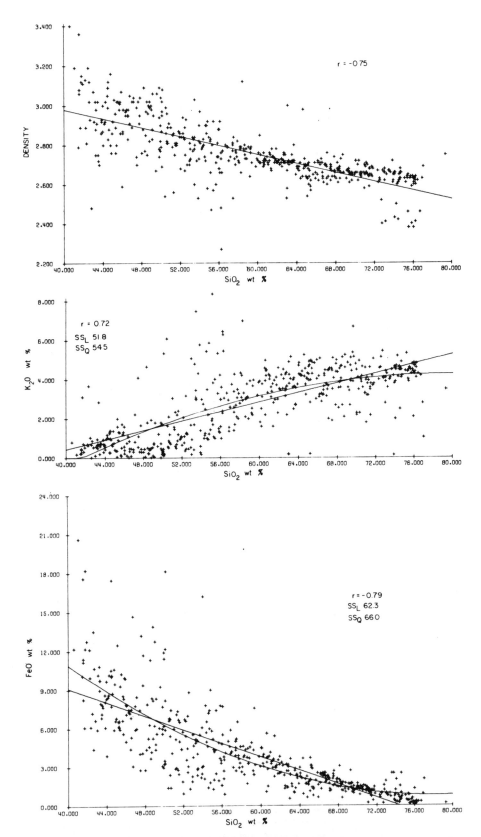

by the percentage of the corrected sum of squares (SS_Q); the percentage of the corrected sum of squares accounted for by the linear regression line is SS_L, and r is the linear correlation coefficient.

using factor analysis for spatially distributed data (Lebart, 1969), Figure 12 shows the similarity of Q-mode factor-analysis maps for the Malsburg Granite of southwest Germany, based on all major oxides. In both cases, the map patterns and the gradients across the maps are only slightly different when grams per 100 cc are used instead of weights percent.

CLOSED-NUMBER SYSTEMS

Chayes (1971) drew attention to the restraints involved when basing conclusions on percentage data or on ratio data in general. Initially, chemical analyses expressed as weights per unit volume may appear to be released from such restraints, because the total weight of each sample is different. Unfortunately, because the volume analyzed is constant, individual oxides or cations are not wholly independent (or completely "open"). Hence, the restraints of the closed-number system are not removed by using weights per unit volume. However, the numerical "null values" for hypothesis testing (computed with the equations developed by Chayes, 1971) are necessarily different from those for weight-percentage data.

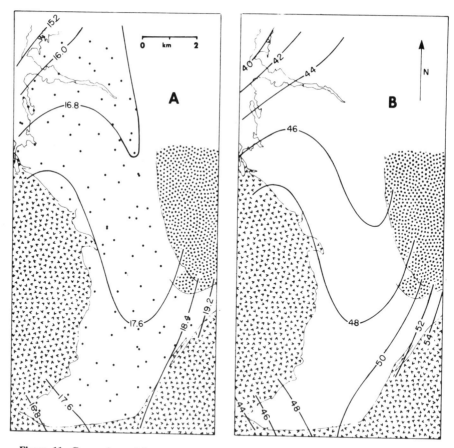

Figure 11. Comparison of degree-3 polynomial trend surfaces for Al_2O_3 in 114 samples of the older granite of Donegal, Éire. Sample sites are indicated by dots; close stipple = older metasedimentary rocks; v-patterns = younger granitic rocks. A, Al_2O_3 weight-percent surface, associated with 41.9 percent of the total corrected sum of squares; B, Al_2O_3 grams-per-100-cc surface, associated with 45.0 percent of the total corrected sum of squares.

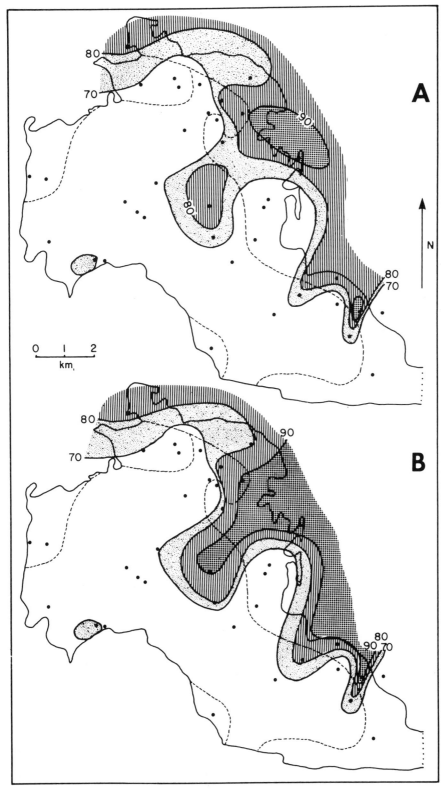

Figure 12. Comparison of manually contoured Q-mode factor-loading maps based on actual values of major oxides in 37 samples of the Malsburg Granite, southwest Germany; sample sites indicated by dots. Patterns used solely to make contour values distinct. A, based on weight-percent data; B, based on grams-per-100-cc data.

CONCLUSIONS

In petrology, modern methods of analysis permit new data to be amassed rapidly and in large quantities. This can lead to the assembly of large sets of traditional data for numerous variables in an arbitrary manner, and then to manipulating the information until an interesting conclusion emerges that can be called the objective of the study. Some variables are more significant than others in the analysis of well-defined, specific, geological problems; in general, the significance of each measurable variable depends upon the purpose of the research. Unfortunately, weight percent has come to be the accepted unit for expressing chemical composition in petrology, and in the literature little attention has been given to whether it is the most appropriate unit. Standard petrographic diagrams (for example, Harker diagrams) are commonly used in an attempt to make sense of petrological data collected in an essentially arbitrary manner. In many cases, the objectives involved in preparing a Harker diagram are unstated and far from obvious. As a result, the variables and sampling plans that should be used also remain obscure. Such procedures can introduce strong biases leading to erroneous petrographic and petrogenetic conclusions. In order not to be burdened with limitless data that have less than optimal usefulness, it is desirable to collect information for those variables that are demonstrably significant for the prespecified model under test (Whitten, 1974).

It has been demonstrated that weight-percentage and weight-per-unit-volume data commonly yield dissimilar results. Figures 1 through 12 provide some measure of the magnitude of the differences. Comparison of gradients shown by Harker diagrams based on samples of unequal size (as provided by weight percentages) can lead to significant biases. In almost all cases, quantitative comparison of samples and the elucidation of geochemical trends for petrological and petrogenetic studies necessitate correlating chemical components in equal volumes of rocks, and thus the use of cation or oxide weights per unit volume. Results for the Palisades and Lugar Sills show that use of grams per unit volume (instead of the traditional weight percentages) involves significant corrections and modifications to standard procedures and conclusions. In economic and engineering projects, the changed gradients and absolute values could be of considerable importance when problems near critical threshold values are involved.

ACKNOWLEDGMENTS

Many fruitful discussions with colleagues at Northwestern University (and particularly suggestions made by R. C. Speed, A. L. Howland, W. C. Krumbein, and M.E.V. Koelling) during the development of these ideas have been greatly appreciated. The work was supported by the U.S. Army Research Office, Durham (Grant DA-ARO-D-31-124-72-G54).

REFERENCES CITED

Albers, J. P., and Robertson, J. F., 1961, Geology and ore deposits of East Shasta copper-zinc district, Shasta County, California: U.S. Geol. Survey Prof. Paper 338, p. 1-107.

Babcock, R. S., 1973, Computational models of metasomatic processes: Lithos, v. 6, p. 279-290.

Baragar, W.R.A., 1967, Wakuach Lake map-area, Quebec-Labrador (23-0): Canada Geol. Survey Mem. 344, p. 1-174.

Barth, T.F.W., 1962, Theoretical petrology (2d ed.): N.Y., John Wiley & Sons, Inc., 416 p.
Bayly, B., 1968, Introduction to petrology: Englewood Cliffs, N.J., Prentice-Hall, 371 p.
Bowen, N. L., 1928, The evolution of the igneous rocks: Princeton, N.J., Princeton Univ. Press, 332 p.
Boyle, R. W., 1961, The geology, geochemistry, and origin of the gold deposits of the Yellowknife district: Canada Geol. Survey Mem. 310, p. 1-193.
―――1965, Geology, geochemistry, and origin of the lead-zinc deposits of the Keno Hill-Galena Hill area, Yukon Territory: Canada Geol. Survey Bull. 111, p. 1-302.
Byers, F. M., Jr., 1959, Geology of Umnak and Bogoslof Islands, Aleutian Islands, Alaska: U.S. Geol. Survey Bull. 1028L, p. 267-369.
Card, K. D., 1968, Geology of Denison-Waters area, District of Sudbury: Ontario Dept. Mines Geol. Rept. 60, p. 1-63.
Chayes, F., 1956, Petrographic modal analysis: N.Y., John Wiley & Sons, Inc., 113 p.
―――1971, Ratio correlation: A manual for students of petrology and geochemistry: Chicago, Ill., Univ. Chicago Press, 99 p.
Coleman, R. G., and Lee, D. E., 1963, Glaucophane-bearing metamorphic rock types of the Cazadero area, California: Jour. Petrology, v. 4, p. 260-301.
Cornwall, H. R., and Kleinhampl, F. J., 1964, Geology of Bullfrog quadrangle and ore deposits related to Bullfrog Hills caldera, Nye County, Nevada, and Inyo County, California: U.S. Geol. Survey Prof. Paper 454, p. J1-J25.
Cross, C. W., Iddings, J. P., Pirsson, L. V., and Washington, H. S., 1903, Quantitative classification of igneous rocks: Chicago, Ill., Univ. Chicago Press, 286 p.
Espenshade, G. H., and Boudette, E. L., 1967, Geology and petrology of the Greenville quadrange, Piscataquis and Somerset Counties, Maine: U.S. Geol. Survey Bull. 1241, p. F1-F60.
Exley, C. S., 1959, Magmatic differentiation and alteration in the St. Austell Granite: Geol. Soc. London Quart. Jour., v. 114 (for 1958), p. 197-230.
Faessler, C., 1962, Analyses of rocks of the Province of Quebec: Quebec Dept. Mines Geol. Rept. 103, p. 1-251.
Fenner, C. N., 1926, The Katmai magmatic province: Jour. Geology, v. 34, p. 673-772.
Hamilton, W., 1965, Diabase sheets of the Taylor Glacier Region, Victoria Land, Antarctica: U.S. Geol. Survey Prof. Paper 456, p. B1-B71.
Hotz, P. E., and Willden, R., 1964, Geology and mineral deposits of the Osgood Mountains quadrangle, Humboldt County, Nevada: U.S. Geol. Survey Prof. Paper 431, p. 1-128.
Huber, N. K., and Rinehart, C. D., 1967, Cenozoic volcanic rocks of the Devils Postpile quadrangle, eastern Sierra Nevada, California: U.S. Geol. Survey Prof. Paper 554, p. D1-D21.
Jakeš, P., and White, A.J.R., 1972, Major and trace element abundances in volcanic rocks of orogenic areas: Geol. Soc. America Bull., v. 83, p. 29-40.
Jenness, S. E., 1966, The anorthosite of northern Cape Breton Island, Nova Scotia, a petrological enigma: Canada Geol. Survey Paper 66-21, p. 1-25.
Jones, W. R., Hernon, R. M., and Moore, S. L., 1967, General geology of Santa Rita quadrangle, Grant County, New Mexico: U.S. Geol. Survey Prof. Paper 555, p. 1-144.
Larrabee, D. M., Spencer, C. W., and Swift, D.J.P., 1965, Bedrock geology of the Grand Lake area, Aroostock, Hancock, Penobscot, and Washington Counties, Maine: U.S. Geol. Survey Bull. 1201, p. E1-E38.
Lebart, L., 1969, Analyse statistique de la contiguite: Pubs. Inst. Stat. Univ. Paris, v. 18, p. 81-112.
Lee, D. E., and van Loenen, R. E., 1971, Hybrid granitoid rocks of the southern Snake Range, Nevada: U.S. Geol. Survey Prof. Paper 668, p. 1-48.
MacKevett, E. M., Jr., 1963, Geology and ore deposits of the Bokan Mountain uranium-thorium area, southeastern Alaska: U.S. Geol. Survey Bull. 1154, p. 1-125.
Maxwell, J. A., Dawson, K. R., Tomilson, M. E., Pocock, D.M.E., and Tetreault, D., 1965, Chemical analyses of Canadian rocks, minerals, and ores: Canada Geol. Survey Bull. 115, p. 1-476.

Nelson, A. E., 1966, Significant changes in volcanism during the Cretaceous in north-central Puerto Rico: U.S. Geol. Survey Prof. Paper 550, p. D172-D177.

Pye, E. G., 1968, Geology of Lac des Iles area: Ontario Dept. Mines Geol. Rept. 64, p. 1-47.

Ragland, P. C., and Butler, J. R., 1972, Crystallization of the West Farrington pluton, North Carolina, U.S.A.: Jour. Petrology, v. 13, p. 381-404.

Ratté, J. C., and Steven, T. A., 1967, Ash flows and related volcanic rocks associated with the Creede caldera, San Juan Mountains, Colorado: U.S. Geol. Survey Prof. Paper 524, p. H1-H58.

Rhodes, J. M., 1969, The geochemistry of a granite-gabbro association [Ph.D. thesis]: Canberra, A.C.T., Australian National Univ., 315 p.

Rinehart, C. D., and Ross, D. C., 1964, Geology and mineral deposits of the Mount Morrison quadrangle, Sierra Nevada, California: U.S. Geol. Survey Prof. Paper 385, p. 1-106.

Sainsbury, C. L., 1969, Geology and ore deposits of the central York Mountains, western Seward Peninsula, Alaska: U.S. Geol. Survey Bull. 1287, p. 1-101.

Shand, S. J., 1947, Eruptive rocks: N.Y., John Wiley & Sons, Inc., 488 p.

Sims, P. K., and Gable, D. J., 1967, Petrology and structure of Precambrian rocks, Central City quadrangle, Colorado: U.S. Geol. Survey Prof. Paper 554, p. E1-E56.

Smedes, H. W., 1966, Geology and igneous petrology of the northern Elkhorn Mountains, Jefferson and Broadwater Counties, Montana: U.S. Geol. Survey Prof. Paper 510, p. 1-116.

Speed, R. C., 1963, Layered picrite-anorthositic gabbro sheet, West Humboldt Range, Nevada: Mineralog. Soc. America Spec. Paper 1, p. 69-77.

Thompson, G. A., and White, D. E., 1964, Regional geology of the Steamboat Springs area, Washoe County, Nevada: U.S. Geol. Survey Prof. Paper 458, p. A1-A52.

Turner, F. J., and Verhoogen, J., 1960, Igneous and metamorphic petrology (2d ed.): N.Y., McGraw-Hill Book Co., 694 p.

Tyrrell, G. W., 1917, The picrite-teschenite sill of Lugar (Ayrshire): Geol. Soc. London Quart. Jour., v. 72 (for 1916), p. 84-131.

——1948, A boring through the Lugar Sill: Geol. Soc. Glasgow Trans., v. 21, p. 157-202.

——1952, A second boring through the Lugar Sill: Geol. Soc. Edinburgh Trans., v. 15, p. 374-392.

Wager, L. R., 1929, Metasomatism in the Whin Sill of the north of England. Pt. I: Metasomatism by lead vein solutions: Geol. Mag., v. 64, p. 97-110.

Walker, F., 1940, Differentiation of the Palisade Diabase, New Jersey: Geol. Soc. America Bull., v. 51, p. 1059-1106.

Walker, K. R., 1969, The Palisades Sill, New Jersey: A reinvestigation: Geol. Soc. America Spec. Paper 111, 175 p.

White, W. H., 1973, Flow structure and form of the Deep Creek Stock, southern Seven Devils Mountains, Idaho: Geol. Soc. America Bull., v. 84, p. 199-210.

White, D. E., Thompson, G. A., and Sandberg, C. H., 1964, Rocks, structure, and geologic history of Steamboat Springs thermal area, Washoe County, Nevada: U.S. Geol. Survey Prof. Paper 458, p. B1-B63.

Whitten, E.H.T., 1972, Enigmas in assessing the composition of a rock unit: A case history based on the Malsburg Granite, SW. Germany: Geol. Soc. Finland Bull., v. 44, p. 47-82.

——1974, Scalar and directional field and analytical data for spatial variability studies: Internat. Assoc. Math. Geol. Jour., v. 6, p. 183-198.

Whitten, E.H.T., and Boyer, R. E., 1964, Process-response models based on heavy-mineral content of the San Isabel Granite, Colorado: Geol. Soc. America Bull., v. 75, p. 841-862.

Wilshire, H. G., 1967, The Prospect Alkaline Diabase-Picrite Intrusion, New South Wales, Australia: Jour. Petrology, v. 8, p. 97-163.

Manuscript Received by the Society May 23, 1973
Revised Manuscript Received November 12, 1973

STATISTICAL METHODOLOGY

Segmentation of Discrete Sequences of Geologic Data

D. M. HAWKINS

Department of Statistics
University of Witwatersrand
Johannesburg, South Africa

AND

D. F. MERRIAM

Department of Geology
Syracuse University
Syracuse, New York 13210

ABSTRACT

Sequential data can be segmented by the techniques of (1) numeric differentiation, (2) split moving window, (3) maximum level variance, and (4) piecewise regression. Each method is compared and contrasted with the others for performance with different types of data strings. In geology, data strings could be electric logs (or other borehole logs), seismic traces, x-ray curves, or traverses measuring some property with distance. Segmenting traces into like parts with uniform characters facilitates comparison and correlation.

INTRODUCTION

Several techniques have been proposed recently for the zonation of digitized sequential data. Geologists and other earth scientists frequently work with data taken along traverses or measured through time; these data are usually analyzed by "time-series" methods. Correlation from one series to another is by cross-correlation techniques if the data are numeric and by cross-association techniques if the data are nonnumeric (Merriam, 1971). These techniques, however, are not intended for segmenting the traces into like parts; it is this problem that is considered here. The determination of segments that are uniform in character may be important in its own right, as well as adjunct to the correlation of segments with similar patterns.

The techniques that are available for zoning discrete sequential data we have termed (1) numeric differentiation, (2) split moving window, (3) maximum level variance, and (4) piecewise regression. The advantages of each technique for analyzing different types of data strings are compared.

TECHNIQUE DESCRIPTION

Numeric-Differentiation Method

This technique, proposed by Kulinkovich and others (1966), is based on the idea of inferring a discontinuity at any point at which the trace has a numerically large derivative. From the digitized trace values x_1 to x_n, successive differences $\Delta x_i = x_{i+1} - x_i$ are produced, and a segment boundary is inferred whenever $|\Delta x_i|$ exceeds some preset value. This technique is sensitive to noise.

Split Moving-Window Method

Webster (1973) used a window width w which is passed along the trace. At each point, the window is split at its center value, and the difference in mean between the two halves is found. A segment boundary is inferred wherever this difference attains a local maximum exceeding some preset value. The window width w is a controllable parameter, and it is recommended that the trace be reanalyzed using a number of different widths. If the width is such that only two points are contained in the window, then this technique is essentially equivalent to numeric differentiation. With wider windows, the method is less affected by noise, but is also less sensitive to short segments in the trace.

Maximum Level-Variance Method

In contrast with the two preceding methods, this method allows examination of the proposed segment as a whole, rather than just its boundary. The trace is broken into k segments by choosing k-1 segment boundaries in order to maximize the between-segment sum of squared deviations or, equivalently, to minimize the within-segment sum of squared deviations. The technique is discussed in Testerman (1962), Gill (1970), and Hawkins and Merriam (1973). The last paper gives an algorithm for locating the segment boundaries. The advantage of this method is that it insures the homogeneity of each segment. The major disadvantage is that its computational requirements are higher than those of the first two methods.

Piecewise-Regression Method

This method is an extension of the maximum level-variance method. In each k segment, the trace is approximated by fitting a functional form such as a low-order polynomial by least squares. The segment boundaries are chosen in order to minimize the total residual variance of the composite approximation. An algorithm for the procedure is given by Hawkins (1972), and the two-segment situation is discussed by Quandt (1958).

In both the piecewise-regression and maximum level-variance methods, the number of k segments is a controllable parameter. It is recommended that the value of k be in one-step increments from 1 until the decrease in residual variance caused by adding an additional segment is small. The computational requirements are marginally higher than for the maximum level-variance method.

TECHNIQUE PERFORMANCE

The practical effectiveness of the various techniques can best be illustrated by a discussion of their performances on the stylized traces shown in Figure 1.

Type *a* Trace

This trace shows isolated peaks with low background noise. All four techniques will be effective in locating the peaks. The numeric-differentiation method will isolate the breaks on both sides of the peak, as will the split moving-window method applied with narrow windows. The maximum level-variance method will separate each peak as a separate segment, whereas the piecewise-regression method isolates both sides of each peak. With suitable choice of parameters, all the techniques will keep the noisy background zones as single segments.

Type *b* Trace

A number of sharp changes in level are present in this trace, together with superposed noise. The sharp level changes on the left will be isolated by all four methods. On the right, in the high-noise area, the numeric-differentiation method will isolate every local peak. The split moving window will locate the peaks if the window is narrow, but will merge them if the window is wide. The maximum level-variance and piecewise-regression methods will not split this area if the number

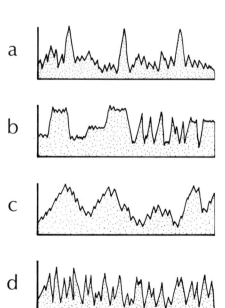

Figure 1. Four basic types of traces: *a*, isolated narrow peaks and low background noise; *b*, sharp changes in level and variable amounts of noise; *c*, "hill and valley" with gradual changes in level; *d*, few obvious breaks and high level of background noise.

of segments allowed is small; however, as the number of segments is increased, the major plateaus will be isolated.

Type c Trace

In this type of trace, no sharp breaks occur, and the numeric-differentiation and split moving-window techniques will yield poor results. The maximum level-variance method is effective in splitting the "hill" tops and "valley" floors from one another. In traces similar to this one, however, the piecewise-regression method excels. The trace may be approximated by a number of line segments in which the segment boundaries correspond to the peaks and valleys in the trace. These boundaries will be located accurately by piecewise regression. This is significant because, when traces of types a and b are filtered, type c traces result. Thus, if any trace is to be smoothed, it should be segmented by piecewise regression.

Type d Trace

This type of trace creates great difficulties, because little visible structure is present. The numeric-differentiation and split moving-window methods will tend to generate a large number of segments corresponding to each local peak; however, this adds little to one's visual impression of the trace. A trace such as this may be filtered to reduce the visual effect of noise on the trace. The disadvantage in this smoothing is that it blurs and displaces zone boundaries. The maximum level-variance method applied to the raw trace can be effective in locating relatively small shifts in mean that are not visually noticeable because of the noise. Similarly, if there is a trend in the segments, then the piecewise-regression method is indicated.

Table 1 summarizes the type of break that is best detected by each of the techniques, and Table 2 lists examples of geologic data of various types.

CONCLUSION

In many situations of practical importance, it is required to break a discrete digitized trace into a number of segments so that each segment exhibits some internal unity and is distinct from its neighbors. Four standard techniques for breaking a trace into such segments have been examined. It emerges that the numeric-differentiation and split moving-window techniques, which are computationally simple, are adequate for traces with low noise and sharp breaks between segments. Other traces should be segmented by the maximum level-variance method if the segments are trend free, and by the piecewise-regression method if the segments contain trend.

All four techniques may be implemented easily on a computer. For the maximum

TABLE 1. Technique Description

Technique	Data type	References
Numeric differentiation	Sharp breaks and spikes	Kulinkovich and others, 1966; Webster, 1973
Split moving window	Sharp breaks with plateaus	Webster, 1973
Maximum level variance	Change in mean level	Gill, 1970; Hawkins and Merriam, 1973
Piecewise regression	Differing slopes	Hawkins, 1972

TABLE 2. EXAMPLES OF DATA TYPES

a	b	c	d
X-ray curves	Electric logs	Variation in sea level through time	Tree rings
Traverses across mineralized veins	Gamma-ray logs	Magnetic profiles	Seismograms
	Drilling-time logs	Gravity profiles	Varve thickness

Note: Graphic representation of the different types a, b, c, and d is shown in Figure 1.

level-variance and piecewise-regression methods, main storage requirements are greater than for the numeric-differentiation and split moving-window techniques, and these techniques can be applied to long traces only by breaking the trace into more manageable subsections.

ACKNOWLEDGMENTS

We thank J. W. Harbaugh of Stanford University and J. E. Robinson of Union Oil Company of Canada for reading the manuscript and making suggestions for its improvement.

REFERENCES CITED

Gill, D., 1970, Application of a statistical zonation method to reservoir evaluation and digitized-log analysis: Am. Assoc. Petroleum Geologists Bull., v. 54, p. 719-729.
Hawkins, D. M., 1972, On the choice of segments in piecewise approximation: Jour. Inst. Math. Applications, v. 9, p. 250-256.
Hawkins, D. M., and Merriam, D. F., 1973, Optimal zonation of digitized sequential data: Jour. Internat. Assoc. Math. Geol., v. 5, p. 389-395.
Kulinkovich, A. Ye., Sokhranov, N. N., and Churinova, I. M., 1966, Utilization of digital computers to distinguish boundaries of beds and identify sandstones from electric log data: Internat. Geology Rev., v. 8, p. 416-420.
Merriam, D. F., 1971, Computer applications in stratigraphic problem solving, in Decision-making in the mineral industry: Canadian Inst. Mining and Metallurgy Spec. Vol. 12, p. 139-147.
Quandt, R. E., 1958, The estimation of the parameters of a linear regression system obeying two separate regimes: Jour. Am. Stat. Assoc., v. 53, p. 873-880.
Testerman, J. D., 1962, A statistical reservoir-zonation technique: Jour. Petroleum Technology (Aug.), p. 889-893.
Webster, R., 1973, Automatic soil boundary location from transect data: Jour. Internat. Assoc. Math. Geol., v. 5, p. 27-37.

MANUSCRIPT RECEIVED BY THE SOCIETY SEPTEMBER 28, 1973
REVISED MANUSCRIPT RECEIVED FEBRUARY 14, 1974

Printed in the U.S.A.

Determination of Important Parameters in a Classification Scheme

Thomas A. Jones
and
Robert A. Baker

Esso Production Research Company
P.O. Box 2189
Houston, Texas 77001

ABSTRACT

Classification and categorization, particularly the numerical classification of samples into categories, are common aspects of geological investigations. However, variables that are irrelevant to the classification or investigation are commonly included. Relevant variables are often difficult to isolate, and objective methods for pointing them out are lacking.

This paper describes a semiobjective scheme for identifying important variables; the scheme can be reduced to basic statistical computations, with some repetition of the numerical classification method. The steps are as follows: (1) Classify or cluster the data and determine the best classification; (2) select a set of potentially important parameters by using the methods described in the paper; (3) reclassify the samples, but use only the parameters selected in step 2; and (4) if the results of step 3 agree with step 1, the procedure is ended; otherwise, go to step 2 and modify the list of selected parameters.

The procedures are illustrated in detail with an analysis of foraminiferal data collected from the Gulf of Mexico. A classification of 38 samples into five categories, based on 252 species, is shown to be obtainable with only 14 species. Reduction in the number of species to be considered may result in future savings in time, both in interpretation and data gathering.

INTRODUCTION

Most geologic data, whether samples or measurements on samples, may be classified or categorized. The categories, for example, may delineate lithologies, depositional environments, biological zones, and so forth. A convenient method of determining categories is cluster analysis, a computerized classification method that places samples into groups on the basis of similarity. We have found cluster analysis to be worthy of application to a variety of problems; the results presented here illustrate an extension beyond its normal usage.

A search for order in a set of data commonly reveals that only part of the data is relevant to the order. It is usually desirable to reduce the number of variables or parameters in an analysis to simplify interpretation and data handling. The purpose of this paper is to demonstrate a procedure whereby significant parameters may be identified.

We assume that the basic categorization of the data is known, either a priori or by use of some classification technique such as cluster analysis. The problem is to answer the question, Are some variables more important than others in defining the classification? We wish to find those parameters that are most useful. The following discussion deals with the analysis of biologic data, but the same reasoning may be applied to other types of data.

The technique we used is not difficult (with the help of a computer) and should be applicable to many geological problems. We used cluster analysis initially to classify our samples semiobjectively. We then computed a quality statistic for each species in each of the sample groupings. This statistic, discussed below, was used to select the best species for defining the sample groupings. The best species were then selected, and the cluster analysis was rerun to determine if a good set had been found.

Our analysis has been applied to bottom samples of foraminifers from the northern Gulf of Mexico (traverse 1, Fig. 1). The samples were grouped by cluster analysis, and the clusters seem to reflect water depth. The original analysis involved 252 species. However, we were able to obtain the same classification by use of only 14 selected species. This required four trials through the procedure.

METHOD OF ANALYSIS

Quality Statistic

A good geological parameter must be (1) easy to detect, (2) consistent, and (3) characteristic. The selection of important parameters lies in quantifying these or other qualities which may be considered essential to a good parameter. In our solution we combine these qualities into a single measure of geological "goodness."

Ease of Detection. In paleontological samples, an easily detected species is usually one that occurs in abundance because it is most likely to occur in the samples. The same relations hold for other geological parameters; for instance, large quantities of a chemical compound are more easily detected than small quantities, and a thick sandstone bed is easier to detect than a thin one. We have chosen the mean value of the parameter as a measure of "ease of detection" because it is a reasonable measure and it can be obtained readily from most data.

Consistency. Consistency is also required of a good geological parameter. If a particular species occurs only in a given facies, but in only a few of the samples from that facies, it cannot be depended upon for identification of the facies. If the species is found, we may conclude that the sample is from the given facies; however, if the species is not found, we do not know if the sample is from the given facies or not. The inconsistency of a parameter may be measured by the standard deviation of the values of the parameter within the category (group of samples). If the parameter varies greatly within a suite of samples, the standard deviation will be large, indicating a high degree of inconsistency. We have taken the *inverse* of the standard deviation as the measure of consistency.

Characterization. A characteristic parameter is one that does not occur in samples outside the specified category or group of samples, or it is one that has a definite functional relation to the sample. We have attempted to quantify this quality by comparing abundance and standard deviation of the species *in other groups*. If a species is markedly more abundant and consistent within a given group than outside of that group, it may be called characteristic of that group.

We have often been successful in our efforts to identify key species by computer, but there remains a definite need for human judgment. Using various formulas combining mean abundance and standard deviation in different ways, the computer will list the parameters (that is, species) objectively according to their usefulness in each of the clusters of samples. But the process of selecting the key parameters requires human judgment and some trial and error.

General Procedure

The three steps of the analysis are as follows: (1) Cluster the data and determine the best classification, (2) determine the important parameters, and (3) test and modify the list of important parameters until satisfactory. These steps are described in more detail below.

Cluster analysis places samples into hierarchical groups, commonly in the form of dendrograms. The geologist determines from these diagrams which samples are to be considered as a natural group. We can recommend no statistical technique for objectively determining the level at which the hierarchy contains more than one significant cluster. This step is done strictly by eye and human judgment. The uses and interpretations of cluster analyses have been discussed by a number of authors, including Sokal and Sneath (1963), Mello and Buzas (1968), Demirmen (1969), Wishart (1969), Hazel (1970), and Blackith and Reyment (1971).

There are many different ways of computing numerical similarity and of clustering samples, and part of the process is to decide which of several methods shows

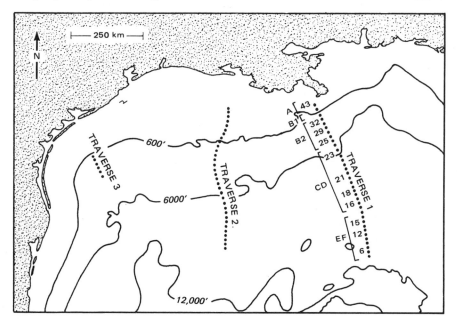

Figure 1. Northern Gulf of Mexico deep-water ecology traverses, with selected sample and cluster locations shown for traverse 1. Contours indicate water depth.

the best groupings. This can be determined only by comparing the numerical groupings with any geological knowledge of the samples. Knowledge of which method or coefficient works best for a specific problem will simplify further processing, such as zonation of other wells or traverses in the area.

After the groups have been satisfactorily determined, the important parameters that generated them must be found. We developed two programs that perform elementary operations on the data and reorder the species according to their goodness, as judged by the algorithm chosen by the user. PIP1 is used to inspect the make-up of each group or cluster of samples, and PIP2 is used to get an overall comparison of the groups.

The most important quality overlooked by PIP1 and PIP2, and to be noted by the analyst, is characterization. A species may be abundant and consistent in every

TABLE 1. ABUNDANCES OF SPECIES, ORDERED BY QUALITY, FOR CLUSTER A

Species	42	41	40	39	38
P56B	34.0	34.0	30.0	31.0	34.0
P53F	15.0	18.0	11.0	10.0	15.0
P86A	10.0	8.0	7.0	7.0	5.0
P58H	7.0	4.0	14.0	13.0	9.0
P55A	2.0	2.0	5.0	4.0	4.0
C23F	6.0	5.0	4.0	4.0	11.0
P56C	8.0	13.0	7.0	2.0	9.0
P58D	6.0	7.0	5.0	4.0	1.0
P0FA	3.0	2.0	8.0	12.0	8.0
C1YG	4.0	26.0	22.0	23.0	1.0
P58A	1.0	2.0	1.0	3.0	2.0
P53D	1.0	1.0	5.0	7.0	9.0
C42B	3.0	2.0	1.0	0.5	4.0
C27E	2.0	10.0	17.0	52.0	43.0
C21B	9.0	13.0	4.0	2.0	1.0
C0VB	0.0	0.5	10.0	5.0	9.0
C1FA	2.0	1.0	10.0	1.0	7.0
P56A	3.0	1.0	0.5	3.0	0.5
P91A	1.0	1.0	1.0	1.0	1.0
C23C	4.0	4.0	0.5	0.5	0.5
P53A	1.0	1.0	5.0	1.0	1.0
C62B	3.0	0.5	0.5	0.5	3.0
C21N	11.0	23.0	3.0	1.0	0.5
C1ZA	0.5	1.0	10.0	5.0	0.5
P54A	6.0	2.0	0.5	0.5	1.0
C27B	12.0	6.0	1.0	0.5	0.0
C21C	3.0	2.0	0.5	0.5	0.0
C89A	0.5	0.5	2.0	1.0	0.5
P53B	1.0	0.5	0.5	2.0	0.5
C28B	0.5	1.0	1.0	0.5	1.0
C30D	0.5	0.5	7.0	1.0	1.0
C30K	4.0	0.5	0.5	0.5	0.5
C75A	4.0	0.5	0.5	0.5	0.5
C0EA	1.0	0.5	0.5	0.5	1.0
C28A	1.0	1.0	1.0	0.5	0.0
C75C	4.0	0.5	0.0	0.5	0.5
C23A	0.0	0.0	0.5	1.0	7.0
A82B	0.5	0.5	0.5	0.5	1.0
C68A	0.5	1.0	0.5	0.5	0.5
P1XA	0.5	1.0	0.5	0.5	0.5
P53E	0.5	0.5	1.0	0.5	0.5
P58I	0.5	2.0	0.5	0.5	0.0

group, but may not be characteristic. Species that are not characteristic will blur rather than sharpen the classification.

We compute the quality of each species within each sample grouping to help find the important parameters. Quality here is measured for each species by $\bar{x}^2/(1 + s^2)$, where $\bar{x}$ and s^2 are the mean and variance of the species counts for the samples in each group. The species are ordered in each grouping by this statistic, so that those with the highest quality appear at the top of the list. The faunal lists thus organized by PIP1 are compared with one another, as shown in part by Tables 1 through 5; the tabled values show the data for each sample in the cluster, with the species ordered by the quality statistic. By this comparison, we are able to estimate how characteristic each species is to each group. Table 7 shows a list from PIP2 that is helpful in making the selections; the species are

TABLE 2. ABUNDANCES OF SPECIES, ORDERED BY QUALITY, FOR CLUSTER B1

Species	37	36	35	34	33	32	31
P56B	26.0	22.0	35.0	22.0	34.0	24.0	24.0
P53F	11.0	10.0	9.0	5.0	13.0	10.0	8.0
P53D	8.0	10.0	4.0	6.0	9.0	5.0	8.0
P58H	19.0	11.0	16.0	11.0	6.0	11.0	10.0
C1FA	14.0	9.0	5.0	7.0	13.0	18.0	15.0
P86A	8.0	13.0	5.0	13.0	10.0	5.0	6.0
C23A	20.0	18.0	20.0	8.0	8.0	9.0	7.0
P0FA	11.0	10.0	6.0	16.0	5.0	7.0	3.0
C57A	5.0	6.0	2.0	3.0	5.0	2.0	2.0
P56C	3.0	13.0	7.0	4.0	5.0	10.0	16.0
P58D	2.0	1.0	2.0	4.0	5.0	8.0	6.0
C27F	18.0	16.0	10.0	4.0	9.0	2.0	2.0
P91A	1.0	2.0	3.0	1.0	1.0	4.0	3.0
C95A	0.0	0.5	3.0	8.0	6.0	6.0	6.0
C1YH	1.0	3.0	7.0	25.0	13.0	11.0	7.0
C41A	4.0	3.0	2.0	1.0	2.0	2.0	9.0
C29A	0.5	1.0	3.0	1.0	2.0	6.0	3.0
C23D	1.0	1.0	3.0	2.0	2.0	9.0	7.0
P54A	3.0	0.5	2.0	1.0	1.0	1.0	4.0
C23F	3.0	6.0	3.0	11.0	5.0	0.0	0.5
C21B	1.0	3.0	1.0	12.0	2.0	4.0	7.0
C30D	1.0	1.0	3.0	1.0	5.0	1.0	7.0
A1WD	0.5	0.5	3.0	2.0	5.0	3.0	0.5
C28B	5.0	3.0	3.0	1.0	2.0	0.0	0.5
A34G	0.5	1.0	2.0	1.0	2.0	3.0	0.5
C0EA	1.0	1.0	1.0	1.0	1.0	1.0	1.0
A1*B	2.0	1.0	2.0	0.5	1.0	0.5	1.0
P58A	1.0	3.0	1.0	1.0	1.0	5.0	0.5
P53A	3.0	0.5	1.0	1.0	2.0	1.0	0.5
C62B	0.5	1.0	3.0	2.0	0.5	1.0	0.5
P56A	1.0	0.5	0.5	1.0	2.0	1.0	1.0
A64A	0.0	0.5	3.0	1.0	5.0	7.0	0.0
C0VB	11.0	7.0	3.0	1.0	0.5	0.0	1.0
P55A	2.0	1.0	4.0	4.0	2.0	23.0	6.0
C21H	2.0	2.0	1.0	1.0	0.5	0.5	0.0
C68A	2.0	0.5	3.0	1.0	0.5	0.0	0.5
A0*A	0.5	3.0	1.0	0.5	1.0	0.5	0.5
A35A	1.0	2.0	1.0	0.5	0.5	0.5	0.5
C95B	1.0	2.0	1.0	0.5	0.5	0.5	0.5
C62C	0.5	0.5	0.5	0.5	1.0	1.0	4.0
P58I	0.0	0.5	0.5	5.0	0.5	3.0	0.5
P55B	0.5	1.0	3.0	0.5	0.5	0.5	0.5

not ordered by any statistic, but their average abundance in each cluster is shown, facilitating comparison of cluster contents.

Our quality statistic is by no means unique or all-inclusive. The use of the programs with the statistic provides a convenient and useful ordering of the samples for further study. Other methods of analysis, such as principal components, factor, and canonical variates analysis (for example, see Seal, 1964; Buzas, 1972; Blackith and Reyment, 1971) can be used to aid in selecting species at this step in the procedure. We feel that applications of these techniques have only partly completed the analysis and that the third step in our procedure is necessary.

The samples are next reclustered, using only those parameters (species) considered important above. If the resulting classification matches the original, the process is ended. If there is a difference in classifications, the selected-species list must be examined, and possibly altered, as in step 2. In this example, and in other

TABLE 3. ABUNDANCES OF SPECIES, ORDERED BY QUALITY, FOR CLUSTER B2

Species	30	29	28	27	26	25	24
P56B	31.0	27.0	24.0	23.0	17.0	28.0	27.0
P56C	10.0	14.0	13.0	6.0	16.0	17.0	12.0
P0FA	5.0	6.0	10.0	4.0	7.0	5.0	6.0
P86A	4.0	8.0	10.0	6.0	9.0	5.0	5.0
C23A	31.0	29.0	29.0	19.0	15.0	13.0	13.0
P58D	6.0	7.0	10.0	3.0	7.0	9.0	5.0
P58H	8.0	13.0	14.0	7.0	18.0	18.0	6.0
C1YH	9.0	12.0	9.0	3.0	4.0	10.0	6.0
P53F	6.0	6.0	5.0	9.0	14.0	15.0	6.0
C62C	6.0	3.0	3.0	3.0	6.0	4.0	12.0
P54A	2.0	1.0	3.0	2.0	2.0	2.0	5.0
C95A	4.0	4.0	12.0	2.0	9.0	3.0	4.0
C30L	0.5	1.0	6.0	4.0	7.0	11.0	9.0
P91A	3.0	1.0	1.0	2.0	1.0	2.0	2.0
C68A	2.0	1.0	2.0	1.0	2.0	5.0	2.0
C21B	6.0	2.0	3.0	1.0	4.0	3.0	11.0
P58I	1.0	0.5	3.0	1.0	2.0	3.0	2.0
C0EA	3.0	2.0	0.5	1.0	6.0	3.0	2.0
P55A	3.0	6.0	2.0	9.0	0.5	1.0	6.0
C74A	1.0	1.0	0.5	3.0	6.0	7.0	4.0
P53D	6.0	5.0	2.0	23.0	5.0	2.0	12.0
C23D	4.0	1.0	0.5	13.0	4.0	1.0	8.0
C30D	6.0	3.0	7.0	0.5	2.0	0.5	0.5
C23F	3.0	3.0	12.0	0.5	2.0	7.0	0.5
P56A	2.0	2.0	2.0	1.0	0.5	0.5	0.5
P53B	2.0	0.5	0.5	0.5	3.0	0.5	2.0
C41A	5.0	6.0	4.0	0.5	0.5	0.0	0.5
C30B	3.0	1.0	1.0	0.5	0.5	0.5	4.0
C0EC	2.0	2.0	0.5	1.0	2.0	0.5	0.0
A1WC	0.0	1.0	2.0	8.0	3.0	2.0	0.5
A0XA	0.5	0.5	3.0	3.0	1.0	0.5	0.5
A35A	0.5	0.5	0.5	2.0	1.0	2.0	0.5
C57A	2.0	4.0	3.0	0.5	0.5	0.0	0.0
P58G	1.0	1.0	0.5	1.0	5.0	1.0	0.5
M0GD	0.0	0.0	0.0	13.0	4.0	1.0	8.0
P53E	2.0	0.5	0.5	2.0	0.5	0.5	0.5
P58A	0.5	2.0	0.5	0.5	0.5	0.5	2.0
C93E	0.5	0.5	1.0	1.0	1.0	1.0	0.5
A34G	0.5	2.0	1.0	1.0	3.0	0.0	0.0
C27G	0.5	0.5	0.5	0.5	6.0	8.0	0.5
M1BA	0.5	0.5	0.0	0.5	0.5	4.0	5.0
A01A	0.5	0.5	0.5	2.0	1.0	0.5	0.5

analyses not described here, we have found that the selection and reclustering step may be repeated three or four times before settling on the classification and parameters which appear to be satisfactory. Specialized knowledge, such as that of a paleontologist, should be used to criticize the results and aid in correlation or environmental interpretation during and after the selection process.

Simply doing an R-mode cluster analysis (clustering species into groups) will be inadequate in most cases (for example, Maddocks, 1966; Mello and Buzas, 1968). Any clustering method *must* place an entity into a cluster and only that cluster. Hence, a species that actually occurs in one or more clusters will not be grouped correctly because it will appear that the species is limited in extent.

CLUSTER ANALYSIS OF ECOLOGY DATA

Gravity and Ewing cores were taken at points along three traverses in the northern Gulf of Mexico (Fig. 1), and live foraminifers in each of the samples were identified and counted. Traverse 1 consists of 38 sampling locations, numbered 6 through 43 (deepest to shallowest). These 38 samples contained 252 identified species of foraminifer. For analytical purposes the foraminifers were given four-character code names. The data are discussed here in terms of these numbers and codes. We will discuss only the zonation of traverse 1; zonation and species selection were also done for the other two shorter traverses with results similar to those of traverse 1.

We desired to cluster the samples only on the basis of their faunal content with the objective of defining a sound ecological classification. The nearest-neighbor-fusion procedure was used to cluster the samples (Wishart, 1969; Blackith and Reyment, 1971), and the product-moment (correlation) coefficient indicated the similarity between samples. The resulting dendrogram was analyzed and the clusters coded as A, B1, B2, CD, and EF. Their locations along the traverse are shown in Figure 1. A possibly more convenient form for this classification is shown in Table 8, column a. The samples in B1 and B2 are actually subdivisions of a larger cluster B; CD and EF were each combined from smaller clusters.

It may be noted that samples 13 and 43 are not assigned cluster codes. The data show that neither of these two samples is very similar to any of the other 36. Because these samples appear to be misfits, they are not considered further in the clustering procedure. The biologist or paleontologist would be expected to examine these samples in further detail on an individual basis.

The next step was a test of the classification. The samples and clusters are plotted versus water depth in Table 8. There is no overlap, and Figure 1 and Table 8 suggest that the clusters may be influenced by an environmental factor associated with or related to water depth, distance from shore, or both. (The samples along traverse 2 also show a similar relation to water depth, though unrelated to distance from shoreline.) External geologic data should be used to test the results of cluster analysis. If there had been significant overlap on these two figures, we would have had to define the clusters more coarsely or search for other explanatory external variables.

With the stations now grouped on the basis of similar fauna, we wish to know which species are important in defining the various zones. One method by which one could select the key species is by tabulating species occurrence against clusters. This is merely a distribution chart ("scramble sheet") for clusters rather than samples. A more automated procedure of selecting key species involves the use of the methods described in the following section.

DETERMINATION OF KEY PARAMETERS

Table 8, column a, shows a plot of the 38 samples and their clusters versus water depth for traverse 1. A total of 252 species of foraminifer was used for this classification. Table 8, column b, shows the water depth plot of the same samples but with the clusters defined by use of only 14 selected species. Remarkable agreement was obtained between the original data and the selected data.

The 14 species we selected should not be regarded as unique; it may well be that another list could be found which would do equally well or possibly better. Our technique is adequate, but possibly not optimal. However, to show that our method is useful, we randomly selected 30 species from the total list of 252. These 30 species were used to cluster the samples; the zonation obtained is shown in Table 8, column c. Note that five clusters were found but that they are not in agreement with the original zonation and are themselves internally inconsistent.

TABLE 4. ABUNDANCES OF SPECIES, ORDERED BY QUALITY, FOR CLUSTER CD

Species	23A	23	22	21	20	19	18	17	16
P56B	40.0	33.0	31.0	58.0	48.0	42.0	21.0	31.0	25.0
C30L	15.0	13.0	9.0	16.0	3.0	9.0	6.0	14.0	16.0
P53F	16.0	16.0	11.0	8.0	13.0	11.0	12.0	4.0	4.0
P86A	6.0	5.0	5.0	3.0	5.0	9.0	11.0	9.0	11.0
P56C	19.0	14.0	12.0	11.0	8.0	8.0	24.0	23.0	28.0
P58H	2.0	3.0	3.0	1.0	2.0	5.0	6.0	5.0	2.0
P58D	4.0	4.0	3.0	3.0	4.0	3.0	13.0	6.0	10.0
C1YH	2.0	5.0	4.0	2.0	2.0	5.0	0.5	2.0	6.0
C27G	9.0	12.0	12.0	12.0	11.0	3.0	0.5	5.0	1.0
C68A	0.5	8.0	6.0	22.0	10.0	7.0	6.0	28.0	24.0
P58A	1.0	3.0	2.0	1.0	1.0	0.5	3.0	3.0	2.0
C0EA	5.0	5.0	4.0	6.0	3.0	0.5	0.0	0.0	3.0
C30B	3.0	3.0	6.0	5.0	6.0	2.0	19.0	3.0	8.0
C62C	4.0	6.0	8.0	3.0	5.0	1.0	0.0	2.0	0.5
P0FA	2.0	23.0	4.0	3.0	2.0	6.0	6.0	8.0	8.0
C93E	0.5	0.5	1.0	2.0	2.0	5.0	0.5	5.0	4.0
C23A	4.0	4.0	2.0	2.0	2.0	0.5	6.0	0.5	0.0
P53D	2.0	8.0	18.0	5.0	8.0	6.0	1.0	2.0	1.0
C02A	5.0	9.0	15.0	9.0	20.0	0.0	6.0	0.0	0.5
P54A	1.0	0.5	2.0	1.0	2.0	2.0	0.5	1.0	1.0
P91A	0.5	1.0	0.5	1.0	1.0	2.0	1.0	1.0	3.0
P58I	0.5	0.5	2.0	0.5	0.5	0.5	2.0	2.0	2.0
C74A	13.0	9.0	5.0	0.5	0.5	2.0	0.5	3.0	2.0
P55A	2.0	6.0	2.0	1.0	3.0	0.5	0.5	0.5	0.5
A1WC	1.0	0.5	1.0	0.5	0.5	2.0	0.0	3.0	4.0
A39A	0.5	0.5	2.0	0.5	0.5	2.0	6.0	3.0	0.5
C95A	3.0	3.0	2.0	2.0	3.0	0.0	13.0	2.0	0.5
C30F	0.5	0.5	0.5	0.5	2.0	2.0	0.5	0.5	4.0
M18A	0.5	1.0	2.0	0.5	0.5	0.5	0.0	5.0	2.0
A64A	0.5	0.5	0.5	2.0	0.5	0.5	0.0	2.0	4.0
A71A	1.0	0.5	0.5	2.0	2.0	0.5	0.0	0.0	2.0
C57A	0.5	0.5	0.5	0.5	0.0	0.5	0.5	6.0	3.0
C30I	0.5	0.5	0.5	2.0	2.0	0.5	0.0	0.0	1.0
C62A	0.5	1.0	2.0	4.0	0.5	0.0	0.0	1.0	0.0
C42A	0.5	0.5	0.5	0.5	0.5	0.5	0.5	2.0	0.5
C1YA	0.5	0.5	0.5	5.0	6.0	0.0	0.0	0.0	0.5
P53E	0.5	0.5	1.0	0.5	1.0	0.5	0.5	0.5	0.5
C41A	1.0	3.0	2.0	0.5	0.5	0.5	0.0	0.0	0.0
C42D	3.0	2.0	0.5	0.0	5.0	0.0	0.0	0.0	0.0
A0XA	0.5	0.5	0.5	0.0	0.5	1.0	6.0	1.0	0.5
C23F	0.5	0.5	0.5	0.5	0.5	3.0	0.0	0.5	0.5
C0EC	1.0	0.0	0.5	0.0	0.5	0.0	0.0	2.0	2.0

Therefore, although we may significantly reduce the parameter list, we must do so carefully.

Columns d and e of Table 8 represent, in effect, intermediate steps between the random grouping of column c and the well-ordered grouping of column b. Column d is the result of using PIP1 and PIP2 without human judgment. Column e is the result of using human judgment along with the two computer programs. Column b is the result of reconsideration of column e and the data provided by PIP1. The steps involved are discussed in detail below.

First Trial Solution

Let us attempt to redefine the original classification with from 10 to 15 species. Tables 1 through 5 give the computer code numbers of the important species for each cluster as determined by PIP1. If we select the first few species in each

TABLE 5. ABUNDANCES OF SPECIES, ORDERED BY QUALITY, FOR CLUSTER EF

Species	15	14	12	11	10	8	7	6
P53F	22.0	26.0	23.0	23.0	28.0	30.0	23.0	25.0
C30L	25.0	35.0	44.0	28.0	28.0	24.0	31.0	29.0
P56B	28.0	17.0	17.0	21.0	31.0	15.0	11.0	15.0
P56C	16.0	12.0	14.0	19.0	6.0	7.0	7.0	16.0
P58D	7.0	6.0	6.0	4.0	0.5	7.0	5.0	8.0
C83B	6.0	2.0	8.0	6.0	1.0	8.0	4.0	8.0
P86A	7.0	5.0	6.0	4.0	2.0	2.0	3.0	2.0
P53D	5.0	21.0	6.0	6.0	3.0	13.0	8.0	8.0
C57A	4.0	4.0	5.0	5.0	6.0	0.5	1.0	0.5
P0FA	3.0	5.0	3.0	4.0	2.0	0.5	0.5	2.0
C68A	10.0	20.0	10.0	17.0	6.0	0.0	0.0	9.0
P58H	3.0	2.0	2.0	3.0	2.0	0.5	1.0	0.5
P53E	0.5	0.5	3.0	2.0	4.0	9.0	6.0	4.0
C42A	0.5	6.0	0.0	4.0	0.5	9.0	4.0	6.0
P54A	2.0	0.5	2.0	2.0	1.0	1.0	1.0	4.0
P55A	2.0	3.0	5.0	5.0	4.0	13.0	28.0	11.0
C02A	1.0	2.0	1.0	4.0	7.0	13.0	20.0	22.0
C30B	6.0	4.0	4.0	1.0	10.0	0.5	0.0	1.0
A01A	1.0	0.0	0.0	2.0	0.5	2.0	3.0	2.0
C93E	3.0	3.0	3.0	6.0	0.5	0.5	0.0	0.0
C30E	6.0	2.0	0.5	0.5	0.5	1.0	3.0	1.0
C42D	1.0	0.0	0.0	3.0	11.0	17.0	6.0	2.0
M1BA	2.0	2.0	0.5	2.0	0.5	0.5	0.5	0.5
P58G	0.5	0.5	1.0	0.5	9.0	5.0	4.0	0.5
C42C	1.0	0.0	2.0	0.5	1.0	7.0	3.0	1.0
C27G	7.0	3.0	0.0	2.0	2.0	0.0	0.5	1.0
C74A	0.5	3.0	7.0	0.5	1.0	0.5	0.5	1.0
P58A	1.0	0.5	0.5	1.0	2.0	0.5	0.5	0.5
P58I	1.0	1.0	2.0	0.5	0.5	0.5	0.5	0.5
C93C	3.0	2.0	0.0	1.0	3.0	0.0	0.0	0.0
A39A	0.5	0.5	1.0	0.5	1.0	0.5	0.5	1.0
A71A	2.0	2.0	0.0	0.5	0.5	0.0	1.0	0.5
P55B	0.5	0.5	0.5	2.0	0.5	0.5	0.5	0.5
P91A	2.0	0.5	2.0	2.0	21.0	1.0	2.0	1.0
C62A	0.0	3.0	4.0	0.5	0.0	0.0	0.0	1.0
C0ED	1.0	0.0	4.0	1.0	1.0	0.0	0.0	0.5
M0GE	0.0	0.0	0.0	0.5	2.0	0.5	0.5	2.0
C1YK	1.0	0.5	0.0	1.0	0.5	0.5	0.5	0.5
C62C	0.0	0.0	2.0	2.0	1.0	0.0	0.0	0.5
C27A	0.0	0.0	0.0	0.0	1.0	2.0	1.0	1.0
C0EA	0.0	0.0	0.0	0.5	1.0	3.0	0.5	0.5
C1ZA	0.5	0.5	0.5	3.0	1.0	0.0	0.0	0.0

list, we would expect to have a reasonable set. Table 6 shows the five species in each cluster with the highest value of the goodness coefficient. There are duplications, so a combined list of these species is also given. The combined list contains 12 species, and these species seem to be the relevant ones.

The stations were clustered by using these 12 species. The dendrogram showed only four clusters, designated R, S, T, and U; these are plotted in the water-depth chart of Table 8, column d. Cluster T corresponds to EF, but this is about the only match in the comparison. Clusters R and S do not correspond to the original zonation at all and, indeed, are not even self-consistent; they overlap each other. It would thus appear that merely taking those species in each cluster with the highest PIP1 coefficient is not much better than random selection.

There are two possible causes for the failure of the 12 species to duplicate the classification. The abundance of the planktonic species (first code digit, P) were computed separately from the benthic species; inasmuch as there are fewer planktonic species than benthic, the abundance percentages tend to be higher, and the goodness coefficient unduly influenced. This is equivalent to using different scales of measurement for the two types of foraminifers. If the percentages had all been computed together, the planktonic percentages would have been lower, and the output list from PIP1 would probably not have been weighted so heavily in favor of planktonic species.

The second reason for the failure to duplicate the classification is that we did not take into account that a parameter must be characteristic of the cluster under consideration. For instance, study of Tables 1 through 5 shows that P56B has approximately the same abundance throughout all the stations in the traverse; this species is not characteristic of any single cluster because it occurs abundantly in all of them.

We could compare Tables 1 through 5 for each species and determine which species are characteristic for each cluster; however, this would be tedious. A computer program (PIP2) has therefore been written to accomplish this task. For each species occurring in the first half of the lists of Tables 1 through 5, Table 7 (output from PIP2) provides the mean abundance of the various clusters. Use of Table 7 simplifies the task of selecting the key species because it shows, among those species already found worthy of further consideration, which ones are characteristic of the individual clusters.

Ideally, we would wish to find a set of species such that species 1 occurs only in cluster A, species 2 only in cluster B1, and so on. When this situation applies, the species are each characteristic of their own cluster and none else; therefore,

TABLE 6. SPECIES WITH HIGHEST COEFFICIENT VALUES FOR EACH CLUSTER AND COMBINED LIST

A	B1	B2	CD	EF	Combined
P56B	P56B	P56B	P56B	P53F	P56B
P53F	P53F	P56C	C30L	C30L	P53F
P86A	P53D	P0FA	P53F	P56B	P86A
P58H	P58H	P86A	P86A	P56C	P53D
P55A	C1FA	C23A	P56C	P58D	P56C
					P0FA
					C30L
					P58H
					P55A
					C1FA
					C23A
					C58D

the original zonation could be duplicated easily. This is a rare situation. A given species may occur in many clusters but with varying abundances. Generally, it is necessary to use pairs of species to distinguish a given cluster from another. This implies that at least two species are usually necessary to distinguish each cluster. We therefore attempt to duplicate the five zones on the basis of 10 species.

Second Trial Solution

Looking at Table 7, we find that species C27B occurs in cluster A and nowhere else. Table 1, however, shows that there is a wide range of abundances, ranging

TABLE 7. AVERAGE SPECIES ABUNDANCE FOR EACH CLUSTER

Species	A	B1	B2	CD	EF
A1WD	0.30	2.07	0.64	0.0	0.0
C02A	0.0	0.0	0.50	7.17	8.75
C21B	5.80	4.29	4.29	0.28	0.13
C21N	7.70	0.21	0.0	0.0	0.19
C23A	1.70	12.86	21.29	2.33	0.13
C23C	1.90	0.07	0.0	0.06	0.06
C23D	0.30	3.57	4.50	0.67	0.44
C23F	6.00	4.07	4.00	0.72	0.19
C27B	3.90	0.0	0.0	0.0	0.0
C27E	24.80	8.71	0.14	0.11	0.19
C27G	0.10	2.43	2.36	7.28	1.94
C28B	0.80	2.07	0.0	0.0	0.0
C30B	0.0	0.43	1.50	6.11	3.31
C30F	0.0	0.0	0.14	1.22	1.81
C30D	2.00	2.71	2.79	0.28	0.0
C30L	0.0	0.14	5.50	11.22	30.50
C41A	0.0	3.29	2.36	0.83	0.13
C42A	0.0	0.07	0.29	0.67	3.75
C42B	2.10	0.29	0.0	0.22	0.0
C57A	0.40	3.57	1.43	1.33	3.25
C62C	0.20	1.14	5.29	3.28	0.69
C68A	0.60	1.07	2.14	12.39	9.00
C74A	0.0	0.36	3.21	3.94	1.75
C83B	0.0	0.0	0.0	1.67	5.38
C93E	0.0	0.29	0.79	2.28	2.00
C95A	0.0	4.21	5.43	3.17	0.06
C0EA	0.70	1.00	2.50	2.94	0.69
C0VB	4.90	3.36	0.36	0.11	0.0
C1FA	4.20	11.56	1.00	0.28	0.0
C1YB	0.30	0.0	0.0	0.06	0.06
C1YH	0.50	9.57	7.57	3.17	0.50
P53A	1.80	1.29	0.50	0.22	0.06
P53D	4.60	7.14	7.86	5.67	8.75
P53F	13.80	9.43	8.71	10.56	25.00
P55A	3.40	6.00	3.93	1.78	8.88
P56B	32.60	26.71	25.29	36.56	19.38
P56C	7.80	8.29	12.57	16.33	12.13
P58D	4.60	4.00	6.71	5.56	5.44
P58G	0.10	0.71	1.43	0.39	2.63
P58H	9.40	12.00	12.00	3.22	1.75
P86A	7.40	8.57	6.71	7.11	3.88
P91A	1.00	2.14	1.71	1.22	3.94
P0FA	6.60	8.29	6.14	6.89	2.50
No. in cluster	5	7	7	9	8

from 12 percent in the shallowest sample (42) to zero in sample 38. We would expect this species to tie stations 40, 41, and 42 together, but stations 38 and 39 may tend to be grouped with all the other stations. However, we shall tentatively use this species. Further study of Table 7 shows that C1YE is abundant in cluster A, much less so in B1, and absent in the other clusters. Table 2 shows that C1YE has large variation, but it may be useful to separate A from B1.

Considering cluster B1, we see that C28B is found in B1, less so in A, and nowhere else. However, comparison of Tables 1 and 2 shows that there is much overlap between the values in A and B1, and thus this species may not be useful. Species A1WD has a similar situation. We are thus forced somewhat from the optimal characterization and must go to the pair approach. Species C1YH occurs strongly in B1, somewhat less so in B2, weakly in CD, and only in trace amounts

TABLE 8. SAMPLES AND CLUSTERS VERSUS WATER DEPTH, TRAVERSE 1

Station no.	Water depth (m)	a Clusters, all species	b Clusters, final species list	c Clusters, 30 random species	d Clusters, first 5 from PIP1	e Clusters, species selected by PIP1 and PIP2
43	151	..	..	V	..	A
42	230	A	A	V	R	A
41	297	A	A	W	R	A
40	373	A	A	X	R	A
39	427	A	A	X	R	A
38	521	A	B1	X	R	B
37	594	B1	B1	X	R	B
36	661	B1	B1	X	S	B
35	715	B1	B1	X	S	B
34	800	B1	B1	X	S	B
33	897	B1	B1	X	R	B
32	991	B2	B1	Y	R	B
31	1,103	B2	B1	Y	R	B
30	1,239	B2	B2	W	S	B
29	1,315	B2	B2	W	S	B
28	1,388	B2	B2	V	S	B
27	1,448	B2	B2	X	..	B
26	1,554	B2	CD	Y	S	B
25	1,648	B2	CD	Y	S	B
24	1,670	B2	B2	X	R	B
23A	1,782	CD	CD	Z	R	CD
23	1,870	CD	CD	Z	R	CD
22	1,871	CD	CD	Z	R	CD
21	2,039	CD	CD	Z	R	CD
20	2,079	CD	CD	Z	R	CD
19	2,112	CD	..	W	R	..
18	2,300	CD	CD	V	U	EF
17	2,318	CD	CD	Y	U	EF
16	2,427	CD	CD	V	U	EF
15	2,524	EF	EF	Y	T	EF
13	2,642	..	..	W	..	..
14	2,688	EF	EF	X	T	EF
12	2,788	EF	EF	W	T	EF
11	2,879	EF	EF	W	T	EF
10	2,958	EF	EF	Z	T	EF
8	3,167	EF	EF	Z	T	EF
7	3,230	EF	EF	Z	T	EF
6	3,467	EF	EF	Z	T	EF

in A and EF. Hence, this species cannot distinguish between B1 and B2 but should be able to separate B1 and B2 from the other clusters. Let us use C1YH and C28B.

Cluster B2 presents further problems. Here no unique species exists and we must use a pair distinction. Note that C23A is more abundant in B2 than elsewhere, though it is common in B1. Although the means are different, the variation in B1 and B2 is such that the abundances overlap; thus, the stations in these clusters may not be correctly separated. C62C has similar characteristics. The variance is much less here though; therefore, overlap is not exceedingly large.

In CD we find that C27G has abundances greater than those of the surrounding clusters, and we use this species. Species C30B and C68A also seem to be reasonable choices. We find in Table 4 that C30B is less variable than C68A, and we shall use C30B and C27G at this point.

Finally, it is apparent that C30L can be used to distinguish EF from the other clusters and that this species also can be used to separate CD from B2 and B2 from B1. Because of its wide range of values, C30L is a useful species. Looking at Table 7 alone would seem to indicate that C42A may be useful. However, Table 5 shows a great variation in abundance at the stations in EF, and C42A is rejected. C83B appears to be characteristic of EF; thus, it will be used.

We have thus selected the following 10 species as being useful in defining the zonation: C1YE, C23A, C30B, C27B, C62C, C30L, C1YH, C27G, C83B, and C28B. These species were used and the samples were clustered as usual. The dendrogram showed four clusters, designated as A, B, CD, and EF; these are plotted on the water-depth chart (Table 8, column e).

This classification is quite similar to that obtained by use of all of the species. Perhaps the most significant deviation is that B1 and B2 are not distinct; they form a large cluster, called B, which has the same range as B1 and B2 combined, except for station 38, which is placed into B at the expense of A. The boundary between B and CD is the same as originally, but the CD-EF boundary places three of the original CD stations into EF.

The task now confronting us is to modify—either substitute in or add to—the species list. These changes must be made with specific goals in mind, that is, to split B1 and B2 and move the CD-EF boundary to a more reasonable position. The modifications must be made carefully, so as not to change an acceptable boundary while attempting to correct an unacceptable one. The A-B boundary will not be studied because a single misclassification at a boundary is not of great importance, particularly when one considers that we have split a continuous graduation into distinct zones.

Third Trial Solution

Let us look first at B. Although C1YH is not useful for separating B1 from B2, it is useful in defining the CD-EF boundary, and so we will not replace it. C28B and C62C are good discriminators between B1 and B2, and they will be retained. Close examination of C23A (Tables 2, 3, and 7) shows that, although the mean abundances are different, the variation is so great as to provide no discrimination. Further study of the tables shows that C1FA will distinguish B1 from B2 quite well, although some loss in the A-B resolution may result. We replace C23A by C1FA.

Looking in a similar manner at the CD-EF discrimination, we see that C30L and C83B are excellent discriminators (low variation and different means). C27G is only partly successful in defining the CD-EF boundary; stations 16, 17, and

18 are much like EF, even though they are in CD. This was doubtless a cause of the error in placing the boundary. However, C27G is very good in defining the B2–CD boundary, and it will be retained. Similarly, C30B will not be removed even though the abundances in CD and EF are similar for most samples; it is needed for B2–CD discrimination.

Finally, we add species to the selected list. Table 7 shows that the mean abundances of P55A in CD and EF are greatly different, and B1 and B2 also have different means. In addition, the within-cluster variability is not great. This species should help to distinguish CD from EF and B1 from B2. Similarly, C42A should help determine the CD–EF boundary.

The zonation resulting from these changes was not completely satisfactory. The B1–B2 boundary was distinguished. However, the B2–CD boundary was altered in the process. The error in the CD–EF classification was unchanged.

Fourth Trial Solution

It was decided to add C68A to the selected list in order to help define the B2–CD and CD–EF boundaries; C57A has low variation within the individual clusters, even though the cluster means are not greatly different. The final list of selected species is as follows: C1YE, C27B, C1YH, C28B, C62G, C27G, C30B, C30L, C83B, C1FA, P55A, C42A, C57A, and C68A.

The classification resulting from the two additions to the species list is shown in Table 8, column b. It can be seen that the CD–EF boundary has been defined perfectly and the B2–CD boundary less well. Samples 25 and 26, previously in B2, were placed in CD by the selected species. This is not too surprising because the content of samples 25 and 26 differs slightly in some respects from the other samples in cluster B2.

The minor misclassifications at the boundaries are not serious, particularly when one considers that the faunal content should actually be considered a more or less continuously varying function of depth, rather than distinct, discrete groups. Only a few species were used to detect subtle differences.

These statements point up an important fact concerning the reduction of the faunal list. If the list is to be shortened radically, then one might expect a loss in precision. (On the other hand, enhanced precision could be achieved if random, confusing elements were removed; each case would have to be judged on its own merits.) Care must be taken, however, in assuring that results from one area can be used safely in another; it is up to the investigator to decide how universal the results may be. This problem is probably less serious in nonbiologic studies in which it is commonly less difficult to identify the important parameters. But in any case, the procedures offer the potential of pointing out which species are characteristic and perhaps of ecological significance.

The list of selected species should not be considered unique; it is merely a subset of the original data that fairly well approximates the original classification. One or more different species may have done as well or even better. Also, the selected species may not be paleontologically significant. They were picked on the basis of duplication of the grouping and without paleontologic consideration. It is at this point that the paleontologist must decide whether the selected list is useful, and if not, how it should be altered.

CONCLUSIONS

The water-depth zonation along the traverse is interesting because it was produced by the computer without knowledge of biology or paleontology. Even by this mechanical method, a classification that seems to make geologic or ecologic sense has resulted. The results obtained by determining key species are also important. If the same classification using only a few species can be obtained as one using more than 200 species, then in some instances we need only consider the few. In addition, similar lists have been obtained by use of presence-absence data. As Buzas (1972) pointed out, quantitative analysis need not be abandoned because of a cursory examination of a fauna. These results can save considerable time in further analysis, both in data gathering and interpretation.

APPENDIX 1. GOODNESS COEFFICIENTS

The coefficient used in the previous example to measure the goodness of a parameter is not the only possible coefficient. Other similar statistics have properties that may be useful with other types or sets of data. In some cases, the data type suggests which coefficient may be most useful; in other situations, it is wise to use PIP1 with different coefficients in order to determine which best fits the requirements.

The coefficient used for species i was computed by $C_{ij} = \bar{x}_i^2/(1 + s_i^2)$ for the samples within the given cluster j; no consideration was taken of the abundance values of samples from outside cluster j. However, if we compute

$$C_{ij}^* = C_{ij} / [\bar{x}_i^2 / (0.1 + s_i^2)]$$

total sample, then we are using the mean and standard deviation of the total sample to standardize the cluster statistic, C_{ij}. Therefore, we denote C_{ij} as a "nonstandardized" coefficient and C_{ij}^* as a standardized coefficient.

A similar statistic uses sums and sums of squares of the parameter values: $C_{ij} = (\sum_m x_{im})^2 / (0.01 + \sum_m x_{im}^2)$. This coefficient may also be standardized by dividing by the corresponding statistic for the total sample, giving C_{ij}^*.

If the data are in a presence-absence form and not in terms of counts or measurements, then a different set of coefficients is needed. Here, we are concerned with the locations at which species occur and not in their abundance. Assume that there are n samples in a cluster, of which k_i of these samples have the particular species i present. As previously, the mean variance ratio for clustering is given by $C_{ij} = k_i/(1 + n - k_i)$. This coefficient can also be standardized by dividing by the corresponding statistic for the total traverse, giving C_{ij}^*.

Another presence-absence coefficient deals simply with the number of samples in a cluster that contain the species of interest, k_i; the more occurrences, the more valuable that species is for defining the cluster. However, if that species is also found in many of the samples in other clusters, it is not as useful. Thus, we also wish to standardize this statistic in some cases.

REFERENCES CITED

Blackith, R. E., and Reyment, R. A., 1971, Multivariate morphometrics: London, Academic Press, 412 p.

Buzas, M. A., 1972, Biofacies analysis of presence or absence data through canonical variate analysis: Jour. Paleontology, v. 46, p. 55-57.

Demirmen, Ferruh, 1969, Multivariate procedures and Fortran IV program for evaluation and improvement of classifications: Kansas Geol. Survey Computer Contr. 31, 51 p.

Hazel, J. E., 1970, Binary coefficients and clustering in biostratigraphy: Geol. Soc. America Bull., v. 81, p. 3237-3252.

Maddocks, R. F., 1966, Distribution patterns of living and subfossil podocopid ostracodes in the Nosy Be area, northern Madagascar: Kansas Univ. Paleont. Contr. 12, 72 p.

Mello, J. F., and Buzas, M. A., 1968, An application of cluster analysis as a method of determining biofacies: Jour. Paleontology, v. 42, p. 747-758.

Seal, H. L., 1964, Multivariate statistical analysis for biologists: New York, John Wiley & Sons, 207 p.

Sokal, R. R., and Sneath, P.H.A., 1963, Principles of numerical taxonomy: San Francisco, W. H. Freeman and Co., 359 p.

Wishart, David, 1969, Fortran II programs for eight methods of cluster analysis (CLUSTAN I): Kansas Geol. Survey Computer Contr. 38, 112 p.

MANUSCRIPT RECEIVED BY THE SOCIETY JULY 9, 1973
REVISED MANUSCRIPT RECEIVED NOVEMBER 21, 1973

Variograms and Variance Components in Geochemistry and Ore Evaluation

A. T. MIESCH

U.S. Geological Survey
Denver, Colorado 80225

ABSTRACT

Investigations of spatial variability in ore deposits and other rock bodies have been approached through classical analysis of variance methods and through the theory of regionalized variables, as developed by the French school of geostatistics. The analysis of variance approach leads to estimates of variance components that can be used to form the variogram employed in geostatistics. The equivalence of the two approaches allows a possibly useful interchange of methods for sampling and statistical estimation.

INTRODUCTION

One of the most fundamental properties of a rock body is the type of chemical and mineralogic variability within it. Many volcanic units, for example, are notably uniform in chemical and mineralogic composition in both vertical and lateral directions. Granitic plutons tend to vary laterally in composition toward their contacts with older country rock, but vertical variations over the limited intervals that we commonly see are typically small. In sedimentary units, the stratigraphic variability is indicative of changes in the sedimentary environment or source area with time, whereas lateral variability reflects the configuration of the sedimentary basin.

Variability within a rock body is also important in sampling, and efficient sampling plans depend on a knowledge of the degree and type of variability present. A compositionally uniform volcanic unit or ore deposit, for example, can be adequately represented for most purposes by relatively few samples, whereas adequate representation of a more variable deposit can be much more difficult. It is well known that the efficiency of sampling can be significantly improved through a knowledge of the type of variability; the most familiar example is the common practice of sampling sedimentary units across the bedding, but significant improvement in sampling efficiency can also be achieved where directional variability, or other types of nonuniform variability, is far more subtle.

Although the degree and type of compositional variability in a rock body are always observed by geologists, they are only rarely measured in any rigorous way. Most computed standard deviations in the geochemical literature suffer from biases brought about by selective sampling, and only a few investigators have bothered to assess the directional and spatial properties of the variability in designing sampling programs. Some notable exceptions, which have involved estimation of variance components, are investigations by Youden and Mehlich (1937), Krumbein and Slack (1956), Baird and others (1967), and Connor and others (1972).

The practical advantages of employing a knowledge of the type of variability in a rock unit in sampling are well recognized in the mining industry, and interest in the subject has prompted intensive research during the past decade into mathematical methods of ore evaluation. The outstanding research appears to have been centered in France and led by Professor Georges Matheron of the École Nationale Supérieure des Mines de Paris. Matheron's work has led to a theory of regionalized variables. The theory is based on the concept of a variogram; an estimated variogram summarizes the information available on the type of spatial variation present in an ore deposit. My own knowledge of the theory comes principally from papers by Matheron (1963), Blais and Carlier (1968), David (1969, 1970), and Olea (1972). A comparison of the two approaches to the problem of assessing spatial variation, one using variograms and the other using variance components, has been a subject of interest, however, and I find that the approaches are fundamentally the same. The purpose of this paper is to support this contention in the hope of bringing proponents of the two approaches closer together.

VARIANCE COMPONENTS

The first application of classical analysis of variance methods to a problem in field geochemistry that I have been able to find was by Youden and Mehlich (1937), who investigated the variability in soil pH over two areas in New York and New Jersey. Their principal interest was in determining the type of variation present in order to design an efficient sampling program. This was done by estimating components of the total variance associated with various sampling intervals. Krumbein and Slack (1956) used a similar approach in the study of radioactivity in a thin shale bed that overlies Coal No. 6 throughout much of the Illinois Basin. They explained the general philosophy and model, the sampling scheme, and the computational procedures in sufficient detail to provide a basic reference for future work. According to the model they used, a radioactivity measurement (actually the square root of the measurement after transformation) is viewed as being determined by the grand mean for the entire area studied, plus a deviation (α) characteristic of the supertownship (a group of nine townships) from which the sample came, plus a deviation (β) characteristic of the township within the supertownship, plus a deviation (γ) characteristic of the mine within the township, plus a deviation (δ) characteristic of the sample from within the mine. Thus, in the regional phase of their study, the total variance was viewed as containing four components:

$$\sigma_x^2 = \sigma_\alpha^2 + \sigma_\beta^2 + \sigma_\gamma^2 + \sigma_\delta^2. \tag{1}$$

The first variance component (σ_α^2) is a function of the differences among supertownships; the second (σ_β^2), a function of differences among townships within supertownships; the third (σ_γ^2), a function of differences among mines within townships;

and the last (σ_δ^2) a function of differences among measurements on individual samples from within the same mine.

Krumbein and Slack (1956) also estimated the components of variance of radioactivity in the shale bed within a single mine and, as in the regional study, viewed the total variance (σ_m^2) as consisting of four components associated with varying sampling intervals. A fifth component consisted of variance arising from measurement errors. Thus,

$$\sigma_m^2 = \sigma_\epsilon^2 + \sigma_\theta^2 + \sigma_\lambda^2 + \sigma_P^2 + \sigma_\phi^2, \qquad (2)$$

where σ_ϵ^2 is the variance arising from differences among sections across the shale bed (called "grids") that are about 1.02 m thick and 3.05 m wide, σ_θ^2 is the variance among major units of the grids that are about 51 cm by 1.53 m, σ_λ^2 is the variance between minor units that are about 25 cm by 76 cm within the major units, and σ_P^2 is the variance among samples from within minor units. The final variance component (σ_ϕ^2) was estimated by making duplicate measurements of the radioactivity on all samples.

The estimated variance components are given in Table 1 along with the average spacings between the various sampling units. The estimated total variance, which includes variance on all local and regional scales as well as variance due to measurement error, is 0.1906. The estimated components indicate that if sampling were confined to any single supertownship, the total observed variance would tend to be less than this amount by only 1.8 percent. Similarly, if sampling were confined to any single township, only an additional 7.2 percent of the variance would be lost, and only an additional 6.2 percent would be lost if the sampling were confined to a single mine. Thus, most of the variance in radioactivity tends to be associated with small scales, and the variance within a single mine tends to be almost 85 percent of that found for the entire area, which is about 320

TABLE 1. AVERAGE SPACINGS BETWEEN SAMPLING LOCATIONS AND ESTIMATES OF VARIANCE COMPONENTS FOR RADIOACTIVITY IN A SHALE BED IN SOUTHERN ILLINOIS*

Source of variation	Average spacing	Estimated variance component	Estimated variance component (% of total)
Regional study			
Between supertownships	..	0.0035	1.8
Between townships within super-townships	15.0 km (9.3 mi)	0.0137	7.2
Between mines within townships	5.0 km (3.1 mi)	0.0119	6.2
Between samples within mines	0.8 km (0.5 mi)	0.1615	84.7
Total		0.1906	99.9
Local study			
Between grids within a mine	1.13 km (0.7 mi)	0.0000†	0.0
Between major units within grids	1.40 m (4.6 ft)	0.0106	8.9
Between minor units within major units	0.67 m (2.2 ft)	0.0561	47.3
Between samples within minor units	0.15 m (0.5 ft)	0.0482	40.6
Between replicates within samples	..	0.0038	3.2
Total		0.1187	100.0

*From Krumbein and Slack (1956).
†Computed estimate is negative, but is set equal to zero as is conventional.

km long. Within the single mine, about 47 percent of the total variance tends to be associated with sampling intervals of 15 to 67 cm, and another 40 percent with intervals of less than 15 cm. Only about 3 percent of the total variance in radioactivity arises from measurement error.

The purpose in reviewing the experimental results of Krumbein and Slack here is to point out, prior to the following discussion of variograms, that application of the classical analysis of variance approach leads to estimation of the variance and the percent of the total variance that is associated with various sampling intervals. The estimates are used, as shown by Krumbein and Slack (1956, p. 749-753), both to improve efficiency in sampling and, therefore, statistical estimation, and to assess the relative importance of geologic processes that act over local and regional scales.

VARIOGRAMS

Although the theory of regionalized variables is only about ten years old, its application to problems of ore evaluation, at least, appears to be further developed than the classical techniques of analysis of variance. Analysis of variance, on the other hand, has received more application in geochemical investigations not directly related to economic matters. Application of the theory of regionalized variables begins with sampling at regular intervals, if possible, down a drill hole, along a traverse or a grid, or within some three-dimensional network. The sample data are then used to construct the estimate of a variogram, which is a curve representing the degree of continuity of mineralization within an ore body (Matheron, 1963, p. 1250). The variogram is defined by

$$\gamma(h) = \frac{1}{2V} \int \int \int_V [f(M+h) - f(M)]^2 \, dV, \qquad (3)$$

where $\gamma(h)$ is the value of the variogram (the ordinate) for sampling interval h (the abscissa). The triple integration is over the volume of the deposit, V, and $f(M+h)$ and $f(M)$ are values of the compositional variable at two points separated by the interval h. Variograms are similarly defined for areas and for traverses, with corresponding decreases in the number of integrals. In practice, most variograms are initially estimated for drill holes or traverses across the deposit in various directions in a search for any anisotropy that may be present. The variograms are then averaged as found to be appropriate.

The variogram for a traverse is estimated by

$$\gamma(h) = \frac{1}{2n} \sum_i (X_{i+h} - X_i)^2, \qquad (4)$$

where n is the number of pairs of measurements (X_{i+h} and X_i) made at points separated by the interval h. Equation (4) is identical to one commonly used by chemists and others to estimate the analytical variance from duplicate analytical measurements on n specimens (Youden, 1951, p. 17) and is obviously a measure of variance, as Matheron (1963) pointed out.

Ore deposits that display completely random or chaotic compositional variation would be represented by variograms that are generally parallel to the abscissa over the entire range of h; that is, the value of $\gamma(h)$ would be the same for each possible sampling interval. Thus, a given number of samples from any small region are just as representative of the deposit as the same number of samples from widely separated points throughout the deposit. More commonly, the variogram increases with increasing h, thereby indicating that the differences are greater for samples taken far apart than for those collected close together. However, most variograms show a plateau or flattening beyond some value of h, indicating that a maximum variance has been attained. Large values of $\gamma(h)$ at small sampling intervals (for example, $h = a$) are said to reflect a nugget effect; that is, samples collected only "a" units apart tend to differ greatly, reflecting an erratic distribution of the ore on a small scale.

Typically, $\gamma(h)$ is based on a large number of sample pairs where the interval h is small and on many fewer pairs where h is large. Because of this, the form of the variogram is generally erratic at large values of h and may contain some negative slopes between points where it has been estimated. The negative slope is caused by the same properties of the data that lead to negative estimates of variance components in classical analysis of variance.

Following the estimation of the variogram, the theory of regionalized variables calls for the fitting of one of several types of models that have been developed (Blais and Carlier, 1968). Selection of the model is not entirely objective, and considerable experience in the theory and its application appears to be required. After the estimate of the variogram has been obtained and an appropriate model has been fitted, the parameters of the model can be used to determine weighting factors for computing moving averages, with confidence intervals, for the various ore blocks throughout the deposit. The method is referred to as *kriging*, and the confidence intervals are determined from the *kriging* variance (David, 1970). The models are also used to estimate the volume of a deposit, as well as its overall value and the associated confidence intervals.

COMPARISON OF METHODS

Proponents of the classical analysis of variance approach and of the variogram approach to the determination of spatial variation in rock bodies view the total variance to exist at a range of scales. That is, any rock body may exhibit variation on a regional scale, a very local scale, or any scale in between. The difference between the two approaches is merely one of variance measurement. If variance components are estimated for the sampling intervals h and $2h$, the component σ_h^2 is a measure of all the variance at scales less than h, and σ_{2h}^2 is a measure of all the variance on scales between h and $2h$. If a variogram is estimated at the points h and $2h$, the value of $\gamma(h)$ is a measure of all the variance at scales less than h ($\gamma(h) = \sigma_h^2$), but the value of $\gamma(2h)$ is a measure of all variance at scales less than $2h$ ($\gamma(2h) = \sigma_h^2 + \sigma_{2h}^2$).

The equivalence of the two approaches as described here is more or less intuitive; the notation becomes cumbersome when an attempt is made to demonstrate the equivalence mathematically. The equivalence might be verified if we were able to compute both variance components and a variogram for the same set of real data, but the methods are normally based on different sampling plans, and no single set of data is satisfactory for both. However, it is possible to generate some hypothetical data for a sampling plan that is suitable for both approaches. The data are given and the sampling plan is illustrated in Figure 1. Computed values of the variogram, using equation (4), and of the variance components, using the procedure given in detail by Krumbein and Slack (1956, p. 754), are given

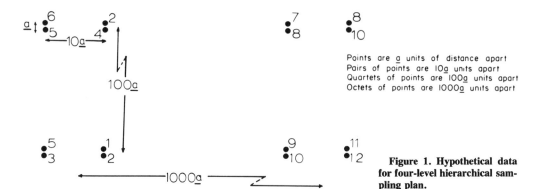

Figure 1. Hypothetical data for four-level hierarchical sampling plan.

in Table 2. The final column of Table 2 shows the cumulative sums of the variance components, which are exactly equal to the corresponding values of the variogram.

It is possible now to construct variograms for radioactivity in the shale bed investigated by Krumbein and Slack from the variance components given in Table 1. The variograms are shown in Figure 2, but inexperience in the theory of regionalized variables prohibits any attempt to fit a model or to carry the procedure further. It may suffice for the present to point out that the flatness of the regional variogram is in accord with Krumbein and Slack's finding that most of the variation is local.

CONCLUDING REMARKS

The apparent equivalence of variograms and cumulative sums of variance components estimated by use of the classical analysis of variance methods should be of interest and benefit to the proponents of both approaches. At least one limitation of the variogram approach is the requirement that observations be made at intervals that are at least approximately equal over the ore deposit or rock body. This is possible in much exploration drilling if the requirement is known beforehand, but can present a difficulty where the variogram is to be constructed from previous data. Hierarchical sampling plans of the type used for the analysis-of-variance approach, on the other hand, may be easy to simulate using previous data and might be used in situations where regular spacing of drill holes or sampling localities is not possible. The hierarchical sampling designs used by Youden and

TABLE 2. VARIOGRAM VALUES AND VARIANCE COMPONENTS*

Sampling interval (h)	Variogram $\gamma(h)$[†]	Variance component (σ_h^2)[§]	Cumulative sum of variance components
a	1.06250	1.06250	1.06250
$10a$	2.87500	1.81250	2.87500
$100a$	3.53125	0.65625	3.53125
$1,000a$	19.87500	16.34375	19.87500

*Computed from data of Figure 1.
[†] Values of $\gamma(h)$ based on comparison of 8, 16, 32, and 64 pairs of values, respectively.
[§] Variance components represent, respectively, differences between points within pairs (8 d.f.), between pairs within quartets (4 d.f.), between quartets within octets (2 d.f.), and between octets (1 d.f.).

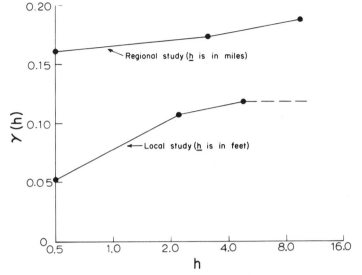

Figure 2. Variograms for radioactivity in shale bed overlying Coal No. 6 in Illinois constructed from variance component estimates of Krumbein and Slack (1956) given in Table 1.

Mehlich (1937) are particularly adaptable to some situations where regular-interval sampling cannot be used, such as in areas where outcrops are sparse and irregularly distributed. If the variograms constructed on the basis of hierarchical sampling plans are satisfactory, the possibility of using variograms can be extended to many investigations where they are currently impractical.

The theory of regionalized variables appears to offer a great deal to those who, heretofore, have approached investigations of spatial variability solely through conventional analysis-of-variance methods. The variance components may be more easily interpreted if cast into a variogram or some similar graphical device. Where variograms are used, it may be possible to employ all or at least some of the theory that has been developed for interpreting them and to derive soundly based confidence intervals for geochemical abundance estimates and for estimates of the mean values that are commonly contoured or otherwise displayed on geochemical maps. Application of the theory of regionalized variables to contouring procedures was shown by Olea (1972).

REFERENCES CITED

Baird, A. K., McIntyre, D. B., and Welday, E. E., 1967, Geochemical and structural studies in batholithic rocks of southern California: Pt. II, Sampling of the Rattlesnake Mountain pluton for chemical composition, variability, and trend analysis: Geol. Soc. America Bull., v. 78, p. 191-222.

Blais, R. A., and Carlier, P. A., 1968, Applications of geostatistics in ore evaluation: Canadian Inst. Mining and Metallurgy, Spec. Vol. 9, p. 41-68.

Connor, J. J., Feder, G. L., Erdman, J. A., and Tidball, R. R., 1972, Environmental geochemistry in Missouri—A multidisciplinary study: Internat. Geol. Cong., 24th, Montreal 1972, Symp. 1, p. 7-14.

David, M., 1969, The notion of "extension variance" and its application to the grade estimation of stratiform deposits, in Weiss, Alfred, ed., A decade of digital computing in the mineral industry: New York, Am. Inst. Mining, Metallurgy, and Petroleum Engineers, p. 63-81.

———1970, Geostatistical ore estimation—A step-by-step case study: Canadian Inst. Mining and Metallurgy, Spec. Vol. 12, p. 185-191.

Krumbein, W. C., and Slack, H. A., 1956, Statistical analysis of low-level radioactivity of Pennsylvanian black fissile shale in Illinois: Geol. Soc. America Bull., v. 67, p. 739–762.

Matheron, G., 1963, Principles of geostatistics: Econ. Geology, v. 58, p. 1246–1266.

Olea, R. A., 1972, Application of regionalized variable theory to automatic contouring: Univ. Kansas Center for Research, Inc., Spec. Rept. to Am. Petroleum Inst., Research Proj. 131, 191 p.

Youden, W. J., 1951, Statistical methods for chemists: New York, John Wiley & Sons, Inc., 126 p.

Youden, W. J., and Mehlich, A., 1937, Selection of efficient methods for soil sampling: Yonkers, N.Y., Contr. Boyce Thompson Inst., v. 9, p. 59–70.

MANUSCRIPT RECEIVED BY THE SOCIETY SEPTEMBER 10, 1973
REVISED MANUSCRIPT RECEIVED FEBRUARY 4, 1974

Effect of Closure on the Comparison of Means in Ternary Systems

P. G. Sutterlin

Department of Geology
University of Western Ontario
London, Ontario N6A 3K7
Canada

R. W. May[*]

Department of Geography
Wilfred Laurier University
Waterloo, Ontario N2L 3C5
Canada

ABSTRACT

Closed-array data result in linear correlation between pairs of variables that are based on the properties of the closed array rather than on any geological or other relation between them. The comparison of means between the variables of two sets of three-component, closed-data arrays also is influenced by closure.

The usual procedure of comparing means is by the utilization of a t test. If the value of t when the means of one pair of variables are compared is greater than twice the critical value, the comparison of the means of at least one other pair of variables must yield the result that they are statistically different. Under these conditions, it is difficult to attach geological significance to more than one of the differences because any others may be a result of the closed-array properties of the two sets of data.

---

[*]Present address: Alberta Research, Edmonton, Alberta, Canada T6G 2C2.

INTRODUCTION

Chayes (1960) defined as "closed" any table of data containing observations such that

$$\sum_{i=1}^{M} x_{ij} = \sum_{i=1}^{M} \bar{x}_i = K, \text{ for all } j, \quad (1)$$

where x_{ij} = the amount of the ith variable in the jth individual, $\bar{x}_i$ = the mean value of the ith variable, M = the number of variables, and K = a constant. The constant term K usually will have a value of 100, because most closed arrays are percentage data. Chayes (1960, 1962, 1971) and Vistelius and Sarmanov (1961) have shown that the conditions in (1) can result in linear correlations between pairs of variables that are based on the properties of the closed array rather than on any geological or other relation between the variables. Krumbein (1972) and Krumbein and Watson (1972) have shown that results of trend surface analysis in three-component systems also can be influenced by closure. This study examines the comparison of means between two sets of three-component, closed-data arrays.

USE OF t TEST

The normal procedure for comparing means is use of a t test (Hoel, 1962). A null hypothesis (H_0) is formulated that presupposes no significant difference between the means, μ_1 and μ_2, of a particular variable in two different populations. At the same time, an alternate hypothesis (denoted by H_A) also is formulated. Symbolically, these are represented as

$$H_0 : \mu_1 = \mu_2 \quad (2)$$

and

$$H_A : \mu_1 \neq \mu_2. \quad (3)$$

Because it is rarely, if ever, possible to analyze an entire population and determine the mean value directly (μ_1 or μ_2), estimates are made by taking samples of the two populations and calculating the means of the variable under consideration. Then, the statistic

$$t = \frac{(\bar{x} - \bar{y}) - (\mu_x - \mu_y)}{\sigma \sqrt{\dfrac{1}{N_1} + \dfrac{1}{N_2}}} \quad (4)$$

is calculated, where $\bar{x}$ = the mean of variable x in sample 1, $\bar{y}$ = the mean of variable y in sample 2, μ_x = the population mean of variable x, μ_y = the population mean of variable y, N_1 = the number of individuals in sample 1, N_2 = the number of individuals in sample 2, and σ = the common standard deviation of the two populations. The probability of obtaining this value of t by chance is found by comparison to values of Student's t distribution obtained from tables such as those given by Pearson and Hartley (1966).

The common standard deviation, σ, because it is generally unknown, is estimated by

$$s = \sqrt{\frac{(N_1 - 1)s_x^2 + (N_2 - 1)s_y^2}{N_1 + N_2 - 2}} \tag{5}$$

where N_1 and N_2 are defined under (4) above, s_x^2 = the variance of variable x in sample 1, and s_y^2 = the variance of variable y in sample 2. The denominator of equation (4) can then be written as

$$\sqrt{\frac{(N_1 - 1)s_x^2 + (N_2 - 1)s_y^2}{N_1 + N_2 - 2} \left(\frac{1}{N_1} + \frac{1}{N_2}\right)} = s^*. \tag{6}$$

Under the usual assumption of the null hypothesis that the difference $(\mu_1 - \mu_2)$ is zero, the formula for the t statistic given in (4) reduces to

$$t = \frac{(\bar{x} - \bar{y})}{s^*}. \tag{7}$$

This calculated value then is compared to a predetermined value, t_α, which may be termed the critical value. The statistic t_α depends on two variables: (1) the value of α termed the level of significance that is conventionally chosen as either 0.05 or 0.01 at the time H_0 and H_A are made, and (2) the degrees of freedom given by the quantity $N_1 + N_2 - 2$, which is necessary because t_α depends on the number of individuals in each of the two samples. If the value of t calculated in (7) is less than t_α, H_0 is accepted; otherwise, H_0 is rejected. In the following analysis, attention is focused on the cases in which H_0 would be rejected.

CASE 1. EQUAL VARIANCES

Assume sample set 1 of size N_1 and sample set 2 of size N_2 and that the variances of the three variables in both samples are equal. The closed-array property gives

$$\bar{x}_{11} + \bar{x}_{21} + \bar{x}_{31} = 100 \tag{8}$$

and

$$\bar{x}_{12} + \bar{x}_{22} + \bar{x}_{32} = 100 \tag{9}$$

where $\bar{x}_{11}$ = the mean of variable x_1 in sample 1; $\bar{x}_{21}$ = the mean of variable x_2 in sample 1; $\bar{x}_{31}$ = the mean of variable x_3 in sample 1; and where $\bar{x}_{12}$, $\bar{x}_{22}$, and $\bar{x}_{32}$ are the means of the analogous variables in sample 2. The homoscedastic assumption gives

$$s_{11}^2 = s_{21}^2 = s_{31}^2 = s_{12}^2 = s_{22}^2 = s_{32}^2. \tag{10}$$

A t test of x_{11} against x_{12} gives, for variable x_1

$$t_1 = \frac{\bar{x}_{11} - \bar{x}_{12}}{s_1^*}, \qquad (11)$$

where

$$s_1^* = \sqrt{\frac{\{(N-1)s_{11}^2 + (N-1)s_{12}^2\}}{N_1 + N_2 - 2}\left(\frac{1}{N_1} + \frac{1}{N_2}\right)}.$$

Because the variances of the variables are equal, $s_1^* = s_2^* = s_3^*$. Consider the outcome, as mentioned previously, where

$$|t_1| > t_\alpha. \qquad (12)$$

It is obvious that $\bar{x}_{11} \neq \bar{x}_{12}$ and that

$$\frac{|\bar{x}_{11} - \bar{x}_{12}|}{s_1^*} > t_\alpha. \qquad (13)$$

Equations (8) and (9) can be rewritten as

$$\bar{x}_{11} = 100 - \bar{x}_{21} - \bar{x}_{31} \qquad (8A)$$

and

$$\bar{x}_{12} = 100 - \bar{x}_{22} - \bar{x}_{32}, \qquad (9A)$$

which upon substitution into (13) gives

$$\frac{|(\bar{x}_{22} - \bar{x}_{21}) - (\bar{x}_{31} - \bar{x}_{32})|}{s_1^*} > t_\alpha. \qquad (14)$$

Assume now that $\bar{x}_{21}$ is not significantly different from $\bar{x}_{22}$ and also that $\bar{x}_{31}$ is not significantly different from $\bar{x}_{32}$. Symbolically,

$$\frac{|\bar{x}_{22} - \bar{x}_{21}|}{s_2^*} < t_\alpha \qquad (15)$$

and

$$\frac{|\bar{x}_{32} - \bar{x}_{31}|}{s_3^*} < t_\alpha. \qquad (16)$$

Combining (15) and (16) gives

$$\frac{|\bar{x}_{22} - \bar{x}_{21}|}{s_2^*} + \frac{|\bar{x}_{31} - \bar{x}_{32}|}{s_3^*} < 2t_\alpha. \qquad (17)$$

Consideration of basic inequalities provides the result:

$$\frac{|(\bar{x}_{22} - \bar{x}_{21}) - (\bar{x}_{32} - \bar{x}_{31})|}{s_1^*} \leq \frac{|\bar{x}_{22} - \bar{x}_{21}|}{s_1^*} + \frac{|\bar{x}_{31} - \bar{x}_{32}|}{s_1^*} . \qquad (18)$$

Combining (14), (17), and (18) gives a lower and upper bound for the quantity on the left-hand side of (15) as follows:

$$t_\alpha < \frac{|\bar{x}_{22} - \bar{x}_{21}) - (\bar{x}_{32} - \bar{x}_{31})|}{s_1^*} < 2t_\alpha . \qquad (19)$$

There is, however, no reason why $|t|$ as given in (11) should be bounded above. Thus it can be concluded that one or both of the assumptions made in (15) and (16) is invalid if $|t| > t_\alpha$. This completes the main argument in the case of equal variances.

CASE 2. UNEQUAL VARIANCES

The situation to be considered here is when

$$s_{11}^2 > s_{21}^2 > s_{31}^2, \qquad (20)$$

$$s_{12}^2 > s_{22}^2 > s_{32}^2, \qquad (21)$$

and

$$s_{i1}^2 = s_{i2}^2; \quad i = 1, 2, 3. \qquad (22)$$

These relations imply that

$$s_1^* > s_2^* > s_3^*, \qquad (23)$$

where these latter variables are analogous to the s^* stated previously in equation (6). The subscripts refer to the value calculated for the means of each sample; for example, s_2^* is the estimate of the standard deviation of the difference when the means of variable 2 are compared—that is, for the quantity $(\bar{x}_{21} - \bar{x}_{22})$. The argument is similar to that for the homoscedastic case. In a comparison of $\bar{x}_{11}$ and $\bar{x}_{12}$,

$$t_1 = \frac{\bar{x}_{11} - \bar{x}_{12}}{s_1^*} . \qquad (24)$$

Assuming that this gives a result showing these two means are significantly different, then $|t_1| > t_\alpha$. Making assumptions analogous to (16) and (17) gives

$$\frac{|\bar{x}_{22} - \bar{x}_{21}|}{s_2^*} < t_\alpha \qquad (25)$$

and

$$\frac{|\bar{x}_{32} - \bar{x}_{31}|}{s_3^*} < t_\alpha. \qquad (26)$$

Consideration of the relations shown in (23) gives

$$\frac{|\bar{x}_{22} - \bar{x}_{21}|}{s_1^*} < \frac{|\bar{x}_{22} - \bar{x}_{21}|}{s_2^*} < t_\alpha \qquad (27)$$

and

$$\frac{|\bar{x}_{32} - \bar{x}_{31}|}{s_3^*} < \frac{|\bar{x}_{32} - \bar{x}_{31}|}{s_1^*} < t_\alpha. \qquad (28)$$

Combining (27) and (28) gives an analogue to (18):

$$\frac{|\bar{x}_{22} - \bar{x}_{21}|}{s_1^*} + \frac{|\bar{x}_{32} - \bar{x}_{31}|}{s_1^*} \leq \frac{|\bar{x}_{22} - \bar{x}_{21}|}{s_2^*} + \frac{|\bar{x}_{32} - \bar{x}_{31}|}{s_3^*} \leq 2t_\alpha. \qquad (29)$$

The end result is then an inequality similar to (19):

$$t_\alpha < \frac{|(\bar{x}_{22} - \bar{x}_{21}) - (\bar{x}_{32} - \bar{x}_{31})|}{s_1^*} < \frac{|\bar{x}_{22} - \bar{x}_{21}|}{s_1^*}$$

$$+ \frac{|\bar{x}_{32} - \bar{x}_{31}|}{s_1^*} < \frac{|\bar{x}_{22} - \bar{x}_{21}|}{s_2^*} + \frac{|\bar{x}_{32} - \bar{x}_{31}|}{s_3^*} \leq 2t_\alpha. \qquad (30)$$

As before, there is no reason why $|t|$ given in (24) should be bounded above, and thus it can be concluded that one or both of the assumptions made in (27) and (28) concerning the differences between $\bar{x}_{21}$ and $\bar{x}_{22}$ and, $\bar{x}_{31}$ and $\bar{x}_{32}$ must be invalid if the test of $\bar{x}_{11}$ against $\bar{x}_{12}$ yields a value of t such that $|t|$ is greater than $2t_\alpha$. This becomes somewhat more critical than the homoscedastic case because the implication from (30) is that the difference between $\bar{x}_{11}$ and $\bar{x}_{12}$ can be smaller than $2t_\alpha$ and still result in a rejection of the null hypothesis when comparing $\bar{x}_{21}$ and $\bar{x}_{22}$ or $\bar{x}_{31}$ and $\bar{x}_{32}$ or both. Qualitatively, if $|t_1|$ is large, then the possibility is greater that one of the other sets of means is different.

VARIANCE RESTRICTIONS

In both case 1 and case 2, certain restrictions were placed on the variances of the three variables involved. These assumptions given in equations (10) [case

1] and (20) to (23) [case 2] were felt necessary to establish the inequalities of equations (17) and (18) [case 1], and (27) and (28) [case 2]. It can be shown, however, that it is only the relation given in (23) concerning the relative sizes of s_1^*, s_2^*, and s_3^* that is necessary to the proof of (30) and not the magnitude of the individual variances. It is a simple matter to arrange the variables under consideration so that the relation given in (23) holds. Thus, the proof of (19) and (30) is not directly dependent upon the variances of the three variables in the two samples.

DISCUSSION

Consider the case if the sample sizes N_1 and N_2 are both sufficiently large to give a value that is greater than 30 for the degrees of freedom. The critical value of t at a 0.05 percent level of significance is 1.96. There are five sets of possible results, depending on the value of t_1, as shown in Table 1. If the calculated value of t found when comparing $\bar{x}_{11}$ and $\bar{x}_{12}$ is greater than twice the critical value (3.92), then, as has already been shown, a comparison of the means of the other two variables must yield at least one other calculated t value greater than 1.96. This is the situation in results 1 and 2 of Table 1. Results 3, 4, and 5 are concerned with the situation where t_1 is between 1.96 and 3.92.

Result 1

In this situation, the null hypothesis of equality of means must be rejected for two of the three values. Because the value of t_1 is greater than twice the critical value of t (3.92), the value of t_2, as has been proven, could be due solely to the closed-array properties of the data. The problem then is to decide which of the differences has geological significance.

Result 2

This result is similar to the previous one except that the null hypothesis must be rejected for all three comparisons. Because t_1 is greater than 3.92, the values of both t_2 and t_3 could be the result of closure. Again, this makes it difficult to attach geological significance to more than one of the three differences.

Result 3

In this situation, it would be valid to attach geological significance to the difference in the means of variable 1.

Results 4 and 5

These are analogs of results 2 and 3. In either of these situations, the possibility exists that any of the differences found between means may have geological significance and is not simply due to the closed-array properties of the data.

An example for illustrative purposes is two different unconsolidated clastic units that are sampled and the percentages of sand, silt, and clay determined as shown in Table 2. This example falls into the category of case 2, in which variable 1 is sand, variable 2 is clay, and variable 3 is silt. The outcome of the t tests is that of result 1 (Table 1). Comparison of the means of the sand percent gives a t value of -5.41, which has an absolute value greater than twice the critical value. Thus it is not surprising that a t test of the mean silt content of the two units yields a t value that is significant at the 95 percent level of confidence. Geological significance can only be attached to one of these differences because the other one is due to closure. Two examples from the literature serve to illustrate further the problem of comparing means between three-variable, closed-data arrays. The choice of these examples should not be construed as implying a criticism of the geological aspects of the works cited.

In a study of tills of the Allegheny Plateau by Gross and Moran (1971, Table 3), a progressive change in the granulometric composition of six tills was shown. Each of the stratigraphic units shows a sand content that is significantly different from that of the till above or below it. It is not surprising, considering the results of case 2 given previously, that in all of the five comparisons shown in Table 2, either the silt or clay mean values also show a significant difference in these stratigraphically adjacent units.

Aseez (1972) illustrated the use of certain sieve data in the determination of depositional environments. The three variables used were coarse and very coarse sand, medium sand, and fine plus very fine sand plus silt. Table 4 of Aseez showed the results of t test comparisons of these three variables in the different environments sampled. Because Aseez did not give the raw data or the means and standard deviations on which the t tests were based, it is not possible to consider fully the results in the light of the previously developed theory. It is not surprising, however, that in the six sets of tests, five show differences between two of the three means compared.

SUMMARY

The chief purpose of this paper is to point out a definite problem that is encountered if the means of two or more three-variable, closed-data arrays are compared using a t test. The examples cited have come from the field of sedimentary petrology; however, the same problem could easily arise in describing the geochemistry of igneous rocks with ternary compositional diagrams. Depending on the value of t calculated, only one of the three comparisons may have geological significance. Differences in the other two sets of means could simply be due to the closed-array nature of the data, a limitation that is not present in open-data arrays.

TABLE 1. POSSIBLE RESULTS OF t TEST FOR COMPARISON OF MEANS BETWEEN THREE-VARIABLE CLOSED ARRAYS

Result No.	Test and hypothesis	Variable 1	Variable 2	Variable 3
1	t_i	$t_1 > 3.92$	$t_2 > 1.96$	$t_3 < 1.96$
	H_0	Reject	Reject	Accept
2	t_i	$t_1 > 3.92$	$t_2 > 1.96$	$t_3 > 1.96$
	H_0	Reject	Reject	Reject
3	t_i	$1.96 < t_1 < 3.92$	$t_2 < 1.96$	$t_3 < 1.96$
	H_0	Reject	Accept	Accept
4	t_i	$1.96 < t_1 < 3.92$	$t_2 > 1.96$	$t_3 < 1.96$
	H_0	Reject	Reject	Accept
5	t_i	$1.96 < t_1 < 3.92$	$1.96 < t_2 < 3.92$	$1.96 < t_3 < 3.92$
	H_0	Reject	Reject	Reject

TABLE 2. HYPOTHETICAL CLASTIC UNITS

	Unit 1 ($N = 30$)		Unit 2 ($N = 40$)		t value 1 versus 2
	$\bar{x}$	s	$\bar{x}$	s	
Sand	30	7.6	40	7.5	−5.41
Silt	46	5.0	38	5.5	6.17
Clay	24	7.0	22	6.0	1.27

ACKNOWLEDGMENTS

We acknowledge the assistance of S. Rinco, Department of Applied Mathematics, University of Western Ontario, for critically reading the manuscript. Financial support for this study came from National Research Council of Canada Grants A4387 and A8714 to P. G. Sutterlin and R. W. May, respectively.

REFERENCES CITED

Aseez, L. O., 1972, Triangular presentation of textural data in the interpretation of depositional environments: Jour. Sed. Petrology, v. 42, p. 729-731.

Chayes, F., 1960, On correlation between variables of constant sum: Jour. Geophys. Research, v. 65, p. 4185-4193.

——1962, Numerical correlations and petrographic variation: Jour. Geology, v. 70, p. 440-452.

——1971, Ratio correlation: A manual for students of petrology and geochemistry: Chicago, Ill., Chicago Univ. Press, 99 p.

Gross, D. L., and Moran, S. R., 1971, Grain-size and mineralogical gradations within tills of the Allegheny Plateau, in Goldthwait, R. P., ed., Till: A symposium: Columbus, Ohio State Univ. Press, p. 251-274.

Hoel, P. G., 1962, Introduction to mathematical statistics: New York, John Wiley & Sons, 427 p.

Krumbein, W. C., 1972, Areal variation and statistical correlation in open and closed number systems, Internat. Statistical Inst., 38th, Washington, D.C., 1971, Proc.: Book 1, Invited Papers, p. 551-556.

Krumbein, W. C., and Watson, G. S., 1972, Effect of trends on correlation in open and closed three-component systems: Internat. Assoc. Math. Geol. Jour., v. 4, p. 317-330.

Pearson, E. S., and Hartley, H. O., 1966, Biometrika tables for statisticians, Vol. 1: New York, Cambridge Univ. Press, 238 p.

Vistelius, A. B., and Sarmanov, O. V., 1961, On the correlation between percentage values; major component correlation in ferromagnesian micas: Jour. Geology, v. 69, p. 145-153.

MANUSCRIPT RECEIVED BY THE SOCIETY SEPTEMBER 10, 1973
REVISED MANUSCRIPT RECEIVED JANUARY 21, 1974

Printed in the U.S.A.

Geological Society of America
Memoir 142
© 1975

Comparison of Fan-Pass Spatial Filtering and Polynomial Surface-Fitting Models for Numerical Map Analysis

MICHAEL D. WILSON

Cities Service Exploration and Production Research Laboratory
Box 50408
Tulsa, Oklahoma 74150

ABSTRACT

A comparison is made of the conceptual and mathematical models of two numerical map analysis techniques, polynomial surface fitting and fan-pass spatial filtering. Fan-pass spatial filtering is designed to isolate trends with a specified orientation. In this respect it is complementary to polynomial surface fitting and band-pass spatial filtering, which delineate trends on the basis of size.

Spatial filtering techniques are superior to polynomial surface fitting because they more accurately preserve the position, orientation, length, width, and relative amplitude of local anomalies. However, two major disadvantages of most forms of spatial filtering are the requirement that the input data be gridded and the presence of edge effects in filter output. Gridding procedures may alter the input data in an undesirable manner by removing, reducing, or creating local anomalies or random noise. Edge effects, which are inherent products of the spatial filtering process, can be minimized in several ways. Development of spatial filtering techniques that utilize irregularly spaced data and establishment of accepted criteria for determining the quality of polynomial trend surfaces would improve the usefulness of both techniques.

INTRODUCTION

In the last decade, the advent of widespread computer use by geologists has fostered numerous attempts at numerical map analysis. The purposes of such map analysis, usually termed trend analysis, are twofold: (1) to predict values of an areally mapped variable between data points or even beyond the limits of data control, and (2) to separate variation in a mapped variable into two or more components, generally referred to as regional trends and local anomalies. Only the second purpose is of concern in this discussion.

Numerical techniques used to define the regional trends or local anomalies in mapped data include moving average, linear regression (trend fitting), nonlinear regression, autocovariance analysis, spectral analysis, and spatial filtering. Agterberg (1969) summarized the theory and application of several of these techniques. Further discussion will be limited to a comparison of the model most frequently utilized—polynomial trend fitting—and a method which has been applied only recently, fan-pass spatial filtering.

POLYNOMIAL MODEL

Conceptual Model

Conceptually, raw data are separable into three components (Miesch and Connor, 1967; Parsley, 1971): (1) regional trends, (2) local anomalies, and (3) random "noise." A model also applied in trend-analysis applications resolves the observed data into regional and residual components where the residual includes both local anomalies and random noise. The regional trend is approximated by a polynomial function, and the local and noise components are obtained by subtraction of the regional from the observed data values. In this article the term polynomial trend is restricted to those functions that fit the linear regression model as described by Krumbein and Graybill (1965).

Regional trends are assumed to have been produced by geologic processes that operated at a level greater than that covered by the map area (systematic measurement error also could produce a significant bias or component in regional trends). Local anomalies are produced by processes that operated on a level smaller than the map area but greater than individual data points. Random noise is usually attributed to the effects of sampling and measurement errors. Such errors are controlled by processes that operated on a level equivalent to or below that of the individual observation.

Although it is conceptually possible to distinguish between regional trends and local anomalies, in practice, the geologic processes that controlled the variable under investigation may have operated on a wide range of geographic levels. In such situations attempts to distinguish regional and local components of variation may not be meaningful or possible, except by arbitrary definition (McIntyre, 1967; Tinkler, 1969, p. 117; Parsley, 1971).

If the regional and local components are relatively obvious in the raw data, application of trend analysis is of little advantage (assuming the geologist can recognize these trends). This situation occurs when the random noise component is insignificant. Trend analysis should be most valuable if the noise component is relatively large. However, if the noise component dominates the regional and local components, it may not be possible to obtain meaningful results from trend analysis or even to determine whether regional trends and local anomalies exist (see Howarth, 1967; Norcliffe, 1969). Consequently, for a set of raw data, the proportion of the variability accounted for by sampling and measurement error should be evaluated prior to embarking on an extensive trend-analysis study.

Mathematical Model

The polynomial trend model is a form of multiple linear regression. The general linear model can be expressed mathematically as

$$Y = a_0 + \sum_{i=1}^{k} (a_i X_i) + e,$$

where Y is the dependent variable, X_i are independent predictor variables measured without error, e is a random component with mean equal to zero and finite variance, and the a_0 and a_i are unknown parameters. The polynomial trend model can be written:

$$Z = a_0 + a_1 X + a_2 Y + a_3 X^2 + a_4 XY + a_5 Y^2 + \ldots + e,$$

where Z is the dependent (predicted) variable, the X and Y terms are geographic coordinates measured without error along orthogonal axes, e is a random component with mean equal to zero and finite variance, and a_0, a_1, ..., a_n are unknown coefficients. The unknown coefficients are estimated by a least-squares solution for the polynomial function of interest. Note that the mathematical model permits separation of the raw data into two components, a random-noise component (e) and a regional plus local component (a_0 and X and Y terms). Separation of regional trend and local anomalies can be accomplished by defining arbitrarily the highest degree (the highest power of X or Y is called the degree of the polynomial fit) terms present in the regional component.

Criteria of Best Fit

Methods of selecting the best polynomial fit are numerous, as described by Agterberg (1964). They include testing F-ratios for significance, stepwise regression using partial correlation coefficients, testing for autoregression of residuals, establishing confidence intervals on the trend surface, and duplicate trend fitting using interpenetrating subsamples. The criterion most frequently applied is the F-ratio test. It provides for successive fitting of higher degree surfaces until the calculated F-ratio falls below the table F value at a given confidence level (Krumbein and Graybill, 1965; Chayes, 1970). The calculated F-ratio is formed by dividing the mean squares accounted for by the addition of a set of polynomial terms (after removing the sum of squares accounted for by all lower degree terms) by the remaining unexplained mean squares. A defect of the F-ratio test, as it has been applied, is its failure to establish the significance of individual terms in the full set of terms of a given degree. The F-test treats each set of terms as a group, and thus one that significantly improves the fit may overshadow the ineffectual nature of the associated terms. Baird and others (1971) noted that the addition of nonsignificant terms in the trend function artificially lowers the F-ratio. Baird and others (1971) and Parsley (1971) also criticized several of the applied standards for use of the F-ratio to determine when trend fitting should cease.

Whitten (1970) described an application of orthogonal polynomial trend fitting of irregularly spaced data, which allows the sum of squares contribution of each coefficient to be computed independent of the effects of the other coefficients. The problem of significance of individual terms also can be eliminated by use of the forward stepwise regression procedure (Draper and Smith, 1966, p. 171-172). This method successively adds individual terms to the surface fit until a partial F-test indicates that addition of further terms does not account for a significant amount of the variation above that removed by previously included terms. A check also is made to determine if addition of a new term has reduced the significance of terms already in the equation to a point where they should be deleted.

If autoregression of the residuals occurs, an F-ratio test is not valid (Krumbein and Graybill, 1965, p. 337). Agterberg (1964) demonstrated that autocorrelation of residuals increased the F-ratio. Parsley (1971) and Draper and Lawrence (1970) described techniques that permit testing for uncorrelated random noise in the residuals. To simplify the processing of the data and interpretation of the results, both the serial correlation technique used by Parsley and the two-dimensional run analysis of Draper and Lawrence require that the raw data be gridded before trend fitting. The effects of gridding a set of raw data are impossible to assess without full knowledge of the interpolation procedures used in the gridding process.

The use of confidence-interval comparisons as a criterion of best fit has been limited to low-order polynomial fits by the complexity of the calculations. However, confidence interval calculations assume that the residual values are uncorrelated. This assumption may not be valid for most low-degree trend fits.

Agterberg (1964) favored use of duplicate trend analysis as a criterion for selection of the most reliable fit, inasmuch as it is not bound by the assumption of independent residuals. Trends are fitted to subsets of data obtained by random sampling of the original data. The degree of surface fit that exhibits the least difference between the two subsets is considered the most reliable. This approach does not eliminate confusion produced by the presence of random noise in the residual maps.

Data-Point Distribution and Independence of Predictor Variables

The requirement by the linear regression model that independent variables be uncorrelated (Jones, 1972) presents a further difficulty in applying polynomial trend surface fitting to areally distributed data. This assumption has been violated in most applications of trend surface analysis. Correlations may be introduced by inadequate data-point distribution and are particularly high between powers of the same coordinate, for example, between X^3 and X^4. Jones recommended use of a ridge regression technique to remove the effects of correlated independent variables.

If the correlation between the independent X and Y terms for a given set of data is relatively strong, the matrix of X and Y terms required for estimating the unknown coefficients in the trend equation may be "ill-conditioned," as described by Miesch and Connor (1967). An ill-conditioned matrix may result in an unstable least-squares solution. Miesch and Connor recommended calculation of a "condition value" that will give an indication of the effects of data-point distribution and interrelationships of X and Y terms on the matrix condition. Norcliffe (1969) suggested use of a "quadrat" method to determine if the observations exhibit a degree of clustering large enough to prevent meaningful trend analysis.

Miesch and Conner (1967) claimed that translation and scale changes of the coordinate system may result in extensive changes in the terms selected by stepwise-regression surface fitting. Such effects may be related to inadequate data-point distribution, and if common, would greatly increase the difficulty of selecting the best surface fit for a given set of data.

Discontinuities in the Mapped Surface

Polynomial surfaces are continuous functions over the mapped area. Discontinuities (faults, for example) in the raw data surface are inadequately accounted for by continuous surfaces where such discontinuities are large, numerous, or poorly defined by the control points. In areas where discontinuities are few and their location relatively well known, polynomial trend surface analysis incorporating the presence of discontinuities may be practical (James, 1970). James noted that if faults are not taken into consideration, the regional trends may be distorted, spurious residuals may occur away from the fault, and residuals near the fault may be obscured.

SPATIAL FILTERING

Conceptual Model

Spatial filtering techniques are of two basic types: band-pass filtering and fan-pass (also referred to as pie-slice) filtering. Band-pass filtering is designed to isolate trends of specified size in a set of data. Conceptually the observed data are separated into a trend component with the frequency range to be retained (passed) by the filter, a trend component with frequency range to be rejected by the filter, and a random-noise component. In practice the noise component is lumped with the rejected frequencies component.

Band-pass spatial filtering is directly analogous to polynomial surface fitting because both are designed to delineate trends of specific size (Fig. 1). As it is generally applied, polynomial fitting has been used to separate regional trends from the combined local and noise components. Band-pass filtering retains those features in the input data that have frequencies (widths) within a specified range. This range can be defined to include features of regional or local extent.

Fan-pass spatial filtering screens for orientation rather than size (Fig. 1) and thus is complementary to both band-pass filtering and polynomial surface fitting. The observed data are separated into a trend component with a range of orientations to be retained, a trend component with orientations to be rejected, and a random-noise component that is consolidated into the rejected orientation component.

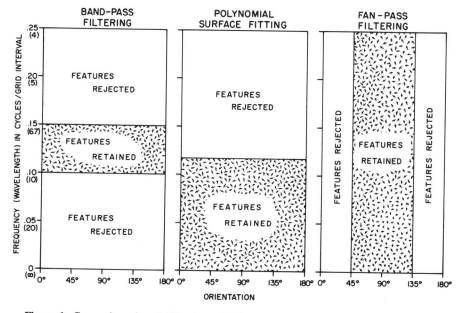

Figure 1. Comparison of spatial filtering and polynomial surface-fitting techniques with respect to size and orientation of features retained in filtered output and trend maps. Width (frequency or wave-length) of smallest feature resolved is dependent on spacing of control points.

Previous Work

The theory of spatial filtering as applied to geophysical data was developed by Swartz (1954) and Dean (1958). Fuller (1967) summarized the many types of grid operators (filters) that can be applied to potential field data. Fan-pass filtering of geophysical data (velocity filtering) was described initially by Fail and Grau (1963) and Embree and others (1963). Treitel and others (1967) described a procedure for significantly reducing the computer calculations involved in fan-pass filtering. Fan-pass filtering also can be accomplished by using optical techniques involving coherent and incoherent light (Dobrin and others, 1965; Patau and others, 1970; Bourrouilh and Bourrouilh, 1972).

As previously applied to geologic data, spatial filtering has primarily been of the band-pass type. Robinson and others (1968), Robinson (1969), Robinson and Charlesworth (1969), and Robinson and others (1969) referred to a study in which Devonian and Cretaceous stratigraphic tops in Alberta are band-pass filtered to delineate structural trends with widths between 13 and 130 km. Robinson and Merriam (1972) discussed the application of band-pass and combined band-pass and fan-pass filtering to stratigraphic data in Kansas and compared the advantages and disadvantages of spatial filtering with polynomial surface fitting. Robinson

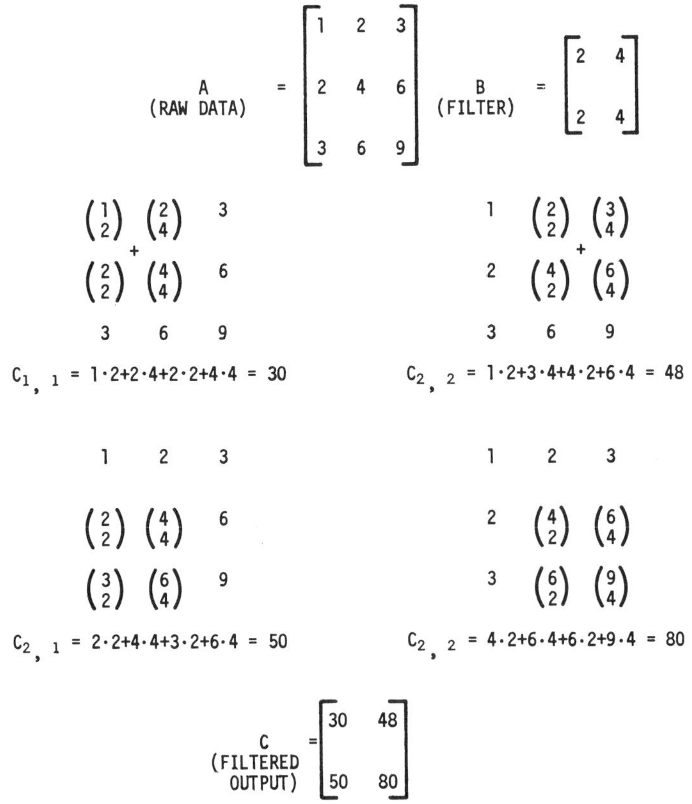

$$A \text{ (RAW DATA)} = \begin{bmatrix} 1 & 2 & 3 \\ 2 & 4 & 6 \\ 3 & 6 & 9 \end{bmatrix} \quad B \text{ (FILTER)} = \begin{bmatrix} 2 & 4 \\ 2 & 4 \end{bmatrix}$$

$$\begin{array}{ccc} \binom{1}{2} & \binom{2}{4} & 3 \\ \binom{2}{2} & \binom{4}{4} & 6 \\ 3 & 6 & 9 \end{array} \qquad \begin{array}{ccc} 1 & \binom{2}{2} & \binom{3}{4} \\ 2 & \binom{4}{2} & \binom{6}{4} \\ 3 & 6 & 9 \end{array}$$

$$C_{1,1} = 1\cdot 2 + 2\cdot 4 + 2\cdot 2 + 4\cdot 4 = 30 \qquad C_{2,2} = 1\cdot 2 + 3\cdot 4 + 4\cdot 2 + 6\cdot 4 = 48$$

$$\begin{array}{ccc} 1 & 2 & 3 \\ \binom{2}{2} & \binom{4}{4} & 6 \\ \binom{3}{2} & \binom{6}{4} & 9 \end{array} \qquad \begin{array}{ccc} 1 & 2 & 3 \\ 2 & \binom{4}{2} & \binom{6}{4} \\ 3 & \binom{6}{2} & \binom{9}{4} \end{array}$$

$$C_{2,1} = 2\cdot 2 + 4\cdot 4 + 3\cdot 2 + 6\cdot 4 = 50 \qquad C_{2,2} = 4\cdot 2 + 6\cdot 4 + 6\cdot 2 + 9\cdot 4 = 80$$

$$C \text{ (FILTERED OUTPUT)} = \begin{bmatrix} 30 & 48 \\ 50 & 80 \end{bmatrix}$$

Figure 2. Filtering of gridded data matrix (A) using grid operator (B). Filtered output is matrix C. Multiplication of input data point by its corresponding filter value indicated by enclosing pair of values in parentheses. Output values not defined where portion of filter falls beyond limits of input data.

(1967) presented the results of fan-pass filtering of aeromagnetic data from central Nevada. Bhattacharya and Raychaudhuri (1967) used band-pass filtering to delineate aeromagnetic anomalies of various sizes in the Appalachian Mountain belt of Canada.

Merriam and Robinson (1970) and Robinson and Merriam (1971) presented simplified fan-pass and band-pass filter maps (Z-trend maps) that display only the presence or absence of trends. Robinson and Ellis (1971) have written a FORTRAN IV computer program for spatial filtering of geologic data.

Filtering Techniques

The spatial filter, also referred to as an operator, consists of a series of weighting factors sequentially applied to (convolved with) subsets of the gridded data (Fig. 2). An elementary discussion of digital filtering procedures can be found in Robinson and Treitel (1964). A fan-pass filter is one that will effectively pass all features with directional properties within a given range of azimuth. The minimum azimuth range is limited to 90° (45° either side of a central azimuth) if the grid spacings are equal in both X and Y directions. Ideally the filter response should resemble as closely as possible that shown in Figure 3; that is, all features in the pass region should be passed with their amplitudes unaffected, whereas those in the reject region should be completely attenuated. The ideal response can only be obtained with a filter of infinite length. Close approximations to the ideal filter can be obtained by use of a 12-row by 25-column matrix.

The two-dimensional spatial fan-pass filter $g(X, Y)$ is obtained by evaluation of the two-dimensional Fourier transform

$$g(X, Y) = \int\int_{-\infty}^{\infty} e^{-2\pi i (f_X X + f_Y Y)} f_X f_Y \, dXdY,$$

where f_X is frequency in the X direction and f_Y is frequency in the Y direction (Embree and others, 1963). Integration of this equation for a fan-pass filter of $2m$ rows and $2n + 1$ columns gives

$$g(X, Y) = \frac{1}{\pi^2 \left[\left(\frac{Ym}{\Delta_Y} \right)^2 - \left(\frac{Xn}{K\Delta_X} \right)^2 \right]},$$

where Δ_X and Δ_Y are grid intervals and K, m, and n are integers (Embree and others, 1963; Robinson, 1967). For $K = 1$, the filter will pass trends within a ±45° range for a specified angle. Larger values of K increase the width of the fan-pass angle.

To filter a set of gridded data at a series of angles one can (1) create a specific filter for each fan-pass angle, (2) maintain a constant filter angle and rotate the coordinate system prior to gridding, or (3) maintain a constant filter angle and shift (lag) the data so that trends with the specified orientation are realigned to correspond with the filter angle. Listings of fan-pass filter weights are presented by Embree and others (1963), Fuller (1967), and Robinson and Merriam (1972). The larger the filter, the more accurate the results. However, increasing the length of the filter increases the processing time required and enlarges the border areas in which unwanted edge effects are present.

Gridding

The major drawback to spatial filtering is the requirement that input data be gridded. Gravity, magnetic, and seismic data usually meet this requirement, but geologic data do not. The procedures used in gridding a set of irregularly spaced data may tend to "smooth" the data by decreasing the effects of local and noise components. Gridding also may destroy regional or local components—or even create additional features. Consequently, it is essential that gridding techniques be examined critically before applying them to data.

The size of the smallest anomaly that spatial filtering (or polynomial fitting) can detect is dependent on the spacing of the observed data and the interval on which the data is gridded. As noted by Robinson and others (1969), the width of a map feature must be greater than about twice the average well spacing or grid spacing before it will be detected.

Filtered Output

Unlike polynomial trend fitting, spatial filtering techniques produce trends whose widths, lengths, positions, orientations, and relative amplitude match closely those of trends present in the raw data. Polynomial regression techniques fit the observed data in one operation (for an example of surface fitting of low-order fixed-degree polynomials to subsets of data see Czeglédy, 1972). If clustering of data points is extreme, the fit in areas of sparse control may be overly influenced by clusters of points in relatively remote portions of the area. Spatial filtering operates on only a small portion of the gridded data at one time and thus approximates local trends uninfluenced by the nature of the data in more distant parts of the map area. Moving-average (Davis, 1973, p. 222-230, 374-390) and spline (Whitten and Koelling, 1973) techniques are similar to spatial filtering in this respect.

Amplitudes of features on spatially filtered maps cannot be compared directly to the amplitudes of the raw data. This is because the filters utilized are approximations to the required filters. Such imperfect filters emphasize or attenuate some trends to a greater degree than is desired. However, relative amplitudes of features within a narrow range of size or orientation are preserved.

A further problem associated with fan-pass filter techniques is their inability to preserve correctly the shape of equidimensional features (domes) in the filter output. Equidimensional features appear on filtered maps as elongated trends with

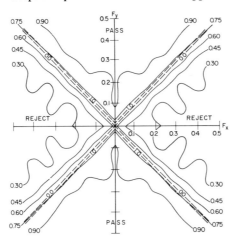

Figure 3. Amplitude response characteristics of ideal north-south fan-pass filter (dashed lines) and actual north-south 12-row by 25-column fan-pass filter (solid lines). F_x represents frequency in cycles per grid interval in X direction and F_Y cycles per grid interval in Y direction. Contours indicate ratio of output to input amplitude for any given frequency in the X or Y directions.

the direction of elongation parallel to the fan-pass angle. Such features can be distinguished from normal elongate trends by the shift of orientation of the trend if the fan-pass angle is changed.

Filtering is not defined at the edges of a map area when a portion of the filter extends beyond the existing data. Therefore a zone one-half the length (and width) of the filter at the map edges can be lost. The anomalies in this unfiltered border zone may be of critical interest in many applications. Methods of avoiding the loss of data at the map edges include (1) extension of the raw data beyond the map area by addition of grid points with a preset numerical value(s), (2) combined modification of the original map data and extension beyond the map edges, and (3) modification of the filter at the map edges. Techniques (1) and (2) have been discussed by Foster and others (1968) and method (3) by Johnson (1971). These techniques do not overcome edge effects completely. Consequently, it should be recognized that filtering programs utilizing such techniques will produce maps that exhibit at least slight edge effects. A better solution is to collect data well beyond the area of interest and then discard the results at the map edges. Unfortunately in many situations this may not be possible or practical.

Application to Test Data

Several of the advantages and drawbacks of fan-pass spatial filtering are illustrated by filtering a set of contrived data. The test data consist of anomalies of varying shape, amplitude, and orientation (Fig. 4) to which random noise (mean of 0.0 and standard deviation of 12.5) and a small vertical fault (offset down to the east 25.0) have been added (Fig. 5). The data were collected at the intersections of a square grid, a portion of which is shown in the lower left corner of Figure 4. The addition of the random noise effectively obscures the anomalies whose amplitudes (before addition of random noise) are less than 25. The data are filtered for southwest-northeast trends (45°) and southeast-northwest trends (135°). The filter used is a 12-row by 25-column fan-pass filter whose amplitude response characteristics are shown in Figure 3. The filter in Figure 3 is actually a north-south (90°) fan filter. In order to filter for trends at 45° and 135°, the gridded data have been shifted (lagged) so that the trends of interest occur at 90°.

Examination of the 45° filter map (Fig. 6) indicates that only the anomalies at 45° and those within a 45° angle from this, with the exception of the anomalies at 180°, are passed by the filter. Trends at 165° to 180° are eliminated as a result of the inability of east-west lagging of data points to realign these trends. I recommend avoiding lagging procedures for most fan-pass filtering applications.

The positions, orientations, shapes, and widths of trends within the 90° fan (45° ± 45°) are retained with the exception that the dome-shaped anomalies in the upper right portion of the area appear as elongate trends. Even the obscure low-amplitude trends are relatively obvious in the filter output. The relative amplitudes of features within 30° of the fan-pass angle are maintained, whereas those at greater angles are progressively attenuated. Filtering over the full extent of the map in the east-west direction is made possible by addition of data points on both edges. The additional points in each row are set equal to the first value (left border) or last value (right border) in the row. A loss of three rows of output occurs at the top and bottom of the 45° filter map. The length of the columns was not extended to permit filtering over the entire north-south extent of the map.

The 135° filter map (Fig. 7) retains trends within 45° of the specified filter angle and effectively eliminates those features with trends outside this range. Comparison of the 45° and 135° filter maps demonstrates the effectiveness of spatial filtering in separating divergent trends within a map area. As in the 45° filter map, trends at 180° are missing as a result of the lagging operation. Note that the dome-shaped anomalies appear as elongate trends with an orientation of 135° (Fig. 7).

The filter output for both 45° and 135° indicates that the presence of the fault in the raw data does not alter the ability of the filtering process to delineate features cut by the fault. However, when the distribution of control points is not as dense as in the present example, the ability of the filtering process to delineate trends in the vicinity of a fault is considerably diminished.

Map Selection Criteria

Statistical criteria for evaluating the effectiveness of spatial filters in isolating local anomalies have not been developed. Fan-pass and band-pass filtering screen input data for trends whose orientation or size has been preselected. The charac-

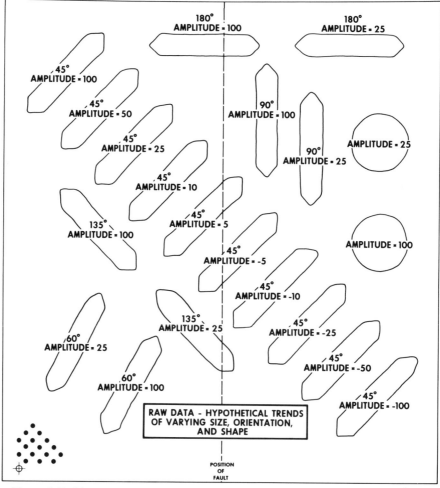

Figure 4. Positions, orientations, and amplitudes of anomalies in gridded test data. Part of grid system is shown in lower left corner along with X-Y coordinate origin (circle with cross).

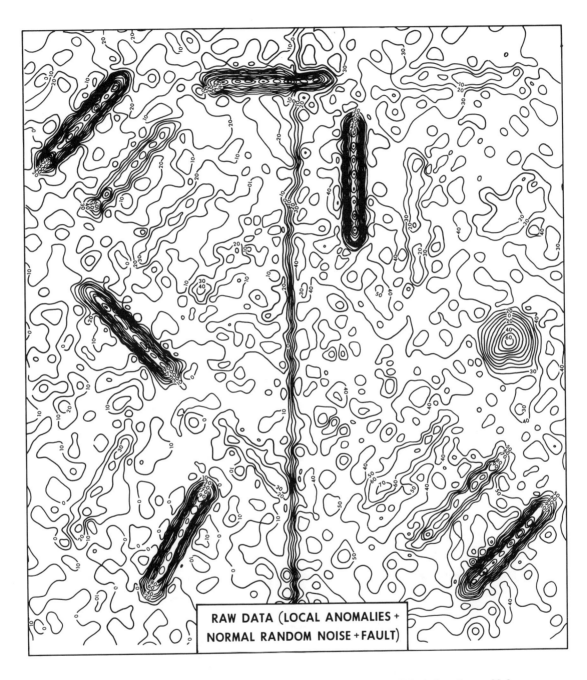

Figure 5. Test data of Figure 4 to which random noise and a small fault have been added. Anomalies with amplitudes less than 25 (southwest-northeast trends only) are obscured by addition of noise component.

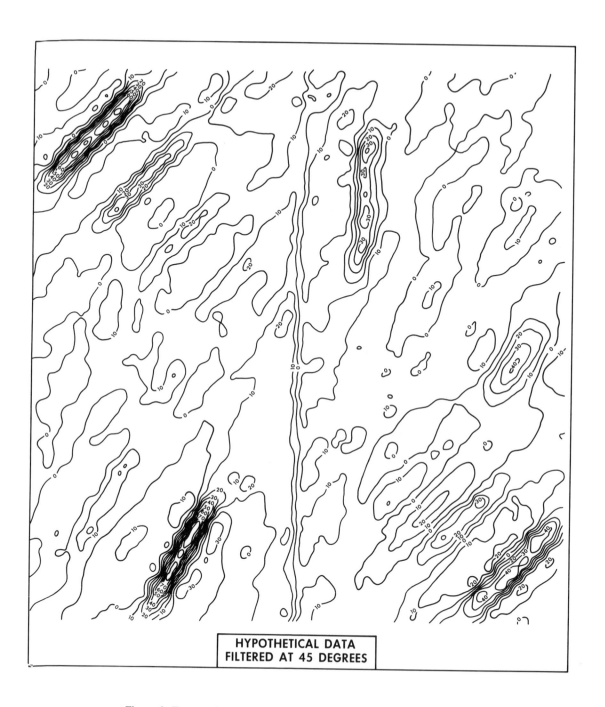

Figure 6. Fan-pass filter at 45° (southwest-northeast) of test data in Figure 5.

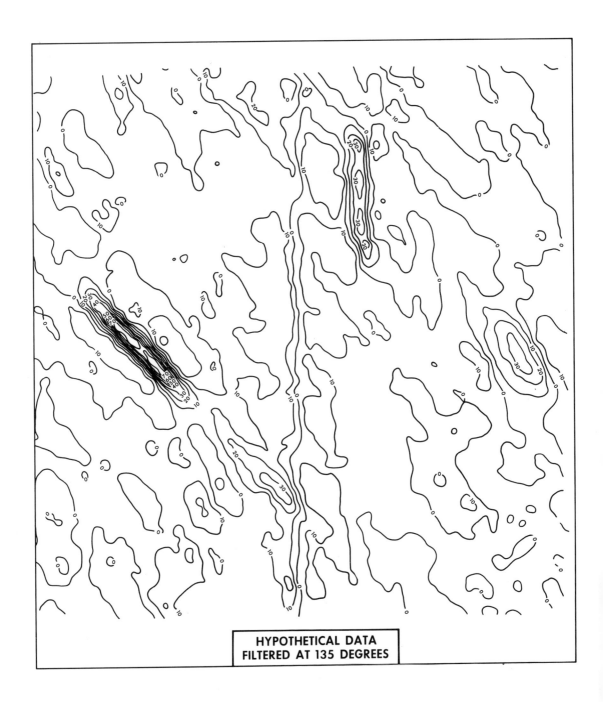

Figure 7. Fan-pass filter at 135° (southeast-northwest) of test data in Figure 5.

teristics of the filter (ideal filter) that best emphasize a given trend can be rigorously defined by means of an amplitude response diagram (Fig. 3). Therefore the degree to which a given filter approximates the ideal filter can be determined by measuring the difference between their amplitude responses.

CONCLUSIONS

The applied forms of spatial filtering and polynomial surface fitting have drawbacks that may severely limit their ability to define accurately regional trends and local anomalies in geologic data. The most severe restraint on spatial filtering is the requirement of gridded input data. Polynomial fitting is affected adversely by the use of intercorrelated independent variables and by the presence of discontinuities (faults) of relatively large offset.

Numerical criteria for determining the similarity of spatial filter response to the desired filter response are needed in addition to a method of filtering nongridded data. A best-fit criterion for polynomial surfaces that includes tests for both significance of predictor terms and degree of autocorrelation in the residual components also needs to be developed.

ACKNOWLEDGMENT

I thank Cities Service Oil Company for permission to publish this article.

REFERENCES CITED

Agterberg, F. P., 1964, Methods of trend surface analysis: Colorado School Mines Quart., v. 59, p. 111-120.

——1969, Interpolation of areally distributed data: Colorado School Mines Quart., v. 64, p. 217-237.

Baird, A. K., Baird, K. W., and Morton, D. M., 1971, On deciding whether trend surfaces of progressively high order are meaningful: Discussion: Geol. Soc. America Bull., v. 82, p. 1219-1234.

Bhattacharya, B. K., and Raychaudhuri, B., 1967, Aeromagnetic and geological interpretation of a section of the Appalachian belt in Canada: Canadian Jour. Earth Sci., v. 4, p. 1015-1037.

Bourrouilh, R., and Bourrouilh, B., 1972, Analyse spectrale et filtrage en éventail des cartes structurales de l'île de Minorque et de l'Est de Majorque; consequences tectoniques: Acad. Sci. Comptes Rendus, v. 275, ser. D, p. 1335-1338.

Chayes, F., 1970, On deciding whether trend surfaces of progressively higher order are meaningful: Geol. Soc. America Bull., v. 81, p. 1273-1278.

Czeglédy, P. F., 1972, Efficiency of local polynomials in contour mapping: Internat. Assoc. Math. Geologists Jour., v. 4, p. 291-305.

Davis, J. C., 1973, Statistical analysis of geological data: New York, John Wiley & Sons, Inc., 525 p.

Dean, W. C., 1958, Frequency analysis of gravity and magnetic interpretation: Geophysics, v. 23, p. 97-127.

Dobrin, M. B., Ingalls, A. F., and Long, J. A., 1965, Velocity and frequency filtering of seismic data using laser light: Geophysics, v. 30, p. 1144-1178.

Draper, N. R., and Lawrence, W. E., 1970, A note on residuals in two dimensions: Technometrics, v. 12, p. 394-398.

Draper, N. R., and Smith, H., 1966, Applied regression analysis: New York, John Wiley & Sons, Inc., 407 p.

Embree, P., Burg, J. P., and Backus, M. M., 1963, Wide band velocity filtering—The pie slice process: Geophysics, v. 28, p. 948-974.

Fail, J. P., and Grau, G., 1963, Les filtres en éventail: Geophys. Prosp., v. 7, p. 131-163.

Foster, M. R., Sengbush, R. L., and Watson, R. J., 1968, Use of Monte Carlo techniques in optimum design of the deconvolution process: Geophysics, v. 33, p. 945-949.

Fuller, B. D., 1967, Two dimensional frequency analysis and design of grid operators: Mining Geophysics, v. 2, p. 658-708.

Howarth, R. J., 1967, Trend surface fitting to random data—An experimental test: Am. Jour. Sci., v. 265, p. 619-625.

James, W. R., 1970, Regression models for faulted structural surfaces: Am. Assoc. Petroleum Geologists Bull., v. 54, p. 638-646.

Johnson, B. D., 1971, Convolution filtering at ends of data sets: Australian Soc. Exploration Geophysicists Bull., v. 2, no. 4, p. 11-24.

Jones, T. A., 1972, Multiple regression with correlated independent variables: Internat. Assoc. Math. Geologists Jour., v. 4, p. 203-218.

Krumbein, W. C., and Graybill, F. A., 1965, An introduction to statistical models in geology: New York, McGraw-Hill Book Co., 475 p.

McIntyre, D. B., 1967, Trend-surface analysis of noisy data, in Merriam, D. F., and Cocke, N. C., eds., Computer applications in the earth sciences: Colloquium on trend analysis: Kansas Geol. Survey Computer Contr. 12, p. 45-49.

Merriam, D. F., and Robinson, J. E., 1970, Trend analysis in geologic and geophysical exploration, in Symposium on mathematical methods in geology and geophysics: Příbram, Czechoslovakia, Hornická Příbram ve vědě a technice, Paper M11, 41 p.

Miesch, A. T., and Connor, J. J., 1967, Stepwise regression in trend analysis, in Merriam, D. F., and Cocke, N. C., eds., Computer applications in the earth sciences: Colloquium on trend analysis: Kansas Geol. Survey Computer Contr. 12, p. 16-18.

Norcliffe, G. B., 1969, On the use and limitations of trend surface models: Canadian Geographer, v. 13, p. 338-348.

Parsley, A. J., 1971, Application of autocorrelation criteria to the analysis of mapped geologic data from the Coal Measures of central England: Internat. Assoc. Math. Geologists Jour. v. 3, p. 281-295.

Patau, J. C., Lesem, L. B., Hirsch, P. N., and Jordon, J. A., Jr., 1970, Incoherent filtering using kinoforms: IBM Jour. Research and Devel., p. 485-491.

Robinson, E. A., and Treitel, S., 1964, Principles of digital filtering: Geophysics, v. 29, p. 395-404.

Robinson, E. S., 1967, Use of fan filters in computer analysis of magnetic-anomaly trends: U.S. Geol. Survey Prof. Paper 575D, p. D113-D119.

Robinson, J. E., 1969, Spatial filtering for geological data: Oil and Gas Jour., v. 67, p. 132-134, 140.

Robinson, J. E., and Charlesworth, H.A.K., 1969, Spatial filtering illustrates relationship between tectonic structure and oil occurrence in southern and central Alberta, in Merriam, D. F., ed., Symposium on computer applications in petroleum exploration: Kansas Geol. Survey Computer Contr. 40, p. 12-18.

Robinson, J. E., and Ellis, M. J., 1971, Spatial filters and FORTRAN IV program for filtering geologic maps: Geocom Programs 1, Geosystems, London, 21 p.

Robinson, J. E., and Merriam, D. F., 1971, Z-trend maps for quick recognition of geologic patterns: Internat. Assoc. Math. Geologists Jour., v. 3, p. 171-181.

———1972, Enhancement of patterns in geologic data by spatial filtering: Jour. Geology, v. 80, p. 333-345.

Robinson, J. E., Charlesworth, H.A.K., and Kanasewich, E. R., 1968, Spatial filtering of structural contour maps: Internat. Geol. Cong., 23d, Prague 1968, sec. 13, p. 163-173.

Robinson, J. E., Charlesworth, H.A.K., and Ellis, M. J., 1969, Structural analysis using spatial filtering in interior plains of south-central Alberta: Am. Assoc. Petroleum Geologists Bull., v. 53, p. 2341-2367.

Swartz, C. A., 1954, Some geometrical properties of residual maps: Geophysics, v. 19, p. 46–70.

Tinkler, K. J., 1969, Trend surfaces with low "explanations"; the assessment of their significance: Am. Jour. Sci., v. 267, p. 114–123.

Treitel, S., Shanks, J. L., and Frasier, C. W., 1967, Some aspects of fan filtering: Geophysics, v. 32, p. 789–800.

Whitten, E.H.T., 1970, Orthogonal polynomial trend surfaces for irregularly spaced data: Internat. Assoc. Math. Geologists Jour., v. 2, p. 141–152.

Whitten, E.H.T., and Koelling, M.E.V., 1973, Spline-surface interpolation, spatial filtering, and trend surfaces for geological mapped variables: Internat. Assoc. Math. Geologists Jour., v. 5, p. 111–126.

MANUSCRIPT RECEIVED BY THE SOCIETY SEPTEMBER 10, 1973
REVISED MANUSCRIPT RECEIVED JANUARY 21, 1974

Texture Analysis

Geoffrey S. Watson

*Department of Statistics
Princeton University
Princeton, New Jersey 08540*

ABSTRACT

G. Matheron, J. Serra, and their colleagues at the Centre de Morphologie Mathématique of the Ecole Nationale Supérieure de Paris have developed a novel theory and device for studying geometric aspects of two-dimensional sets which is of considerable interest to statisticians and petrographers (especially sedimentary petrographers). A stationary random set, regarded as the union of subsets, is scanned at discrete points in a finite region **D**, and the results are stored in a computer. The theory indicates how programs should be written to estimate quantities such as the size distribution of subsets (granulometric properties) and their mutual relations (structural properties). This paper provides an introduction to their work.

INTRODUCTION

Geological problems have led to two major developments of statistical theory. Krumbein's (1939) paper on the preferred orientation of pebbles in sedimentary deposits began the statistics of orientation data. This development was accelerated by paleomagnetic studies (for a survey of the field, see Watson, 1970). Krumbein and co-workers also described the texture of sedimentary rock as the totality of descriptors of grains and pores in a rock. (Krumbein and Pettijohn, 1938; Pettijohn, 1957). This paper is a survey of more recent observational and mathematical methods that have been developed for this problem, particularly in the Paris School of Mines.

Other sciences (for example, metallurgy) also face the problem of describing and measuring the shape, size, and mutual relations of particles of different materials suspended in space. The observational techniques usually involve passing a "probe" (a line or plane) through the body and examining, with various devices, the intersections of the probe and the particles. The relevant literature is large and scattered; the same facts are often rediscovered, and different sciences use different terms that reflect more the possible physical measurments than the inherent mathematical notions. The nature and limits of this survey are described below.

Mathematically, particles of a given material distributed in space may be described by a set **A** of the points **x** belonging to these particles. The set **A** is made up of nonintersecting subsets or particles which will be called grains. All points not belonging to **A** belong to the pores, thus maintaining the sedimentary motivation. An equivalent description is given by a function $I_A(x)$, which is equal to unity where **x** is a point in **A** and is zero otherwise. The chaotic local and overall homogeneous nature of a sediment suggests that the function $I_A(x)$ should be regarded as a random function whose properties, however, are the same over all space, that is, a function generated by a random process whose mechanism is the same everywhere. This was first suggested by Fara and Scheidegger (1961) to describe a linear traverse through a rock. Matheron (1967) developed this idea in great generality and much of this survey will be devoted to his work.

Any such theory must, if it is to be useful, be closely allied with possible observing and measuring devices. For years, petrographers have made manual examinations of parallel traverses of a plane section of rock. In the 1960s flying spot scanners were developed for studying two-dimensional areas. At the National Bureau of Standards, metallurgical problems motivated the early work of Kirsh (1957), which is further described in Moore and others (1968). In France, Serra developed the "texture analyser" (see Klein and Serra, 1972). The technology varies, but for all these devices it should suffice here to have the following picture in mind. Imagine a beam of light of very small cross section traversing, by parallel straight lines, a rectangular area. A voltage signal proportional to the gray level of the spot lighted is sampled at regular intervals. If high voltage corresponds to solid and low voltage to void, time points (and thus grid points) when the voltage is above a discrimination level will be given the value 1, and when it is below that level, the value 0.

These values could produce, on a cathode-ray tube, an image that would be a digitized approximation of the area examined. In most devices (for example, Argonne National Laboratory's ALICE) the grid is rectangular. In Serra's machine the grid is hexagonal, which is better for his purposes. Because of noise in the circuits, problems in preparation of the surface examined, and possible drifts in the voltage level, errors will be made that can be discussed statistically. This topic will not be treated here, although it is vital to applied work. It will be assumed here that the image has been "cleaned up" as much as possible by two-dimensional filtering.

Given that it is possible to process a digitized image of the part of set **A** in region **D** (that is, an image of **A** $\cap$ **D**), geometric information can be derived from it in the following manner. The proportion of the area of region **D** occupied by **A** is estimated easily by the proportion of grid points in **D** having the value 1. Since the area of **D** is known, the area of the scanned set **A** $\cap$ **D** is determined in the following way. The trick is to reduce other problems to finding the area of some set. For example, the perimeter of **A** in **D** is roughly the area of a band around the periphery of **A** divided by the width of the band. To get this set we could, in theory, use a circular disc **B** of small radius r, place its center at every point **x** inside **A**, and record the points where **B** intersects the complement of **A**, A^c. An approximation of set **C** of points so obtained can be found digitally and shown on the cathode-ray tube. Hence, the perimeter can be estimated from set **C**, which is **A** transformed by set **B** (that is, $C = \phi(A, B)$). **B** will be called the structural element. Hence, it is necessary to know some operations that transform sets; these are given in the next section, after which they are used in the above manner to obtain geometric information about set **A**.

In practice, the sequence of zeros and ones from the scanner is fed into a computer, and the programs that realize the above operations (and others such as counting grains) depend on many things. The elegance of the French approach described above is that it provides a general mathematical method for generating measuring procedures, and one does not have to think in terms of programs, although these must be written eventually. Its strength comes from the fact that it suggests a wealth of ways in which texture might be described and investigated. Serra has even suggested that it forms the basis of a theory of mathematical morphology (for example, see Serra, 1969).

In practice, set **A** will lie in three dimensions. There are stereological problems, mentioned briefly below, in reconstructing **A** from information on lines and planes. Serra (1969) gave an account suitable for his approach. A less mathematical and more materials-science point of view may be found in DeHoff and Rhines (1967). Weibel (1973) also provided an elementary review of stereological techniques and emphasized the petrographic origins of many of the principles involved.

This survey ends with an account of size distributions of grains and pores in a homogeneous sediment. One of the many related topics avoided in this paper is pattern recognition, because our interest here is morphology rather than taxonomy. There are strong connections between Matheron's theory and geometrical probability (see Kendall and Moran, 1963) and, through stereology, with integral geometry (see Hadwiger, 1957). The theory of random sets is the subject of Matheron (1975) and Harding and Kendall (1974).

SET TRANSFORMATIONS

The set transformations of most value in morphology may be generated by an operation due to Minkowski. Let **A** and **B** be two sets in an n-dimensional Euclidean space $\mathcal{R}^n$. In practice $n = 1, 2,$ or 3. Points in the space will be denoted by their position vectors **x**, **y**, and so forth. The Minkowski sum of **A** and **B**, $\mathbf{A} \oplus \mathbf{B}$, is then defined by

$$\mathbf{A} \oplus \mathbf{B} = \{\mathbf{x} + \mathbf{y} \mid \mathbf{x} \in \mathbf{A}, \mathbf{y} \in \mathbf{B}\}, \tag{1a}$$

$$= \bigcup_{\substack{\mathbf{x} \in \mathbf{A} \\ \mathbf{y} \in \mathbf{B}}} (\mathbf{x} + \mathbf{y}),$$

where $\cup$ is the sign for union. If **B** consists of a single point **y** only, $\mathbf{A} \oplus \mathbf{B}$ is the set **A** translated by the vector **y**, which may be written more simply as $\mathbf{A}_\mathbf{y}$. Thus

$$\mathbf{A}_\mathbf{y} = \bigcup_{\mathbf{x} \in \mathbf{A}} (\mathbf{x} + \mathbf{y}),$$

with the notation

$$\mathbf{A} \oplus \mathbf{B} = \bigcup_{\mathbf{x} \in \mathbf{A}} \mathbf{B}_\mathbf{x} = \bigcup_{\mathbf{y} \in \mathbf{B}} \mathbf{A}_\mathbf{y}. \tag{1b}$$

Examples of $\mathbf{A} \oplus \mathbf{B}$ are shown in Figure 1.

Also, the reflection $\check{\mathbf{B}}$ of a set $\mathbf{B}$ in the origin of $\mathscr{R}^n$ is defined as

$$\check{\mathbf{B}} = \bigcup_{\mathbf{x} \in \mathbf{B}} (-\mathbf{x}).$$

Therefore, a point $\mathbf{z}$ belongs to $\mathbf{A} \oplus \mathbf{B}$ if and only if (iff) the intersection of $\mathbf{A}$ and $(\check{\mathbf{B}})_\mathbf{z}$ is not empty; that is,

$$\mathbf{z} \in \mathbf{A} \oplus \mathbf{B} \rightleftarrows (\check{\mathbf{B}})_\mathbf{z} \cap \mathbf{A} \neq \emptyset. \tag{1c}$$

If $\mathbf{z} \in \mathbf{A} \oplus \mathbf{B}$, $\mathbf{z} = \mathbf{x} + \mathbf{y}$, where $\mathbf{x} \in \mathbf{A}$, $\mathbf{y} \in \mathbf{B}$ by equation 1a, then $\mathbf{x} = \mathbf{z} - \mathbf{y} \in \mathbf{A}$. Since $(\check{\mathbf{B}})_\mathbf{z} = \bigcup_{\mathbf{y} \in \mathbf{B}} (\mathbf{z} - \mathbf{y})$, we have $(\check{\mathbf{B}})_\mathbf{z} \cap \mathbf{A} \neq \emptyset$. The converse is proved similarly. Matheron (1967) defined the dilatation of $\mathbf{A}$ by $\mathbf{B}$ to be

$$\mathbf{A} \oplus \check{\mathbf{B}} = \{\mathbf{z} \mid \mathbf{B}_\mathbf{z} \cap \mathbf{A} = \emptyset\}. \tag{2}$$

Thus, $\mathbf{z}$ belongs to the dilatation of $\mathbf{A}$ by $\mathbf{B}$ iff $\mathbf{B}_\mathbf{z}$ is entirely contained in the complement $\mathbf{A}^c$ of $\mathbf{A}$, so that $\mathbf{B}_\mathbf{z} \subset \mathbf{A}^c$. Figure 2 shows by example how $\mathbf{A} \oplus \mathbf{B}$ and $\mathbf{A} \oplus \check{\mathbf{B}}$ may differ. If $\mathbf{B}$ is a circle centered at the origin, then $\mathbf{B} = \check{\mathbf{B}}$ and they are identical.

In the introduction, the set $\mathbf{C} = \{\mathbf{z} \mid \mathbf{B}_\mathbf{z} \subset \mathbf{A}\}$ arose. By the remark following equation (2), $\mathbf{C} = \mathbf{A}^c \oplus \check{\mathbf{B}}$. To obtain $\mathbf{C}$ as a transformation of $\mathbf{A}$, we define, by analogy with equation (1b),

$$\mathbf{A} \ominus \mathbf{B} = \bigcap_{\mathbf{y} \in \mathbf{B}} \mathbf{A}_\mathbf{y}. \tag{3a}$$

The analogue of equation (1c) is

$$\mathbf{z} \in \mathbf{A} \ominus \mathbf{B} \rightleftarrows (\check{\mathbf{B}})_\mathbf{z} \subset \mathbf{A}. \tag{3b}$$

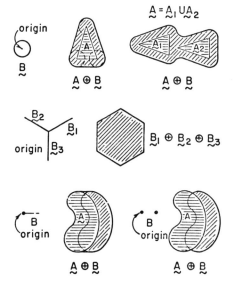

Figure 1. Illustrations of Minkowski sums of sets.

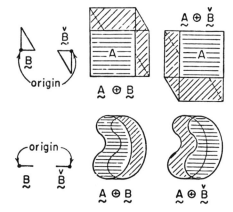

Figure 2. Comparisons of $\mathbf{A} \oplus \mathbf{B}$ and of $\mathbf{A} \oplus \check{\mathbf{B}}$, the dilatation of $\mathbf{A}$ by $\mathbf{B}$.

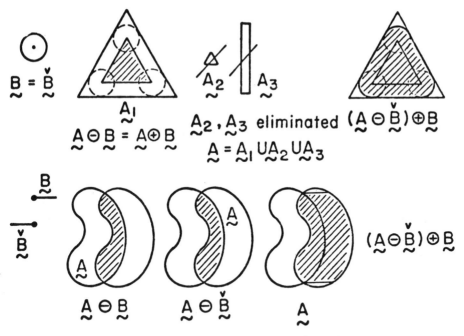

Figure 3. Illustrations of the erosion and opening of a set A by a set B.

This leads to the definition of the erosion of **A** by **B** as

$$\mathbf{A} \ominus \check{\mathbf{B}} = \{\mathbf{z} \mid \mathbf{B}_z \subset \mathbf{A}\}, \tag{4}$$

which should be contrasted with equation (2). Parts of **A** are removed when **A** is eroded by **B**, and these may not be replaced when **B** is added to **A** ⊖ **B**. The resulting set, however, is of great interest in the study of granulometry (the "sizes" of grains), as will be seen in the last section of this survey. It is called the opening of **A** by **B** and is denoted by **A** ω **B**, with the definition

$$\mathbf{A} \, \omega \, \mathbf{B} = (\mathbf{A} \ominus \check{\mathbf{B}}) \oplus \mathbf{B}. \tag{5}$$

It may be shown that **A** ω **B** is the set of all points swept out by **B** as it moves to all positions inside **A** and so is a smoothed-off version of **A**. Figure 3 illustrates the operations defined in equations (3), (4), and (5).

MEASUREMENT OF GEOMETRIC ASPECTS OF A

In these applications the arbitrary set **A** is a model of a collection of particles. Matheron originally worked with grains in a sedimentary rock, so that $\mathbf{A}^c$ is the space between them or pores. In studying the flow of fluid through rocks or the power of a sediment to store oil, $\mathbf{A}^c$ is more important than **A**. In this case, the grains have a fairly regular shape by mathematical standards. If **A** is the set of fractures in a rock body—a set of surfaces in $\mathscr{R}^3$ or lines in $\mathscr{R}^2$—the same is true. (If these methods are applied to the analysis of photographs of cloud cover as seen from a satellite or to the growth and disappearance of blood vessels around a wound, the sets become more complex. Significant aspects of the pattern will differ from case to case.)

To avoid mathematical difficulties, our arguments will be intuitive and based on diagrams. Although the grains shown in the figures have very simple shapes, the results will hold true for realistic irregular grains. It is also important to illustrate the relation of the method of measurement and the geometric features of **A**. Different assumptions about **A** will be appropriate on different occasions and required for different aspects. In this section, **A** is a bounded set whose points **x** are either interior points or boundary points. At almost all boundary points **x**, the external normal **n**(**x**) exists and the total perimeter is finite. The boundary of **A** is denoted by $\partial \mathbf{A}$. **A** may be a union of disjoint-connected subsets called grains.

It is assumed that all needed quantities exist, in particular the measure of any set **C**, which will be denoted by *mes* **C**, the length of **C** in $\mathcal{R}$, the area of **C** in $\mathcal{R}^2$, and its volume in $\mathcal{R}^3$. If $I_\mathbf{C}(\mathbf{x})$ is the indicator function of set **C** then

$$mes\ \mathbf{C} = \int I_\mathbf{C}(\mathbf{x})\, d\mathbf{x}, \tag{6}$$

where $d\mathbf{x}$ is the volume element, $dx_1 \ldots dx_n$ in $\mathcal{R}^n$. The texture analyser produces, in $\mathcal{R}^2$, discrete approximations to *mes* **C** for all sets considered below.

If $\underset{\sim}{\alpha}$ is a unit vector, the width of **A** in the direction $\underset{\sim}{\alpha}$, $b(\mathbf{A}, \underset{\sim}{\alpha})$ is the distance between two hyperplanes with normal $\underset{\sim}{\alpha}$ that just enclose **A**. The diameter $D(\mathbf{A})$ and thickness $d(\mathbf{A})$ of **A** are defined by

$$\begin{aligned} D(\mathbf{A}) &= \sup_{\underset{\sim}{\alpha}} b(\mathbf{A}, \underset{\sim}{\alpha}), \\ d(\mathbf{A}) &= \inf_{\underset{\sim}{\alpha}} b(\mathbf{A}, \underset{\sim}{\alpha}), \end{aligned} \tag{7}$$

where $\sup_{\underset{\sim}{\alpha}}$ and $\inf_{\underset{\sim}{\alpha}}$ are the maximum (supremum) and minimum (infimum) for all values of $\underset{\sim}{\alpha}$.

Let **B** be a ball of radius r with center at origin, and consider $mes\ (\mathbf{A} \oplus \check{\mathbf{B}})$ (which in this example, is the same as $mes\ (\mathbf{A} \oplus \mathbf{B})$) a function of r. This is the typical method—to consider the measure of a family of sets as a function of the parameter defining the family. Here the measure of the dilatation of **A** by **B**, $(\mathbf{A} \oplus \check{\mathbf{B}})$, clearly increases with r and is equal to *mes* **A** at the origin $r = 0$. The shape at the origin may be interpreted geometrically since, for small r, $mes\ \mathbf{A} \oplus \check{\mathbf{B}} = mes\ \mathbf{A} + mes$ (layer around **A**, thickness r). Thus, in $\mathcal{R}^2$ the area of this layer is r times the perimeter of **A**, and in $\mathcal{R}^3$ the volume of this layer is r times the surface area. Therefore,

$$\left[\frac{d}{dr} mes\ \mathbf{A} \oplus \check{\mathbf{B}} \right]_{r=0} = \begin{cases} \text{perimeter of } \mathbf{A}\, (\mathcal{R}^2) \\ \text{surface area of } \mathbf{A}\, (\mathcal{R}^3). \end{cases} \tag{8}$$

If **A** is a collection of particles, we obtain the total of all perimeters or surface areas. Notice that in $\mathcal{R}^1$ one finds twice the number of intervals, these being the grains of **A**. Equation (8) could be used practically in $\mathcal{R}^2$ by measuring *mes* $\mathbf{A} \oplus \check{\mathbf{B}}$ for various values of r and fitting a curve through the results. The slope of the fitted curve at the origin is then an estimate of the perimeter of **A**.

Again using the ball, the family $\mathbf{A} \ominus \check{\mathbf{B}}$ yields similar results. The value of *mes* $\mathbf{A} \ominus \check{\mathbf{B}}$ decreases with r and vanishes when r is the radius of the largest inscribed ball. The negative slope at $r = 0$ has the above properties.

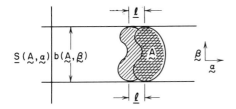
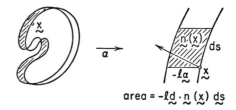

Figure 4. Illustration of how the measure of a dilatation of A can be used to get its width or apparent projected area.

Figure 5. In the left-hand drawing, the new area has been shaded, and in the right, a magnified view is given of the region about x when l is small.

Suppose now that **B** is a segment of length l and direction $\underset{\sim}{\alpha}$, issuing from the origin. Because the structural element **B** is not isotropic like the ball, directional features of **A** should be revealed. Examples show the behavior of *mes* $\mathbf{A} \oplus \check{\mathbf{B}}$, which should increase as l increases. For simple **A**, as in Figure 4, we find

$$mes\, \mathbf{A} \oplus \check{\mathbf{B}} = mes\, \mathbf{A} + lb(\mathbf{A}, \beta), \quad (\mathcal{R}^2)$$
$$= mes\, \mathbf{A} + lS(\mathbf{A}, \underset{\sim}{\alpha}), \quad (\mathcal{R}^3),$$

where in $\mathcal{R}^2$, $S(\mathbf{A}, \underset{\sim}{\alpha}) = b(\mathbf{A}, \beta)$ is the length of the projection of **A** on a line with normal $\underset{\sim}{\alpha}$, and in $\mathcal{R}^3$, $S(\mathbf{A}, \underset{\sim}{\alpha})$ is the area of the projection onto a plane with normal $\underset{\sim}{\alpha}$. However, Figure 5, where ds is the perimeter element on $\partial \mathbf{A}$, shows that new volume is only gained where $-l\underset{\sim}{\alpha} \cdot \mathbf{n}(\mathbf{x})$ is positive and the volume element is $sup\{0, -l\underset{\sim}{\alpha} \cdot \mathbf{n}(\mathbf{x})\, ds\}$. More precisely,

$$\left[\frac{d}{dl} mes\, \mathbf{A} \oplus \check{\mathbf{B}}\right]_{l=0} = -\int_{\partial \mathbf{A}} inf\{0, \underset{\sim}{\alpha} \cdot \mathbf{n}(\mathbf{x})\}\, ds;$$

this reduces to the measure of the projection in simple cases.

If **B** is a pair of points $(\mathbf{0}, \mathbf{h})$, then $\mathbf{A} \oplus \check{\mathbf{B}} = \mathbf{A} \cup \mathbf{A}_{-\mathbf{h}}$. It is of more interest to study $\mathbf{A} \ominus \check{\mathbf{B}} = \mathbf{A} \cap \mathbf{A}_{-\mathbf{h}}$. We define the covariance[1] of **A** as

$$K(\mathbf{h}) = mes\, \mathbf{A} \cap \mathbf{A}_{-\mathbf{h}},$$
$$= \int I_{\mathbf{A}}(\mathbf{x}) I_{\mathbf{A}}(\mathbf{x} + \mathbf{h})\, d\mathbf{x}, \qquad (9)$$
$$= \int I_{\mathbf{A}}(\mathbf{y} - \mathbf{h}) I_{\mathbf{A}}(\mathbf{y})\, d\mathbf{y},$$
$$= K(-\mathbf{h}).$$

It will sometimes be useful to indicate the set involved by writing $K(\mathbf{h}; \mathbf{A})$. It is easy to verify that

$$K(\mathbf{0}) = mes\, \mathbf{A}, \qquad \int K(\mathbf{h})\, d\mathbf{h} = (mes\, \mathbf{A})^2. \qquad (10)$$

---

[1] This is *not* the statistical covariance but a geometrically defined quantity. In later sections, **A** will be a random set, and this quantity will correspond to a statistical covariance, except that it is not corrected for the mean.

If **A** is an isolated grain, $K(\mathbf{h})$ will decrease to zero as $|\mathbf{h}|$ increases; if **A** is composed of several grains, however, the curve will reflect the overlapping of different grains and the behavior is more subtle. As usual, most interest lies in the behavior of $K(\mathbf{h})$ for small $\mathbf{h}$; that is, with $\mathbf{h} = l\underset{\sim}{\alpha}$,

$$K'_{\underset{\sim}{\alpha}}(0) = \lim_{l \to 0} \frac{K(l\underset{\sim}{\alpha}) - K(0)}{l}.$$

Figure 6 shows that, with infinitesimal l,

$$2K(0) = 2 \ \text{mes (shaded region)} + \text{mes (boundary region)};$$

that is,

$$2K'_{\underset{\sim}{\alpha}}(0) = -\int_{\partial A} |\mathbf{n}(\mathbf{x}) \cdot \underset{\sim}{\alpha}| \, ds.$$

Thus,

$$K'_{\underset{\sim}{\alpha}}(0) = -\int_{\partial A} \sup\{0, -\mathbf{n}(\mathbf{x}) \cdot \underset{\sim}{\alpha}\} \, ds,$$

$$= \int_{\partial A} \inf\{0, \mathbf{n}(\mathbf{x}) \cdot \underset{\sim}{\alpha}\} \, ds, \qquad (11)$$

$$= -\tfrac{1}{2} \int_{\partial A} |\mathbf{n}(\mathbf{x}) \cdot \underset{\sim}{\alpha}| \, ds.$$

For simple boundaries, as we saw above, $K'_{\underset{\sim}{\alpha}}(0)$ gives the total measure of the projection of **A**.

If $K'_{\underset{\sim}{\alpha}}(0)$ is estimated from observations in many equally spaced directions $\underset{\sim}{\alpha}$ and the results are averaged, their result will be an estimate of

$$\underset{\underset{\sim}{\alpha}}{\text{ave}} \, K'_{\underset{\sim}{\alpha}}(0) = -\tfrac{1}{2} \Sigma \int \underset{\underset{\sim}{\alpha}}{\text{ave}} |\mathbf{n}(\mathbf{x}) \cdot \underset{\sim}{\alpha}| \, ds, \qquad (12)$$

where the sum is over the disjoint grains A_i composing **A**. For each grain, $\underset{\underset{\sim}{\alpha}}{\text{ave}} |\mathbf{n}(\mathbf{x}) \cdot \underset{\sim}{\alpha}|$ is a dimensional constant. Thus,

$$\underset{\underset{\sim}{\alpha}}{\text{ave}} \, K'_{\underset{\sim}{\alpha}}(0) = -\tfrac{1}{2} \underset{\underset{\sim}{\alpha}}{\text{ave}} |\mathbf{n} \cdot \underset{\sim}{\alpha}| \, \Sigma S(A_i), \qquad (13)$$

where $S(A_i)$ is the measure of ∂A_i. In $\mathscr{R}^2$,

$$\underset{\underset{\sim}{\alpha}}{\text{ave}} |\cos \theta| = (1/2\pi) \int_0^{\pi} |\cos \theta| \, d\theta$$

$$= 2/\pi.$$

In $\mathcal{R}^3$,

$$\underset{\alpha}{ave}|\cos\theta| = (1/4\pi) \iint |\cos\theta| \sin\theta \, d\theta \, d\phi$$

$$= 1/2.$$

Hence,

$$\mathcal{R}^3: \underset{\alpha}{ave} - K'_\alpha(0) = S/4, \quad S = \text{total surface area}, \tag{14}$$

$$\mathcal{R}^2: \underset{\alpha}{ave} - K'_\alpha(0) = 2L/\pi, \quad 2L = \text{total perimeter},$$

$$\mathcal{R}^1: \quad -K'(0) = \text{number of intervals in } \mathbf{A}, \tag{15}$$

where in $\mathcal{R}^1$, $\mathbf{A}$ is the union of disjoint intervals.

Since our treatment is intuitive (being dependent on using diagrams), one may question whether equation (13) applies when grains $\mathbf{A}_i$ of $\mathbf{A}$ have "holes" in them. Figure 7 illustrates this situation and is analogous to Figure 6, where the boundary region is dotted. The deduction of equation (11) is unchanged if $\partial \mathbf{A}_i$ stands for both the internal and external boundaries. Thus, equation (14), which will be of practical use, does not carry topological restrictions on its validity.

In $\mathcal{R}^2$ we may use the covariance of $\mathbf{A}$, $K(\mathbf{h}; \mathbf{A})$ to determine its connectivity, $\nu(\mathbf{A})$, which is the number of grains minus the number of holes. Let $\mathbf{A}$ be the union of n grains $\mathbf{A}_i$, and let $\mathbf{y}$ be an infinitesimal directed line segment orthogonal to α. If

$$-K'_\alpha(0; \mathbf{A}_i \oplus \mathbf{y}) + K'_\alpha(0; \mathbf{A}_i) = \delta[-K'(0; \mathbf{A}_i)],$$

then

$$\delta[-K'(0; \mathbf{A}_i)] = \tfrac{1}{2} \int_{\partial(\mathbf{A}_i \oplus \mathbf{y})} |\mathbf{n}(\mathbf{x}) \cdot \underset{\sim}{\alpha}| \, ds - \tfrac{1}{2} \int_{\partial \mathbf{A}} |\mathbf{n}(\mathbf{x}) \cdot \underset{\sim}{\alpha}| \, ds.$$

From Figure 8

$$\int_{\partial(\mathbf{A} \oplus \mathbf{y})} |\mathbf{n}(\mathbf{x}) \cdot \underset{\sim}{\alpha}| \, ds = \int_{\partial \mathbf{A}} |\mathbf{n}(\mathbf{x}) \cdot \underset{\sim}{\alpha}| \, ds + 2|\mathbf{y}|,$$

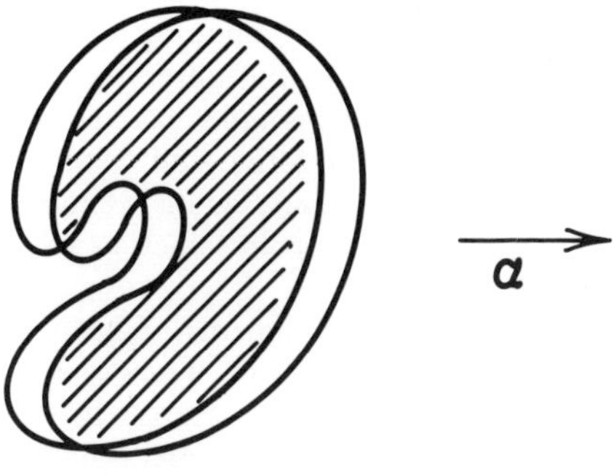

Figure 6. The area in common is shaded, rather than the new area as in Figure 5.

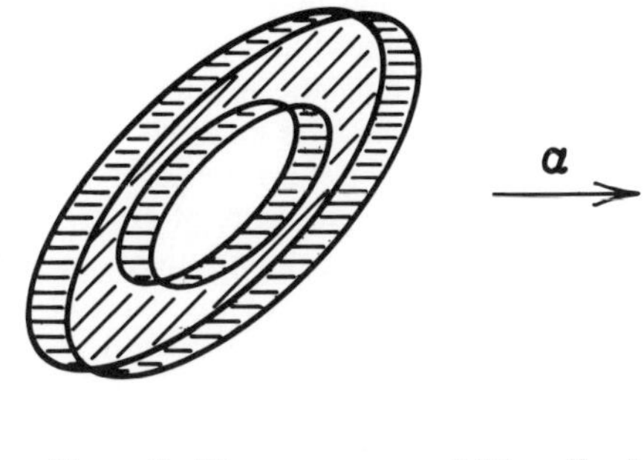

Figure 7. The common area (oblique lines) and boundary area (horizontal lines) for grain containing a hole.

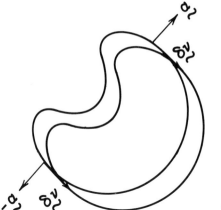

Figure 8. Illustration to show the boundaries of A_i and $A_i \oplus y$ are related. $\delta\underline{v} \equiv y$ of the text.

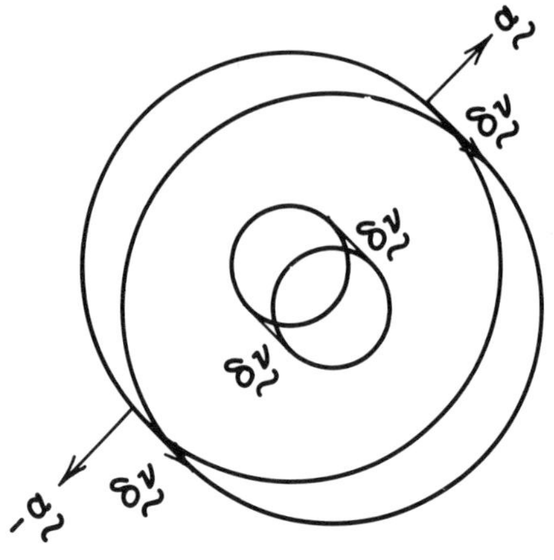

Figure 9. Illustration to show the boundaries of A_i and $A_i \oplus y$ when the grain contains a hole. $\delta\underline{v} \equiv y$ of the text.

where $|y|$ is the length of the line segment y. Hence,

$$\delta\left[-K'(0; A_i)\right] = |y|,$$

so that

$$\delta\left[-K'(0; A)\right] = n|y|.$$

The effect of indentations and holes must then be checked. Diagrams show that the former have no effect, but a hole leads to a contribution of $-|y|$. As seen in Figure 9, the internal boundary decreases; evaluating the integrals over it leads to $-|y|$. Thus, for a set A in $\mathscr{R}^2$, we have

$$\begin{aligned}\delta\left[-K'_\alpha(0)\right] &= |y| \text{ (number of grains minus number of holes)}, \\ &= \nu(A)|y|,\end{aligned} \quad (16)$$

where $\nu(A)$ is the connectivity number of A. The usual definition of connectivity of a single body is 1 + the number of holes. (A disc is singly connected, a donut is two-ply connected, and so forth.)

Other procedures may be used to count the number of grains, such as the program developed by Moore and others (1968). Perimeters may also be computed in other ways. Determining the most appropriate program depends on the observing and measuring device. The methods suggested above are merely examples of how the symbolism works; Haas and others (1967a, 1967b) used it to define and compute a measure of the convexity of grains, measures of angularity, and size distributions of many kinds. Size distributions, as well as several technical points such as edge effects, will be discussed in a later section.

We conclude this section with some brief stereological comments. What can be reconstructed of A in $\mathscr{R}^2$, given data on all lines through it? What can be reconstructed of A in $\mathscr{R}^3$, given data on *all* planes through it? On lines one sees intervals, knows the positions of their ends, their lengths and their number (number of ends divided by two). Putting parallel lines together in the place where we will see cross sections of the grains (the areas of the cross sections being built up of the intervals), the end points of the intervals make up the perimeters of the cross sections. Putting parallel planes together, the cross-sectional areas compose the volume of grains, and cross-sectional perimeters compose the surface area of grains. This is reflected in Table 1, which shows by arrows what may be proved possible from Crofton's theorem (for example, see Hadwiger, 1957).

In practice, one has data only on a finite selection of lines or planes so that additional difficulties arise. To extend this discussion to cope with set transformations made with the Minkowski-type operations used above, it is necessary to introduce Minkowski functionals. This is beyond the scope of this paper but is treated fully in Hadwiger (1957) and sketched in Serra (1969). More practical information may be found in DeHoff and Rhines (1967).

TABLE 1. STEREOLOGICAL POSSIBILITIES

$\mathcal{R}^3$	Volume	Surface area	Mean curvature	Number
	↑	↑	↑	
$\mathcal{R}^2$	Surface area	Perimeter	Number	
	↑	↑		
$\mathcal{R}^1$	Length	Number		

RANDOM 0-1 FUNCTIONS

The mathematical tools developed above are appropriate for a finite number of bounded grains. To describe the fabric of rocks and other practical textures, it is natural to embed the finite part we observe in an infinite texture. Thus, **A** and its indicator function $f(\mathbf{x})$ are supposed to be defined over all $\mathcal{R}^n$. If **D** is the part of $\mathcal{R}^n$ examined, $\mathbf{A} \cap \mathbf{D}$ is the set that is available for study, which might be a block of rock or a photograph, for example. Using the methods of the preceding section, we need to determine the relation of the geometric quantities that will be calculated for the entire infinite texture. To assume that $\mathbf{A} \cap \mathbf{D}$ is typical of **A** is to assume that **A** is "homogeneous." A block of rock is often cut so that two-dimensional sections may be studied to get information about three dimensions; the plane sections are samples from the block, just as the block is a sample from the formation. By assuming a random model, remarks can be made about the geometric quantities as estimators and about various sampling schemes that might be proposed.

If the random set **A** is to be homogeneous, its probability structure must be the same everywhere, that is, invariant under translation. The indicator function $f(\mathbf{x})$ of **A** should thus be a stationary random function. Strict stationarity will be assumed for simplicity. Explicit probability models for **A** (Poisson points in $\mathcal{R}^2$ being the prototype) are discussed below. The probability law for **A** will be known if we have a self-consistent way of computing $prob(\mathbf{B}' \subset \mathbf{A}, \mathbf{B}'' \subset \mathbf{A}^c)$ for all sets $\mathbf{B}', \mathbf{B}''$.

The treatment is again intuitive. Assuming probability space $(\Omega, \mathcal{S}, P)$, a set $\mathbf{A}_\omega$ of $\mathcal{S}^n$ corresponds to all $\omega \in \Omega$, and all $\mathbf{A}_\omega$ should have the simple structure demanded above (consisting only of interior and boundary points and having a simple boundary). The σ-algebra $\mathcal{S}$ should contain events of interest such as "the open set $\mathbf{B}^0$ is contained in **A**." The indicator function f should be written $f(\mathbf{x}, \omega)$, have the values one if $\mathbf{x} \in \mathbf{A}_\omega$ and zero if $\mathbf{x} \notin \mathbf{A}_\omega$, and for fixed **x**, be a random variable on Ω. Once a probability on (Ω, β) is constructed, it is necessary to prove the self-consistent computation of $prob\,(\mathbf{B}' \subset \mathbf{A}, \mathbf{B}'' \subset \mathbf{A}^c)$ for all sets $\mathbf{B}'$, $\mathbf{B}''$. This program was carried out in Annexe II of Matheron (1967).

If $\mathbf{B}_\mathbf{x}$ is $\mathbf{B}$ translated by $\mathbf{x}$, and $\mathbf{A}$ is a stationary random set, $prob(\mathbf{B}_\mathbf{x} \subset \mathbf{A})$ is the same for all $\mathbf{x}$; that is,

$$prob(\mathbf{B}_\mathbf{x} \subset \mathbf{A}) = prob(\mathbf{B} \subset \mathbf{A}), \quad \text{all } \mathbf{x}$$
$$= P(\mathbf{B}). \tag{18}$$

In particular, for a point $\mathbf{x}$,

$$P(\mathbf{x}) = P(\mathbf{x} + \mathbf{h}) = E[f(\mathbf{x})]$$
$$= p. \tag{19}$$

Also,

$$P(\mathbf{x}_1, \mathbf{x}_2) = E[f(\mathbf{x}_1)f(\mathbf{x}_2)]$$
$$= E[f(\mathbf{x})f(\mathbf{x} + \mathbf{x}_2 - \mathbf{x}_1)]$$
$$= E[f(\mathbf{x})f(\mathbf{x} + \mathbf{h})], \quad \mathbf{h} = \mathbf{x}_2 - \mathbf{x}_1.$$

We therefore define the noncentered covariance of $\mathbf{A}$ by

$$C(\mathbf{h}) = E[f(\mathbf{x})f(\mathbf{x} + \mathbf{h})]$$
$$= prob(\mathbf{x}, \mathbf{x} + \mathbf{h} \in \mathbf{A}) \tag{20}$$
$$= prob(\mathbf{x} \in \mathbf{A}_{-\mathbf{h}} \cap \mathbf{A}).$$

Then,

$$P(\mathbf{x}_1, \mathbf{x}_2) = C(\mathbf{x}_2 - \mathbf{x}_1)$$
$$C(\mathbf{h}) = C(-\mathbf{h}), \, C(\mathbf{0}) = P, \, 0 \leq C(\mathbf{h}) \leq P. \tag{21}$$

Since $1 - f(\mathbf{x})$ is the indicator of $\mathbf{A}^c$, the probability of events where some points belong to $\mathbf{A}$ and some to $\mathbf{A}^c$ may be found from P, when the points involved are enumerable. For this reason, it is useful to define

$$Q(\mathbf{B}) = prob(\mathbf{B} \subset \mathbf{A}^c).$$

It seems intuitively clear that if $\mathbf{A}$ is a stationary set, the set transformations described in the second section will lead to new stationary sets, provided $\mathbf{B}$ is compact. The grains of $\mathbf{A}$ are modified in the same way wherever they occur, that is, the operations should not destroy the homogeneity. For example, the erosion of $\mathbf{A}$ by $\mathbf{B}$ is $\mathbf{A} \ominus \check{\mathbf{B}} = \{\mathbf{z} \mid \mathbf{B}_\mathbf{z} \subset \mathbf{A}\}$. However, for any sets $\mathbf{G}'$ and $\mathbf{G}''$ and vector $\mathbf{h}$,

$$prob(\mathbf{G}' \subset \mathbf{A} \ominus \check{\mathbf{B}}'), \mathbf{G}'' \subset (\mathbf{A} \ominus \check{\mathbf{B}}')^c) = prob(\mathbf{G}'_\mathbf{h} \subset (\mathbf{A} \ominus \check{\mathbf{B}}')_\mathbf{h}, \, \mathbf{G}''_\mathbf{h} \subset (\mathbf{A} \ominus \check{\mathbf{B}})^c_\mathbf{h}),$$
$$= prob(\mathbf{G}'_\mathbf{h} \subset \mathbf{A}_\mathbf{h} \ominus \check{\mathbf{B}}, \, \mathbf{G}''_\mathbf{h} \subset (\mathbf{A}_\mathbf{h} \ominus \check{\mathbf{B}})^c),$$
$$= prob(\mathbf{G}'_\mathbf{h} \subset \mathbf{A} \ominus \check{\mathbf{B}}, \, \mathbf{G}''_\mathbf{h} \subset (\mathbf{A} \ominus \check{\mathbf{B}})^c),$$

since the translation of the eroded **A** is the erosion of the translation of **A** (similarly for the complements), and $\mathbf{A_h}$ has the same probability law as **A**.

The beautiful relation in $\mathcal{R}^2$ between the texture analyser, the Minkowski operations, and the stationary random function model can now be made clear. As explained in the Introduction, the analyser scans any set **C** in a two-dimensional region **D** and outputs a discrete approximation to *mes* **C** $\cap$ **D**/*mes* **D**. This comes from counting the proportion of points in **D** that belong to **C**. If **C** is stationary with indicator $g(\mathbf{x})$, an unbiased estimator of $p^* = prob(\mathbf{x} \in \mathbf{C})$ is given by

$$\hat{p}^* = \int_\mathbf{D} g(\mathbf{x}) \, d\mathbf{x} / mes \, \mathbf{D}, \qquad (22)$$

$$= mes \, \mathbf{C} \cap \mathbf{D} / mes \, \mathbf{D},$$

since

$$E(\hat{p}^*) = \int_\mathbf{D} E \, g(\mathbf{x}) \, d\mathbf{x} / mes \, \mathbf{D} = p^*.$$

This result may be restated as

$$prob(\mathbf{x} \in \mathbf{C}) = E \, mes \, \mathbf{C} \cap \mathbf{D} / mes \, \mathbf{D}$$

or (23)

$$E(mes \, \mathbf{C} \cap \mathbf{D}) = prob(\mathbf{x} \in \mathbf{C}) \, mes \, \mathbf{D}.$$

Equation (23) shows that for a stationary **C**, $prob(\mathbf{x} \in \mathbf{C})$ is the expected measure of **C** per unit area or per unit volume. If **C** were not stationary and $\mathbf{x} \in \mathbf{D}$, we would only be able to state that $\mathbf{x} \in \mathbf{C}$, or $\mathbf{x} \notin \mathbf{C}$. Repetition of **C** would be required to obtain a serious estimate of p^*. In effect, stationarity provides replication.

Suppose that from observations of $\mathbf{A} \cap \mathbf{D}$ we wish to estimate $prob(\mathbf{B} \subset \mathbf{A})$ for some **B** with **A** stationary. Since $prob(\mathbf{B_x} \subset \mathbf{A}) = prob(\mathbf{B} \subset \mathbf{A})$ for all $\mathbf{x}$, it is intuitively sensible, by analogy with the derivation of equation (22), to check for each point $\mathbf{x}$ in **D** whether or not $\mathbf{B_x} \subset \mathbf{A}$ and to use the proportion of successes as an estimate. Thus, we obtain the estimator

$$\frac{mes \, \{\mathbf{x} \mid \mathbf{B_x} \subset \mathbf{A}, \mathbf{x} \in \mathbf{D}\}}{mes \, \mathbf{D}}. \qquad (24a)$$

If $\mathbf{B_x} \not\subset \mathbf{D}$, one cannot determine whether $\mathbf{B_x} \subset \mathbf{A}$; this estimator therefore needs slight modification in practice. Ignoring this edge effect, the numerator set is

$$\{\mathbf{x} \mid \mathbf{B_x} \subset \mathbf{A}\} \cap \mathbf{D} = (\mathbf{A} \ominus \mathbf{\check{B}}) \cap \mathbf{D},$$

so that one estimator is essentially

$$mes \, (\mathbf{A} \ominus \mathbf{\check{B}}) \cap \mathbf{D} / mes \, \mathbf{D}. \qquad (24b)$$

This is unbiased because, as we have seen above, $\mathbf{C} = \mathbf{A} \ominus \mathbf{\check{B}}$ is stationary, and by equations (22) and (23) its expectation is $prob(\mathbf{x} \in \mathbf{A} \ominus \mathbf{\check{B}})$. However, the event $x \in A \ominus \check{B}$ is the same as the event $B_x \subset A$, so the expectation is equal to $prob(B_x \subset A) = prob(B \subset A) = P(B)$. Thus,

$$prob(B \subset A) = \frac{E\,mes\,(A \ominus \check{B}) \cap D}{mes\,D},$$
$$= P(B).$$
(25)

Actually, allowance for the edge effects of D is easy, since $B_x \subset D$ iff $x \in D \ominus \check{B}$. Hence, it is only necessary to replace D in equations (24) and (25) by $D \ominus \check{B}$. Since

$$A \ominus \check{B} \cap D \ominus \check{B} = (A \cap D) \ominus \check{B},$$

the practical form of equation (24) is

$$mes\,(A \cap D) \ominus \check{B} / mes\,D \ominus \check{B}. \tag{26}$$

Similarly, to estimate $Q(B_x) = Q(B) = prob(B_x \subset A^c)$, one observes that

$$Q(B) = prob(x \in A^c \ominus \check{B}),$$
$$= 1 - prob(x \in (A^c \ominus \check{B})^c);$$

that is,

$$1 - Q(B) = prob(x \in A \oplus \check{B}). \tag{27}$$

By analogy with equation (26), the unbiased estimator of $1 - Q(B)$ will be

$$mes\,(A \oplus \check{B}) \cap D / mes\,D. \tag{28a}$$

The analyser will compute equation (28) by counting the points in D for which $B_x \subset A^c$, that is, using a version of

$$mes\,\{x \mid B_x \subset A^c\} \cap D / mes\,D, \tag{28b}$$

corrected for edge effects, as was done in equation (26).

Recalling definition (5) of the opening of A by B,

$$A \,\omega\, B = (A \ominus \check{B}) \oplus B,$$

it may be verified that $A \,\omega\, B$ is stationary, and a repetition of the above arguments shows that

$$prob(x \in A \,\omega\, B) = E\,mes\,A \,\omega\, B \cap D / mes\,D. \tag{29a}$$

Delfiner (1971) showed that the edge-corrected form of the estimator of $prob(x \in A \,\omega\, B)$ is

$$mes\,(A \cap D) \,\omega\, B \cap D \ominus (B \oplus \check{B}) / mes\,D \ominus (B \oplus \check{B}). \tag{29b}$$

When the texture analyser is used to compute the estimates (equations (22), (24b), (28b)) of $prob(\mathbf{x} \in \mathbf{C})$, $prob(\mathbf{x} \in \mathbf{A} \ominus \mathbf{B})$, and $prob(\mathbf{x} \in \mathbf{A} \oplus \mathbf{B})$, the points $\mathbf{x}$ where the checks are successful may be lighted on a cathode-ray tube. The sweep speed is so great that persistence of vision yields a picture of $\mathbf{C}$, $\mathbf{A} \ominus \mathbf{\check{B}}$, and $\mathbf{A} \oplus \mathbf{\check{B}}$, the latter by using $\mathbf{A}^c$ rather than $\mathbf{A}$. Thus, the erosion and dilatation of $\mathbf{A}$ by $\mathbf{B}$ may be viewed, even though they are not constructed. Some storage is required for all $\mathbf{B}$ other than a single point. There is another approximation arising from discreteness—a circular $\mathbf{B_x}$ must, for example, be approximated by a polygonal lattice (usually a hexagon) of points. The device has the capability of adjusting the lattice size to the scale of the phenomenon to reduce these difficulties.

The noncentered covariance function of a stationary set $\mathbf{A}$ is defined by equation (20) as

$$C(\mathbf{h}) = prob(\mathbf{x} \in \mathbf{A} \cap \mathbf{A_h}),$$
$$= E\, mes\, (\mathbf{A} \cap \mathbf{A_h}) \cap \mathbf{D}/mes\, \mathbf{D}, \qquad (30)$$

since $\mathbf{A} \cap \mathbf{A_h}$ for fixed $\mathbf{h}$ is also a stationary set (by equation (23)). (This latter fact may be shown by using the strict stationarity of the indicator $f(\mathbf{x})$ of $\mathbf{A}$ with the indicator $f(\mathbf{x})\, f(\mathbf{x} - \mathbf{h})$ of $\mathbf{A} \cap \mathbf{A_h}$.) Equation (30) is estimated by the proportion of points $\mathbf{x}$ in $\mathbf{D}$ such that $\mathbf{x}$ and $\mathbf{x} - \mathbf{h}$ lie in $\mathbf{A}$ and therefore corresponds with equation (9).

It was shown above that $\mathbf{A} \ominus \mathbf{\check{B}}$ is stationary. By analogy with equation (30), its covariance is given by

$$E\, mes\, (\mathbf{A} \ominus \mathbf{\check{B}}) \cap (\mathbf{A} \ominus \mathbf{\check{B}})_\mathbf{h} \cap \mathbf{D}/mes\, \mathbf{D}. \qquad (31)$$

From equation (3)

$$mes(\mathbf{A} \ominus \mathbf{\check{B}}) \cap (\mathbf{A} \ominus \mathbf{\check{B}})_\mathbf{h} = \left(\bigcap_{\mathbf{y} \in \mathbf{B}} \mathbf{A_y}\right) \cap \left(\bigcap_{\mathbf{y} \in \mathbf{B}} \mathbf{A_y}\right)_\mathbf{h},$$

$$= \left(\bigcap_{\mathbf{y} \in \mathbf{\check{B}}} \mathbf{A_y}\right) \cap \left(\bigcap_{\mathbf{y} \in \mathbf{\check{B}}} \mathbf{A_{y+h}}\right),$$

$$= \left(\bigcap_{\mathbf{y} \in \mathbf{\check{B}}} \mathbf{A_y}\right) \cap \left(\bigcap_{\mathbf{z} \in \mathbf{\check{B}_h}} \mathbf{A_z}\right),$$

$$= \bigcap_{\mathbf{x} \in \mathbf{\check{B}} \cup \mathbf{\check{B}_h}} \mathbf{A_x},$$

$$= \mathbf{A} \ominus (\mathbf{\check{B}} \cup \mathbf{\check{B}_h}).$$

Thus, equation (31) may be written as

$$E\, mes\, \mathbf{A} \ominus (\mathbf{\check{B}} \cup \mathbf{\check{B}_h}) \cap \mathbf{D}/mes\, \mathbf{D}. \qquad (32)$$

This will be estimated, by analogy with equations (24) and (25), by the proportion of points $\mathbf{x}$ in $\mathbf{D}$ for which

$$(\mathbf{\check{B}} \cup \mathbf{\check{B}_h})_{\check{\mathbf{x}}} \subset \mathbf{A}, \quad \text{or } (\mathbf{B} \cup \mathbf{B_h})_\mathbf{x} \subset \mathbf{A}.$$

The variance of p^* in equation (22) is

$$\sigma^2(\hat{p}^*) = (1/(mes\,\mathbf{D})^2) \int_{\mathbf{D}} \int_{\mathbf{D}} E[g(\mathbf{x}_1)g(\mathbf{x}_2)]\,d\mathbf{x}_1\,d\mathbf{x}_2 - p^{*2}$$
$$= (1/(mes\,\mathbf{D})^2) \int_{\mathbf{D}} \int_{\mathbf{D}} [C(\mathbf{x}_1 - \mathbf{x}_2) - p^2]\,d\mathbf{x}_1\,d\mathbf{x}_2, \qquad (33)$$

if $C(\mathbf{h})$ is the covariance function of $\mathbf{C}$. However, it will not be satisfactory to evaluate this numerically using crude estimates of $C(\mathbf{h})$ and $p = C(\mathbf{0})$ obtained from equation (30). Some plausible, positive definite function should be fitted to $C(\mathbf{h})$. In other words, the probability model needs to be more explicit so that it includes a covariance function probably containing some parameters requiring estimation. The same problem arises in estimating the variance of the estimate of $P(\mathbf{B})$ in equation (26).

In an earlier section, we derived geometric formulas such as equation (14) in $\mathscr{R}^3$:

$$\underset{\alpha}{ave} - K'_\alpha(0) = S(\mathbf{A})/4.$$

What is the meaning of this when $\mathbf{A} = \mathbf{A} \cap \mathbf{D}$ and $\mathbf{A}$ is a stationary set? First, all realizations of $\mathbf{A} \cap \mathbf{D}$ must satisfy the requirements for the derivation of equation (14) (that is, consist only of interior and boundary points and have a simple boundary). $K(\mathbf{h}) = mes\,\mathbf{A} \cap \mathbf{A}_\mathbf{h}$ of equation (11) corresponds to $C(\mathbf{h})$ in equation (30). Since in $\mathscr{R}^3$, $mes\,\mathbf{D} = $ volume of $\mathbf{D}$, equation (30) is the expected covariance per unit volume. Hence,

$$\underset{\alpha}{ave} - C'_\alpha(0) = \tfrac{1}{4}\,E\,mes(\mathbf{A} \cap \mathbf{D})/mes\,\mathbf{D}$$
$$= \tfrac{1}{4}\,(\text{expected surface area per unit volume}) \qquad (34)$$
$$= \tfrac{1}{4}\,(\text{specific surface area}).$$

Similarly, in $\mathscr{R}^2$

$$\underset{\alpha}{ave} - C'_\alpha(0) = \text{specific perimeter}/\pi. \qquad (35)$$

These quantities would be estimated by averaging the estimated initial slopes of $\hat{C}(\mathbf{h})$, where $\mathbf{h}$ is given a number of equally spaced directions $\underset{\sim}{\alpha}$.

If the probability law of the random set is invariant under rotation as well as translation, it is "isotropic." Thus, $C'_\alpha(0)$ is independent of $\underset{\sim}{\alpha}$. In practice, different values of $\underset{\sim}{\alpha}$ are used since this assumption would usually need checking.

It was remarked after equations (22) and (23) that for stationary $\mathbf{C}$, $prob(\mathbf{x} \in \mathbf{C})$ is the expected measure of $\mathbf{C}$ per unit area or volume. The estimator $mes\,\mathbf{C} \cap \mathbf{D}/mes\,\mathbf{D}$ of $p^* = prob(\mathbf{x} \in \mathbf{C})$ is unbiased, and its variance is given by equation (33). Since the integrand will tend to zero as $|\mathbf{x}_1 - \mathbf{x}_2| \to \infty$, $\sigma^2(\hat{p}^*) \to 0$ as $mes\,\mathbf{D} \to \infty$. Thus, $\hat{p}^* \to p^*$ in probability as $mes\,\mathbf{D} \to \infty$. As $mes\,\mathbf{D} \to \infty$, the edge effects caused by $\mathbf{D}$ will tend to zero. These facts provide an intuitive justification for procedures used in the next section where probabilities will be calculated by finding measures and ignoring **D**. Rigorous proofs require that $\hat{p}^* \to p$ almost surely, and a formal proof requires that the edge effects tend to zero. These proofs may be obtained by assuming that $C(\mathbf{x}) \to 0$ as $|\mathbf{x}| \to \infty$ and that **D** is a simple region like a square.

This section concludes with a few specific models for random sets **A**. Let points $\underline{\xi}_i$ be distributed over the whole space $\mathscr{R}^n$ so that the number of points in nonoverlapping regions is independent, and the expected number in any region **V** is given by θ *mes* **V**. Then, $\mathbf{A} = \cup \underline{\xi}_i$, the set of all such points, is the Poisson distribution in $\mathscr{R}^n$, and it is the basis of the models that follow and many others. It is clearly spatially homogeneous, that is, stationary; it is also isotropic because it has no directional features.

Suppose we have a way of generating a sequence of independent random sets $\mathbf{C}_1, \mathbf{C}_2, \ldots$, all with the same probability properties. For example, in $\mathscr{R}^2$ all $\mathbf{C}_i$ might be circular discs with independent radii drawn from some distribution. Using the Poisson points $\underline{\xi}_i$ from above and $\mathbf{C}_i$ translated by $\underline{\xi}_i$ which yields $\mathbf{C}_{\underline{\xi}_i}$, consider $\mathbf{A} = \cup \mathbf{C}_{\underline{\xi}_i}$. This **A** is a stationary random set. If the $\mathbf{C}_i$ were circular discs, **A** would also be isotropic; if they were elliptical discs, however, it might be anisotropic. Given the probability law of **C**, we could calculate for any fixed set **B**,

$$\chi(\mathbf{B}) = prob(\mathbf{B} \cap \mathbf{C} = \emptyset). \tag{36}$$

To calculate $prob(\mathbf{B} \cap \mathbf{A} = \emptyset)$ and the covariance function $C(\mathbf{h})$ of **A**, consider some large region **D** containing **B**. The number of $\underline{\xi}_i$ points in **D** is the Poisson mean θ *mes* **D**. Given that n points $\underline{\xi}_i$ are in **D**, they are independently and uniformly distributed over **D**. Thus, for any one point known to be in **D**,

$$prob(\mathbf{B} \cap \mathbf{C}_{\underline{\xi}} = \emptyset \mid \underline{\xi} \text{ in } \mathbf{D}) = prob(\mathbf{B}_{-\underline{\xi}} \cap \mathbf{C} = \emptyset \mid \underline{\xi} \text{ in } \mathbf{D}),$$

$$= \int_{\mathbf{D}} \frac{\chi(\mathbf{B}_{-\underline{\xi}}) \, d\underline{\xi}}{mes \, \mathbf{D}}, \tag{37}$$

using definition (36) and ignoring edge effects. Denoting this probability by q, the probability that none of the $\mathbf{C}_{\underline{\xi}_i}$, $\underline{\xi}_i \in \mathbf{D}$ intersect **B** is then $E(q^N)$, where N is the random number of $\underline{\xi}_i$ in **D**. This is the Poisson mean $\lambda = \theta$ *mes* **D**; hence,

$$E(q^N) = \sum_0^\infty e^{-\lambda} \lambda^n q^n / n!,$$

$$= \exp\{-\lambda + \lambda q\},$$

$$= \exp\left\{-\theta \, mes \, \mathbf{D} \int_{\mathbf{D}} \frac{[1 - \chi(\mathbf{B}_{-\underline{\xi}})]}{mes \, \mathbf{D}} \, d\underline{\xi}\right\},$$

$$= \exp\left\{-\theta \int_{\mathbf{D}} [1 - \chi(\mathbf{B}_{-\underline{\xi}})] \, d\underline{\xi}\right\}.$$

It is intuitively clear that as **D** increases to include all space, this probability becomes $prob(\mathbf{B} \cap \mathbf{A} = \emptyset)$, so that

$$prob(\mathbf{B} \cap \mathbf{A} = \emptyset) = \exp\left\{-\theta \int_{\mathscr{R}^n} [1 - \chi(\mathbf{B}_\xi)] \, d\xi\right\}. \tag{38}$$

Since

$$1 - \chi(\mathbf{B}_\xi) = prob(\mathbf{B}_\xi \cap \mathbf{C} \neq \emptyset),$$
$$= prob(\mathbf{B}_\xi \text{ intersects } \mathbf{C}),$$
$$= prob(\xi \in \mathbf{C} \oplus \check{\mathbf{B}}),$$
$$= E I_{\mathbf{C} \oplus \check{\mathbf{B}}}(\xi),$$

where $I_{\mathbf{C} \oplus \check{\mathbf{B}}}(\xi)$ is the indicator function, then

$$\int [1 - \chi(\mathbf{B}_\xi)] \, d\xi = \int E[I_{\mathbf{C} \oplus \check{\mathbf{B}}}(\xi)] \, d\xi,$$
$$= E \int I_{\mathbf{C} \oplus \check{\mathbf{B}}}(\xi) \, d\xi,$$
$$= E \, mes \, \mathbf{C} \oplus \check{\mathbf{B}}.$$

Thus, equation (38) may be rewritten as

$$prob(\mathbf{B} \cap \mathbf{A} = \emptyset) = \exp(-\theta \, mes \, \mathbf{C} \oplus \check{\mathbf{B}}). \tag{39}$$

Finally, to find the covariance function of **A**, we define

$$K(\mathbf{h}) = E \, mes \, \mathbf{C} \cap \mathbf{C}_{-\mathbf{h}} \tag{40}$$

so that $K(0) = E \, mes \, \mathbf{C}$. Let set **B** in equation (39) be the pair of points **x**, $\mathbf{x} + \mathbf{h}$ so that $\mathbf{C} \oplus \check{\mathbf{B}} = \mathbf{C} \cup \mathbf{C}_{-\mathbf{h}}$. Since

$$mes \, \mathbf{C} \cup \mathbf{C}_{-\mathbf{h}} = 2 \, mes \, \mathbf{C} - mes \, \mathbf{C} \cap \mathbf{C}_{-\mathbf{h}},$$

then

$$prob\{(\mathbf{x}, \mathbf{x} + \mathbf{h}) \cap \mathbf{A} = \emptyset\} = \exp(-2\theta \, E \, mes \, \mathbf{C} + \theta E \, mes \, \mathbf{C} \cap \mathbf{C}_{-\mathbf{h}}).$$

Thus,

$$C(\mathbf{h}) = prob(\mathbf{x} \in \mathbf{A}, \mathbf{x} + \mathbf{h} \in \mathbf{A}),$$
$$= 1 - prob(\mathbf{x} \notin \mathbf{A}) - prob(\mathbf{x} + \mathbf{h} \notin \mathbf{A}) + prob(\mathbf{x} \notin \mathbf{A}, \mathbf{x} + \mathbf{h} \notin \mathbf{A}),$$
$$= 1 - 2\exp[-\theta K(0)] + \exp[-2\theta K(0) + \theta K(\mathbf{h})];$$

that is,

$$C(\mathbf{h}) = 1 - 2q + q^2 \exp \theta K(\mathbf{h}), \tag{41}$$

where $q = \exp -\theta K(\mathbf{0})$. If $K(\mathbf{h}) \to 0$ as $|\mathbf{h}| \to 0$, $C(\mathbf{h})$ decreases from $1 - q$ to $(1 - q)^2$ as $|\mathbf{h}|$ goes from 0 to ∞.

Stationary random functions are often analysed spectrally. When they are indicator functions, as here, they must have appreciable power in the high-frequency range to produce the sharp boundaries to the grains. The high-frequency end of the spectral density corresponds to the covariance function near the origin. It is not surprising then that the calculations above depended on the slope of the covariance function at the origin.

GRANULOMETRY

In many sciences, attempts are made to describe texture. In metallurgy, petrography, and biology, for example, many words and definitions have been coined to describe the roundness, size, preferential orientation, and so forth of particles in a plane section. Such properties do not depend on the relative positions of the grains or particles. For our purpose, these will be considered aspects of the granulometry of the set of grains **A**. By contrast, other properties, such as the covariance function, depend on the relative positions of the grains; these are termed structural properties.

Let **A** be a set in $\mathcal{R}^2$ and **D** the region viewed, for example, a slide prepared by a petrographer. The petrographer customarily makes traverses and records the lengths of intersection of the traverse with the grains of a specific mineral. These measurements (in a fixed set) give an idea of grain size. Even if all grains were spheres so that there would be no argument about the meaning of "size," it is clear that traverses would hit larger spheres more often. The average size so obtained is thus weighted by the size itself. If the diameters of all circular sections were measured and averaged, there would be no weighting. These are, respectively, simple examples of granulometry *in measure* and granulometry *in number*, by the definitions of Matheron (1967).

Particles or grains are often sieved. Matheron (1967) has abstracted this process to give a general formulation of all granulometries. Suppose that the grains of **A** that do not pass through a sieve of dimension λ are denoted by $\psi_\lambda(\mathbf{A})$, so that $\psi_\lambda(\mathbf{A}) \subset \mathbf{A}$. If $\lambda \geq \mu$, there should be more left behind on the smaller sieve, that is, $\psi_\lambda(\mathbf{A}) \subset \psi_\mu(\mathbf{A})$. Defining $\psi_0(\mathbf{A}) = \mathbf{A}$, if $\psi_\lambda \bigcirc \psi_\mu$ stands for using the μ sieve first and sieving the residue with the λ sieve, then $\psi_\lambda \bigcirc \psi_\mu = \psi_\mu \bigcirc \psi_\lambda = \psi_{\sup(\lambda,\mu)}$. Finally, if **B** and **C** are subsets of **A** and $\mathbf{B} \subset \mathbf{C}$, then $\psi_\lambda(\mathbf{B}) \subset \psi_\lambda(\mathbf{C})$ for any λ. If **A** is in $\mathcal{R}^n$, so that one can speak of translations and scale changes, ψ_λ should only translate with translations of **A** and $\psi_\lambda(\mathbf{A}) = \lambda\psi_1(\mathbf{A}/\lambda)$. These six properties are an axiomatic definition of a granulometry, although some of the properties are deducible from others. Further, consider *mes* $\psi_\lambda(\mathbf{A})$ as the volume of the residue. When $\lambda = 0$, it is *mes* **A**, since $\psi_0(\mathbf{A}) = \mathbf{A}$. As λ increases, *mes* **A** tends to zero. Thus,

$$G(\lambda) = \frac{\textit{mes } \mathbf{A} - \textit{mes } \psi_\lambda(\mathbf{A})}{\textit{mes } \mathbf{A}} \tag{42}$$

is a distribution function of the grain sizes according to the definition of ψ, that is, of the particular granulometry.

These definitions apply to many situations other than the process of sieving gravel. Every useful class of granulometries is associated with the set operation of opening (equation (5)). Let $\mathbf{B}(\lambda)$ be a family of sets indexed by λ, with the property that

$$\lambda > \mu \Rightarrow \mathbf{B}(\lambda) = [\mathbf{B}(\lambda) \ominus \check{\mathbf{B}}(\mu)] \oplus \mathbf{B}(\mu), \tag{43}$$

that is,

$$\mathbf{B}(\lambda) = \text{opening of } \mathbf{B}(\lambda) \text{ by } \mathbf{B}(\mu), \quad \lambda > \mu.$$

This is the case, for example, if $\mathbf{B}(\lambda)$ is a ball of radius λ centered at the origin or a segment of length λ (leaving the origin in a direction $\underline{\alpha}$). For any such family, if

$$\psi_\lambda(\mathbf{A}) = [\mathbf{A} \ominus \check{\mathbf{B}}(\lambda)] \oplus \mathbf{B}(\lambda),$$
$$= \mathbf{A}\,\omega_\lambda, \quad \text{say} \tag{44}$$

it may be shown that $\psi_\lambda(\mathbf{A})$ satisfies the axioms of a granulometry and equation (42) becomes

$$G(\lambda) = 1 - mes\,\mathbf{A}\,\omega_\lambda / mes\,\mathbf{A}. \tag{45}$$

To transfer this notion of granulometry of a fixed set $\mathbf{A}$ to a random set $\mathbf{A}$ observed in $\mathbf{D}$, it is natural to define

$$G(\lambda) = 1 - (E\,mes\,\mathbf{A}\,\omega_\lambda \cap \mathbf{D})/(E\,mes\,\mathbf{A} \cap \mathbf{D}). \tag{46}$$

If $\mathbf{A}$ is a stationary set, this expression does *not* depend on $\mathbf{D}$, because $E\,mes\,\mathbf{A} \cap \mathbf{D}/mes\,\mathbf{D}$ is $prob(\mathbf{x} \in \mathbf{A})$ for any $\mathbf{x}$ (that is, p), and the numerator $E\,mes\,\mathbf{A}\,\omega_\lambda \cap \mathbf{D}/mes\,\mathbf{D}$ is $prob(\mathbf{x} \in \mathbf{A}\,\omega_\lambda)$ for any $\mathbf{x}$. Since $\mathbf{A}\,\omega_\lambda$ is also stationary, this probability is also a constant. Thus,

$$G(\lambda) = prob(\mathbf{x} \notin \mathbf{A}\,\omega_\lambda \mid \mathbf{x} \in \mathbf{A}). \tag{47}$$

The definition of equation (46) is, however, directly related to the method by which $G(\lambda)$ could be estimated in practice.

Linear granulometry results from choosing $\mathbf{B}(\lambda)$ to be the set of points $\underline{\Lambda} = \{l\underline{\alpha},\, 0 \leq l \leq \lambda\}$, where $\underline{\alpha}$ is a unit vector. $G(\lambda)$ from equation (45) will be related, in this case, to scanning or traversing $\mathbf{A}$ in a direction $\underline{\alpha}$. If $\mathbf{A}$ is scanned on a finely spaced set of lines parallel to $\underline{\alpha}$, and each time a grain is entered the length of the traverse is recorded, it would be possible to sum the lengths of all traverses $\leq \lambda$ and divide it by the sum of the lengths of all traverses. The ratio would certainly be a nondecreasing function of λ. (It will, in fact, be $G(\lambda)$ of equation (45).) In probabilistic terms, if a point of $\mathbf{A}$ is chosen at random, $G(\lambda)$ is the probability that it belongs to a traverse of length $\leq \lambda$. Furthermore, by the definition of the opening of $\mathbf{A}$ by $\underline{\Lambda}$, $mes(\mathbf{A} \ominus \check{\underline{\Lambda}}) \oplus \underline{\Lambda}$ is the measure of the set of all points covered by translates of $\underline{\Lambda}$ that lie inside $\mathbf{A}$; it is the shaded area in Figure 10.

Thus, $mes\,(\mathbf{A} \ominus \check{\underline{\Lambda}}) \oplus \underline{\Lambda}$ is proportional to the total length of all traverses that are $\geq \lambda$. Also, $mes\,\mathbf{A}$ is proportional to the total length of all traverses. In $\mathcal{R}^2$ the proportionality constant is the spacing of the lines. Thus, the distribution function mentioned above is indeed $G(\lambda)$ of equation (45). Grains that are large in directions perpendicular to α will be scanned and recorded many times. Thus, the distribution obtained will be quite different from the distribution function of the widths of all grains in direction α. The latter is an example of granulometry in number; $G(\lambda)$ is a granulometry in measure.

If **A** is a stationary random set, $G(\lambda)$ may be expressed in terms of $P_\alpha(\lambda)$, which is the probability that a line segment Λ of length λ and direction α is covered by **A**. If Figure 11 is typical, this probability equals $prob(\mathbf{x} \in \mathbf{A} \ominus \Lambda)$ (for example, in the derivation of equation (26)). From equation (46)

$$[1 - G(\lambda)] \, prob(\mathbf{x} \in \mathbf{A}) = prob\{\mathbf{x} \in (\mathbf{A} \ominus \check{\Lambda}) \oplus \Lambda\}, \tag{48}$$

$$= E\,[mes\,(\mathbf{A} \ominus \check{\Lambda}) \oplus \Lambda \cap \mathbf{D}]/mes\,\mathbf{D},$$

where

$$prob(\mathbf{x} \in \mathbf{A}) = P_\alpha(0).$$

Writing $P'_\alpha(\lambda)$ for $d/d\lambda\, P'_\alpha(\lambda)$, Matheron (1967) showed by an argument reminiscent of renewal theory that

$$1 - G(\lambda) = [P_\alpha(\lambda) - P'_\alpha(\lambda)]/P_\alpha(0). \tag{49}$$

We will use the argument from ergodic theory mentioned in the last section, which is more intuitive and closer to experimental facts and also more generally applicable.

Suppose **D** in equation (48) is the square $(0, s) \times (0, s) = \mathbf{D}_s$. Then the limit, as $s \to \infty$, of $s^{-2}\,mes\{[(\mathbf{A} \ominus \check{\Lambda}) \oplus \Lambda] \cap \mathbf{D}_s\}$ is almost surely $prob\{\mathbf{x} \in (\mathbf{A} \ominus \check{\Lambda}) \oplus \Lambda\}$. **A** will be a union of sets in general looking like Figure 13, but as will be shown below, we may begin by considering Figure 11 where there are no holes or re-entrant features. In Figure 11, $(\mathbf{A} \ominus \check{\Lambda}) \oplus \Lambda$ is shaded and shown to be the union of two sets. Region ① is $\mathbf{A} \ominus \check{\Lambda}$; region ② has everywhere a width in direction α of λ and a projection ③ (in $\mathcal{R}^2$ a length, in $\mathcal{R}^3$ an area) so that its area is λ times ③. Writing Λ as $\Lambda(\lambda)$, for Figure 12

$$mes\,\mathbf{A} \ominus \Lambda\,(\lambda + \delta\lambda) - mes\,\mathbf{A} \ominus \Lambda\,(\lambda) = -\,mes(\text{horizontally shaded area}$$
$$\text{or volume}),$$
$$= -\delta\lambda\,③ + 0(\delta\lambda^2),$$

or

$$\text{projection } ③ = -\frac{d}{d\lambda}\,mes\,\mathbf{A} \ominus \Lambda\,(\lambda).$$

Thus,

$$\frac{mes\,\{(\mathbf{A} \ominus \check{\Lambda}) \oplus \Lambda\} \cap \mathbf{D}_s}{s^2} = \frac{mes\,(\mathbf{A} \ominus \check{\Lambda}) \cap \mathbf{D}_s}{s^2}$$

$$-\frac{\lambda}{s^2}\frac{d}{d\lambda}\,mes\,[\mathbf{A} \ominus \Lambda(\lambda)] \cap \mathbf{D}_s + \frac{b}{s^2},$$

where b is the measure of the region near the boundary of $\mathbf{D}_s$ in which the above operations need modification. However, b must be of order $4s\lambda$. Thus, the ergodic argument as $s \to \infty$ yields equation (49). If $G(\lambda)$ has a density, it is given by

$$g(\lambda) = -\frac{\lambda\, P_{\alpha}''(\lambda)}{P_{\alpha}(0)}. \qquad (50)$$

As stated before, $1 - G(\lambda)$ is the probability that a random point of **A** lies on a traverse of length $>\lambda$.

To complete the proof of equation (49), we need to observe that if **A** is divided into subsets $\mathbf{A}_{(1)}, \mathbf{A}_{(2)}, \ldots$ by surfaces whose generators are parallel to $\underline{\Lambda}$, then

$$mes(\mathbf{A} \ominus \underline{\check{\Lambda}}) \oplus \underline{\Lambda} = \Sigma_i\, mes\,(\mathbf{A}_{(i)} \ominus \underline{\check{\Lambda}}) \oplus \underline{\Lambda}. \qquad (51)$$

An example in $\mathcal{R}^2$ of **A** divided into three subsets is given in Figure 13. The truth of equation (50) follows by illustrating the opening of **A** and these three parts. The subsets in Figure 13 have the properties assumed in Figure 12—essentially, that on proceeding from any point in the set in the $\underline{\alpha}$ direction, the boundary is crossed only once.

If a line of direction α that cuts **A** is randomly chosen, the chance that the traverse length **A** is greater than λ (from Figs. 12 and 13) is

$$1 - F(\lambda) = P_{\alpha}'(\lambda)/P_{\alpha}'(0) \qquad (52)$$

and

$$f(\lambda) = -P_{\alpha}''(\lambda)/P'(0). \qquad (53)$$

This is the distribution obtained by recording all traverses in direction $\underline{\alpha}$; that is, $F(\lambda)$ is the limit of the number of traverses of length $\leq \lambda$ divided by the number of traverses. By contrast, $G(\lambda)$ is the sum of the number of traverses of each length $\leq \lambda$, multiplied by this length, and divided by the total length of all traverses (also compare equations (49) and (52)). Hence equations (52) and (53) are called the distribution of traverses in number, while equations (49) and (50) are the distribution of traverses in length. The distribution could also be weighted by other quantities than length.

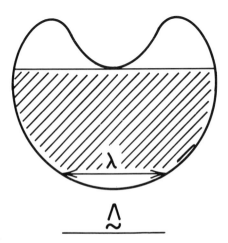

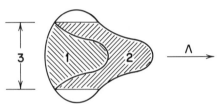

Figure 10. Illustration of $(\mathbf{A} \ominus \underline{\check{\Lambda}}) \oplus \underline{\Lambda}$ which is shaded.

Figure 11. Illustration of the division of $(\mathbf{A} \ominus \underline{\check{\Lambda}}) \oplus \underline{\Lambda}$ into regions 1 and 2.

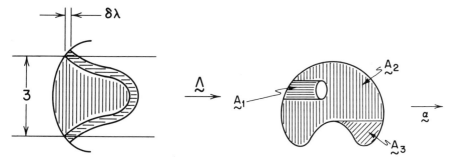

Figure 12. Illustration of the calculation of the area of region 1.

Figure 13. Illustration of the subdivision of A into regions.

Since equation (50) integrates to unity, equation (53) shows

$$\int_0^\infty \lambda f(\lambda)\, d\lambda = -\frac{P_\alpha(0)}{P'_\alpha(0)} \int_0^\infty g(\lambda)\, d\lambda;$$

that is,

$$E(\lambda) = -P_\alpha(0)/P'_\alpha(0) \qquad (54)$$

is the mean traverse in number. $P_\alpha(0) = p$ is the fraction of space covered by A. Hence, equation (54) reads

$$p = E(\lambda)\,[-P'_\alpha(0)].$$

Thus, the granulometry of traverses depends on a special version of $P(\mathbf{B})$ defined by equation (18). There is an interesting and related application of $Q(\mathbf{B}) = prob(\mathbf{B} \subset \mathbf{A}^c)$. Let $\mathbf{x} \in \mathbf{A}^c$, and call all points $\mathbf{y}$ of $\mathbf{A}^c$ the *star of* $\mathbf{x}$ where the line segment $[\mathbf{x}, \mathbf{y}] \subset \mathbf{A}^c$. If $S(\mathbf{y}; \mathbf{x})$ is its indicator, then

$$S_\mathbf{x} = \text{measure of the star of } \mathbf{x} = \int_{\mathcal{R}^2} S(\mathbf{y}; \mathbf{x})\, d\mathbf{y},$$

so that

$$E(S_\mathbf{x}) = \int prob\{(\mathbf{x}, \mathbf{y}) \subset \mathbf{A}^c \mid \mathbf{x} \in \mathbf{A}^c\}\, d\mathbf{y}$$

$$= (1/q) \int Q\{(\mathbf{x}, \mathbf{y})\}\, d\mathbf{y}.$$

If a ball is used instead of a segment, the granulometry is isotropic like the ball. If $\mathbf{B}(r)$ is a ball of radius r, then $P[\mathbf{B}(r)] = prob[\mathbf{B}(r) \subset \mathbf{A}] = prob\{\mathbf{x} \in \mathbf{A} \ominus \mathbf{B}(r)\}$ for any $\mathbf{x}$, since A is, as usual, stationary. Also,

$$prob\{\mathbf{x} \in \mathbf{A} \ominus \mathbf{B}(r) \mid \mathbf{x} \in \mathbf{A}\} = P(\mathbf{B}(r))/P(0)$$

has the following interpretation: if R is the distance from $\mathbf{x}$ in A to the nearest point in $\mathbf{A}^c$ and $F(r) = prob\{R \leq r\}$, then "$\mathbf{B}_\mathbf{x}(r) \subset \mathbf{A}$" is equivalent to "$R > r$." Hence,

$$1 - F(r) = P[\mathbf{B}(r)]/P(0),$$

a result that is very familiar when $\mathbf{A}^c$ is a Poisson field of points. This distribution is not a granulometry; to obtain one, we use $prob\{\mathbf{x} \in [\mathbf{A} \ominus \mathbf{\check{B}}(r)] \oplus \mathbf{B}(r) \mid \mathbf{x} \in \mathbf{A}\}$ from the general definition (see Delfiner, 1971).

ACKNOWLEDGMENTS

I thank M. Georges Matheron for the opportunity to work at the Centre de Morphologie Mathématique and Frederick Almgren, Jr., and Richard Holley for their mathematical advice. This research was partially supported by the Office of Naval Research Contract N00014-67-A-0151-0017 awarded to the Department of Statistics, Princeton University.

REFERENCES CITED

DeHoff, R. T., and Rhines, F. N., 1967, Quantitative microscopy: New York, McGraw-Hill Co., 422 p.

Delfiner, P., 1971, Étude morphologique des milieux poreux automatization des mesures en plaques minces [thesis]: L'Université de Nancy, 241 p.

Fara, H. D., and Scheidegger, A. E., 1961, Statistical geometry of porous media: Jour. Geophys. Research, v. 66, p. 3279-3284.

Haas, A., Matheron, G., and Serra, J., 1967a, Morphologie mathématique et granulométries en place: Annales Mines, v. 11, p. 735-753.

―――1967b, Morphologie mathématique et granulométries en place: Annales Mines, v. 12, p. 767-782.

Hadwiger, H., 1957, Vorlesungen über Inhalt, Oberfläsche und Isoperimetrie: Berlin, Springer Verlag, 312 p.

Harding, E. F., and Kendall, D. G., 1974, Stochastic geometry: New York, John Wiley & Sons, Inc., 416 p.

Kendall, M. G., and Moran, P.A.P., 1963, Geometrical probability: London, Charles Griffin and Co., 125 p.

Kirsh, R. A., 1957, Processing pictorial information with digital computers: Am. Inst. Electrical Engineers, CP 57-878.

Kirsh, R. A., Cahn, L., Ray, C., and Urban, G. H., 1957, Experiments in processing pictorial information with a digital computer: Proc. Eastern Joint Computer Conf., 221 p.

Klein, J. C., and Serra, J., 1972, The texture analyser: Jour. Microscopy, v. 95, p. 349-356.

Krumbein, W. C., 1939, Preferred orientation of pebbles in sedimentary deposits: Jour. Geology, v. 47, p. 673-706.

Krumbein, W. C., and Pettijohn, F. J., 1938, Manual of sedimentary petrography: New York, Appleton-Century Co., 549 p.

Matheron, G., 1967, Eléments pour une théorie des milieux poreux: Paris, Masson et Cie, 166 p.

―――1972, Ensembles fumes aléatories, ensembles semi-markoviens et polyèdres poissoniens: Advances in Applied Probability, v. 4, p. 508-541.

―――1975, Random sets and integral geometry: New York, John Wiley & Sons, Inc. (in press).

Moore, G. A., Wyman, L. A., and Joseph, H. M., 1968, Comments on the possibilities of performing quantitative metallographic analysis with a digital computer, *in* DeHoff, R. T., and Rhines, F. N., eds., Quantitative microscopy: New York, McGraw-Hill Book Co., p. 381-400.

Pettijohn, F. J., 1957, Sedimentary rocks (2d ed.): New York, Harper and Brothers, 718 p.

Serra, J., 1969, Introduction à la morphologie mathématique: Fascicule 3, Les Cahiers du Centre de Morphologie Mathématique de Fontainebleu, L'Ecole Nationale Supérieure des Mines de Paris, 160 p.

Watson, G. S., 1970, Orientation statistics in the earth sciences: Uppsala Univ. Geol. Inst. Bull., new ser., v. 2, p. 73-89.

Weibel, E. R., 1973, Stereological techniques for electron microscopic morphometry, *in* Hayat, M. A., ed., Principles and techniques of electron microscopy, Vol. 3: New York, Van Nostrand Reinhold Co., p. 237-296.

MANUSCRIPT RECEIVED BY THE SOCIETY NOVEMBER 30, 1973
REVISED MANUSCRIPT RECEIVED MARCH 18, 1974

Printed in the U.S.A.

Author Index

Abbott, R. T., 203
Abe, K., 176, 182
Aberdeen, E. J., xiv
Ager, D. V., 4, 34
Agnew, A. F., 7, 26, 34–35
Agterberg, F. P., 352–354, 364
Albers, J. P., 299, 306
Alexander, R. R., 19, 34
Allan, J. F., 87
Allen, J. A., 65, 69
Allen, P., xvii
Armstrong, R. L., 82, 87
Arntson, R. H., 220, 235
Aseez, L. O., 348–349
Asquith, G. B., 7, 34
Atchison, J., 42, 59
Atwater, T., 254–255
Aurelia, M., 65, 69
Ayala, F. J., 36

Baars, D. L., 77, 87
Babcock, R. S., 285, 306
Back, W., 228–231, 235–236
Backus, M. M., 365
Badiozamani, K., 7, 9, 26, 34, 220, 223, 235
Bagnold, R. A., 140, 145, 147
Baird, A. K., 334, 339, 353, 364
Baird, K. W., 364
Banks, P. O., 256
Baragar, W. R. A., 299, 306
Barazangi, M., 256
Barksdale, H. C., 203
Barnard, B. J. S., 152, 161
Barnes, C. R., 36
Barnes, I., 228, 235
Barth, T. F. W., 284, 307
Bassett, M. G., 4, 34
Bayly, B., 284, 307
Bays, C. A., 7, 34
Bear, J., 193, 203
Beckman, W. A., Jr., xvii
Behre, C. H., Jr., 34, 35
Behrens, W. W., III, 217
Bell, W. C., 7, 8, 36
Belyea, H. R., 77, 87
Ben-Manahem, A., 176, 182–183

Bennett, J. G., 39, 59
Benson, B. T., xvii
Bent, D. H., 70
Berg, D. W., 142, 147
Berg, R. R., 55, 59
Berner, R. A., 188, 200, 203
Bhattacharya, B. K., 357, 364
Bischoff, J. L., 200, 203
Blackith, R. E., 21, 34, 319, 322–323, 332
Blais, R. A., 334, 337, 339
Blanc-Lapierre, A., 190, 203
Blatt, H., 42, 59
Boas, M. L., 210, 217
Bogli, A., 220, 235
Bogoliubov, N. N., 152, 162
Bolin, B., 208–209, 217
Bollinger, G. A., 176, 182
Bolt, B. A., 176, 182
Bonham-Carter, G., 83, 87
Bookstrom, A. A., 256
Boudette, E. L., 285, 291, 307
Bourrouilh, B., 356, 364
Bourrouilh, R., 356, 364
Bowen, A. J., 152, 162
Bowen, N. L., 262, 267, 282, 286, 307
Bowen, Z. P., 4, 34
Boyer, R. E., 241, 256, 284, 308
Boyle, R. W., 285, 299, 307
Bretsky, P. W., 7–8, 29, 32, 34
Broecker, W. S., 196, 203, 208–210, 217
Brown, J. A. C., 43, 59
Bruno, R. O., 143, 147
Brush, B. M., 123, 136
Burg, J. P., 365
Burnham, C. W., 258, 277–281
Butler, J. R., 284, 308
Buzas, M. A., 319, 322–323, 331–332
Byerly, G., 281
Byers, F. M., Jr., 299, 307

Cahn, L., 390
Cain, J. A., 256
Caldwell, L. T., xiv
Caner, B., 172–173
Canitez, N., 176, 182
Card, K. D., 299, 307
Carlier, P. A., 334, 337, 339

Author Index

Carnahan, B., 210, 217
Case, J. E., 242, 256
Charlesworth, H. A. K., 356, 365
Chayes, F., 247, 253, 255, 259, 262, 266, 281, 284, 304, 307, 342, 349, 364
Chesser, S. A., 148
Childs, C. W., 210, 217
Christ, C. L., 221, 235
Christiansen, R. L., 255
Churinova, I. M., 315
Clarke, J. M., 35
Cline, L. M., 35
Cole, J. D., 157, 162
Coleman, R. G., 299, 307
Connor, J. J., 334, 339, 352, 354, 365
Conover, W. J., 105-107, 113, 119
Cooper, G. A., 7, 18-19, 34
Copper, P., 4, 34
Corey, A. T., 141, 147
Cornell, J. R., 36
Cornwall, H. R., 286, 299, 307
Courtois, G., 122, 136
Craig, H., 210, 217
Crawford, R. D., 241, 255
Crickmore, M. J., 122, 125, 236
Croneis, C. G., xiv
Cross, C. W., 284, 307
Culkin, F., 227, 230, 235
Culling, W. E. H., 105, 119
Czamanske, G. K., 244, 255
Czeglédy, P. F., 358, 364

Dacey, M. F., xviii, 119, 125, 134, 136, 258, 263, 281
Dapples, E. C., xv-xvii, 35, 39, 53, 59, 88
David, M., 334, 337, 339
Davis, J. C., 240, 255, 358, 364
Davis, R., Jr., 222, 236
Davis, R. A., Jr., 35
Dawson, K. R., 240, 255, 307
Dean, W. C., 356, 364
Deevey, E. S., Jr., 208-209, 217
De Hoff, R. T., 369, 377, 390
Deike, R. G., 236
Delfiner, P., 390
Delwiche, C. C., 208, 217
Demirmen, F., 319, 332
DeMott, L. L., 7, 34
Denbigh, K. G., 210, 217
Deninger, R. W., 7, 34
De Vore, G. W., 263, 281
Dickinson, W. R., 254-255
Dickson, F. W., 249, 255
Dixon, W. J., 197-198, 203
Dobrin, M. B., 356, 364
Doe, B. R., 241, 255
Doeglas, D. J., 38, 59

Domenico, P. A., 193, 203
Domrachev, S. M., 77, 81, 87
Doveton, J. H., 266-268, 281
Draper, N. R., 353-354, 364-365
DuBois, H. M., 4, 34
Dunstan, W. M., 208-209, 218

Eaton, J. P., 176, 182
Ehrlich, R., 263, 281
Einstein, H. A., 124, 127, 136, 139, 147
Ellis, M. J., 357, 365
Embree, P., 356-357, 365
Emery, K. O., 55, 59
Emigh, G. D., 213, 217
Epstein, B., 39, 40, 59
Erdman, J. A., 339
Eriksson, E., 210, 217
Ernst, L., 220, 235
Espenshade, G. H., 285, 291, 307
Ethington, R. L., 36
Exley, C. S., 284-285, 307

Faas, A. V., 258, 282
Faessler, C., 286, 299, 307
Fail, J. P., 356, 365
Fara, H. D., 368, 390
Feder, G. L., 339
Fehrenbacher, J. B., xviii
Feller, W., 124-126, 136
Fenner, C. N., 286, 298, 307
Fenton, C. L., 7, 18-19, 34
Fessenden, F. W., 60
Fisher, F. G., 241, 256
Fisher, R. A., 90, 92, 101
Fleming, R. H., 218
Flinn, D., 258, 281
Flint, R. F., 197, 203
Foerste, A. F., 28, 35
Folk, R. L., 38-39, 42, 59, 231, 235
Fortet, R., 190, 203
Foster, M. R., 359, 365
Fournier, R. O., 258, 281
Fraisier, C. W., 366
Frazee, C. J., xviii
Frear, G. L., 220, 235
Frey, R. W., 65, 70
Friedman, G. M., 38, 49-50, 59
Frolova, E. V., 220, 236
Frost, A. A., 210, 217
Fukao, Y., 176, 182
Fuller, A. O., 38, 59
Fuller, B. D., 356-357, 365
Fuller, W. H., 208, 217
Fulton, R. J., 165, 170, 173
Fyfe, W. S., 200, 203

Gable, D. J., 299, 308

Gardner, W. C., 77, 81, 87
Garrels, R. M., xv, 209, 217, 221, 235-236
Garrett, C. J. R., 152, 157-158, 160, 162
Gaudin, A. M., 41, 46, 48-49, 59
Geer, M. R., 39, 59
Gibbs, R. J., 142, 148
Gill, D., 312, 315
Gilluly, J., 259, 277, 279, 281
Gingerich, P. D., 263-264, 267-268, 281
Goddard, E. N., 241, 255
Goodell, H. G., 138, 148
Gorsline, D. S., 138, 148
Gould, S. J., 29, 35, 203
Grable, D. J., 308
Graf, W. H., 139, 141, 148
Grant, F., 240, 255
Grau, G., 356, 365
Graybill, F. A., xvii, 197, 203, 244, 255, 352-354, 365
Grayston, L. D., 77, 87
Greenwood, H. J., 201, 203
Griffith, J. S., xiv
Gross, D. L., 348-349
Guilbert, J. M., 258, 281
Gulbrandsen, R. A., 212, 217

Haas, A., 376, 390
Hadwiger, H., 369, 377, 390
Hall, J., 35
Hamilton, W., 299, 307
Hansen, D. L., 36
Hanshaw, B. B., 228, 230-231, 235-236
Harbaugh, J. W., 83, 87, 240, 255
Harding, E. F., 369, 390
Harms, J. C., 140, 148
Harned, H. S., 222, 236
Harrison, W., xvii, 143, 148
Hart, S. R., 242, 255
Hartley, H. O., 342, 349
Hatherton, T., 254-255
Hawkins, D. M., 312, 315
Hazel, J. E., 319, 332
Hedgecock, D., 36
Heiskanen, K. I., 105, 119
Hempkins, W. B., xvii
Hernon, R. M., 307
Heyl, A. V., 7, 34, 36
Hirsch, P. N., 365
Hoel, P. G., 342, 349
Hopkins, E. M., 138, 148
Hotz, P. E., 299, 307
Howard, A. D., 220, 236
Howarth, R. J., 352, 365
Hower, J., 244, 255
Hubbell, D. W., 122, 136
Huber, N. K., 299, 307
Hubert, J. F., 90, 101

Hughes, C. P., 8, 36
Hull, C. H., 70
Hunt, C. A., 217
Huntley, D. A., 152, 162
Huxley, J. S., 4, 35
Hyndman, D. W., 242, 249, 254-255

Iddings, J. P., 307
Ikeuti, M., 148
Imbrie, J., xvii
Ingalls, A. F., 364
Inman, D. L., 38, 60, 123, 136, 138, 148, 152, 162
Ippen, A. P., 138, 148
Irving, E., 90, 96, 102
Israel, M., 182
Ivanov, D. N., 264, 281

Jacobson, R. L., 222, 226, 236
Jahns, R. H., 258, 277-281
Jakeš, P., 298, 307
James, W. R., xvii, 84, 87, 355, 365
Jenness, S. E., 299, 307
Johnson, B. D., 359, 365
Johnson, J. G., 82, 87
Johnson, M. W., 218
Johnson, N. I., 42-43, 60
Johnson, T., 176, 182
Johnston, J., 220, 235
Jones, B. F., 221-222, 236
Jones, T. A., xviii, 38, 42, 60, 354, 365
Jones, W. R., 299, 307
Jonson, D. C., 256
Jordon, J. A., Jr., 365
Joseph, H. M., 391
Joyner, B. F., 227, 230, 236

Kamilli, D. C., 281
Kanamori, H., 175, 182
Kanasewich, E. R., 365
Karklins, O. L., 7, 35
Kauffman, E. G., 65, 70
Kay, G. M., 7, 35
Keller, W. D., 38, 60
Kemeny, J. G., 112, 114, 116, 119
Kendall, D. G., 369, 390
Kendall, M. G., 85, 88, 369, 390
Ketchum, B. H., 218
Keulegan, G. H., xv
King, C. A. M., 122, 136
Kirsh, R. A., 368, 390
Kittleman, L. R., 39, 40, 60
Klein, J. C., 368, 390
Kleinhampl, F. J., 286, 299, 307
Koch, G. S., Jr., 101, 240-241, 244, 255
Koelling, M. E. V., 358, 366
Koizumi, M., 142, 148

Koldijk, W. S., 38, 60
Kolmogorov, A. N., 43, 60, 83, 87
Komar, P. D., 122, 133, 136, 140, 145, 148
Korzina, G. A., 208-209, 213, 218
Kotz, S., 42-43, 60
Kretz, R., 258, 281
Krog, M., 61
Krueger, W. C., 61
Krumbein, W. C., xiv-xviii, 35, 38-40, 42, 51, 60, 88, 125-126, 134, 136, 197, 203, 240, 244, 255, 258, 262-263, 266, 281, 334-340, 342, 349, 352-354, 365, 367, 390
Kruskal, W., 266, 281
Kuenzler, H., 183
Kulinkovich, A. Ye., 312, 315

Lal, D., 119, 210, 217
Lamb, H., 151, 153, 162
LaMonica, G. B., xvii
Land, L. S., 192-193, 196, 203, 231, 235-236
Landers, W. S., 39, 60
Langbein, W. B., 105, 119
Langmuir, D., 222, 225, 226, 228, 236
Larochelle, A., 96, 101
Larrabee, D. M., 299, 307
Larsson, I., 90, 101
Lawrence, W. E., 354, 364
Lean, G. H., 122, 125, 136
Lebart, L., 304, 307
Lee, D. E., 285, 291, 299, 307
Lemon, E. R., 208-209, 217
Lerman, A., 106, 113, 118-119, 210, 217
Lesem, L. B., 365
Levinton, J. S., 29, 35
Lévy, P., 190, 203
Libby, W. G., xvi
Lieblein, J., xvi
Lilliefors, H. W., 197-198, 203
Link, R. F., 101, 240-241, 244, 255
Lipman, P. W., 241-242, 254-255
Long, J. A., 364
Looff, K. M., 90, 101
Lorenz, D. M., 29, 32, 34
Lovering, T. S., 241, 255
Lowell, J. D., 258, 281
Luther, H. A., 217
Lyons, E. J., 34-35

Machta, L., 210, 217
Mackenzie, F. T., 203-204, 209, 217, 222, 226, 228, 236
MacKenzie, W. B., 256
MacKevett, E. M., Jr., 299, 307
Maddocks, R. F., 323, 332
Madsen, O. S., 122-123, 136
Magnus, W., 157, 162
Mahony, J. J., 152, 162

Malaika, J., 141, 148
Mandelbaum, H., 244, 255
Manohar, M., 145, 148
Martin, C. S., 122, 136
Mashima, Y., 139, 142, 145, 148
Massey, F. J., Jr., 197-198, 203
Matalas, N. C., 105-107, 113, 119
Mather, J. R., 194, 203
Matheron, G., 334, 336, 340, 368-370, 378, 385, 387, 390
Mathews, G. W., 241-242, 247, 251, 256
Mathews, R. K., 220, 236
Mathews, W. H., 165, 173
Maxwell, J. A., 286, 299, 307
May, J. P., 148
McCammon, H. M., 4, 35
McCammon, R. B., 38, 60
McEwen, M. C., 39, 48-49, 60
McGoldrick, L. F., 152, 160, 162
McIntyre, D. B., 339, 352, 365
McKerrow, W. S., 4, 35
McLachlan, N. W., 158, 162
McNown, J. S., 141, 148
Meadows, D. H., 209, 213, 217
Meadows, D. L., 217
Mehlich, A., 334, 339, 340
Mello, J. F., 319, 323, 332
Merriam, D. F., 240, 255, 311-312, 315, 356-357, 365
Middleton, G. V., 38, 40-43, 54, 59, 60
Miesch, A. T., 352, 354, 365
Mikumo, T., 176, 182
Millard, R. C., 255
Miller, M. C., 140, 145, 148
Miller, R. L., xvi
Mitra, K. C., 4, 35
Mitropolsky, Y. A., 152, 162
Miyake, Y., 142, 148
Moiola, R. J., 38, 60
Monaco, A., 122, 136
Monk, G. D., xv
Moore, G. A., 368, 376, 391
Moore, S. L., 307
Moore, W. J., 258, 281
Morales-Alamo, R., 143, 148
Moran, P. A. P., 369, 390
Moran, S. R., 348-349
Morgan, J. J., 209, 218
Morgan, R. E., 210, 217
Mörner, N. A., 196, 203
Morton, D. M., 364
Moss, A. J., 38, 60
Muncaster, N. K., 256
Murray, R., 59

Nagel, F. G., xvi
Nason, R., 181, 183

Nelson, A. E., 299, 308
Nie, N., 64, 70
Nielsen, A. E., 188, 203
Nielsen, R. L., 258, 281
Norcliffe, G. B., 352, 354, 365
Norton, O. A., 65, 70

Ohsiek, L. E., xv
Oldham, C. H. G., 240, 256
Olea, R. A., 334, 339-340
Olson, J. S., 90, 101
Orme, A. R., xviii
Ostrom, M. E., 25, 35, 53, 60
Otto, G. H., 39, 42, 53, 60

Parsley, A. J., 352-354, 365
Passega, R., 38, 60
Patau, J. C., 356, 365
Patten, B. C., 210, 218
Pearson, E. S., 342, 349
Pearson, R. C., 241, 256
Pearson, R. G., 210, 217
Peikert, E. W., 240-241, 244, 256
Penman, H. L., 194, 203
Perry, L. M., 65, 70
Peters, B., 210, 217
Pettijohn, F. J., xiv, 39, 42, 60-61, 83, 87, 367, 390-391
Phair, G., 241, 256
Pincus, H. J., 90, 101
Pirsson, L. V., 307
Plummer, L. N., 194, 203, 222, 226, 228, 236
Poché, D., 138, 143-144, 148
Pocock, D. M. E., 307
Poldervaart, A., 208, 218
Pope, J. K., 9, 35
Potter, P. E., 83, 87, 89-90, 101
Press, F., 175, 183
Pritchard, W. G., 152, 161
Prostka, H. J., 255
Pryor, W. A., 91, 101
Purdy, E. G., 194, 200, 203
Putman, J. A., 122, 136
Pye, E. G., 299, 308
Pytkowicz, R. M., 210, 218

Quandt, R. E., 312, 315

Raasch, G. O., 7, 18-19, 34-35, 61
Ragland, R. C., 284, 308
Rammler, E., 39, 61
Randers, J., 217
Randolph, J. R., 203
Rao, C. R., 64, 69-70
Rao, J. S., 90, 101
Rasmussen, W. C., xiv
Ratté, J. C., 299, 308

Ray, C., 390
Raychaudhuri, B., 357, 364
Read, W. A., 264, 268, 281
Reardon, E. J., 225, 236
Redfield, A. C., 208-209, 218
Reid, W. T., 39, 60
Remson, I., 194, 203
Reyment, R. A., 21, 34, 319, 322-323, 332
Rhines, F. N., 369, 377, 390
Rhodes, J. M., 291, 308
Richards, F. A., 218
Richards, H. G., 196, 203
Richards, R. P., 9, 19-20, 32, 35
Richter, H., 281
Rinehart, C. D., 299, 307-308
Ristvet, B. L., 193-196, 202-204
Roberson, C. E., 212, 217
Robertson, J. F., 299, 306
Robinson, E. A., 357, 365
Robinson, E. S., 356-357, 365
Robinson, J. E., 356-358, 365
Rogers, J. J. W., 38, 40, 60-61
Rohlf, F. J., 21, 35
Ronov, A. B., 208-209, 213, 218
Rosenman, M., 182
Rosin, P., 39, 61
Ross, D. C., 299, 308
Rudwick, M. J. S., 19-20, 32, 35
Runnells, D. D., 220, 236
Russell, E. W., 213, 218
Ryther, J. H., 208-209, 218

Sainsbury, C. L., 299, 308
Salmon, E. S., 7, 9, 18, 35
Sandberg, C. H., 308
Sanders, H. L., 29, 32, 35
Sarmanov, O. V., 342, 349
Savage, J. C., 176, 183
Sayre, W. W., 122, 136
Sbar, M. L., 256
Scheidegger, A. E., 105, 119, 200, 203, 368, 390
Scherer, W., xviii, 84, 86, 87
Scholes, S. R., Jr., 222, 236
Scholz, C. H., 182, 254, 256
Schuchert, C., 18, 36
Schuenemeyer, J. H., 90, 101
Schwengel, J. S., 65, 70
Scott, R. J., 7, 35
Seal, H. L., 322, 332
Sengbush, R. L., 365
Sengupta, S., 90, 101
Serra, J., 368-369, 377, 390-391
Shand, S. J., 284, 308
Shanks, J. L., 366
Shepard, F. P., 38, 61
Sherwin, D. F., 87

Shields, A., 140, 148
Shigemura, T., 148
Shreve, R. L., xviii
Shrock, R. R., 18-19, 35
Shternina, E. B., 220, 236
Siever, R., 90, 101
Simpson, G. G., 4, 35
Sims, P. K., 242, 256, 299, 308
Skjelbreia, L., 142, 148
Skymer, T., 203
Slack, H. A., xvi, 334-340
Sleath, J. F. A., 122, 136
Sloan, R. E., 7, 35
Slobodkin, L. B., 29, 32, 35
Sloss, L. L., xv-xvii, 25, 29, 35, 74, 76, 87-88
Smedes, H. W., 299, 308
Smith, H., 353, 365
Smith, S. R., 256
Smith, W. O., 194, 203
Sneath, P. H. A., 21, 35, 319, 332
Snell, J. L., 112, 114, 116, 119
Sokal, R. R., 21, 35, 319, 332
Sokhranov, N. N., 315
Soulé, M. E., 29, 35
Speed, R. C., 74, 87, 291, 308
Spencer, C. W., 307
Spry, A., 263, 281
Stanley, S. M., 65, 66, 70
Starkey, J., 276, 282
Stein, J., 210, 218
Steinmetz, R., 90, 92, 102
Stern, T. W., 256
Steven, T. A., 299, 308
Stevenson, R. E., 55, 59
Stewart, D. B., 256
Strahler, A. N., 105, 119
Stumm, W., 208-209, 218
Sulima, J. H., 7, 26, 36
Surface, V. E., 256
Sutcliffe, H., Jr., 227, 230, 236
Sutherland, D. B., 240, 256
Suzuki, Y., 247, 255
Sverdrup, H. U., 208, 218
Swain, F. M., 7, 36
Swartz, C. A., 356, 366
Sweet, W. C., 7, 36
Swift, D. J. P., 307

Tait, R. J., 152, 162
Taylor, S. R., 209, 218
Temple, J. T., 4, 21, 36
Testerman, J. D., 312, 315
Tetreault, D., 307
Thomas, H. H., 256
Thompson, G. A., 286, 299, 308
Thompson, M. E., 221, 236
Thompson, W. H., 7, 36

Thomson, K. C., 183
Thornthwaite, C. W., 194, 203
Thorstenson, D. C., 193, 205
Thrailkill, J., 193, 204, 220, 228, 236
Thwaites, F. T., 61
Tidball, R. R., 339
Tikhii, V. N., 77, 81, 87
Tinkler, K. J., 352, 366
Tisdel, F. W., xiv, 39-40, 51, 60
Toksöz, M. N., 183
Tomilson, M. E., 307
Trask, P. D., 38, 40, 61
Treitel, S., 356-357, 365-366
Trueman, E. R., 64, 70
Truesdell, A. H., 221-222, 236
Tukey, J. W., xvi
Turner, F. J., 287, 308
Tuttle, O. F., 262, 267, 278, 281-282
Twenhofel, W. H., 53, 61
Tweto, O., 241-242, 256
Tyrrell, G. W., 285, 287-289, 308

Udden, J., 38, 55, 61
Upchurch, S. B., 195, 204
Urban, G. H., 390
Ursell, F., 152-153, 162

Vaccaro, R. F., 208-209, 218
Vacher, H. L., 192-193, 195-196, 204
Valentine, J. W., 29, 36
Van Donk, J., 196, 203
Van Loenen, R. E., 285, 291, 307
Van Wazer, F., 208, 218
Verhoogen, J., 287, 308
Vernon, R. O., 231, 234, 236
Vistelius, A. B., 258, 262-263, 267, 282, 342, 349
Vogel, T. A., 281
Von Platen, H., 267, 282

Wadell, H. A., 141, 148
Wadsworth, W. B., 259, 271, 276, 280, 282
Wager, L. R., 285, 308
Wahlstrom, E. E., 241-242, 256
Walcott, R. I., 164, 170-171, 173
Walker, F., 284, 287, 289-290, 308
Walker, K. R., 284, 287-288, 308
Wallace, S. R., 241, 256
Ward, W. C., 38, 42, 59
Washington, H. S., 307
Watson, G. S., xviii, 90, 93, 96-97, 102, 342, 349, 367, 391
Watson, R. J., 365
Waye, I., xvii
Webbers, G. F., 7, 36
Webster, R., 312, 315
Weertman, J., 175-176, 178, 180-181, 183

Weertman, J. R., 176, 180, 183
Weibel, E. R., 369, 391
Weiler, M. G. W., 35
Weinberg, B., 281
Weinberg, R., 210, 217
Weiser, D., 38, 60
Weiss, M. P., 7, 8, 36
Welday, E. E., 339
Wentworth, C. K., 38, 61
White, A. J. R., 298, 307
White, D. E., 286, 299, 308
White, W. H., 284, 308
Whitten, E. H. T., xvii, 240-241, 255-256, 263, 282-283, 291, 306, 308, 353, 358, 366
Whittington, H. B., 8, 36
Wickens, A. J., 172-173
Wilkes, J. O., 217
Willden, R., 299, 307
Williams, A., 8, 9, 36
Wilshire, H. G., 299, 308

Wilson, W. S., xvii
Winchell, N. H., 18, 36
Winkler, S., 157, 162
Winter, J., 4, 36
Wishart, D., 319, 323, 332
Wu, F. T., 175-176, 182-183
Wurster, P., 90, 102
Wyllie, P. J., 242, 256
Wyman, L. A., 391

Yancey, H. F., 39, 59
Yang, S. Y., 35
Yoder, H. S., 249, 256
Youden, W. J., 334, 336, 338, 340
Young, R., 38, 61
Yule, G. U., 85, 88

Zingg, T., 139, 141, 142, 148
Zumwalt, G. S., 36

Subject Index

Activation energy, 189-190
Analytical error, 243
Angularity, measures of, 376
Anondonita, 65
Aquifers
  fresh water, 220
  limestone, 220, 231
Aragonite, 192-202, 220
Arizona, 257-258, 270
Audubon-Albion Stock, 239, 241-246, 250-254
Autocorrelation in residual, 364
Autoregression of residuals, 353-354

Basin
  model, 85
  parameters, 74-76
  slope, 78, 80, 86-87
  width, 78-80
Basin and Range province, 172
Basins, Devonian, 77
Beaverhill Lake Formation, 79
Benioff zone, 254
Bermuda, 189, 192-193, 196, 199-201
Bessel function, 124, 156, 158
Binomial distributions, 266
Bivariate-normal
  distribution, 71, 85-86
  function, 78
  model, 76, 79, 81, 83-84
Boulder zone, Florida, 231, 234
Brachiopod, 3-4, 7-9, 18-19, 29, 32
British Columbia, 163-172
Brownian motion, 105
Bryozoa, 7-9
Burger's vector, 178, 181
Burrowing depth, 63-67

Calcarenites, 192-195
Calcite, 194-196, 219-220, 224-235
  high-magnesium, 193-201
  ion-activity product, 225
  low-magnesium, 193-194, 200
  magnesian, 192, 202
  saturation index, 225
California, 133, 249
Caloosahatchee Formation, 64
Canonical variates analysis, 322

Cation
  percent, 285
  weight percent, 283, 292
  weight per unit volume, 283, 286, 292
Chapman-Kolmogorov theorem, 114
Chione, 65
Chi square, 259, 266-267
  goodness of fit, 197
Classification, 318, 329-331
Clinal maps, 3, 24, 27-28, 31-32
Clines, 4, 9, 24, 26-27
Closed arrays, 342-343, 347
Closed-data arrays, 3-component, 341, 348
Closed-number systems, 304
Closure, 264, 266, 341
Cluster analysis, 317-319, 323
  R-mode, 323
Colorado, 239, 241-242, 254
Color index, 284
Composition of igneous rocks, 283
Confidence intervals, 337
Connectivity, 375
Conodont, 7
Contamination
  beach, 123
  coastal, 121
  duration, 131-133
Continuous exponential distribution, 128
Coralline algae, 195-196
Corals, 195-196
Cordillera
  Canadian, 172
  ice load, 163
  ice sheet, 171
Corey shape factor, 137-143
Cornelia pluton, 257-258, 261, 266, 270, 277, 280
Correlation coefficient, 64, 287, 289, 299
  linear, 29, 100, 292, 295, 301, 303
Covariance, 373, 375, 378, 381
  function, 382-385
Cratonic basins, 72
Crinoid, 8
Crofton's theorem, 377
Cross-bed
  orientations, 89
  variability, 89

vector, 93-94
Crushing law, 37, 39, 43

Debye-Hückel theory, extended, 221
Decorah Formation, 7, 18
Dendrogram, 319, 323, 326, 329
Density, 285-286, 289, 291
  concentration profile, 112-114
  function, 43
Deviations, 244
Devonian time, 79
  Middle, 78
Diagenesis, 188-189, 195-200, 219-220
Diagenetic
  environment, 187, 193, 195
  process, 192
  reactions, 187, 193, 201
Difference matrix, 268
Differentiation, 287
  numeric, 311-317
  trends, 288
Diffusion
  gradients, 189
  processes, 125
Dilatation, 370, 372-373, 381
Dolomite, 231, 234
Dolomitization, 220, 231-235
Donegal Older Granite, 291, 299, 304
Doomsday, 205, 212-214
Dosina, 65
Drag, coefficient of, 139-145

Earthquake, 179
  dislocations, velocity of, 175
  supersonic rupture velocity, 176
Edge dislocations, 175-177, 180
Edge-wave
  perturbation, 156
  standing perturbation, 155
  subharmonic standing, 161
Edge waves
  resonant, 153-155, 160
  standing, 152-153
Eifelian time, 77-78, 80-81
Eigenvalue, 22-23
Elk Point Basin, 72, 77, 79-82
Emsian time, 77, 80-82
Eolianite, 193-197, 201
  Bermudian, 187, 198-199
  calcareous, 189
  carbonate, 192
Erosion of a set, 371, 378-379, 381
Eucrassatella, 65

F statistic, 96-97, 198, 246, 353-354
Factor analyses, 322
  maps, Q-mode, 304

Famennian time, 80, 82
Fault, 176-177, 180-181, 354-355, 359-361
Faulting, 277
Filtering
  spatial, 351-352, 355-359, 362-364
  two-dimensional, 368
Florida, 64, 138, 219, 221, 230-234
  beach, 137
Flow, oscillatory, 139
Foraminifera, 317-318, 323-324, 326
Fourier transform, two-dimensional, 357
Fractionation index, 288
Fraser Glaciation, 163, 165
Frasnian time, 77-78, 81-82
Friction
  law, dynamic, 175, 180-181
  stress, 177, 179
Front Range, 254

Gabbro, orthoclase, 239, 242-251
Gastropod, 8, 195
Gaussian distributions, 77, 84, 259
Gedinnian time, 79
Geochemical
  cycle, 205-207, 214
  reservoirs, 207-208, 211, 216
Geometric
  distribution, 128-129, 134
  probability, 369
Givetian time, 77, 80-81
Glacial lakes, 163, 165
Glauconite, 53-58
Goodness coefficient, 197, 326, 331
Grain
  contact frequencies, 258
  preferential orientation, 385
  roundness, 385
  sequence classes, 272, 274, 276, 280
  sequence types, 268, 270-271
  shape, 137-138, 143-144
  shape index, 137
  size, 385
  transition probabilities, 257-258
  transition sequences, 259
  transition tallies, 265
  transport, 137
Grains
  convexity of, 376
  surface area of, 377
  volume of, 377
Granite, 241, 247, 258, 269, 291, 297, 304-305
Granodiorite, 259-261, 267, 270, 275, 277
Granularmetric analysis, 142
Granulometric properties, 367
Granulometry, 371, 385-386, 389-390
  in measure, 385, 387
  in number, 385, 387

Green River Basin, 90
Gulf of Mexico, 323

Halimeda, 195-196
Harker diagram, 286-288, 290-298, 300-302, 306
Harker-type
  diagrams, 289-290
  plots, 291
Hartley Complex, 291
Hill equation, 151, 157, 160
Homogeneity, 267, 378
Homogeneous, 377
Homoscedastic, 343, 345-346
Homotrema, 195-196
Hydrothermal fluids, 257
Hypersurface, 244, 247-254

Idaho Springs Formation, 241, 247
Illinois, 3-8, 25-26
  basin, 83, 90
Independent trials matrix, 268
Inviscid model, 153
Ionic strengths, 230-231, 235
Iowa, 3-7
Isoline maps, 299
Isopach maps, 29, 74-76, 79, 84, 100
Isophene, 25, 27-28, 32
Isopleth map, 99
Isopleths, 248-250

Kaskaskia
  cycle, 78
  Formation, 82
  unconformity, 77
Kolmogorov-Smirnov test, 197-198
Kriging, 337
Kurtosis, 39, 42-43, 53, 55, 58

Lake Hamilton, 165-171
Lake Merritt, 165, 167, 169
Lake Michigan, 51-52
Lake Quilchena, 164-165, 168-169
Laramide, 241
Laurentide
  ice load, 171
  ice sheet, 164, 170, 172
Lithosphere
  flexure, 163
  thickness, 172
Local anomalies, 351-352
Log-normal
  density functions, 40
  distributions, 37, 40, 42-45, 51-58, 197
  frequency distribution, 52
Lone Rock (Franconia) Formation, 55-57
Louisiana, 50

Lucina, 65
Lugar Sill, Scotland, 283, 285, 287-289, 306

Macrocallista, 65
Magnolia Member, 7
Mahalanobis distance, 63-69
Malsburg Granite, 291, 304-305
Map analysis, numeric, 351
Marine limestones, 193
Markov
  chain, 112-116, 124-126, 258, 262, 264, 267
  chain, absorbing, 112
  chain, embedded, 258, 262, 264-267
  chain, finite, 125
  chain, tridiagonal, 121, 133
  model, 126
  processes, 124, 258, 267
  properties, 267
Mashima transport ratio, 141-142, 146
Mass-action equations, 221
Mass-balance equations, 221
McGregor Member, 18
Michigan Basin, 72, 77-83
Mifflin Member, 3-8, 18, 24
Minkowski
  functionals, 377
  operations, 379
  sum, 369-370
Minnesota, 3-7, 29
Misclassification, 329-330
Mississippi River, 50-51
Missouri, 90
Mixing models, multilayer, 124
Modes, 283-284
Mohorovičíc discontinuity, 172
Mollusks, 8-9
Montana, 284
Monzonite, 241-254, 259-261, 267, 270-271, 276-280
Morphological variability, 3-4
Morphology, theory of mathematical, 369
Morphometrics, multivariate, 21
Moscow Basin, 72, 77, 79-82
Moving average, 358
Moving window, split, 311-315
Moxie pluton, 291

Nearshore zone processes, 138
Nebraska, 50-51
New York, 284, 287
Nonsteady-state cycle, 210
Normal distribution, 42-43, 77, 84
Norms, 287
Null
  hypothesis, 266, 347
  values, 304
Numerical taxonomic system, 21

Oepikina, 3-4, 6-13, 18-33
Opening of a set, 371, 386, 388
Ordovician, Middle, 3, 7-9, 18, 33, 44
Ostracode, 7-9
Oxide weights
 percent, 283
 per unit volume, 283

Paget Formation, 194, 196
Palisades sill, 283-284, 287-290, 306
Pegmatite genesis, 257-258, 277-279
Pennsylvanian, 90
Petrogenetic model, 262
 conceptual, 258
Phenotypic variability, 4
Phi
 diameter, 45-52, 54, 56-57
 normal frequency distribution, 50, 52-58
 normal scale, 53
 scale, 38
 size distribution, 57
 transformation, 38, 42
Phosphorus, 206-208, 211-216
 cycle, 205, 207, 210-216
 mining, 214-215
Photosynthesis, 212
Plate tectonics, 254
Platte River, 50-51
Platteville
 Formation, 3-4, 7-9, 18-19
 Limestone, 5
Pleistocene, 192-193, 196, 201
Poisson
 distribution, 383
 field of points, 390
 process, 124
Pollutant concentration, 124, 127
Polynomial
 model, 353
 trend analysis, 3-dimensional, 239, 241-244, 247, 249, 254
Population variability, 9
Principal components, 322
 analysis, 21-24, 64-65, 68-69
 axes, 21-24, 28
Probability, 108, 110, 124-129, 134, 139, 144, 189, 264, 268-269, 377-378, 383-388
 conditional, 108-109, 135
 density function, Bessel, 124
 density functions, 191-192, 197
 distributions, 189
 fixed, 124
 functions, 43
 geometrical, 369
 law, 379
 law, log-normal, 43, 45
 law, normal, 43

 matrix, transition, 113, 127, 264, 268
 matrix, tridiagonal transition, 126
 model, 258, 382
 transition, 109-110, 121, 125-129, 133, 135
Product moment (correlation) coefficient, 323
Proglacial lakes, 164

Q-mode factor loading maps, 305

Rafinesquina, 9
Random
 function, 187-190, 201, 368
 model, 377
 noise, 352-355, 361
 process, 368
 sets, 383, 386
 variable, 111, 114, 124, 188, 199-201
 walk model, 106
 walk model, recurring, 110-113, 117
 walk, recurring, 105, 107, 111, 114, 118
 walks, 105, 107, 112-115, 125
Ratio data, 304
Rayleigh velocity, 175-176, 180-181
Regional
 component, 353
 trend, 244, 351-352
Regionalized variables, 333-334, 336, 338-339
Regression
 analysis, 244
 lines, 286, 292, 298
 lines, degree-two, 296-298, 302
 lines, linear, 291, 294-297, 300-303
 multiple linear, 352
 piecewise, 311-315
 sequential multiple, 259
 step-wise, 353-354
Relevant variables, 317
Resonance, 153-154, 157, 160
 nonlinear subharmonic, 151
 parametric, 152
 subharmonic, 161
Resonant instability, 159
Reynolds number, 142
Rock density, 288, 293-295
Rosin's
 crushing law, 37, 39-52, 58
 distribution, 39-40, 44-51, 57-58
 scale, 46-52
Russian Platform, 77, 83

Sample correlation coefficients, 66
Sampling, 242
 design, 263
 design, nested, 89, 91
 efficient, 333
 plan, 293, 306

Sand
    beach, 51-52
    cratonic, 52
    river, 49-50
Screw dislocations, 175-177, 180
Sea water, 219-235
Sediment
    tracer, 122, 134
    transport, 105
Sequential data, 311
Settling velocity, 143
Set transformations, 369
Shape factor, 141
Shear
    stress, 140, 176
    wave velocity, 80, 175
Siegenian time, 77, 80-81
Significant parameters, 318
Silver Plume Granite, 241, 247
Size
    distribution, 38, 40, 49, 55-58, 367, 376
    frequency, 49
Skewness, 39, 42-43, 49, 57-58
Spatial filtering, 355, 364
    band pass, 355-357
    fan pass, 351-352, 355-359, 362-363
Spatially homogeneous, 383
Spatial variability, 333-334
Specific gravity, 285, 291-292, 295, 298
Spline, 358
Standard deviations, 53, 55, 66, 68
Stationarity, 379, 381
Stationary, 383, 386, 389
    cycle, 207
    random function, 377, 379, 385
    random set, 367, 378, 387
    set, 386
Steady-state
    cycle, 207
    model, 206, 216
Stereology, 369, 376-377
Stick-slip propagation velocity, 176
Stillwater Complex, 284
Stochastic
    matrix, 264
    model, 189
    model of crystallization, 258
    process, 187-188, 201
    process model, 124
Strandline
    Algonquin, 163, 171-172
    deformation, 163-164
    lake, 172
Strophomena, 3-4, 6, 9, 15-33
Subduction zone, 254
Supersonic velocity, 175
Syenite, 241-246, 250, 252-253

Synecology, 7

t-test, 341-342, 344, 347-348
Tagelus, 65
Tally matrix, 263-265
Target population, 299
Taylor series, 157
Tellina, 65
Tertiary, 258
Texture
    Analyser, 368, 372, 379, 381
    analysis, 367
    attributes, 258
    classes, 270
    maturity, 40, 58-59
    varieties, 267
Time-stratigraphic unit, 71-72
Trachycardium, 65
Traction transport, 139
Transient cycle, 210
Transient-state model, 206, 216
Transition
    frequencies, one-step, 263
    tally matrix, 262
Transportation laws, 39
Transport rate, 139
Trask's sorting coefficient, 40
Trend component, 355
Trend fitting, orthogonal polynomial, 353
Trend-surface, 248, 250
    analysis, 76, 252, 342, 354
    analysis, polynomial, 240, 304, 351-352, 355, 358, 364
    analysis, three-dimensional, 239, 241-244, 247, 249, 254
    map, 299
    orthogonal polynomial, 240
Trilobite, 7
Turbulent flow, 139

Upper Elk Point unit, 79, 81

Vadose
    environment, 194
    zone, 187, 193-196, 199
Van't Hoff equation, 221
Variability, morphological, 3-4
Variance, 129-130, 134-135, 285, 321, 329, 331, 333-336, 343-346, 353, 382
    analysis of, 89, 96-97, 243-246, 259, 333-338
    components, 333-335, 337-338
    covariance matrices, 29
    kriging, 337
    maximum level, 311-313
Variation
    chemical, 288
    clinal pattern of, 5

coefficients of, 30-31
phenetic, 3
Variogram, 333-334, 336-339

Walsingham Formation, 196
Wasatch Formation, 90
Water
    carbonate ground, 230-234
    fresh carbonate, 220
    ground, 219, 221, 228-229
    mixing natural, 230
    saline, 231, 233, 235
    saline subsurface, 230, 235
    sea, 219-235
Wave, standing primary, 155
Waves, cross, 152
Weibull
    distributions, 43-44, 50-51
    equation, 58
    functions, 44, 57
    probability law, 42
Weight
    percent, 285-300, 306
    per unit volume, 286-288, 293-300, 306
Wisconsin, 3-8, 26, 29, 52-56, 59
    Arch, 26
    Dome, 25
Wyoming, 89-92

Xenolith, 49, 247, 261, 275, 280

Yucatan Peninsula, 219, 221, 230, 233

Zingg shape classification, 138